# Schumpeter's Evolutionary Economics

**The Anthem Other Canon Series**

The **Anthem Other Canon Series** is a collaborative series between Anthem Press and The Other Canon Foundation. The Other Canon – also described as 'reality economics'– studies the economy as a real object rather than as the behaviour of a model economy based on core axioms, assumptions and techniques. The series publishes classical and contemporary works in this tradition, spanning evolutionary, institutional, and Post-Keynesian economics, the history of economic thought and economic policy, economic sociology and technology governance and works on the theory of uneven development and in the tradition of the German historical school.

**Forthcoming Titles**

*Ragnar Nurkse (1907–2007): Classical Development Economics and its Relevance for Today,*
Edited by Rainer Kattel, Jan A. Kregel & Erik S. Reinert

*Ragnar Nurkse: Trade and Development,*
Edited by Rainer Kattel, Jan A. Kregel & Erik S. Reinert

*Techno-Economic Paradigms: Essays in Honor of Carlota Perez,*
Edited by Wolfgang Drechsler, Rainer Kattel, & Erik S. Reinert

# Schumpeter's Evolutionary Economics

## A Theoretical, Historical and Statistical Analysis of the Engine of Capitalism

Esben Sloth Andersen

Anthem Press

Anthem Press
An imprint of Wimbledon Publishing Company
*www.anthempress.com*

This edition first published in UK and USA 2009
by ANTHEM PRESS
75-76 Blackfriars Road, London SE1 8HA, UK
or PO Box 9779, London SW19 7ZG, UK
and
244 Madison Ave. #116, New York, NY 10016, USA



*British Library Cataloguing in Publication Data*
A catalogue record for this book is available from the British Library.

*Library of Congress Cataloging in Publication Data*
A catalog record for this book has been requested.

ISBN-13: 978 1 84331 334 2 (Hbk)
ISBN-10: 1 84331 334 0 (Hbk)

ISBN-13: 978 1 84331 335 9 (Ebk)
ISBN-10: 1 84331 335 9 (Ebk)

1 3 5 7 9 10 8 6 4 2

# Contents

# List of Figures

# List of Tables

# Preface

In a previous book called *Evolutionary Economics: Post-Schumpeterian Contributions* (Andersen, 1994, 1), I emphasised the need for "a 'dialogue' with older, verbal studies of economic evolution". Although several of the great economists of the past can be seen as contributors to such verbal studies, the title of the book suggested that I was primarily thinking of Joseph A. Schumpeter. It seemed clear to me that Schumpeter, early in life, had chosen what can today be classified as evolutionary economics as the field in which he would make a great contribution to the progress of his chosen science. Actually, he seems to have wanted to do for economics what Darwin had done for biology: to develop a theory that supports the study of the historical process of the evolution of economic life. However, he rejected to work along Darwinian lines. Instead, he combined broad evolutionist perspectives with the tools of neoclassical economics and the German historical school. Furthermore, he recognised that the full explanation of economic evolution is dependent on the explanation of socio-political evolution, so he also worked to promote evolutionary sociology. Although his attempts to develop these areas and to include them into the body of economics and other social sciences cannot be described as unconditional successes, his unique efforts were not made in vain. His vision of the evolutionary process and the concepts that he used to implement this vision are still able to challenge and inspire. His old-fashioned works are still able to influence the research agenda because we are far from having established solid forms of evolutionary analysis within economics and the other social sciences.

Since Schumpeter is widely, and increasingly, cited in practically all social sciences, one might expect that thorough treatments of all of his works are available. This is not the case. For instance, there exists no systematic evaluation of his first and crucial book on the essence of theoretical economics at the centennial of its publication. The book is not even translated into English. We are also approaching the seventieth anniversary of his *Business Cycles* without having a systematic exposition of its confusing but rich materials. Instead, much of the discussion of Schumpeter relies on selected parts of his books *The Theory of Economic Development* and *Capitalism, Socialism, and Democracy*. One of the reasons is that Schumpeter's works are peak results of a type of wide-ranging and verbose scholarship that was becoming old-fashioned during his lifetime and that has presently become replaced by the type of focussed research communication that has always characterised the natural sciences. The basic reason for the limited exploitation of his verbose works, however, is that Schumpeter did not succeed in establishing a cumulative research tradition.

Ever since Schumpeter died in 1950, a significant number of economists and other social scientists have been both fascinated and confused about the Schumpeterian world of ideas. This confused fascination has led to attempts to explore his scientifically suggestive ideas in the midst of the many misunderstandings that easily emerge from the study of his complex works in the style of a long-past epoch. Actually, quite a few of these researchers have started their personal Schumpeter projects. His many friends started the efforts immediately after his death, but they were constrained by their engagement in the development of Keynesianism and econometrics. Around the centennial of his birth in 1983, Schumpeter's work was interpreted in the light of the problems of innovation and structural change that seemed important complements to studies of the short-term functioning and simple growth of economic systems. Furthermore, sociologists started to pay systematic attention to Schumpeter's work. The resulting studies have greatly enhanced our understanding of the characteristics of Schumpeter's ideas and their present-day relevance. However, most of the contributors lacked an appreciation of the complex problems of studying economic and socio-political evolution and the extent to which these difficulties influenced Schumpeter's work. The increasing emphasis on these issues is a major background for a new wave of Schumpeter studies—to which the present book is a contribution.

The idea of developing the present book crystallised while I, in 1993, wrote a short review of Richard Swedberg's biography of Schumpeter for the *Journal of Economic Literature* (Andersen, 1993). Here I remarked that "Swedberg has produced a unique and inspiring book which is a must for anyone interested in Schumpeter's personality as well as in his rich set of ideas". I also noted that Swedberg's book obtained its coherence by emphasising Schumpeter's contribution to economic sociology and its inspiration from Max Weber's wide-ranging "social economics" (Sozialökonomik). This theme is not the only one that may serve to uncover the core of Schumpeter's efforts, and I emphasised one of the alternatives:

> "One possibility is to consider Schumpeter as the initiator of an evolutionary-economic research program. Given this theme, we would explore his life-long search for a precise model of capitalist evolution and his attempts to apply his evolutionary vision to sociology and the theory of science. We would also focus less on his general approach to *Sozialökonomik* than on his attempts to develop an evolutionary version of it in *Business Cycles* and elsewhere. Finally, we would explore his special blend of elitism, marginalism, and institutionalism as a reflection of his ideas of the jerky and punctuated form of the evolutionary process." (Andersen, 1993, p. 1970)

The suggested evolutionary interpretation of Schumpeter's work can be developed in two ways. On the one hand, the biographies produced by Swedberg and others (Loring Allen and Thomas McCraw) suggest a historical reconstruction of the emergence and development of Schumpeter's evolutionary theory in close connection to his academic life. On the other hand, the evolutionary interpretation can be developed in terms of a rational reconstruction of his evolutionary contributions to the social sciences—especially his evolutionary economics. Although my preferences point at rational reconstruction, the difficulties of handling the Schumpeterian type of evolutionary social science suggest that historical reconstruction is also needed. Furthermore, this historical reconstruction seems to be supported by detailed knowledge of his life and work. Consequently, my Schumpeter project grew in complexity—as demonstrated by several of my papers and working papers. However, a more systematic treatment did not begin to emerge before I received a request to produce a small Schumpeter book in Danish language. I had no difficulty in combining chapters on Schumpeter's major works with chapters on his life in the resulting unambitious book (Andersen, 2004a). My forthcoming expansion of this small book into a larger intellectual biography in English language demonstrates that the systematic combination of Schumpeter's work and core aspects of his life is difficult (Andersen, forthcoming). The present book is different; it focusses on Schumpeter's evolutionary economics and primarily bases its historical and rational reconstruction on his major books.

# Acknowledgements

The author has received support and critical comments from a large number of researchers. Much of this support is acknowledged in his papers and reports, which have not least been commented upon at different conferences. Thanks should especially be given to many of the participants in the conferences of the International Joseph A. Schumpeter Society (founded in 1986) and the Danish Research Unit for Industrial Dynamics (DRUID, established in 1995). Furthermore, a draft manuscript, which is in several ways is very different from the present book, has been commented upon by most members of the IKE group of innovation studies at the Department of Business Studies, Aalborg University. The draft was also distributed to a significant number of international experts from whom important comments have been given. Since the book is the outcome of a very complex and personal research project and since it has only recently obtained its final shape, the usual disclaimer has to be emphasised more than usually.

The author and the publishers wish to thank the following who have kindly given permission for the use of copyright material.

Duncker & Humblot for the translations of excerpts from: *Das Wesen und der Hauptinhalt der theoretischen Nationalökonomie* by Joseph Schumpeter, 1908, 1998. Alle Rechte vorbehalten.

Duncker & Humblot for the translations of excerpts from: *Theorie der wirtschaftlichen Entwicklung by Joseph A. Schumpeter,* 1912, 2006. Alle Rechte vorbehalten.

HarperCollins Publishers for the excerpts from: *Capitalism, Socialism and Democracy* by Joseph A. Schumpeter, 3rd edn, 1942, 1950. [300 very brief quotations totalling 2300 words from CAPITALISM SOCIALISM AND DEMOCRACY, 3RD ED. by Joseph A. Schumpeter. Copyright 1942, 1947 by Joseph A. Schumpeter. Reprinted by permission of HarperCollins Publishers.]

Harvard University Archives for the cover photo of Joseph A. Schumpeter, Harvard University Archives, call # HUGBS 276.90p (2).

Harvard University Press (President and Fellows of Harvard College) for the excerpts from: *The Theory of Economic Development* by Joseph A.

Schumpeter and translated into English by Redvers Opie, 1934.

The McGraw-Hill Companies, Inc. for the excerpts from: *Business Cycles* by Joseph A. Schumpeter, 1939.

Oxford University Press for the excerpts from: *History of Economic Analysis* by Joseph A. Schumpeter, 1954.

Princeton University Press for the excerpts from: The *Economics and Sociology of Capitalism* by Joseph A. Schumpeter and edited by Richard Swedberg, 1991. [SWEDBERG, RICHARD; *ECONOMICS AND SOCIOLOGY.* © 1990 Princeton University Press. Reprinted with permission of Princeton University Press.]

Taylor & Frances Books UK for the excerpts from: *Capitalism, Socialism and Democracy,* Copyright © 1942, 1950, Joseph A. Schumpeter and Allen & Unwin. Reproduced with permission by Taylor & Frances Books UK.

Taylor & Frances Books UK for the excerpts from: *History of Economic Analysis,* Copyright © 1954, Joseph A. Schumpeter and Allen & Unwin. Reproduced with permission by Taylor & Frances Books UK.

Transaction Publishers for the excerpts from: *Essays on Entrepreneurs, Innovations, Business Cycles, and the Evolution of Capitalism* by Joseph A. Schumpeter, edited by Richard V. Clemence and republished by Richard Swedberg, 1951, 1989.

Every effort has been made to trace all the copyright holders, but if any have been inadvertently overlooked the publishers will be pleased to make the necessary arrangements at the first opportunity.

# Abbreviations

The present book concentrates on Schumpeter's major books. They are referred to by means of abbreviated titles. The major books in English are:

- *Development* = *The Theory of Economic Development* (1934; largely a translation of *Entwicklung II*)
- *Cycles* = *Business Cycles* (1939)
- *Cycles Abr.* = *Business Cycles*, abridged edition (1964)
- *Capitalism* = *Capitalism, Socialism and Democracy* (1942; the 1950 edition is used)
- *History* = *History of Economic Analysis* (1954)

One of Schumpeter's major books is only available in German. Furthermore, the first German edition of *Development* comes close to a separate book. We also have to consider its second German edition. The abbreviations are:

- *Wesen* = *Das Wesen und der Hauptinhalt der theoretischen Nationalökonomie* (1908)
- *Entwicklung I* = *Theorie der wirtschaftlichen Entwicklung*, first edition (1912)
- *Entwicklung II* = *Theorie der wirtschaftlichen Entwicklung*, second edition (1926)

Although the present book focusses on the above-mentioned books, it also refers to Schumpeter's other books and to his many papers. Full information is found in "Schumpeter's Works" (starting on page 447). The entries are organised in three lists:

1. Schumpeter's books. They are always cited by words from their titles. This list includes three books that have not been mentioned above: *Doctrine* = *Economic Doctrine and Method*; *Vergangenheit* = *Vergangenheit und Zukunft der Sozialwissenschaften*; and *Geldes* = *Das Wesen des Geldes*.

2. Collections of Schumpeter's papers. They are cited by capital letters taken from their titles. For instance, *TGE* = *Ten Great Economists*; and *BL* = *Briefe/Letters*.

3. Schumpeter's papers. They are referred to by an '*S*' followed by the year of publication—or the year of his finishing of the manuscript. To take an example: *S1932c* = "Development" is a paper published in 2005, which is based on a manuscript from 1932.

Four comments on the citing and quoting of Schumpeter's works seem relevant. First, the above abbreviations are typeset with *slanted Roman font* (like *Development*) to distinguish them from the titles of books that are typeset with *italic font* (like *The Theory of Economic Development*). Second, if a quotation refers to a non-English text, it has without special notice been translated by the present author. Third, any changes of italics in quotations are explicitly mentioned. Fourth, further information on the availability and subjects of Schumpeter's works is found Appendix C.

# 1
# Introduction

When the Austrian-American economist Joseph A. Schumpeter died in 1950, he had received many forms of recognition. He was Professor of Economics at the famous Harvard University in the USA; he was the most cited scholar in the whole field of economics (Samuelson, 1981a, 1); he had recently served as the President of the American Economic Association and had just been elected to become the first President of International Economic Association. This exceptional status was based on Schumpeter's contributions to all major parts of economics and to other social sciences, his all-encompassing network of scholarly contacts, and his ardent support for the new generation of ambitious economists. However, as pointed out by Paul Samuelson, Schumpeter was not satisfied with the status he received. According to Samuelson, who considered himself to have been both Schumpeter's friend and pupil, he was sceptical about his "Popeship" because this was not what he had strived for. From his youth, Schumpeter's main ambition had been to become one of the great economists, and he thought that such economists are not orthodox Popes, but radical scientific innovators. Since he did not succeed in his attempt to renew the science of economics, Samuelson (1981a, 1) suggested that Schumpeter "would have traded his Popeship for a Keynesian revolution". This does not mean that Schumpeter would have liked to promote an arbitrary scientific revolution. Instead, he wanted to trade his position within the economic establishment for an evolutionary-economic breakthrough, which he had tried to obtain throughout his academic life.

The decades since Schumpeter's death have clearly proven the low value of his Popeship. Like John Maynard Keynes, he was born in 1883, and they died within few years of each other; but there are crucial differences in the way posterity reacted to them. Keynes is commonly remembered as the major initiator of macroeconomic theory. The large extent to which modern economic theory has integrated and developed his problems and tools means that we seldom have to refer to him directly. Schumpeter obtained a quite different posthumous status. We economists have forgotten his Popeship of equilibrium economics, but we still maintain folklore about him as the most romantic and paradoxical of all economists. More importantly, we remember him for his vision of innovation and structural change within the economic system. Although we largely base our remembrance on a few striking formulations from the most reader-friendly parts of his works, like a couple of chapters from *Capitalism, Socialism and Democracy*, he still serves important functions. First, Schumpeter represents our bad consciousness of not having treated in a convincing way the grand questions of economic evolution and the transformation of capitalism. Second, his incomplete analytical contributions serve as a challenge to apply the much-improved mathematics of evolution to reconsider his visionary accounts. Finally, since the great questions of the process of economic evolution are unlikely to become answered through isolated economic analysis, he points us to the hitherto largely unsolved problem of establishing a systematic collaboration between the different social sciences.

The evolutionary-economic breakthrough that Schumpeter hoped for is presently within sight; and similar breakthroughs might come about in sociology, political science, and elsewhere. This situation allows us to consider his wish of trading a Popeship for a scientific revolution in a new light. Actually, it has, for many researchers, become obvious that Schumpeter was an important pioneer. However, the more precise study of his contributions to the extension of the social sciences towards the analysis of evolutionary processes has not been easy. The difficulties, which will become clear from the present book, are largely related to the fact that he produced an immensely complex work. Furthermore, his work applies old-fashioned terminology and confronts the problems of past generations of researchers. Finally, he wrote in two languages, and some of his core contributions have not yet been translated from German. For these and other reasons, there is still much confusion about the nature of his work and about the degree to which it still represents a challenge for modern researchers. This is the background for the present account for the emergence and elaboration of Schumpeter's evolutionary economics, which is mainly based on reconstructive readings of his books. The main proposition is that he was a pioneer of a special form of evolutionary economics and that he wanted to add analogous forms of evolutionary sociology and evolutionary political science. Although his form of evolutionary analysis is not identical to the forms that are presently emerging, there is a large degree of compatibility (Fagerberg, 2003). Furthermore, the remaining incompatibility cannot only be considered the result of Schumpeter's idiosyncrasies. It also represents a challenge for present-day research.

## 1.1 The name of the game: 'evolutionary economics'

The core propositions of the present book are: (1) that Schumpeter's basic ambition was to complement equilibrium economics with an evolutionary economics that analyses capitalist economic evolution; and (2) that his major contributions to economics relate to his attempts to fulfil his ambition. The arguments for the two propositions will bring us back to the works that he produced before the First World War. However, the fact that some of these works are still only available in German language is only one of the minor difficulties in developing the argument. The major obstacle is the terminological and conceptual problems that relate to the terms 'evolutionary' and 'evolution'. Although the clarification of these terms is a major theme throughout the present book, it is convenient to reach a minimum level of clarification from the very beginning. The works of the early Schumpeter do not supply us with such clarification. A major reason is that he developed his basic contribution at a time when the evolutionary approach, after an upswing in the nineteenth century, was facing an eclipse. The reaction against the inflated application of the evolutionary approach meant that during "the first few decades of the twentieth century evolution was a dirty word" and that "[e]volutionism as a theoretical approach ... was practiced or endorsed only at risk to one's intellectual career" (Sanderson, 1990, 45–6). This reaction was especially strong while Schumpeter used German language as his primary means of presenting his scientific contributions. It is in this context that we should interpret a remark that Schumpeter made in the radically revised and shortened second edition of his *Theorie der wirtschaftlichen Entwicklung*. Here he emphasised that

> "we must be careful in dealing with the phenomenon of evolution [Entwicklungsphänomen] that we observe, still more with the concept in which we comprehend it, and most of all with the word by which we designate the concept[.] ... [A]ll the over-hasty and insufficiently founded generalisations in which the word [soziale Entwicklung] plays a part have led many of us to lose patience equally with the word, the concept, and the issue." (*Entwicklung II*, 88–9; cf. *Development*, 57–8)

The worries that Schumpeter, in 1926, formulated about the German word 'Entwicklung' re-emerged when he helped Redvers Opie in translating his magnum opus into English. Even the chosen title of the book—*"The Theory of Economic Development"*—was problematic. Schumpeter, quickly after the publication of *Development*, began to switch from the term 'economic development' to 'economic evolution'; and he used the latter term consistently in *Business Cycles* from 1939. Moreover, even while *The Theory of Economic Development* was being printed, he in May 1934 wrote a letter in which he called his book "The Theory of Economic Evolution" (*BL*, 267); and he accepted the French translation with the title *Théorie de l'évolution économique*, that is, without using the French word 'développement' (*S1935e*). Nevertheless, modern economists and sociologists still use term 'economic development' as the primary label for his theory. On this background, it is interesting to note that it is not the necessary translation of his German text. In contrast to the situation in France and the English-speaking countries, speakers of German language in the first decades of the twentieth century largely used a single word for two concepts. 'Entwicklung', which etymologically means 'unwinding', covered both 'development' and 'evolution'.

The English word 'evolution' originates from the Latin 'evolutio'. It is related to the verb 'evolvere' that means 'unwinding' or 'unrolling'. The book of antiquity was a rolled volume of writing and its reading required a process of unrolling. The noun 'evolutio' was used to denote the process of reading through the unrolling of such a book. This background meant that the original usage of the English 'evolution' referred to goal-directed and pre-programmed processes. The same was the case for 'development'. This word arrived to England from France, where 'développement' referred to the process of unfolding. Then Darwin published his *Origin of Species* with its radically different account for change. As a result, the word 'evolution' started to obtain a new meaning; and it ended up denoting the unplanned process of the irreversible change of biological species, human languages, and the routines of social life. In contrast, 'development' kept the meaning that it had originally shared with 'evolution'. Although this division of labour between the two words arrived slowly, the distinction was relatively clear in the early 1930s. At that time, the best translation of the title of "*Theorie der wirtschaftlichen Entwicklung*" was "The Theory of Economic Evolution", that is, the title Schumpeter used in the mentioned letter from 1934. By the way, this letter commented on the fact that he—a foreigner—had got a paper (*S1934a*) included in a collection intended for courses in English style and composition.

The confusion that has emerged from the two possible translations of a single German word has today largely been overcome. It has become clear that the unplanned process of change described in Schumpeter's *The Theory of Economic Development* has nothing to do with the goal-directed and programmable processes of old-style developmental thinking about social change. Nevertheless, the choice of the words "Economic Development" for the title of this book has retarded the formulation of the evolutionary interpretation of Schumpeter's work significantly. In the 1930s, Schumpeter and his translator (Redvers Opie) missed an important opportunity for labelling the book in accordance with modern scientific terminology. The present book tries to undo their decision by translating all occurrences of 'Entwicklung' as 'evolution'—unless Schumpeter was clearly exploiting the ambiguity of German language to use the word in the meaning of 'development'. The importance of this decision is emphasised by the fact that Schumpeter predominantly wrote in German from the decade before World War I until he moved to the USA in 1932.

Schumpeter's early writings are characterised by another terminological problem that had largely been overcome in the 1926 edition of *The-*

*orie der wirtschaftlichen Entwicklung*, which was used for the production of *Development*. However, the terminological problem is obvious in the first edition of this book (from 1912) and in his programmatic account of the essence and limits of equilibrium economics (from 1908). In these two books and elsewhere, he used the term "Economic Statics" for what can roughly be called equilibrium economics and "Economic Dynamics" for the theory of economic evolution. Although the distinction between these branches of economics is crucial for his argument, he largely applied the original terms—Statics and Dynamics—in quotation marks and he pointed out that the terminology is "very unfortunate" (*Wesen*, 182). One of the problems is that the terminology was likely to create confusion, and this problem was to an overwhelming extent confirmed by a large literature that in the next decades tried to clarify the Statics–Dynamics dichotomy—but instead created increasing confusion. The problem was that the words 'statics' and 'dynamics' were used to denote a large number of concepts and that these concepts have often been very loosely defined. Therefore, Fritz Machlup (1959, 109) characterised them as "kaleidoscopic words". Just as children have used the old-fashioned tube with mirrors and coloured glass to produce a huge number of different patterns, economists have used the "kaleidoscope" of the static–dynamic dichotomy to develop a surprisingly large number of meanings. According to Machlup, the problem "is not that the division of economic analysis into Statics and Dynamics makes *no* sense, but that it makes *too many* senses". Although the clarification of the senses in which Schumpeter used the terms "Economic Statics" and "Economic Dynamics" is an important theme of the present book, their translation into 'equilibrium economics' and 'evolutionary economics' may serve as a preliminary solution.

It is only after we have solved the major problems of terminology that it becomes clear that the evolutionary interpretation of Schumpeter's works has been provided by himself. Actually, we can derive most of this interpretation from a couple of sentences in his first book from 1908—the untranslated *Wesen und Hauptinhalt der theoretischen Nationalökonomie*. In this book on the essence and main contents of theoretical economics, he emphasised the need of a strict division of labour between works in the two fundamental fields of economics:

> "Statics [equilibrium economics] and Dynamics [evolutionary economics] are completely different fields; they concern not only different problems but also different methods and different materials. They are not two chapters of one and the same theoretical building but two completely independent buildings. Only Statics [equilibrium economics] has hitherto been somewhat satisfactorily worked up and we essentially only deal with it in this book. Dynamics [evolutionary economics] is still in its beginnings, is a 'land of the fu-

ture'." (*Wesen*, 182–3)

Although Schumpeter's first book focussed on equilibrium economics, its purpose was not just to propose a reform plan for this branch of economics. On the contrary, the problems, methods, and materials that he reserved for evolutionary economics were those that, from the very beginning, engaged him as a researcher. He actually throughout *Wesen* pointed at this branch. He did so by repeatedly emphasising what concepts equilibrium economics cannot really contain and what problems it cannot solve. This negative definition of evolutionary economics as covering parts of the residual of concepts and problems left over by equilibrium economics was followed by a positive definition. This definition is most clearly stated in *Entwicklung I*—where the analysis of capitalist economic evolution is performed in terms of the concepts of capital, entrepreneurial profit, interest, credit, and business cycles. Together, these two books (*Wesen* and *Entwicklung I*) provided Schumpeter with the research programme that would engage him for the rest of his life. They demonstrate that he did not want to obtain a "Popeship" by working within the already-defined equilibrium economics but strived for the honour of being the pioneer of a complementary and very important branch of economics.

Schumpeter's research programme depended on his decomposition of theoretical economics into two logically distinguishable branches: equilibrium economics and evolutionary economics. This distinction was easy to make in the first decade of the twentieth century. At that time evolutionary economics consisted only of loose sketches; and the predominant neoclassical economics could largely be reduced to an equilibrium economics that included the processes of re-establishing equilibrium after exogenous disturbance. During Schumpeter's academic life the distinction apparently became more blurred because equilibrium economics became extended with analyses of economic growth and business cycles. However, Schumpeter tended to reject these extensions as unrealistic substitutes for the evolutionary economic analysis of the long-term transformation of capitalist economies. This topic provided the main contents of his scientific contribution.

## 1.2 Schumpeter's evolutionary pivot

The study of Schumpeter's work is made difficult by the way it was published. We nowadays normally present the results of our research in scientific journals; so it is an obvious strategy to browse his 200 papers (see the list starting on page 449) in order to find accounts for his core scientific contributions. This strategy largely fails, however. Contrary to today's habits, he followed the old-fashioned rule that the size of a publication should reflect its scientific importance. His smaller papers are normally made for the occasion, his longer papers are presenting more ambitions research, and his books present basic scientific contributions. Since these books are large and complex, it is also natural to browse them to find a few core pages that summarise his main contributions. Unfortunately, this strategy also fails. Schumpeter designed the books of his youth in a complementary way; and his later books contain extensions, corrections and perspectives. Therefore, we can to some extent reduce the reconstructive task to the provision of an understanding of his books as a more or less integrated whole.

An obvious strategy is to study Schumpeter's books in chronological order. This strategy can be improved by Yuichi Shionoya's (1997, 16, 23) distinction between Schumpeter's "early trilogy" and his "later trilogy". The early trilogy consists of the untranslated *Das Wesen und der Hauptinhalt der theoretischen Nationalökonomie* (1908); *Theorie der wirtschaftlichen Entwicklung* (1912), which in modified form was published in 1934 as *The Theory of Economic Development*; and *Economic Doctrine and Method* (1914). The later trilogy consists of *Business Cycles* (1939); *Capitalism, Socialism and Democracy* (1942); and *History of Economic Analysis*, which was published posthumously in 1954. Although the treatments of these books in several biographical accounts follow this approach, it is neither used by Shionoya nor by the present book. Shionoya's (1997) *Schumpeter and the Idea of Social Science* assumes that his work can be treated as a whole; and this assumption allows Shionoya to arrange the analysis of the different parts of the work freely. The present account is more constrained by chronology. The background for combining "rational reconstruction" and "historical reconstruction" is that Schumpeter's evolutionary economics can hardly be understood sufficiently without considering the way in which it emerged and was later developed. One of the implications is that Mark Blaug's (1996) strategy of reconstructing the works of the great economists of the past by means of the tools of present-day economics is insufficient for handling the case of Schumpeter.

The present book's combination of historical and rational reconstruction suggests that it has to be based on a grouping of Schumpeter's major books. A possible grouping is provided by Shionoya's idea that *Wesen, Entwicklung I/Development*, and *Doctrine* are the early trilogy while *Cycles, Capitalism*, and *History* are the later trilogy. However, this grouping does not suit the present purposes for at least three reasons. First, although *Doctrine* has been published as an independent book in English, it is really a huge entry commissioned by Max Weber for an extensive handbook of economics in the broad sense. Furthermore, it can, for most purposes, be considered as having been replaced by *History*. Second, *History* is so different from the other books that it is best treated separately as a kind of postscript to the whole of Schumpeter's work. Third, Shionoya has been effective in demonstrating that *Entwicklung I* and *Development* come close to being two different books. For these and other reasons, the present book does not apply the partitioning in an early trilogy and a later trilogy. Instead, the basic architecture of Schumpeter's work can be described in terms of three sets of, partly unfinished, books:

1. The two programmatic books. Schumpeter presented his evolutionary research programme of *Wesen* and *Entwicklung I.* These two books were originally conceived as a two-volume book; but their basic approaches ended up as being different. *Wesen* had announced Schumpeter's evolutionary theory by studying the narrow limits of the non-evolutionary approach and pointing towards a "dynamic" complement. *Entwicklung I* could not present this complement in terms of a new set of well-defined analytical tools. It instead presented the dichotomy between routinised behaviour and innovative behaviour and sketched out how this dichotomy could be used to explain important economic problems. The book even suggested that the dichotomy could be applied for analysing the evolutionary process of change in any sector of social life.

2. The evolutionary trilogy. Schumpeter elaborated his evolutionary theory in *Development, Cycles,* and *Capitalism.* According to Rendigs Fels (1964, viii), it is this "trilogy setting forth 'the Schumpeterian System' ". We have already noted that *Development* is distinct from *Entwicklung I.* It represents the beginning of Schumpeter's attempt to make his evolutionary theory operational. This attempt is further developed in the first 200 pages of *Cycles.* The rest of this book can be seen as the application of the evolutionary theory for the analysis of the waveform history of capitalist economic evolution. This analysis brought Schumpeter far beyond the limits of the economic system. His solution was to return to his early idea of a general theory of social evolution. The results of this return are presented in *Capitalism.*

3. Works in progress. The present book tries to demonstrate that all Schumpeter's major works can best be considered works in progress. When he died in 1950, he nevertheless left unfinished not only *History,* but also a couple of other research projects that are even more important for the understanding of his evolutionary economics. First, he tried to promote the efforts of *Cycles* to combine theory with the history and statistics of capitalist economic evolution. Second, he wanted to complement *Wesen* with a book on the analytical tools available for the formalisation of evolutionary theory (as well as equilibrium economics). Third, he actually finished most of his *History.* From the perspective of the present book, it

is important to note that the manuscript provides an evolutionary interpretation of the history of economic analysis. Furthermore, an implicit theme of *History* is why the science of economics has not expanded to include a complementary analysis of economic evolution. The explanation is apparently that adequate analytical tools had not been provided by the time of Schumpeter's death.

The hints of the evolutionary interpretation of these three sets of books are rather abstract. There is, however, an easy way of obtaining an impression of major parts of the contents of Schumpeter's books. It is to exploit his tendency to mirror himself in the work of other great researchers. Actually, Schumpeter's description of the essence of the contribution of a great economist and social scientist also serves to characterise major parts of his own contribution:

1. "His topic was capitalist evolution. Everything he ever wrote, even his scheme of a stationary society, was written to elucidate this topic" (*History*, 1131).
2. "[H]e was essentially period-bound as a theoretical technician" (*History*, 391).
3. Nevertheless, "the grand vision of an immanent evolution of the economic process ... remains after the most vigorous criticism has done its worst" (*History*, 441).

Although Schumpeter used these formulations to characterise the contribution of Karl Marx, they can even better be used for his own work. The reason is that Marxian analysis mixes the idea of unfolding development with the idea of unpredictable evolutionary transformation to a much larger degree than Schumpeter did. If we leave the problem of the full clarification of the Schumpeterian meaning of the term 'evolution' to the rest of the present book, the above quotations provide us with three rough propositions about his work.

The first proposition is, of course, exaggerated if we take the phrase "[e]verything he ever wrote" too literally. Many of Schumpeter's minor writings, as well as major parts of *History*, reflect his position as a university professor who, apart from his evolutionary research, had to teach and organise economics in general (see the classification of Schumpeter's works based on subjects, which starts on page 425). We, however, should take seriously the proposition that he designed apparently non-evolutionary parts of works like *The Theory of Economic Development* and *Business Cycles* as an integrated part of his analysis of the capitalist process of economic evolution. For instance, he clearly considered his model of the "circular flow" in a non-changing economy as a part of his large-scale development of evolutionary modelling.

The second proposition concerns the influence of the available analytical tools on Schumpeter's theoretical development. We have already seen that he missed the tool that consists in a solid terminology. This fact had consequences for not only the spread of his theory, but also his development of it. His major tools of thought, however, were provided by neoclassical economics, and they were quite inadequate for the development of his evolutionary modelling. Actually, his research was seriously hampered—just as Marx's work had been constrained by the use of the analytical apparatus of classical economics. To overcome this problem, Schumpeter searched in vain for "a new apparatus [that] poses and solves problems for which the older authors could hardly have found answers even it they had been aware of them" (*History*, 39). The failure of this search largely explains why Schumpeter's evolutionary research programme did not really succeed during his lifetime: He simply lacked the conceptual, mathematical and statistical tools needed for expressing and developing his vision of economic evolution. To be more precise, Schumpeter did formulate simple analytical models for handling innovative change of the economic system; but he never developed them into full-blown models that could open up the realm of economic evolution for systematic research by a large group of researchers.

The third proposition emphasises that it is possible to discern a pre-analytic "vision" of evolutionary change that is only imperfectly represented by formal evolutionary analysis. Although Schumpeter's vision of capitalist economic evolution cannot be counted as a scientific contribution, it inspired both him and subsequent researchers to make such contributions. His core vision seems to have been developed in polite opposition to Léon Walras's model of general economic equilibrium—especially by reinterpreting the concepts of the entrepreneur and economic equilibrium. However, his vision of a "magnificent dynamics" (Baumol, 1959, Ch. 3) can even better be understood in its opposition to Marxism. What Schumpeter responded to was not Marxian economics in the narrow sense, but the evolutionary vision and the broad research problems confronted by Marx. Schumpeter can be described as the sole economist in the first half of the twentieth century whose thought and analysis "*turned* upon evolution" (*History*, 436). Since Marx was the sole author of the past who could be considered as having had a similar evolutionary pivot for his work, he represented an important challenge for Schumpeter's research.

Schumpeter focussed on the mechanisms of economic and social evolution under capitalist conditions. His vision of these mechanisms sees all parts of life in a capitalist society as characterised by two opposite forces. On the one hand, most people wish to enter the different aspects of social life in a routinised manner. Thereby, such people contribute to the functioning of society by giving it an important degree of predictability. They are thus, in a certain sense, the pillars of society. On the other hand, there exist people who do not like fixed routines and who have the capability and will to renew these routines. These pioneers, or innovators, create innovative firms, new art forms, and novel political parties. In economic affairs, the innovative entrepreneurs need credit to obtain profit from their innovative projects. The successful projects give raise to extraordinary incomes for their initiators and many other agents. However, these incomes will sooner or later disappear because of the diffusion of the innovations and/or because of the emergence of new innovations. Both in the economy and in other spheres of social life, these innovations emerge in waves. Schumpeter's basic reasons are that the efforts of the innovators presuppose that most of social life is more or less predictable and that the room for successful innovative activities in any particular period is quite limited. The more innovators, the more difficult was the situation for the next innovator with respect to resources and predictability. Things have to settle down to ordinary routine before further innovation can take place. Thus, the expansion and transformation of the routine economy tends to emerge in waves. Through a series of such innovative bursts, the evolution of the capitalist routine system takes place.

The Schumpeterian "process of industrial mutation" thus includes "revolutions" that "occur in discrete rushes which are separated from each other by spans of comparative quiet"; this means that there "always is either revolution or absorption of the results of revolution" (*Capitalism*, 83n). His formulations suggest that he was envisaging a process of change that is able to stir the imagination and emotions of both economic actors and researchers deeply. He expressed his vision of this dramatic process by means of a basic analytical model or "scheme" that apparently tamed the drama. However, the drama is just below the surface of the scheme. Actually, several of his contemporaries suggested that Schumpeter had constructed a semi-formalised epic of a sequence of heroes whose innovations transform a society that otherwise tends to produce dull routine. Nevertheless, the semi-formalised scheme helped him develop specific theories and models, and they in turn helped him organise the complex facts of economic evolution. To function in this way, his basic scheme and his concrete models needed to make assumptions that were not part of the vision. More specifically, the purpose of Schumpeter's assumptions was to allow an untraditional form of equilibrium analysis. His concept of equilibrium implies that the evolutionary process has come to a temporary halt, and this concept allowed him to study evolution in well-defined steps. In the simplest case, the innovative activities emerge from a routine system characterised by this type of equilibrium, and after the implementation of the innovations, a new equilibrated routine system emerges. Let us consider the case of long-term economic evolution for which we can describe Schumpeter's basic analytical scheme in the following way:

- Initial equilibrium: We start from an economic system in which evolution has come to a halt so that it is based on solid routine behaviour. This system is assumed to have found an equilibrium that allows the economic agents to operate year after year in their accustomed ways.
- Economic innovation: The initial equilibrium breaks down when a minority of innovators renews some of the routines. Under capitalist conditions, a strong credit system helps these innovators—the Schumpeterian entrepreneurs who establish new firms.
- Creative destruction and renewed equilibrium: After a competitive struggle between agents related to old and new routines, a renewed and well-established routine system emerges.
- Long-tern economic evolution: The renewed equilibrium forms the basis for another phase of disturbing innovative activity. The long-term economic evolution of the routine system consists of a series of routinised equilibria and innovative rebellions against these equilibria.

This scheme of punctuated equilibria cannot be interpreted as reflecting a form of developmentalism in Schumpeter's thinking. The reason is that he did not assume that the renewed equilibrium was the deterministic result of the initial equilibrium, not even if we knew the innovation that disturbed it. The scheme is rather reflecting a strategy for analysing the immensely complex process of evolution. This strategy might appear as an extended form of comparative-static analysis, but we should not overlook that Schumpeter is applying a very untraditional concept of equilibrium for his theory of waveform economic evolution. The initial equilibrium has not come about by the deliberations of actors with perfect foresight and flexible behaviour. Instead, it is the outcome of a process of bankruptcy, job destruction, and stressful learning.

According to Schumpeter, the actors with vested interests in an equilibrium—in which evolution has temporarily come to a halt—consider economic evolution as the process of the repeated destruction of their routinised positions. Therefore, they strongly oppose a repetition of this painful process by a new wave of innovation. Their power positions largely determine the speed of replacement of old routines by new routines "in the perennial gale of creative destruction" (*Capitalism*, 84). In the feudal economies of Europe, the opponents of "creative destruction" normally had the upper hand, and the result was that economic evolution for long periods came to a halt. In contrast, capitalism in its classical period did not allow such evolutionary stasis. The Schumpeterian entrepreneurs did not have to consider the social costs of their activities, so there was a relatively unhindered movement of the economic frontier. This lack of power of the carriers of old routines did not hinder their resentment, however, and even classical capitalism included the repeated emergence of social and political movements that were either plainly against "creative destruction" or for a redistribution of the gains from this process.

The short description of the application of Schumpeter's analytical scheme for the analysis of the conflicts of capitalist economic evolution might suggest that he was solely an economic sociologist who did not think in terms of basic economic concepts. This impression is wrong. However, he changed the economic concepts so that they could serve his emphasis on the process of innovative change of the economic system. Actually, he designed the whole of his theory of economic evolution with a single "analytic intention, namely, the intention to make the facts of innovation the basis of our model of the process of economic change" (*Cycles*, 87; *Cycles Abr.*, 62). This is obvious from the following battery of definitions:

- Innovation: "we simply define innovation as the setting up of a new production function" (*Cycles*, 87; *Cycles Abr.*, 62).
- Entrepreneur: "For actions which consist in carrying out innovations we reserve the term Enterprise; the individuals who carry them out we call Entrepreneurs" (*Cycles*, 102; *Cycles Abr.*, 77).
- Profit: "an entrepreneur who ... carries out an innovation ... [receives] Entrepreneur's Profit, or simply Profit" (*Cycles*, 104–5; *Cycles Abr.*, 79).
- Credit "is essentially the creation of purchasing power for the purpose of transferring it to the entrepreneur" (*Development*, 107).
- Capital is "that sum of means of payment which is available at any moment for transference to entrepreneurs" (*Development*, 122; emphasis removed).
- Business cycles: "innovations would suffice to produce alternating prosperities and depressions" (*Cycles*, 223; *Cycles Abr.*, 115).
- Economic evolution: "The changes in the economic process brought about by innovation, together with all their effects, and the response to them by the economic system, we shall designate by the term Economic Evolution" (*Cycles*, 86; *Cycles Abr.*, 61).
- Capitalism "is that form of private property economy in which innovations are carried out by borrowed money, which in general, though not by logical necessity, implies credit creation" (*Cycles*, 223; *Cycles Abr.*, 179–80).

- Capitalist evolution is "economic evolution as conditioning, and being conditioned by, the institutional pattern of bourgeois society" (*Cycles*, 304; *Cycles Abr.*, 202).

This list of definitions serves to emphasise that everything in Schumpeter's evolutionary economics turns on his definition of innovation. This definition—innovation as the introduction of a new production function—presupposes a routine system in which the notion of innovative jumps is well-defined. Schumpeter normally presupposed a very special kind of evolutionary process that has more similarities with the biological theory of "punctuated equilibria" (Gould, 2002) than with gradualist accounts for biological evolution or economic evolution. Such an evolutionary process takes its starting point in an equilibrated system in which novelty is very difficult to introduce; but when a radical innovation succeeds, it promotes a cascade of further innovations before the system settles to a new equilibrated state. Schumpeter's reason for applying this approach was largely analytical, so we may speak of his 'methodological saltationism'. While Marshall's (1961) *Principles* had the motto "*Natura non facit saltum*" (Nature does not make jumps), Schumpeter's motto for his evolutionary analysis was obviously *Natura facit saltus*—Nature does make (sudden) jumps. It was by means of this methodology that he could most easily introduce his novel versions of the concepts of entrepreneurship, profit, credit, capital and capitalism. Furthermore, he could use a generalised version of the methodology when he turned to the study of the evolution of social norms, politics, art and the sciences. In his general theory of evolution in all the areas of social life, he needed broader concepts of innovation, entrepreneurship, and evolution; and he did not use concepts like profit, credit, and capital. We, nevertheless, shall see that the basic features of the different evolutionary processes are the same.

The transformation of Schumpeter's "magnificent dynamics" to solid evolutionary analysis was strongly constrained by the available analytical tools of neoclassical economics. Nevertheless, he hardly considered it paradoxical that he promoted the use of these tools through his teaching. According to the analyses in his two first books (in 1908 and 1912), the full modernisation of the science of economics required a division of labour between two separate and complementary branches of economic analysis. The non-evolutionary branch should promote a reformed neoclassical economics. This branch was clearly teachable and it also provided relatively predictable research projects for graduate students. In contrast, the evolutionary branch of economic analysis was hardly teachable during Schumpeter's academic life. Furthermore, it was characterised by very difficult and risky research topics. Therefore, he saw no paradox in serving one branch by his teaching and the other branch through his research.

Schumpeter's double strategy for economics as a science required a mutual recognition of the peculiarities of the two core branches of economic analysis. On the one hand, an evolutionary researcher should recognise that modern neoclassical economics deals with the core issue of understanding and managing economic systems at given levels of evolution. On the other hand, non-evolutionary researchers should recognise that some of the topics that were traditionally treated by them should be left for evolutionary economics. This mutual recognition was not made easy by Schumpeter's rejection of the possibility of the long and gradual march from statics toward the "Mecca" of evolutionary economics, which had been suggested by Alfred Marshall. Furthermore, Schumpeter argued as if his evolutionary economics was the only way of handling adequately core economic phenomena like profit, credit, capital, and the business cycle. It would probably have been easier for him to obtain serious attention if he had paid more respect to alternative approaches to economic evolution. It might also have helped him to make the more modest claim that his type of evolutionary analysis allows a novel and, perhaps, deeper analysis of central aspects of core economic phenomena.

## 1.3 Alternative images of Schumpeter's work

The starting point for the Schumpeter studies that emerged immediately after his death (see Harris, 1951b) can best be described by the intellectual situation of his many students at Harvard University in the 1930s and 1940s. One of them, Robert Heilbroner (1983, 238), wrote:

> "No one ever knew quite what to make of this small, dark, aristocratic man with a taste for dramatic prose and theatrical gestures. ... Everyone agreed that he was brilliant—and perplexing. His students at Harvard complained that he was never predictable, and they were entirely right. ... [They] had to reconcile the fact that he was the most 'hopeless' of conservatives and at the same time an admirer of Marxist economics; a sarcastic critic of the critics of capitalism, and yet its severest critic himself; a scoffer of those who fussed over every sign of trouble in the economy, and himself a diagnostician of its failing health."

This description points out that his students largely regarded Schumpeter conflicting opinions as reflecting his taste for paradox and conspicuous performance. Thus, his students largely missed the point that the perplexing standpoints are typical of a serious and broadly interested explorer of capitalist evolution. They were not seriously interested in complex evolutionary issues; and Schumpeter did not try to change the situation by teaching about his core topic and its difficulties. His teaching strategy was to stimulate students to work based on the most recent and advanced research results of modern economics; and he never directly included his own theories and their real background. As another former student remarked, the "unforgivable omission" in Schumpeter's otherwise excellent—and very demanding—lectures was that "his students never heard a word of Schumpeterian economics" (Smithies, 1950, 633). Since important parts of his work were difficult to obtain and only available in German language, this meant that most of his contemporaries had difficulties in seeing the Schumpeterian wood for all his individual trees.

The incoherence of the efforts to interpret Schumpeter's work reflects its scale and scope as well as the fact that he did not create a school of economics. Actually, he actively opposed the formation of a closely knit group of followers: "I have never tried to bring about a Schumpeter school. There is none and it ought not to exist" (*S1932b*, 600). His ambition was "not to close doors but to open them". This ambition has proved fruitful for the development of economics as a science since he pushed his many research students and other research contacts in the directions that they had chosen themselves. For instance, he promoted the work of Paul Samuelson on the foundations of modern economics, Nicolas Georgescu-Roegen on even more foundational issues, Ragnar Frisch on the establishment of econometrics, James Tobin on macroeconomics, Paul Sweezy on Marxian economics, and Hans Singer on development economics. He also promoted the collective efforts to modernise non-evolutionary economics and to organise econometrics. However, he hardly ever tried to promote directly his own form of evolutionary economic theory and its potential extension with an evolutionary econometrics.

The search for Schumpeter's core contribution was difficult even for memorialists with mastery of his complex background and of his two main languages (English and German). Furthermore, when they found his evolutionary core, they were sceptical about claiming it the reason for his greatness. For instance, the Austrian-American economist Gottfried Haberler was excellently endowed to characterise the scientific contribution of his long-time friend and Harvard colleague. Nevertheless, it was not in terms of a core contribution that Haberler (1950, 333) substantiated Schumpeter's claim for greatness. Instead, he respected Schumpeter as a highly talented and resourceful personality—and as a man who was not afraid of becoming a member of the smallest possible minority. Thus Haberler (1950, 344) wrote: "His independence was not a pose. One could truly say of him what Nietzsche said about Schopenhauer: 'Seht ihn nur an—Niemandem war er untertan'." The translation is: "Look at him—he was mastered by no one". This formulation was the highest possible praise that could be given in the intellectual culture of the old Austria in the beginning of the twentieth century, which was the background of both Schumpeter and Haberler (see Janik and Toulmin, 1973). The praise was restricted, however. Austrian intellectuals knew that Nietzsche praised Schopenhauer as a great personality while rejecting his work: "Was er lehrte, ist abgetan"—what he taught is past history. Haberler omitted this part of the characterisation from his memorial; but he gave the reference so that everyone could look it up. Therefore, he was implicitly suggesting that the future could not build on Schumpeter's work.

Paul Samuelson had an image of Schumpeter that was close to that of Haberler; but he later added that Schumpeter was not satisfied with combining a great vision with minor contributions to a surprisingly large number of scientific fields. According to Samuelson (1981b, 20),

> "Schumpeter was a universalist in economics. Mention a field in the subject of political economy, and you will find his name already established there: economic theory, macroeconomic business cycles, methodology, econometrics, Marxian economics, economic history, *Dogmengeschichte* [history of economic thought]—the list is only countable finite. ... As he himself might put it, 'This is a remarkable performance.' It is one that ought to have brought him satisfaction and fulfilment. But ... [t]he Wagnerian hero does not strive to be a Jack-of-all-trades".

Schumpeter had definitely not strived for a "Popeship" of economics that was earned by a "Jack-of-all-trades", and both Samuelson and Haberler would probably have agreed that this is not the whole story. The reason why we commemorate and study Schumpeter is not only that he was a strange and great man, but also that what he taught about a scientific 'revolution' has not yet been carried out. This problem has become increasingly recognised because we today seem to have overexploited the non-evolutionary Walrasian paradigm and because new evolutionary concepts—as well as new evolutionary mathematics and statistics—seem to provide the tools needed for moving beyond Schumpeter's evolutionary economics. In this situation, his efforts to analyse capitalist economic evolution become just as fascinating as his life. Schumpeter *was* able to perform many kinds of work, but this versatility largely expressed the quest for *one* great goal that is worthy of a Wagner-style romantic hero.

Any ambitious study of Schumpeter's work has to confront important methodological issues; and these issues are an important theme in Yuichi Shionoya's *Schumpeter and the Idea of Social Science: A Metatheoretical Study*. Shionoya (1997, xi) emphasised that "we must consider all of his work ... rather than reading him in snatches". The assumption is that "Schumpeter's thinking can be reconstructed as a paradigm" (p. xii). The major results of Shionoya's reconstruction is that Schumpeter's research programme was that of a "universal social science" and that he implemented it by means of instrumentalist methodology. In his response, Mário Moura (2002, 819) suggested that this proposition represents the wrong assumption "that there is a Schumpeterian essence". Moura (p. 805) also summarised the result of this assumption:

> "You are somewhat confused. Having consulted Schumpeter's major works and the literature devoted to his writings, you find (a) that it is often the case that interpreters present contrasting views on him and (b) that almost invariably there exists textual support for their contrasting theses. In fact, you are beginning to wonder whether you are ever going to be able to *understand* Schumpeter. This could demand that you study his entire work—a considerable enterprise, and the end of which it could be that you remain confused."

According to Moura, the problem is that the interpreters of Schumpeter base their interpretations on the assumption that Schumpeter produced consistent results. In Shionoya's case these results seems to be interpreted as the outcomes of Schumpeter's consistent use of a consistent instrumentalist methodology. Moura suggests that the real task of metatheoretical reconstruction is to explain why pervasive tensions and inconsistencies emerged in Schumpeter's work. The abstract explanation is that there is an incompatibility between the Schumpeterian world view and the world view underlying the scientific tools that he used. Schumpeter accepted theories that presuppose closed systems, but he wanted to study an open-ended social system characterised by innovation and entrepreneurship. Although this alternative interpretation is a useful complement to Shionoya's reconstruction, it is hardly characterised by a deep understanding of the precise nature of the challenges that Schumpeter had to confront in his evolutionary research. By dramatising his inconsistency through metatheoretical reconstruction, Moura might actually reinforce the opinion of the majority of economists that methodological and ontological discourse on their work is likely to create more heat than light. It might also support their tendency to take refuge in pragmatic instrumentalism—as Schumpeter seems to have done.

Although Schumpeter's instrumentalism is a major theme in Shionoya's book, his main emphasis is put on the interpretation of Schumpeter as having a fundamental idea of a universal social science. This interpretation of Schumpeter's work seems to be different from the interpretation of the present book. Actually, the phrase 'the evolutionary interpretation of Schumpeter' was coined by Shionoya to emphasise that other interpretations are available. His own institutionalist interpretation of Schumpeter becomes especially clear if we consider parts of his recent collection of papers (Shionoya, 2005) as well as his review of *The Contribution of Joseph Schumpeter to Economics*, edited by Richard Arena and Cécile Dangel-Hagnauer (2002). According to Shionoya's (2003, 604–5) review, the mission of this volume "is to resist the currently dominant [evolutionary] interpretation of Schumpeter" by an "institutionalist interpretation" that emphasises his promotion of social science in general. Shionoya is probably right that the predominant viewpoint among modern economists is that Schumpeter can better be characterised as an evolutionist than as an institutionalist. However, although this viewpoint might be the "currently dominant interpretation", it is not dominating the book-length accounts for Schumpeter's work. On the contrary, the dialogue is seriously hampered because it is hardly possible to find a satisfactory exposition of the evolutionary interpretation of this work. The present book tries to remedy this problem. Since the development of the evolutionary interpretation of Schumpeter is difficult, there will be very little room for returning to the alternative images of his work. Therefore, readers interested in exploring other interpretations and in determining the limitations of the present evolutionary interpretation will have to consult the broad Schumpeter literature (see Appendix B).

## 1.4 The structure of the present book

As already suggested, the study of Schumpeter's work can largely concentrate on his major books. If we want to interpret his research efforts through a combination of historical reconstruction and rational reconstruction, we need to group these books. The grouping defined by the list on page 8 seems sufficient for the purposes of the present book. In Part I, we shall confront the two programmatic books that Schumpeter produced before World War I. In Part II, we analyse the evolutionary trilogy that was published between 1934 and 1942. Finally, in Part III we consider the works in progress that Schumpeter left when he died in 1950. Let us quickly consider the contents of these three parts.

Part I studies the emergence of Schumpeter's evolutionary economics and its relation to equilibrium economics. Chapter 2 gives information on his early academic life and on the scientific context in which he developed his programmatic books. Chapter 3 emphasises that *Wesen* not only contains a call for a modernisation of economics in its non-evolutionary form, but also for the development of evolutionary economics as a complementary branch of the science of economics. The next two chapters focus on the parts of *Entwicklung I* that were omitted or rewritten in *Development.* These parts of Schumpeter's second programmatic book are used for two major purposes. Chapter 4 tries to demonstrate how the strange phenomenon of elite theory influenced his theory of the mechanisms of economic evolution and social evolution. Chapter 5 explores how he used the resulting analysis to contribute to contemporary scientific discussions on the major problems of capitalist economic evolution.

Part II turns to Schumpeter's evolutionary trilogy. Chapter 6 discusses the relationships between *Development, Cycles,* and *Capitalism.* The chapter also emphasises that his evolutionary economics is not least characterised by the analysis of the mechanisms of the "capitalist engine". The next three chapters analyse the books of the evolutionary trilogy in reverse order. Chapter 7 analyses core parts of the contents of *Capitalism.*

Although the study of the oligopolistic version of the capitalist engine is emphasised, there is also room for discussing Schumpeter's suggestions on the analysis of political evolution and on the long-term sociological trends that point beyond capitalism. Chapter 8 turns to *Cycles* in order to explore his idea that the evolutionary process within capitalism necessarily takes the form of waves of economic evolution. It also covers *Cycles*'s attempt to integrate the theoretical, statistical, and historical analysis of economic evolution. Chapter 9 returns to Schumpeter's magnum opus in the form that it took in *Development* (and *Entwicklung II*). By adding the closely related account for economic evolution found in the first chapters of *Cycles*, the chapter tries to explore Schumpeter's basic model of the capitalist engine that emphasised innovative entrepreneurs and the conservatism of incumbent firms. The chapter ends by considering the differences between the two models of the capitalist engine found in the evolutionary trilogy: Schumpeter's Mark I and Mark II models.

Part III presents the further development of Schumpeter's programme for the science of economics in terms of works in progress. Chapter 10 steps back from his concrete evolutionary analyses to consider his role during and after what has been called the years of high theory and high econometrics (1926–39). Chapter 11 demonstrates how he in *History* developed his theory of scientific evolution and how he applied it to the history of economic analysis. One of the implications of this theory of the history of economics is that evolutionary economic analysis had failed to take off because of its lack of adequate tools. Chapter 12, the final chapter of the book, discusses how Schumpeter's unfinished works point beyond his personal evolutionary economics. The discussion is organised around the three fundamental fields of evolutionary economics, which is implicitly suggested by *History*. The fields are: evolutionary economic theory, evolutionary economic statistics, and evolutionary economic history. The discussion also points at evolutionary economic sociology and the many fields of applied evolutionary economics.

The book includes four appendices. Appendix A contains a reconstruction of Schumpeter's curriculum vitae. Appendix B shortly presents the literature that takes its starting point in the writing of Schumpeter's biography. Appendix C classifies Schumpeter's works with an emphasis on their English availability. Finally, Appendix D gives a relatively formal account for some analytical tools needed for the ecological and statistical approaches to economic evolution. The bibliography at the end of the book is divided in two parts. The bibliography of "Schumpeter's Works" is relatively complete; and it can be used for many purposes. The bibliography in "Other References" reflects the present book's relatively minimalistic mode of citation.

# Part I

# Equilibrium Economics and Evolutionary Economics

# 2
# The Early Years

It is often assumed that Schumpeter made the first formulation of his evolutionary research programme in *The Theory of Economic Development* (*Development*). This assumption is misleading for at least two reasons. First, the 1934 English edition of this book is the somewhat modified translation of the second German edition of 1926 (*Entwicklung II*). In turn, this is the radically revised version of the first German edition of 1912 (or 1911, if we emphasise the point of time at which it became available). Second, the first edition of *Theorie der wirtschaftlichen Entwicklung* (*Entwicklung I*) is clearly a sequel to Schumpeter's first book in which he—in 1908—announced the research programme that added evolutionary economics to equilibrium economics. This book is *Das Wesen und der Hauptinhalt der theoretischen Nationalökonomie* (*Wesen*)—and it has never been translated into English. In very free translation, this book can be called 'The Essence and Limits of Equilibrium Economics' while the title of his second book could have been 'The Essence of Evolutionary Economics'. Schumpeter implicitly pointed at the importance of these books by emphasising the importance of the "third decade" in the lives of great economists. This decade represents "that period of sacred fertility which, in the case of every thinker, creates what is subsequently worked out" (*S1921*, 87). In his own case, the results of the "period of sacred fertility" from 1903–13 were largely presented in *Wesen* from 1908 and *Entwicklung I* from 1912.

## 2.1 Research programmes for the twentieth century

Schumpeter's based his two programmatic books on a careful and independent study of the state of the art of his chosen science. Economics was in the beginning of the twentieth century dominated by the results of the neoclassical revolution from the 1870s and onward. As he later emphasised, the maturation of this revolution meant that economics had entered "a classical situation in our sense" (*History*, 754). The "leading works" of this classical situation "exhibited a large expanse of common ground and suggest a feeling of repose, both of which created, in the superficial observer, an impression of finality". When Schumpeter spoke of "the finality of a Greek temple that spreads its perfect lines against a cloudless sky" he was probably not least referring to the beautiful architecture of Walras's model of an economic system in general equilibrium. However, he recognised that the "classical situation" was better expressed by Alfred Marshall's (1961) *Principles of Economics*. This book is "the classical achievement of the period, that is, the work that embodies, more perfectly than any other, the classical situation that emerged around 1900" (*History*, 834). It obtained this status by de-emphasising general equilibrium analysis and by promoting a pragmatic form of economic analysis that could relate to the statistical and historical work of the historical school and point towards the analysis of long-term economic evolution. Nevertheless, Schumpeter added that "in the last decade or so before the outbreak of the First World War, even the superficial observer should have been able to discern signs of decay, of new breaks in the offing, of revolutions that had not yet issued into another classical situation" (p. 754).

Schumpeter was no superficial observer, and he immediately focussed on the challenges for economic research. The architecture of the theoretical constructs of the leading neoclassical economists had been obtained at serious costs. Especially, the relationship between crucial facts of economic life and the theories of academic economists had become even more troublesome than before. One of the most fundamental reasons for the discrepancy between neoclassical theory and facts of economic life is that the facts are reflecting a rapid historical process. Marshall (1897, 121, 133) emphasised this problem when he prepared for the turn of the century by addressing "the new generation". He pointed out that social science is "the reasoned history of man" and that "the true analytical study of economics is the search for ideas latent in the facts" that are provided by "the historian and the observer of contemporary life". He especially focussed on the challenge of analysing the long-term social and economic changes. Since he had not made this task sufficiently clear in the first edition of his *Principles of Economics*, he now pointed out that the crucial step forward was to analyse "evolution", or "progress". This is the background for his famous statement that "[t]he Mecca of the economist is economic biology rather than in [mechanical] economic dynamics" (Marshall, 1898, 42, 43). In Schumpeter's semi-centennial appraisal of *Principles*, he pointed out that "Marshall held a definite theory of economic evolution, though true to his habit he did not press it upon the reader's attention" (*S1941a*, 106). Actually, this theory "stood in the very center of his thought."

When the American Thorstein Veblen (1898b; 1899–1900), in a series of papers, confronted the question "Why is economics not an evolutionary science?", he chose to ignore the contribution of Marshall and to confront the neoclassical analysis by John B. Clark, a leading American economist. Veblen argued that although the evolutionary theories of Darwinism and Social Darwinism were spreading rapidly in the last decades of the nineteenth century, they had not influenced the core of economics. His explanation was that economic theorists had been engaged in defending the status quo of the capitalist economy by building static models that showed that a perfect market economy gave optimal results. To build such models, economists had applied unrealistic assumptions of pleasure-seeking hedonism, antisocial atomism, and instantaneous calculations. Veblen's solution was that economic theory should instead apply the assumption of instinctive and habit-based behaviour—since such behaviour explains the emergence and gradual evolution of economic institutions.

For theoretically minded economists, it was Veblen that seemed to be out of touch with the actual scientific development. He suggested abandoning solid non-evolutionary theory in favour of an evolutionary economic sociology that he never specified in a precise way. Furthermore, he ignored the difficulties of evolutionary analysis that several leading neoclassical economists—including Marshall and the members of the Austrian school of economics—had tried to overcome. Finally, he took no notice of the fact that Darwinism was facing a crisis even in biology. These and other characteristics of Veblen's arguments meant that Schumpeter, although he hardly took explicit notice of the arguments, was one of his critics. When Schumpeter started to write in the first decade of the twentieth century, most biologists had accepted the phenomenon of evolution; but they thought—wrongly—that the Darwinian mechanism of natural selection had been outdated by explanations based on the inner characteristics of organisms (learning, laws of organismic functioning, and genetic mutation). At the same time, it had become clear that Social Darwinism had led the social sciences into a quagmire of largely unfounded speculation. On these backgrounds, a strong reaction set in against the very tendency that Veblen tried to promote. This reaction helps to explain why Schumpeter saw hasty evolutionary theorising as part of the problem rather than as part of the solution. Instead, he saw the theoretical core of economics as a structure that should be developed and complemented instead of being thrown away—as Veblen seemed to suggest.

Schumpeter's special brand of evolutionary economics can to some extent be seen as a response to the Marshallian challenge. However, he did not try to develop what has later been called "Marshall's evolutionary economics" (Raffaelli, 2003) and he largely ignored the rudiments of an evolutionary economics of his teachers from the Austrian school of economics. Instead, his response followed anti-Marshallian lines and, to some extent, anti-Austrian lines. Schumpeter rejected not only a loosely defined "economic biology" but also the feasibility of the gradual movement from static analysis to evolutionary analysis. According to Schumpeter's interpretation, those economists engaged in developing the core of neoclassical economics were essentially dealing with the functioning of *given* economic systems. He furthermore thought that the maturation of this type of analysis was hampered by the Marshallian strategy of the gradual expansion towards evolutionary issues. Finally, he seems to have agreed with Veblen that the development of neoclassical economics to modern non-evolutionary economics required behavioural assumptions that hindered the development of evolutionary economics. However, Schumpeter's conclusion in *Wesen* was that a division of labour between two *complementary* types of economic analysis was necessary. Here he applied the idea by Clark that economics consists of two fundamental branches. Schumpeter initially applied Clark's old-fashioned terminology by calling these branches "Economic Statics" (i.e., equilibrium economics) and "Economic Dynamics" (i.e., evolutionary economics). In *Wesen* (182–3), he emphasised that "Statics [equilibrium economics] and Dynamics [evolutionary economics] are completely different fields, they concern not only different problems but also different methods and different materials." He agreed with Marshall that "[o]nly Statics [equilibrium economics] has hitherto been somewhat satisfactorily worked up and we essentially only deal with it in this book. Dynamics [evolutionary economics] is still in its beginnings, is a 'land of the future'." However, in contrast to Marshall, he emphasised that "[t]hey are not two chapters of one and the same theoretical building but two completely independent buildings" (p. 183). As a consequence, the novel field of evolutionary economics had to be created in relative isolation.

Although Schumpeter's first book focussed on equilibrium economics, its purpose was not only to propose a reform plan for this fundamental branch of economics. On the contrary, *Wesen* especially served to point at the problems that cannot be analysed adequately by equilibrium economics. In other words, this book provided a negative definition of evolutionary economics as the " 'land of the future' " where alternative methods and materials were used for the study of some of the residual problems left over by equilibrium economics. Schumpeter developed his positive definition of evolutionary economics in the 1912 book called *Theorie der wirtschaftlichen Entwicklung*, that is, "The Theory of Economic Evolution". This book, which was later shortened and translated as *The Theory of Economic Development*, initiated the evolutionary research programme that would engage Schumpeter for the rest of his academic life. However, his full programme for the development of economics also encompassed the further development of equilibrium economics—and he felt a strong responsibility for promoting this development. The fact that he was committed to both equilibrium economics and evolutionary economics has, in the Schumpeter literature, created the impression of a certain of intellectual schizophrenia. However, it is hardly considered schizophrenic that a biologist covers both the physiological functioning of a biological organism and the evolutionary process that has created this organism. Since Schumpeter believed that equilibrium economics can best be compared with physiology and that evolutionary economics is analogous to evolutionary biology (*Cycles*, 36–7; cf. *Cycles Abr.*, 14–15), the paradoxes related to his work largely dissolve. Nevertheless, the underdevelopment of his chosen branch of economics meant that he tended to apply some of the methods and materials of equilibrium economics for solving the problems of evolutionary economics. Therefore, we cannot totally reject the idea "that Schumpeter's academic schizophrenia was caused by his attempt at unifying two fundamentally incompatible world-views" (Reinert, 2002, 23).

Schumpeter not only considered the strict division of labour between equilibrium economics ("Economic Statics") and evolutionary economics ("Economic Dynamics") from the viewpoint of producing scientific results but also as a means of overcoming scientific conflicts. Since the core conflict that he was facing in his youth was the battle of methods between the neoclassical economics of his native Austria and the German historical school, *Wesen* had to add a third branch of economics to equilibrium economics and evolutionary economics. This was the sociologically, historically and statistically oriented type of economics that in the beginning of the twentieth century was very predominant in German-speaking countries. The importance of the three branches of economics is summarised in Schumpeter's concluding statement that "it shall always stay our principle to be silent—or ... to delimit ourselves to summaries of facts—about things on which we have nothing *exact* or sufficiently *interesting* to say" (*Wesen*, 618–19; emphasis in original). His general imperative of being silent about things "on which we have nothing *exact* ... to say" points at equilibrium economics. The exception that we should deal with things about which we can say something "sufficiently *interesting*"—although we cannot initially make fully formalised propositions—point at evolutionary economics. The exception is that of making propositions that serve as limited "summaries of facts"; and this exception points at the main activity of the historical school. Given Schumpeter's main principle of exactness and his two exceptions, *Wesen* can be read as a reform programme for economics that acknowledges the relative independence of three branches of economic analysis:

1. Equilibrium economics. This branch should be promoted by recognising the crucial alliance between mathematics and neoclassical economics and by cleansing it of everything that does not concern the interdependent system of economic elements.

2. Evolutionary economics. This novel branch should be given a chance to develop separately instead of being hidden and constrained by its superficial inclusion in neoclassical treatises. Furthermore, equilibrium economics should recognise that some of the topics it would like to cover can only be treated in a fundamental manner within evolutionary economics.

3. Historical and statistical economics. These branches of economics should be allowed at least the same degree of independence as the experimental branches of the natural sciences. The branches are especially crucial for evolutionary economics since they provide data on the detailed working of evolutionary mechanisms, the historical change of the mechanisms, and the institutional settings in which they work.

## 2.2 Preparing to become a great economist

The facts that Schumpeter published the 626 pages of *Wesen und Hauptinhalt der theoretischen Nationalökonomie* when he was twenty-five years old and the 548 pages of *Theorie der wirtschaftlichen Entwicklung* at the age of twenty-eight are clearly in accordance with his theory of the importance of the "third decade" in the intellectual lives of great economists. His two programmatic books resulted in his promotion of the professionalisation and mathematisation of non-evolutionary economics as a university teacher and his development of evolutionary economic analysis as a scientific researcher. It is this 'academic double life' that serves to explain why he, for a while, became the 'heretic Pope' of modern economics. Since the present book focusses on his development of an evolutionary economics, the reader largely has to look up accounts of Schumpeter's efforts as a university teacher, international coordinator, and historian of economics elsewhere—especially in the Schumpeter biographies by Loring Allen (1991), Richard Swedberg (1991), and Thomas McCraw (2007). However, these biographies largely ignore aspects of Schumpeter's early life and work that are of importance for the study of the emergence of his evolutionary economics. This is one of the backgrounds for the following sketch, which is partly based on unpublished research by the present author and on documents provided by Kiichiro Yagi (1993).

Joseph A. Schumpeter was born in 1883 in Triesch, a rather remote town in the vast and fragile Austrian-Hungarian Empire (see Figure 2.1 on the facing page). Although his father was a textile manufacturer who died early, his mother's second marriage to a retired lieutenant general assured that he entered 'high society' and obtained the best possible education, both generally and as an economist. This development took place in the centre of the empire, Vienna. This glittering and paradox-ridden city was the cradle of Freud's psychology, Klimt's paintings, Mahler's music, Boltzmann's statistical concept of nature, and the Austrian school of economics. It not only became *Wittgenstein's Vienna* (Janik and Toulmin, 1973) but also 'Schumpeter's Vienna'. The life and work in this city and other parts of Austria-Hungary meant that Schumpeter, as pointed out by his friend Wassily Leontief (1950, 103), "knew that old world at its best and he did not want to see a different one". He "remained to the

**Figure 2.1.: Map of Schumpeter's Europe before 1914**

Comment: Schumpeter was born in Triesch. His primary school was in Graz, the secondary school was the Theresianum of Vienna, and his tertiary studies were at the University of Vienna, 1901–05. Then followed postgraduate studies in Berlin and London, as well as a job in Cairo. He returned to Vienna before obtaining academic posts in Czernovitz and Graz. He then was minister and banker in Vienna before moving to the University of Bonn in 1925. He left Europe in 1932. Source: The map was drawn by Ebbe Sloth Andersen.

end the cultivated Austrian gentleman of the old school ... who found in the succession of events from 1914 onward no very striking evidence of progress" (Haberler et al., 1951, 89).

Schumpeter's education included eight years at the elite grammar school called Theresianum. Although this school had been created for the aristocracy, it also allowed entry of pupils from bourgeois families. Theresianum gave him not only a surprisingly solid knowledge in many areas (including six languages!), but also aristocratic manners and a wish to serve the Austrian-Hungarian Empire in the spirit of a non-partisan civil servant. The efficiency with which Schumpeter exploited his high school years might reflect overcompensation for a feeling of inferiority due to the modest background that he really came from. His efforts to feel at home in an aristocratic environment might even have given him a rudimentary version of his evolutionary theory. According to Robert

Heilbroner (2000, 309), it was at this

> "aristocratic school in Vienna ... he absorbed the values that were to become so important in his own life. Are we mistaken in seeing those values transferred to his own vision of history in which an elite [of entrepreneurs] becomes the central moving force? Certainly this elite is an aristocracy, embodying the belief in the natural superiority of the chosen few that lies at the core of all aristocratic views of society. But notice that the Schumpeterian few are chosen not by blood but by 'intellect and will'. It is thus an aristocracy of *talent*."

Although Heilbroner's suggestion seems plausible, it can hardy be proved. For the purposes of the present book, it is more important to emphasise a fact about the stay at Theresianum that has not been mentioned in the Schumpeter literature. This fact is that the teacher of mathematics, physics, and logic was the philosopher Alois Höfler (whose general activities have been sketched by Blackmore et al., 2001). Apart from the general influence of Höfler's teaching on Schumpeter, it should be noted that Höfler's teaching of logic must have emphasised the *System of Logic* by the economist and philosopher John Stuart Mill. Schumpeter later remembered that this book "occupied a place of honor not much below Plato's in the mind of at least one *philological* philosopher I knew as a boy" (*History*, 449n). This lover of logic, Höfler, must have presented Mill's multifaceted view of scientific methods—and this view must later have led Schumpeter to wonder about the widespread and acrimonious methodological controversies in the social sciences. Höfler also gave a university course on the methodology of mechanics. This course, which Schumpeter attended immediately after moving to the University of Vienna (*S1901–05*, 67), concerned "Discussions on the history and philosophy of mechanics (based on original philosophical texts by Galilei, Newton, d'Alembert, Lagrange, Kirchhoff, Hertz, as well as the history of Dühring und Mach)". The "discussions" of the course (see Höfler, 1899; 1900) were closely related to the instrumentalist philosophy of science developed by famous physicist-philosophers like the Austrian Ernst Mach (who provided the starting point for the development of logical positivism by the Vienna Circle of the 1920s).

Schumpeter's effective use of the stay at Theresianum meant that he was quickly maturing scientifically after starting the nearly five years at the University of Vienna in the autumn of 1901. His partially self-selected syllabus at this university is summarised by Table 2.1 on the next page. This table demonstrates that economics was taught as part of a broad educational programme in law and economics. This programme prepared the students for high positions in the Austrian administration, so Schumpeter was confronted with a surprisingly wide range of topics. In this cacophony of disciplines there was no secure place for economics, and the

**Table 2.1.: Schumpeter's syllabus at the University of Vienna, 1901–05**

| General topics | Specific topics | Sum of weekly semester hours | % of non-law courses |
|---|---|---|---|
| Law courses | | 138 | |
| Other courses: | Philosophy | 14 | 11% |
| | Mathematics | 38 | 29% |
| | History | 33 | 25% |
| | Applied economics | 31 | 23% |
| | Theoretical economics | 15 | 11% |
| | English language | 2 | 1% |
| All other courses | | 133 | 100% |

Comment: During Schumpeter's study of economics and law, more than half of the formal activities dealt with law. He emphasised the other courses, however. He studied mathematics fully on his own initiative. The study of applied economics and history was more standard, but he increased the weight of history. The low weight of theoretical economics reflects the limited supply of courses in this area. Source: The calculations underlying the table are based based on *S1901–05*.

theoretical economists were still engaged in the bitter battle of methods launched by the leader of the Austrian school, Carl Menger, against the leader of the historical school, Gustav von Schmoller (see *Doctrine*, 167–74). Furthermore, there was no place at all for mathematics in the law and economics programme, and this explains the common opinion that "[i]n his university years, he [Schumpeter] never took a course in mathematics" (Allen, 1991:I, 146). This is wrong, since Schumpeter followed many mathematics courses. Furthermore, his preparation for the mathematical economics of Léon Walras and others had been suggested by the most prominent mathematician of the University of Vienna, Leopold Gegenbauer (*S1909a*, 75). Actually, his preparation for handling creatively the mathematical economics also included the studying of the history and methodology of mechanics (including Mach, 1919, Ch. 4). He must thus have sensed, as it has later been pointed out by Jaffé (1983, 132, 275), that the "true *fons et origo* [source and origin] of Walras' multiequational formulation of general equilibrium was Louis Poinsot's once famous textbook in pure mechanics, *Éléments de statique* (1803)". Here "we find virtually the whole formal apparatus that Walras later employed in his *Éléments d'économie politique pure*. ... It was Poinsot's model that Léon [Walras] later imitated and adapted to his portrayal of general economic equilibrium."

Schumpeter, however, started with the historically and statistically oriented economics that was taught by members of the historical school.

Several of these teachers, most importantly Inama-Sternegg and Sigmund Adler, were not only engaged in practical policy issues but also in discussions on the great questions of the development of capitalism. The latter discussions influenced him to start up a large-scale research project on the economic history of the Austrian tax state (Perroux, 1965, 16)—probably already in his second university year! Then he quickly moved to the theoretical economics of the Austrian school as taught by Friedrich von Wieser and Eugen von Böhm-Bawerk, but he did not drop his historical and statistical interests (*S1905c, S1905b, S1905a*). Furthermore, he prepared for the change by an intensive and, at that time and place, unique study of mathematics and mathematical economics (*S1906a*). Thereby, he became a student of theoretical economics that was both brilliant and troublesome for his Austrian teachers. He thinking was much closer to the French-Swiss Walras and the American Clark than to the Austrian school of neoclassical economics. Furthermore, he did not accept that his teacher Böhm-Bawerk's (1921) *Capital and Interest* provided a viable response to Karl Marx's *Capital* and to the neo-Marxists. Actually, he seemed more intellectually associated with those of his fellow students who were becoming important neo-Marxists (Otto Bauer, Rudolf Hilferding, and Emil Lederer) than with the later conservative leader of the neo-Austrian school (Ludwig von Mises). This became clear when they all confronted fundamental issues of the analysis of capitalism in the famous Böhm-Bawerk Seminar of 1905 (see Section 5.2).

After graduation in January 1906, Schumpeter continued his intensive and wide-ranging studies in Berlin and London. During this period he not only visited famous economists like Alfred Marshall and Francis Edgeworth; he also studied under the statistician and evolutionary biologist Karl Pearson. These contacts as well as Schumpeter's own writings demonstrate an unusual maturity that was emphasised (Leontief, 1950, 104) by pointing out that while Keynes's "intellectual progress ... was slow at the beginning", we have to recognise that "Schumpeter attained his full scientific stature very early, before the age of thirty".

His early maturation meant that Schumpeter, at the age of 25, could present a research programme that he largely followed for the rest of his life. He presented major parts of this programme in the more than 600 pages of *Wesen und Hauptinhalt der theoretischen Nationalökonomie*, that is, "The Essence and Principal Content of Theoretical Economics". In this 1908 book, Schumpeter explained what a reformed neoclassical economics can do and what it cannot do. His starting point in *Wesen* was the battle of methods between Menger and Schmoller. His main proposition in this respect was that the two conflicting methods—the theoretical method based on the study of economic equilibria and the historical and statistical method that sought a richer understanding—were both legitimate. Since other researchers, like Max Weber (1903–06; 1904), had pointed at legitimate forms of the historical method, Schumpeter could concentrate on demonstrating the legitimacy of a modernised analysis of economic equilibrium. He considered that analysis crucial for the study of the economic system, but he also emphasised that neoclassical economists could not use their core method to contribute to the study of many of the topics that they usually tried to cover. Actually, some of these topics might only be reachable through the historical method, but he suggested that an important subset of problems is treatable if a new method of theoretical analysis emerges. In modern terms, this subset is clearly discernible as the problems of economic evolution, and they cannot be studied without the development of new methods for evolutionary analysis. Schumpeter gave several examples, like the explanation of the rate of interest and entrepreneurial profits, of the reach of the new type of analysis, and he promised to treat them more fully in a companion volume to *Wesen*.

In 1909, Schumpeter obtained his licence to teach at the Austrian universities by fulfilling several requirements. His "large doctorate" required making a test lecture. This unpublished lecture—on "The Verification of Abstract Theorems by Means of Statistics" (*S1909a*, 73)—might be seen as a forerunner to his later promotion of the Econometric Society (*S1933c*). He also submitted a teaching programme that demonstrated that he upheld his broad range of scientific interests (see Table 2.2 on the following page). For the Schumpeter literature, the most surprising topic is "modern statistics". This type of statistics was not only related to the "verification" of the theorems of economic theory, but also to the more or less evolutionary problems of biometrics and anthropology. It is, furthermore, obvious that Schumpeter wanted to connect to the best experts. Even for the mentioned test lecture, he seems to have consulted the leading Viennese statistician Emmanuel Czuber. However, he largely obtained his Continental-style doctorate by means of the programmatic *Wesen*. This book must have displeased Wieser and Böhm-Bawerk in their roles as evaluators. They were especially sceptical about his proposal to modernise neoclassical economics by means of mathematics and instrumentalist philosophy as well as on the contribution of Léon Walras rather than on the Austrian school. Although Schumpeter's brilliance and productivity secured him the degree and academic positions in the old Austria, his aggressive independence meant that his career did not take place at the wished-for University of Vienna. Instead, it was at lower-ranking universities that he implemented his research programme.

The major result of Schumpeter's work within his research programme was another large book, which came out late in 1911: *Theorie der wirtschaftlichen Entwicklung* (*Entwicklung I*). This book—which was later shortened and translated as *The Theory of Economic Development* (*Development*)—presented his theory of economic evolution as driven by innova-

**Table 2.2.: Schumpeter's 1909 teaching programme**

| Course | Description |
|---|---|
| Fundamental economic analysis | "Introduction to the study of political economy for beginners"—apparently a rather advanced introduction to the analytical tools needed for writing seminar papers. |
| History of economics | "The heritage of the classics and modern economic theory"—including classical ideas that have never been fully exploited. |
| Money | "The theory of the money market"—including descriptive economics and sociological issues. |
| Public finance | "The problems of modern public finance"—including an analysis of expenditures for military, industrial and social purposes. |
| Sociology | "The foundations and contemporary state of sociology"—including the sociology of everyday life and the sociological knowledge of artists. |
| Statistics | "Problems of modern statistics"—of the kind that had been developed in relation to biometrics and anthropology; the course would concentrate on works like those of Pearson and Edgeworth. |
| Economic evolution | "[E]ntrepreneurs and capitalists"—with special emphasis on capitalist concentration and the money market. |

Comment: Schumpeter described the first six potential courses in a letter to the University of Vienna. His list was part of the requirements for obtaining the licence to teach at Austrian universities. The last course was announced at the University of Vienna after Schumpeter obtained his doctorate. It was never held because he moved to the University of Czernovitz. Source: The quotes are from *S1909a* (77–8, 83, emphasis removed), from which the additions have also been derived.

tive entrepreneurs. This evolutionary theory was troublesome for most economists since it seemed to redefine Homo Economicus as a follower of routines and since it added the entrepreneur as something like a Homo Creativus that was not too far from the Nietzschean Superman. However, Schumpeter was not only a troublemaker. He saw his theory as a synthesis between neoclassical economics and the historical school as well as the starting point for a general theory of social evolution. His efforts were well received by one of the leaders of the historical school, Max Weber, who invited him to contribute to a major handbook of economics in the broad sense. Schumpeter's task was to give an account for the historical development of economics that served to overcome the battle of methods. The result was later translated into English as *Economic Doctrine and*

*Method* (*Doctrine*).

At the eve of the First World War Schumpeter had established himself as a major economist not only in Austria, but also internationally. This position was not only achieved through his books, but also through his international networking, including a one-year guest professorship in the USA. However, it was in his writings that he had begun the treatment of the topics that would engage him for the rest of his life. The most important of these topics were defined by *Wesen* and *Entwicklung I*.

## 2.3 Schumpeter's new intellectual combination

The above summary of Schumpeter's early intellectual development suggests that it related to two controversies. Since we have already considered the controversy between equilibrium economics and evolutionary economics, it is the battle of methods between neoclassical economics and the historical school that needs further comment. Marshall had a double solution to this battle. On the one hand, the two principal approaches to economics were depicted as complements in the same way as theoretical physics is complementary to experimental physics. On the other hand, Marshall made clear that the role of formal theory diminishes as we move from equilibrium economics to evolutionary economics, while the role of historical evidence increases. Although Schumpeter's programmatic books demonstrate that he largely agreed with Marshall, he formulated his goals more ambitiously. First, he thought that it is not the Marshallian, or the Austrian, type of neoclassical economics but the Walrasian-style equilibrium economics that should be related to statistical and historical evidence on the functioning of given economic systems. Second, he wanted to create a theoretical evolutionary economics that could complement the historically oriented studies of economic evolution within the historical school. Third, the contributions of the historical school suggested to him that this evolutionary economics needed to be complemented by evolutionary sociology in order to account for the overall process of economic and social evolution within capitalism.

As we shall see in the following three chapters on Schumpeter's programmatic books, his early efforts resulted in a novel and surprising combination of preexisting intellectual material (see Figure 2.2 on the next page). His evolutionary economics includes not only evolutionary economic dynamics but also evolutionary economic statics and evolutionary economic sociology. Furthermore, his general evolutionary vision and theory provided important inputs to his contributions in other areas (like the study of the history and methods of economics). A major question is how he arrived at his "magnificent" vision and theory. The present answer, which is an extended version of Perroux's answer, is that he developed his evolutionary pivot by confronting and combining four very

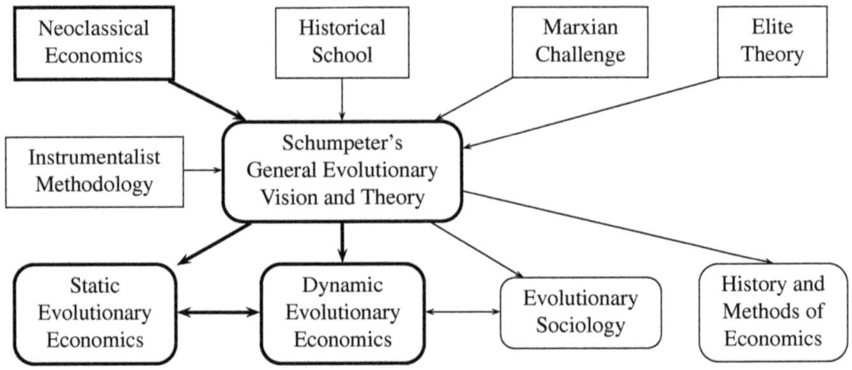


**Figure 2.2.: Schumpeter's evolutionary pivot as well as the main sources and components of his work**

Comment: The main sources of Schumpeter's evolutionary vision and theory were not only neoclassical economics and the German historical school, but also the challenges and inspirations from elite theory, Marxian analysis, and instrumentalist methodology. His main results are contributions to basic evolutionary economic analysis as well as to evolutionary sociology and the history and methodology of economics. The most surprising elements of the figure might be the inclusion of elite theory as a source and of static evolutionary economics as a result of Schumpeter's work. The latter field is characterised by the study of the functioning of states in which evolution has come to a halt and by the comparison between such states.

different contributions to the study of the economic system and its process of change: (1) neoclassical economics, (2) the historical school, (3) Marxism and (4) elite theory. It should also be noted that he applied (5) instrumentalist methodology to interpret the different sources of his work and the main results that he derived from them.

Schumpeter's response to neoclassical economics was influenced by his background in the instrumentalist methodology that he derived from the works of science-oriented philosophers, like the Austrian Ernst Mach (Shionoya, 1997, Ch. 5). This background also seems to explain why he preferred the variant of neoclassical economics developed by Walras as well as how he applied ideas of the American economist John B. Clark. Nevertheless, Schumpeter could not avoid the strong influence from the economists of his native Austria (Menger, Böhm-Bawerk and Wieser), although they subscribed to a very different methodology of science. Furthermore, his development of the evolutionary research programme was influenced by those of his teachers at the University of Vienna who belonged to the historical school as well as by the main German representatives of this school (Schmoller, Sombart, Spiethoff, and Max Weber). These researchers provided what appears to be a combination of a strong emphasis on the static analysis of the functioning of economic and social institutions with rather weak evolutionary perspectives. Furthermore, their evolutionary work was to a large extent related to the challenge provided by Marxism. Finally, some members of the historical school shared with Schumpeter an interest in elite theory. Nevertheless, he was alone in dedicating his research life to evolutionary economics and to the evolutionary analyses within other social sciences.

Schumpeter's creative overcoming of the battle of methods through the development of his evolutionary economics and evolutionary sociology was not easy. Furthermore, he presented his results by means of a confusing terminology. His basic endeavour was to argue for the peaceful coexistence of several fields of scientific studies. On the one hand, he argued for the complementarity of theoretical, historical and statistical studies. On the other hand, he also argued that equilibrium economics and evolutionary economics should coexist. For the latter argument, he defined Statics as equal to equilibrium economics and Dynamics as equal to evolutionary economics. However, he later changed his definitions of the terms statics and dynamics. According to these definitions, dynamics denotes a method that makes explicit use of the concept of time while the static method does not do so. These methods can be used by both equilibrium economics and evolutionary economics. Therefore, it is no contradiction in terms to speak of static evolutionary economics. This term simply denotes the field of evolutionary economics that studies the functioning of states in which evolution has come to a halt and that compares between such states. Since he never fully recognised the need of this term, he tended to merge static evolutionary economics with static equilibrium economics. However, the latter field operates under the assumption that evolution can be excluded from the analysis.

Schumpeter's overcoming of the battle of methods has been noted by several researchers (including Shionoya, 2005). However, it is convenient to compare the present interpretation with an old statement by François Perroux in his book-length introduction to *Théorie de l'évolution économique*—the 1935 French translation of *Entwicklung II*. Perroux's conclusion starts from the fact that

> "economic research is naturally oriented in two general directions: the one historical and sociological, the other mathematical and deductive. In relation to works in German language, the originality of Schumpeter is to have attempted a synthesis of the two types of work 'by means of a new definition of the terms statics and dynamics'. … In relation to the totality of his thought, Schumpeter is actually attempting a [double] synthesis, on the one hand, between the Austrian school and the Lausanne school [Walras and Pareto] and, on the other hand, between these two abstract systems and the historical and sociological system of Werner Sombart and Max Weber [the

historical school]." (Perroux, 1965, 21, emphasis removed)

According to this interpretation, Schumpeter attempted to make a double "synthesis", which can most easily be memorised in terms of the following pseudo-equation:

$$\text{Schumpeter's "synthesis"} = (\text{Austrian School} + \text{Lausanne School}) + \text{Historical School}$$

Perroux's formula covers major characteristics of Schumpeter's work. This work gives much evidence of the inspiration from the neoclassical economics of the Austrian school and the Lausanne school (Walras and Pareto). It is also relatively easy to detect inspirations from the empirically and methodologically oriented efforts of the historical school, that is, Schmoller and his most prominent followers (Sombart, Spiethoff and Max Weber). Perroux (1965, 189, emphasis removed) suggested that Schumpeter's "synthesis" consisted of an attempt to translate "the essence of our sociological and historical knowledge of modern capitalism" into "the language of marginal utility and equilibrium". Since this is an impossible task, Schumpeter had to drop many of the rigorous results of neoclassical economics. Nevertheless, he tended to stick as closely as possible to the Walrasian formalisation of static equilibrium economics instead of explicitly recognising the need for static evolutionary economics as the logical counterpart of his favourite field: dynamic evolutionary economics.

There are two main problems with Perroux's formula. The first problem is that Schumpeter's work does not represent a 'synthesis' in the standard meaning of this term. Schumpeter did not try to synthesise everything found in neoclassical economics and the works of the historical school. On the contrary, he freely used elements of the two opposing lines of research. That is the reason why it is better to use the term 'combination' instead of the term 'synthesis'. The other problem is that Schumpeter included much more in his new intellectual combination than just neoclassical economics and the research of the historical school. We have already seen that he included elite theory and that much of his work appears to be a response to challenges from Karl Marx and the neo-Marxists. Since elitism and the analytical aspects of Marxism also influenced the members of the historical school, this may appear a minor problem with respect to Perroux's formula. Nevertheless, Figure 2.2's specification of all the major components and sources of Schumpeter's intellectual combination seems more operational for the present study.

# 3

# From Walrasian Statics To Evolutionary Dynamics

It is presently 100 years since Schumpeter published his first book. This book is largely about the foundations of analytical economics and it was published when he was 25. Therefore, the following formulations by Stanley Fischer (1987, 235–6) may read as if they describe his book:

> "*Foundations* is the work of a 25-year-old. There are signs of youth in the eagerness to proselytize for the new mathematical faith and overreaching in trying to impose an entirely coherent theme on the material. But the book bears the unmistakable command of the economics of his material, at home with technique, and most remarkably for a young man in a hurry, thoroughly familiar and patient with the literature. It is, as Schumpeter no doubt remarked, a remarkable performance."

Schumpeter's praise, however, was not for himself but for his friend Paul Samuelson. Like Schumpeter, Samuelson finished his PhD thesis at the age of 25. At that age, both of them had ploughed through an enormous literature with an emphasis on the underlying formal structures and analytical tools. They had found widespread confusion and a lack of recognition of the basic unity under the multiform surface of topics and modes of formulation, and they both wanted to overcome confusion and lay the foundations for future research. In these and other respects, there are surprising similarities between their works, and Schumpeter's introductory statement may cover both works: "The following account belongs to the family of purely theoretical works, and it tries to carefully examine the foundation, the methods and the major results of pure economics as well as its nature, value and potential" (*Wesen*, 20).

The similarities between Schumpeter's *Wesen und Hauptinhalt* and Samuelson's *Foundations of Economic Analysis*, however, are dwarfed by two differences. First, although both books concentrate on mathematical economics, they do so in very different ways. Schumpeter argues that mathematics is the natural language of analytical economics, but he develops this argument with hardly any explicit use of mathematics. Samuelson demonstrates the importance of mathematics by applying it. Second, the two books have entirely different strategies of how to extend the reach of theoretical economics. Samuelson (1947, 284–5) considers dynamic analysis as implicit in Walrasian statics and formulates a "Correspondence Principle" according to which the "statical systems are simply degenerate special cases" of dynamical systems. The perspective is the gradual movement towards a "comparative dynamics" that ranges to the problems of "the business cycle, and even to the majestic problems of economic development" (p. 355). In contrast, Schumpeter conjured a basic dichotomy been "Statics" and "Dynamics". In *Wesen*, he considered the intermediate type of "dynamics" as illegitimate because the parameters of the statical system are not conserved over time. While a cautious form of short-term comparative statics is useful, the study of longer periods must include a theory of the evolution of parameters like production functions and consumption functions. Hence, *Wesen* is not only about the 'Foundations of Static Economic Analysis' in the limited sense; it also tries to demonstrate why the 'Foundations of Evolutionary Economic Analysis' has to be treated in an entirely different book. We shall see that some of the ideas for his next book are actually present in *Wesen*. We shall also consider some of the ways in which the study of neoclassical economics promoted Schumpeter's evolutionary vision and analysis.

## 3.1 Different interpretations of *Wesen*

The English title of Schumpeter's *Das Wesen und der Hauptinhalt der theoretischen Nationalökonomie* would have been something like "The Essence and Principal Content of Economic Theory"—but the book has never been translated to English language. Wassily Leontief (1950, 105), probably, gave the best available characterisation of *Wesen* by stating that

> "this remarkable book remains practically unknown in the English-speaking world and yet it contains the statement of his fundamental views which constitute the basis of Schumpeter's whole scientific *weltanschaung* [world view]. Some of these were never restated again as explicitly and with as much *elan* [vivacity]. It is indicative of his turn of mind that the nearest approximation to exposition of the general principles of economics was undertaken by Schumpeter at the very beginning of his career."

The consequence of the widespread ignorance of this book is not only that the best opportunity for understanding Schumpeter has been ignored; even mathematical economists missed an excellent account for their own work. He was able to provide this opportunity because he was not only interested in what could be modelled with the existing analytical techniques, but also in what was left over for novel kinds of analysis. It was his interest in the *limits* of mathematical modelling that, according to Leontief (1950, 105), explained the uniqueness and vivacity of *Wesen*: "The insight into the nature and appreciation of the true significance of a theory or a scientific procedure often finds its most sensitive measure in a clear and unequivocal statement of its inherent limitations." These limits of the prevailing form of mathematical economics became especially clear because Schumpeter not only "delineated the margins of its effective range" but also "designated what he called the process of development as the particular aspect which could not be encompassed by the conceptual schemes of static general equilibrium theory" (p. 105). Thus, *Wesen* combines issues of relevance to mathematical economists as well as to students of "the process of development", which presently is identified as economic evolution. This characteristic of the book gave it a vigour that was noted by other readers than Leontief. One of them was the Austrian-American Oscar Morgenstern, who is known for his contribution to John von Neumann's development of game theory. Morgenstern (1951, 198) got a copy of *Wesen* in the 1920s, and he remembered "what sort of revelation it was to me when I first laid hands on it and, like many others of my generation, I resolved to read everything Schumpeter had written and would ever write."

The majority of economists considered the book very differently. One of them was Friedrich von Wieser, Schumpeter's formed teacher at the University of Vienna. Actually, *Wesen* provoked the only book review that Wieser (1911, 54) ever wrote, probably because he felt that it attacked the very essence of his own work. Presently, we shall only note that he remarked that Schumpeter's "main error is that he wants to master too much; one gets the feeling that the author has not yet reached his equilibrium and still has to learn to delimit himself." Although Wieser phrased his verdict in an urban way, its slightly ironic but precise characteristic of the youthful equilibrium theorist must have hit Schumpeter hard. More than thirty years later, he emphasised his dislike of his first book: "I have no copy and have been trying to atone for this effort of my youth since it was issued" (*S1944a*, 1). Actually, he wanted to produce a radically new edition and this meant that he disallowed a German reprint and rejected proposals for a translation into English language. As a result, only the 1000 copies of the first edition were (until 1970) available for posterity, the only exception being translations into Japanese (*BL*, 263–4) and Italian that somehow avoided the Schumpeterian blockade.

The limited availability of *Wesen* is not sufficient to explain the very limited attention to its arguments. The main explanation is that it is strange for most economists. We may obtain a first impression of the strangeness of its argumentative strategy by comparing it with *Tractatus Logico-Philosophicus* by the Austrian-British philosopher Ludwig Wittgenstein (who was born in Vienna in 1889). Since Wittgenstein's *Tractatus* is notoriously dense while Schumpeter's *Wesen* is very verbose, the similarity is definitely not a matter of style. Instead, it concerns their overall method of analysis, and the reason for the similarity in this respect is not least that mathematicians and physicist-philosophers inspired them both. With such inspiration, it became a major task to clean science from metaphysics and loose talk. Wittgenstein emphasised this task as the all-dominant purpose of *Tractatus*:

> "The whole sense of the book might be summed up in the following words: what can be said at all can be said clearly, and what we cannot talk about we must pass over in silence. ... Philosophy [i.e. the early Wittgenstein's type of philosophy] sets limits to the much disputed sphere of natural science. ... It will signify what cannot be said, by presenting clearly what can be said." (Wittgenstein, 2001, Preface, §4.113, and §4.115)

This sounds like formulations of a hard-core physicist and positivist, and the natural-science-oriented intellectuals of Vienna Circle interpreted Wittgenstein in this vein in the 1920s. These logical positivists were convinced that what we can speak clearly and meaningfully about is all that matters—but Wittgenstein did not agree. On the contrary, he believed that what we cannot with acceptable clarity express is what really matters: "We feel that even if all *possible* scientific questions have been answered, the problems of life have still not been touched at all" (§6.52). The precise mapping of the realm of scientific solutions is thus, at the same time, a mapping of the core aspects of life—like meaning and value—that we have to handle in an intuitive manner. In other words, he designed his *Tractatus* like a painter who is not interested in the figures of his painting but rather in the negatively defined background of the figures. Furthermore, his theory of the limits of scientific modelling told him that he could not precisely fill out this all-important part of the painting. Much later he tried, in his *Philosophical Investigations*, to approach the hitherto untouched areas. This was a natural move; but the distance of thirty years between his two philosophical manifestos made it difficult for his audience to grasp the connection between them.

In contrast to Wittgenstein, Schumpeter was establishing himself in the science of economics with its engagement in the problems of the Austrian-Hungarian Empire and with its battle of methods between empiricists and abstract theorists. There, nonetheless, are a number of similarities. Compared with *Tractatus*, readers of Schumpeter's *Wesen* are spared for Schopenhauerian issues like death and the troublesome meaning of life, but instead they have to endure his elitist suggestions about the differences between routine-based behaviour and the behaviour of innovative entrepreneurs. Nevertheless, we, in both cases, encounter an overwhelming argument for the strict application of logic and mathematics in science—followed by short accounts of the limits of these results with respect to the really interesting issues. A closer analysis demonstrates that the overall method in *Wesen* is surprisingly analogous to that of Wittgenstein, and this methodological congruence is even reflected in Schumpeter's formulations. Thus, *Wesen*, which was published ten years before Wittgenstein finished his *Tractatus*, stated that

> "it shall always stay our principle to be silent—or ... to delimit ourselves to summaries of facts—about things on which we have nothing *exact* or sufficiently *interesting* to say." (*Wesen*, 618–19)

The similarity of formulation is not due to any direct contact between Schumpeter and Wittgenstein. Even indirect relationships are difficult to detect. However, it was not too difficult to develop their approaches from widespread views in Vienna during first decade of the twentieth century. Furthermore, the quotation from *Wesen* demonstrates that their views were not identical. The similarity is found in the general imperative of being silent about things "on which we have nothing *exact* ... to say". Schumpeter, however, made two exceptions to this rule. The first is to allow for "summaries of facts", like the main activity of the historical school. The second exception is that we may deal with things about which we can say something "sufficiently *interesting*"—even though we cannot formulate it in an exact way. Here it is not difficult to detect that he was thinking of the effects of innovative entrepreneurs.

Given Schumpeter's main principle of exactness and his two exceptions, there are at least three interpretations of *Wesen*:

1. Promoting the modernisation of neoclassical economics: We can read it in the mode of the Vienna Circle by simply emphasising exactness. Then the formulation becomes a clear-cut endorsement for the sharpened-up neoclassical economics by means of mathematical modelling.

2. Overcoming the battle of methods: The possibility of non-exact summaries of facts allows us, like Max Weber, to see the book as suggesting the coexistence of the neoclassical approach and the historical approach.

3. Preparing evolutionary economics: The approval of looser sayings about sufficiently interesting phenomena points toward Schumpeter's development of evolutionary analysis beyond the limits of "exact" economics.

Although the first two interpretations give much insight into Schumpeter's academic life and work, it is the third interpretation that best fits the purposes of the present book. According to this interpretation, *Wesen* is an—excessively long—introduction to his *Entwicklung I* and his later works. It largely serves as such an introduction by putting strong emphasis on the *limited* results that can be produced by following the refined neoclassical programme. Furthermore, while Wittgenstein took some thirty years to move to the *Philosophical Investigations*, Schumpeter had already, while writing *Wesen*, begun to explore the realm beyond what was, at least in his days, sayable in an exact manner. This interpretation seems to be the one that is most faithfully reflecting the goals that directed Schumpeter's writing of the 626 pages of *Wesen*. However, two other interpretations are important for understanding his evolutionary research programme. His ultimate goal was to bring as many evolutionary phenomena as possible into the realm of the exactly sayable. Furthermore, this clarified theory should help the summarising of whole classes of historical and statistical facts. Therefore, we shall try out all three interpretations and consider the possibility of combining them into a fourth interpretation:

4. *Wesen* promotes a research programme for economics in the broad sense. Modern economics should be developed to contain two main departments that are complementary. The analysis provided by of the first department assumes a given stage of the economic system, while the second department focusses on economic evolution. This research programme should help to overcome the battle of methods.

## 3.2 Exploring the "Magna Carta" of theoretical economics

Since the first interpretation concentrates on the foreground argument of *Wesen*, it is relatively easy to apply. The "essence" of economic theorising is defined in relation to the model of the economic system by Walras and his successor Pareto, who together formed the core of the Lausanne school. However, Schumpeter's book presents their general equilibrium theory in verbal terms. Therefore, German-speaking students who found themselves handicapped when confronted with the mathematical form, and the level of abstraction, of Walras's *Elements of Pure Economics* could use *Wesen* instead. The book served the same function in the US. Here Wesley Mitchell, an important statistician and institutionalist economist, used it in his lectures at Columbia University in the 1920s. Mitchell (1969, 376) summarised the hard-core contribution of *Wesen* by emphasising that "Schumpeter develops substantially just one important thesis which is the most important result of Walras's speculations." The large size of the book is caused by its "elaborate methodological discussion of what he is going to do, the way in which he is going to do it, the limitations of what he had done, and finally the importance of the results which he set forth." The basic task of "pure economics" is that of "finding how the essential results can be demonstrated with utmost economy of intellectual means" (p. 377).

Although *Wesen* argues for the great advantages of the mathematical method for economists, it does not do so "with utmost economy of intellectual means". This is quickly recognised by comparing it with *Foundations of Economic Analysis.* After four pages of introduction, Samuelson (1947, 7–20) spends 14 pages introducing the formal characteristics of "Equilibrium Systems and Comparative Statics" ranging from systems of a single variable to systems "involving thousands of variables". In contrast, Schumpeter concentrates on the equilibrium of the Walrasian system because it covers, or can be specialised to cover, nearly all of what has traditionally been analysed by economists. This idea is presented on the first 440 pages (see Table 3.1 on the following page). Then follows the account for the bread-and-butter activity of economic theorists: comparative statics. The explanation of this verbose presentation is found in the Preface: *Wesen* wants to make "the German economist" familiar with concepts, propositions, and views that hitherto had been "foreign" (*Wesen*, xxi). One of the most important reasons why this body of knowledge and methods have not been spread is "the question of the 'mathematical method'". Hence, the book will not require any significant level of mathematical skill. Furthermore, *Wesen* is "not a textbook" but rather an introductory treatment of the "very narrow field" within the social sciences that "allows exact treatment" (pp. x–xi). The analysis of the real meaning of the propositions of pure economics provides "something like an epistemology of economics" (p. xii). Let us start by considering *Wesen* along these lines. More specifically, let us recognise that Schumpeter's 'foreground argument' concerned an equilibrium model of the economic system: "In the centre of the book stands the problem of equilibrium, the importance of which is only slight from the viewpoint of practical applications of theory, but which is nevertheless fundamental for science" (pp. xix). In the context of German-speaking countries, he emphasised that economic equilibrium had not received "sufficient consideration". Therefore, he wanted to "emphasise that it is at the foundation for our theoretical system. The theory of exchange, price and money, and its most important application, the exact theory of distribution, is based on it."

Part I of *Wesen* is called "Foundation" and it provides a "criticism ... that is necessary when asking about the foundations of theoretical economics" (*Wesen*, 26, emphasis removed). This criticism suggests a cleansing of the expositions of economic theory from anything that are not essential. This cleansing is needed because nearly all writers of economic textbooks ignored that they were presenting a formal system that, no less than Euclid's *Elements of Geometry*, is based on axiomatic assumptions. For instance, we may think of Marshall's *Principles of Economics* that spends the first 270 pages motivating the basic assumptions underlying the theoretical apparatus and continues to include motivating di-

**Table 3.1.: *Wesen und Hauptinhalt der theoretischen Nationalökonomie* (1908)**

| Translated title | Comments and pages |
|---|---|
| Preface | Background, purpose and analytical tools (pp. v–xxii) |
| I. Foundation | Basic methodological issues relating to the analysis of economic systems (pp. 1–113) |
| II. Problems of Static Equilibrium | Walrasian equilibrium economics and the Statics–Dynamics dichotomy between core neoclassical economics and evolutionary economics (pp. 115–311) |
| III. Theory of Distribution | The neoclassical theory covers adequately wage and rent but not profit and interest; evolutionary accounts for the role of profit and, especially, interest are sketched out (pp. 313–440) |
| IV. The Method of Variation | Comparative statics and some comments on the related form of evolutionary thinking (pp. 441–519) |
| V. The Nature, Cognitive Value, and Development Possibilities of Theoretical Economics | The reform programme for economic science based on the distinction between core neoclassical economics ("Statics") and evolutionary economics ("Dynamics") (pp. 521–611) |

Comment: *Wesen* is a methodologically oriented book that contains an analysis of the structure and explanatory power of neoclassical economics. By studying the "essence" of the models of Walras and other mathematical economists, Schumpeter tried to demonstrate the limits of static analysis and the need for complementing it with evolutionary analysis. The book thus points at a complementary volume on the essence and principal contents of "Dynamics", i.e., evolutionary economics.

gressions. Such a motivation requires intrusion into the realms of other sciences, like psychology, in which economists are "only dilettantes" and this intrusion provides "points of attack for the adversaries" (*Wesen*, 24). Just like the science of mechanics would never have been created if it had continued to be engaged in explaining what "power" and "mass" really are, the science of economics cannot be grounded in this way. Instead, economists have to acknowledge that they presuppose a system of interdependent elements. These elements are agents that possess given quantities of different goods; and the core question is whether the system has an unambiguous state of equilibrium. The task of theoretical economics is to explain this equilibrium and its infinitesimal change without relying on other sciences. The totality of the propositions that can be derived in this way defines the discipline of theoretical economics (pp. 28–9).

The specification of what John Stuart Mill had called "the laws of co-ordination" starts with chapters on "The Exchange Relation". At a given point in time, each good is assumed to be bought or sold at a given price. Hence a unit of this good can be exchanged with fixed quantities of all other goods. The investigation of these relationships is the subject matter of all pure economic theory (*Wesen*, 53). This theory can be constructed in several ways that cannot be judged from their truth but only from their usefulness. For instance, the labour theory of value of classical economics should not be considered as false, but as unfruitful for the treatment of important economic problems (pp. 57–63). In contrast, the solution to the problem of economic interdependence is directly served by focussing on marginal evaluations performed by each economic agent (pp. 71, 105–7). Since these evaluations are reflected in observable behaviour, the "value principle" does not need the psychological reflections of many neoclassical economists.

Part II deals extensively with "The Problem of Static Equilibrium". Since any difference in the evaluations of the marginal contributions to an agent's utility would contradict equilibrium, they have to be equalised according to the "law of the level of marginal utility" (*Wesen*, 129–31). We may also say that the relationship between the marginal utilities of any two goods have to be equal to the reverse relationship between their prices (p. 213). This rule, the "alpha and omega" of pure economics, can also be used for the analysis of the problems of production. This becomes clear when we study a competitive economy in which equilibrium prices have to reflect marginal costs. Here the equilibrium condition is that the marginal costs of a good are equal to its marginal utility (pp. 214–15). Like Walras, Schumpeter recognises that this simply means that productive services are included among the elements of the economic system. Since this solution did not satisfy the consumption-oriented members of the Austrian school, he tries to demonstrate that supply curves can be reinterpreted as an alternative form of demand curves (p. 235). He, nevertheless, focus on the Walrasian theory of price: the definition of a system of equations that simultaneously determines all quantities and all prices (pp. 260–2). He presents the necessary condition for the uniqueness of the solution by counting equations and he emphasises the missing equation for the price of the money good. He also presents the limitations of the solution by rehearsing the poor state of price theory with respect to "limited competition" and by remarking that perfect competition is at best an approximation to reality (pp. 269–72). This approximation, however, is good enough for many purposes. It not least clarifies the theory of money to a surprising degree (pp. 276–7) while the theory of saving seems to be in a poor state (pp. 304–5).

Part III handles the "Theory of Distribution". This is not only "the most important application of price theory" but also a tool for unwarranted answers to highly controversial socio-economic problems (*Wesen*, 315–16). Economists had normally tried to explain (and justify) the incomes derived from labour, land, and physical capital. In addition to wage, rent, and interest on capital, some economists had added entrepreneurial profit as a fourth basic category of income. *Wesen* points out that much of the confusion on these matters is due to the study of the factors of production instead of their productive services. Given this "artifice" (p. 372), the size of wages and rents can easily be determined by including the services of labour and land as elements of the system of economic equations (pp. 330–1, 368). In contrast, entrepreneurial profits cannot be treated in this system because they are expressions of disequilibrium. The concrete explanation of interest on capital was a more controversial matter, but practically everyone agreed that it should be included as a source of income in an equilibrated economic system (p. 392). Schumpeter disagreed for reasons that we shall consider in Sections 5.3. In any case, he argued that when compared with classical economics, his truncated theory of distribution is a major advance because it explains wage and rent in exactly the same way and because it is a more powerful analytical tool (p. 379).

Part IV moves from what John Stuart Mill had called "the laws of coordination" to what Schumpeter, together with Mill, call "the laws of motion" (*Wesen*, 443). This part of the book has the heading of "The Method of Variation". This method starts from an equilibrium state of the economic system and compares it with the new equilibrium that is caused by a change in one of the elements of the system (a quantity or a price of a good or service). The core issue is the changes, or "variations", of the endogenous variables. Thus, we are dealing with what we today, with a phrase that was coined by Oppenheimer (*History*, 855, 965), call comparative statics. Actually, Schumpeter seems to be the first to describe this method systematically—and he does so with utmost care. While the theory of distribution solves one core problem, comparative statics solves another core problem and provides "the second group of results of exact economics" (*Wesen*, 443). Furthermore, the method is heavily used by all economists and it is even underlying the do-it-yourself economics of "politicians" and "historians" (pp. 475–6). The problem with all these applications is that the method is used without sufficient consideration of the underlying methodological problems.

As anyone who has performed physical experiments in high school ought to know, comparative statics presupposes that we have a well-defined system and only change one thing at a time. If we take a "snapshot" (*Wesen*, 142) of an equilibrium state and change one element of the system, then we would like to know in advance what the snapshot in the resultant equilibrium looks like. To make this prediction, we need to take something as given; and in the present case this something is the consumption functions and the production functions of the economic agents. These functions serve to determine the response of economic agents to, for instance, a change in price. A widespread use of this method had developed without a careful study of the underlying problems. The resulting controversies can be avoided by acknowledging the nature of the system of economic equations (pp. 456–75). Since the given functional relations between the elements of the model describe individual reactions to small and continuous change, these functions cannot be used for the analysis of the response of the economic system to major or discrete changes of its elements. Comparisons between equilibria that are separated by long time periods are also problematic since they draw attention away from the major changes of the functional relationship that are likely to occur during such periods. Therefore, the method of variation is, in general, only applicable in the immediate neighbourhood of an equilibrium of the given economic system, measured in time and in the state space of the system. This limitation and the possibilities of overcoming it are illustrated by 40 pages of examples on taxes, import duties, changes of income, and the introduction of machinery (pp. 478–519).

Part V has the long title "Summary of Conclusions on the Essence, Cognitive Value, and Development Possibilities of Theoretical Economics". This long heading precedes one hundred pages of conclusions! Nevertheless, the conclusions are simple with respect to the above argument. First, the demarcation of the domain of theory-based economics should be defined by the set of problems for which its basic model and the underlying "schema of exchange" can be applied (*Wesen*, 582). Second, the "methodological and epistemological essence" of pure economics demonstrates that it is "a 'natural science' and its theorems are 'natural laws' " (p. 536). Third, this science is best served by sticking to its place in the scientific division of labour. Although economic research and the writing of economic textbooks might reflect inspiration from the tools of the natural sciences and the contents of other social sciences, they should be cleansed from the frequent intrusions into the domains of other sciences (pp. 536–53). These three conclusions provide a surprisingly accurate description of much of the later development of modern economics. They suggest that theoretical economics, especially through its Walrasian formulation, forms a "closed and autonomous province within the realm of knowledge" (p. 523). As Schumpeter later said (*History*, 242, 827), Walras had created the "Magna Carta" of this province—both as the first complete map and as the original constitutional document—and this made him "the greatest of all economists". Even for those unfamiliar with the development of the English constitutional law based on the Magna Carta of the year 1215, the meaning of the underlying caveat should be clear: Walras, of course, was not perfect. Nevertheless, he had demonstrated that the "subject matter" of economic theory "is a cosmos and not a chaos" (*Cy-*

*cles*, 41) and he had implicitly determined the borderlines of this theory.

## 3.3 Resolving the battle of methods

The second interpretation of *Wesen* is that it wants to create room for researchers engaged in non-exact summaries of facts. Since this is largely how the early Schumpeter characterised the activities of the German historical school, his first book can be seen as promoting a resolution of the battle of methods between that school and the Austrian school or, more specifically, between Gustav von Schmoller and Carl Menger. This reading is further helped by recognising that the book is addressed to a German audience that was dominated by Max Weber and other "historians" who worked according to the "the Schmoller programme" (*S1926b*, 186). The foreground argument about the narrow domain reserved for pure theory leaves plenty of room for alternative modes of study—and this point is explicitly made. The first signal is found on the first pages of the Preface. Here we meet a programme of reconciliation: "nearly every 'school' ['Richtung'] and every individual author are right in their propositions ... from the standpoint of the purposes for which they are intended". Therefore, the task is to "learn, not criticise; analyse and work out the correct in each proposition, not merely accept or reject" (*Wesen*, v–vi; emphasis removed). Although these statements are general, they are especially intended to cover controversies within theoretical economics as well as the battle of methods. However, Schumpeter immediately adds that he considers the controversy "between pure theory and history to be largely overcome" and that he, for each scientific problem, will "investigate whether the one or the other treatment is most recommendable" (p. vii). In contrast, the "whole history of the battle of methods" can be described by the sentence: "Everyone is convinced of his *exclusive* rights while he only *partially* can demonstrate this, and the beginner does not know which to adhere to" (p. xvi).

Part I continues this story. The battle of methods is hardly surprising since even natural sciences like chemistry and mechanics are characterised by controversies between experimentalists and theorists as well as in the camp of pure theory (*Wesen*, 4–6). The main reason is that each researcher has to specialise, that this specialisation becomes part of his personality, and that he hardly recognises the borderlines of his speciality. The same situation is found in the science of economics, where price theory is beyond the reach of the empirically oriented researchers, while the problem of the overall organisation of the social economy cannot be treated by abstract theorists (p. 7). This problem was overlooked because the historical school started by attacking classical economics at a time when it had entered a period of deep stagnation that crippled even the creativity of the great Stuart Mill (pp. 9–10). Hence, it is understandable that the historical school chose to throw abstract theory overboard and concentrate on facts and practical problems. However, the members of this school never understood the nature of the theorising of the emerging neoclassical economics. Instead, they started "the development of new theories on the basis of historical materials" of which the "most well-known example of this is probably the 'Theory of Modern Capitalism' by W. Sombart" (p. 18). Schumpeter emphasised that this group seemed to be "in a quick upswing and soon will dispose of a significant literature." This literature "does not build an abstract system, but makes individual hypotheses on concrete questions ... [that] relate always to definite historical facts." Such hypotheses have "similarity with the hypotheses of biology"; and this similarity "is strengthened by their dealing mostly with the problems of evolution [Entwicklung]. They are everything else than 'static', ... [b]ut perhaps the area of 'Dynamics' belongs to them! That will have to be seen" (p. 18).

These formulations demonstrate that Schumpeter did not dismiss history-friendly theorising. Actually, he simply considered the "descriptive" method and the theoretical method as two ways of handling facts. Since both methods have inductive and deductive elements, there is no basic difference. The only difference is that the theorist tries to cover whole classes of fact by developing a formal model that is characterised by the utmost economy of thought (*Wesen*, 41–4). This level of formality and simplicity is only applicable to a few areas of social life. Furthermore, the overselling of the descriptive accuracy of abstract theory was quickly recognised by empirically oriented researchers and contributed to the battle of methods (p. 48). The apparent founding of neoclassical economics in a broadly conceived atomism and individualism was also highly provoking for researchers who emphasised altruism and collectivism (p. 82). The problem is that many economic treaties start with a specification of Homo Economicus or other theoretical versions of *Homo sapiens*. This is, however, not the strategy of Walras and Schumpeter (*Wesen*, 85–7). As theoretical economists, we should *not* consider economic agents—households and firms—from their inside through psychology (as Wieser wanted) or from the viewpoint of organisation theory (as members of the historical school did). In contrast, agents should be studied from outside, from what we as researchers can observe about their behaviour. This is all we need if we take seriously that we are dealing with the properties of a system of quantities of goods and exchange relations between them (prices). The theoretical agents are designed to fit the model of this system. To emphasise this point, Schumpeter coined the phrase "methodological individualism" and dedicated a whole chapter to clarifying the issue (pp. 87–98).

Parts II–IV can be browsed quickly since we know their main contents from the above account. With respect to the battle of methods, we simply need to note that Schumpeter, at each major step of his argument, emphasises the limited scope of theoretical economics and thus the room for alternative treatments. For instance, theorists make "formal assumptions" while others might make "theories on the causes of economic action" (*Wesen*, 129). These theories might concern the interdependent influences of "[r]ace, cultural level, social position, education, personality" well as those of "the natural environment and social organisation" (p. 142). In contrast, the theorist only needs assumptions on the preferences that determine short-term choice in the economic system. Furthermore, production technology and industrial organisation are also represented by parameters by the theorist while they are problems for the studies of the historical school (p. 148). The fact that Marshall treated these topics extensively only serves to create confusion about the real structure of the standard model of the economic system (p. 150). Similarly, the basic model fails to support an extension of economic theorising into the long run and thus to the great phenomena of economic evolution (pp. 177, 186). The impression of a limited contribution of theoretical economics and the need for a complementary effort by the historical school becomes stronger and stronger. It also covers money and finance (p. 297), saving behaviour (p. 308) and long-term issues of rent on land (pp. 375–6).

Part V's narrow programme for theoretical economics obviously leaves a very large domain for the historical school. Actually, the members of this school had already occupied much of this domain while they considered the domain of theoretical economics of very little interest. Therefore, they could hardly oppose Schumpeter's conclusion. This conclusion was that the Walrasian Magna Carta allowed him to "preach a kind of Monroe Doctrine of economics" (*Wesen*, 536), that is, a dual principle of foreign policy: no acceptance of intervention from foreigners, no attempt to intervene against outsiders. If this doctrine was accepted, the transgressions of the battle of methods were overcome. Theoretical economists could concentrate on developing and applying their analytical tools within safe borderlines. At the same time, these borders defined the domains in which the historical school should not fear any attack from the theorists.

Schumpeter's argument for staying within the borderlines of economic analysis is largely related to advantages of a strict division of labour between the different fields of a broadly defined economics. For instance, his argument implies that much of what is included in the introductory parts of Marshall's *Principles* should be left out and taken over by the emerging field of economic sociology. This purging of economic theory had largely been completed by Walras, and the main task for theoretical economists is to complete his work on Economic Statics. Another part of his argument was not really developed. However, he implicitly suggested that the phenomenon of business cycles should largely be left over to statistically inclined members of the historical or to the evolutionary economics that he was pointing at. In contrast, dynamical extensions of the Walrasian system would only create confusion. Thus, theoretical economists seem to be trapped within their static model; and *Wesen* can be read as an acceptance of this situation. What are left are simply problems of scientific coordination between the different specialities. For instance, the division of labour creates a problem for those theorists interested in the empirical verification of their theorems that are of no immediate interest for empirically oriented researchers (*Wesen*, 532). Another need felt by some theorists is the collaboration with sociologists, of which many had connections to the historical school. Although sociology has nothing to do with economics in the methodological and epistemological sense, the "development of sociology ... will perform great services to economics" (p. 541). By following their predominant inclinations, the members of the historical school would provide these services.

However, neither the historical school nor the neoclassical economists liked the borderlines needed for upholding the "Monroe Doctrine of economics". Even Walras transgressed the borderlines defined by *Elements of Pure Economics* (Walras, 1954, 380–1). Here he acknowledged that the *tâtonnement* process might be "tending towards equilibrium without ever actually attaining it" because "the market is like a lake agitated by the wind, where the water is incessantly seeking its level without ever reaching it" . He went even further by moving from the "wind" of parameter change to the "storm" of crises: "just as a lake is, at times, stirred to its very depths by a storm, so also the market is sometimes thrown into violent confusion by *crises*, which are sudden and general disturbances of equilibrium." As we shall see in Section 5.4, Schumpeter rejected the possibility of treating such crises within equilibrium economics. Instead, he counted crises with the large number of phenomena that "can only be understood from the standpoint of evolution. Here belong the problems of the formation of capital and others, especially those of economic progress and the crises" (*Wesen*, 587). In this setting, he also changed the Walrasian metaphor of "wind" to the Schumpeterian "perennial gale of creative destruction" (*Capitalism*, 84).

## 3.4 The Statics–Dynamics dichotomy

The third interpretation of *Wesen* concerns the largely evolutionary phenomena that actually were at the centre of the attention of many economists, although they are not covered by the Walrasian Magna Carta. The general opinion was that these phenomena would soon be subsumed under neoclassical economics by its gradual extension towards long-term issues. Economics had in the beginning of the twentieth century reached what Schumpeter later called "a classical situation" in which the leading works "exhibited a large expanse of common ground and suggest a feeling of repose" (*History*, 754). The most dominant of these works was Alfred Marshall's *Principles of Economics*. This book is designed in a way that makes it very difficult to determine the limits of neoclassical economics. However, Schumpeter found that it was ultimately founded on something that came very close to the Walrasian system. Therefore, he considered its promises of leading towards the analysis of the great problems of progress and evolution unfounded. Instead, he was among those who were "able to discern signs of decay, of new breaks in the offing, of revolutions that had not yet issued into another classical situation" (*History*, 754).

Schumpeter focussed on the problem that the beautiful architecture of the theoretical constructs of the leading neoclassical economists had been obtained at serious costs. The architecture of the models excluded a serious treatment of that long-term change in the economic framework that arises from the innovative competition among firms as well as related phenomena like entrepreneurial profit, credit and business cycles. Since American economists were very interested in these phenomena, Schumpeter's careful study of their contributions helped him to recognise signs of "new breaks" (see especially *S1910a*). These novelties suggested a need of complementing the theory of an equilibrating economic system with a theory of the processes that change the economic system. This need is obvious if we accept not only *Wesen*'s positive definition of the domain of theoretical economics but also the definition of what is not included in that domain. In the Preface, he emphasised that his

> "exposition depends on the fundamental separation between economic 'Statics' and 'Dynamics', a point whose importance cannot be overstated. For the time being, the methods of pure economics are only sufficient for the former area, and our results hold only for this area. 'Dynamics' is something that in any respect is completely different from 'Statics', methodologically as well as regarding contents. ... We shall see ... that it [the separation] holds the key to the solution of many controversies and many apparent contradictions" (*Wesen*, xix).

From Schumpeter's perspective, Statics could be defined as the combination of equilibrium analysis and the core phenomena covered by that method. His definition of Dynamics is less clear—mainly because it is "a land of the future" (*Wesen*, 183). His definition appears to be something like: Dynamics covers the phenomena of "Entwicklung" that can be handled theoretically combined with the method that is to be designed for the analysis of these phenomena. This version of the Statics–Dynamics dichotomy was crucial for the arguments in *Wesen* and *Entwicklung I*. However, he largely applied the terms in quotation marks and he pointed out that he applied the terminology "for reasons of convenience" although he considered it "very unfortunate" (*Wesen*, 182). There were good reasons for this caution. Actually, his early definitions of Statics and Dynamics created so much confusion that he later dropped them (see Figure 10.1 on page 319). The terminology is obviously not taken from mechanical physics in which 'dynamics' is the method of analysing the deterministic movement of systems of bodies (like the Solar system). The terminology was, rather, derived from Stuart Mill's (1868:II, 513) *System of Logic*. Here we find the distinction between "Statics" as the study of the conditions of stability and "Dynamics" as the study of the laws of progress. Mill got these concepts from the philosopher Auguste Comte who in turn seems to have got them from the zoologist and anatomist Henri Blainville (*Development*, xi).

A much more obvious source of Schumpeter's dichotomy, however, is the work of the leading American economist John Bates Clark. Clark had been involved in a lengthy and famous controversy with Böhm-Bawerk on the concept of capital. This controversy caught Schumpeter's attention, and he even, in 1907, translated one of Clark's contributions for the scientific journal of the Austrian economists. As a side effect, Schumpeter must have recognised that Clark's work helped him to overcome the exclusion of a real analysis of economic evolution from neoclassical economics. Clark (1898; 1899b; 1899a) suggested his strategy in papers like "The Future of Economic Theory" as well as in his large book on *The Distribution of Wealth* from 1899. Here it became clear that although Clark had developed his own contribution to neoclassical economics, he had also tried to clarify how a movement towards the study of "Economic Dynamics" could take place. Like neoclassical economists in general, Clark had analysed a stylised and stationary economy. His particular contribution was to determine the distribution of income by means of the marginal contribution to production of the different types of agent. He was, however, aware that his elegant analysis did not provide an explanation for the major determinants of income distribution in the real world. Furthermore, neoclassical economics was becoming less productive as it moved into modelling detail. In the perspective of research for the coming century, Clark (1898, 14) argued that it was high time to move to "the science of Social Economic Dynamics". Schumpeter fully accepted the need for moving forward to Economic Dynamics, but he also appreciated Clark's conviction that the starting point was the well-established area of "Economic Statics". Thus, Clark's Statics–Dynamics dichotomy became a crucial tool for the development of the core arguments in *Wesen* and *Entwicklung I*. More specifically, Schumpeter seems to have wanted to define two relatively independent areas of theoretical economics according to the formula:

$$\begin{aligned} \text{Theoretical Economics} &= \text{Statics} + \text{Dynamics} \\ \text{Statics} &\approx \text{Neoclassical Economics} \\ \text{Dynamics} &\approx \text{Evolutionary Economics} \end{aligned}$$

Although Clark did not share Schumpeter's interpretation of Dynamics as what may today be called evolutionary economics, the background of the formula in Clark's work is not difficult to detect. Even in the Preface to his *Distribution of Wealth,* Clark made formulations that must have raised the interest of young Schumpeter. These formulations cover both the non-evolving neoclassical system and the transcendence of this system. Clark (1899a, vi) characterised the former system as "static", that is, as covering a stationary state. Neoclassical economics built on the implicit assumption of this state, and it is even hidden in the concept of "natural" prices of the factors of production used by classical economics: "The term *natural* ... was unconsciously employed as an equivalent of the term *static;* and it is such natural or static standards that this volume undertakes to present." By referring to "static standards", Clark theorised as if "the changes that are going on in the shape of the industrial world and in the character of its activities were to cease." His purpose was "completely to isolate the static forces that act in distribution from the dynamic forces."

Schumpeter seems to have received a first specification of the limitation of the analysis provided by theoretical economists from Clark's formulations. Classical economics had hidden this limitation by verbal accounts for economic progress, but the formal approach of neoclassical economics served to clarify the problem: it makes its analysis of the effects of "static forces" under the assumption that the "forces of progress" have become "paralyzed" (Clark, 1899a, vi–vii). Since change and evolution underlie many of the most conspicuous phenomena of the capitalist economy, this problem is a serious one. The task for the new century, therefore, was to remove the paralysing assumption. According to Clark, this opens up the area of Economic Dynamics—the study of economic growth and evolution:

> "It is already clear that the field for new investigation offered by economic dynamics is an indefinitely fruitful one. It ... [deals with] essentially new problems, because the prevailing mode of economic study has not heretofore isolated them, brought them clearly into view and afforded the data for solving them. ... The mere theory of economic dynamics will enlarge by many fold the scope of political economy; it will lift theory to a new plane. The statement of the pure laws of economic change will open, as it were, the vestibule of the science of the future." (Clark, 1899a, 35, 76)

Clark did not enter this "vestibule" in his *Distribution of Wealth*—he reserved this entering for a later work—but he gave a few indications about its characteristics. In terms of the "standards" of economic life, the task was to analyse both the forces that establish fixed standards and the forces that create new standards. In Clark's (1899a, 32) terms, this meant the introduction of both "static" and "dynamic" forces: "Static forces set the standards, and dynamic forces produce the variations." The complete process of economic evolution covers both the disturbance of the economic standards and their re-establishment. The strategic problem was how to specify the "dynamic forces". Clark did not say much about this problem, and he seems primarily to have thought of forces influencing the economic system from the outside. However, he made remarks that pointed in another direction. Although they are not central to Clark's own argument, Schumpeter must have found his remarks on the entrepreneur of great interest. In these remarks, the term entrepreneur is in italics because it was considered a French word. In any case, the background is that "[t]he prices that conform to the cost of production are, of course, those which give no clear profit to the *entrepreneur*" (p. 70). This situation is changed when "an invention first gives a profit to *entrepreneurs* and then, in the way that we have described, adds something to wages and interest" (p. 405). This process seems to define the area of Dynamics: "Dynamic science deals with profits in their original state, as normally created by improvements in industry, in the proceeds of which the *entrepreneurs* have a share" (p. 410). In contrast, "static science deals with them [profits] in their later and permanent state, as they are transmuted into increments of wages and interest."

In these remarks about innovative renewal and the subsequent transformation emergence of the standards of reformed stationary state, the entrepreneur is obviously at the very centre of the stage (see also Clark, 1899b, 194–8). If we combine these remarks with other sources of Schumpeter's thinking (see the next two chapters), we obtain a glimpse of his grand vision that, in modified form, influenced him for the rest of his life. Based on Clark's hints, he must have waited eagerly for the "dynamic" sequel to *Distribution of Wealth*, which Clark even suggested had some relationship to Böhm-Bawerk's work. This second volume came out in 1907, and it was a disappointment because of Clark's emphasis on simple growth and automatic progress. In his review, Schumpeter (*S1908b*, 655) especially missed the "*energetic* aspects" that are a major "lever of economic evolution". This made him realise that he had to make his own advance towards a type of Economic Dynamics that emphasised the innovative activities of the entrepreneur: "in him lies the most essential difference between the dynamic viewpoint opposite to the Statics that presuppose hedonic equilibrium men" (*S1908b*, 655). In *Wesen*, however, he started by stating the problem more generally. As we have already seen in Chapter 2, he emphasised that

> "Statics and Dynamics are completely different fields, they concern not only different problems but also different methods and different materials. They are not two chapters of one and the same theoretical building but two completely independent buildings. Only Statics has hitherto been somewhat satisfactorily worked up and we essentially only deal with it in this book. Dynamics is still in its beginnings, is a 'land of the future'." (*Wesen*, 182–3)

This formulation is indirectly pointing at an independent treatise of Economic Dynamics. This promise is made more explicit elsewhere in *Wesen*. Thus, Schumpeter remarked that he in relation to how fortunes are created had found "new elements for ... a kind of theory of economic evolution [Theorie der wirtschaftlichen Entwicklung]" that he could not include but that he, perhaps, "later" could present to the public (*Wesen*, 309–10). Furthermore, he pointed out that it is not only the explanation of interest, entrepreneurial profit, and the formation of fortunes that have to rely on the development of Dynamics. This field also has to cover "the problems ... of the economic progress and crises" (p. 587). Even with respect to the price of the services of labour and land he, to some extent, suggested the application of evolutionary economics. With respect to the value of land, he thought that "all that is really interesting in it gravitates toward the great problems of evolution" (p. 374) and that he had found "a seed for a new theory" (p. 588n). Finally, he presented an entrepreneur-based theory of interest on capital in a section called "Prolegomena to a Dynamic Theory of Interest" (pp. 414–30). Obviously, such 'introductory remarks' have to be followed by a more extensive treatment. Although this treatment is found in *Entwicklung I* (see Section 5.3), the development of core aspects of Economic Dynamics are present in *Wesen*. However, although these aspects were "interesting", it was still an open question whether something "exact" could be said about them.

## 3.5 Types of entrepreneurs and parameters of the system

Schumpeter's attempts to develop an "exact" theory about economic evolution probably started from the Walrasian system. Many years after he wrote his *Wesen*, he emphasised his immediate recognition of the limitations of this system due to its focus on Economic Statics. In the Preface to the Japanese edition of *Development* he stated that he discovered that Walras's focus was not made for purely methodological reasons. On the contrary, "Walras would have ... said (and, as a matter of fact, he did say it to me the only time that I had the opportunity to converse with him) that of course economic life is essentially passive" (*S1937*, 166). According to Walras, economic life "merely adapts itself to the natural and social influences which may be acting on it". This implies "that the theory of a stationary process constitutes really the whole of theoretical economics and that as economic theorists we cannot say much about the factors that account for historical change, but must simply register them."

Walras's conception of a passively adapting economy provoked Schumpeter's critical response that was based on his elitist vision of economic change: "I [Schumpeter] felt very strongly that this was wrong, and that there was a source of energy within the economic system which would of itself disrupt any equilibrium that might be attained" (*S1937,* 166). His vision of this disruptive force determined his analytical efforts. He felt confident that "there must be a purely economic theory of economic change which does not merely rely on external factors propelling the economic system from one equilibrium to another." Schumpeter emphasised that "[i]t is such a theory that I have tried to build and I believe now, as I believed then, that it contributes something to the understanding of the struggles and vicissitudes of the capitalist world". The scientific contribution of his theory is that it "explains a number of phenomena, in particular the business cycle, more satisfactorily than it is possible to explain them by means of either the Walrasian or the Marshallian apparatus" (p. 166).

Schumpeter thus developed his version of Economic Dynamics in opposition to Walras (and Marshall). However, there is little doubt that his reading of Walras's *Elements of Pure Economics* directly influenced his evolutionary theory. Since this influence is difficult to detect in *Wesen,* we shall start the study of this issue by the account for the emerging Economic Dynamics that we find in Schumpeter's "prolegomena". The starting point is the theory of interest on capital developed by Böhm-Bawerk. Even Carl Menger had stated: "'The time will come when people will realize that Böhm-Bawerk's theory is one of the greatest errors ever committed'" (*History,* 847n). In contrast, Schumpeter tried to rescue parts of the Böhm-Bawerkian theory by emphasising that is was a contribution to "dynamics" rather than to "statics" (*Wesen,* 428, 408–13). However, his own theory of interest started from the viewpoint of the entrepreneur who created radically new lines of production in a given economic system. Although the presentation of his alternative theory logically should have been postponed to *Entwicklung I,* he considered it so important that he inserted a sketch in *Wesen* (414–30). Here he suggested that his theory might partly be expressed in terms of Böhm-Bawerk's concept of advantageous roundabout methods of production. The main point is that the use of roundabout methods does not generally and permanently create an ability to pay a premium to the capitalist savers. Instead, it is the investment of entrepreneurs in innovations with higher productivities that provides the profit that can be used to pay interest. They do so by solving an inter-temporal problem. The presently available "world of goods" is governed by a system of prices that has no room for the innovative project. However, if some of these goods are brought into a "new combination", then they can be used to produce a future output that can pay a premium when measured in terms of the present price system. The condition is that—although the present state has no room for it—credit money is created for the innovative project. With this money the entrepreneur detracts some of the present goods for his innovative purpose. His demand for "purchasing power" (p. 427) essentially explains the phenomenon of interest.

Given his emerging vision of Economic Dynamics, it is not difficult to detect how Schumpeter developed his criticism of the sketches of a dynamical process that are found in Walras's *Elements*. The driver of this sketchy dynamics of the economic system is an agent that is presently best denoted the Walrasian entrepreneur, or the W-entrepreneur. This W-entrepreneur has a paradoxical role in Walras's essentially static system. On the one hand, the W-entrepreneurs bring the elements of the system together into a coherent whole through an equilibrating process that is motivated by profit opportunities. On the other hand, the holders of the role of W-entrepreneurship have no income when the system has moved to equilibrium. These two characteristics of the Walrasian system become clear as soon as we consider its basic structure. Walras (1954, 222) defined a role list that contains four essential types of agent. The list starts by calling "the holder of land ... a *land-owner*, the holder of personal faculties a *worker* and the holder of capital proper a *capitalist*". Then the list designates "by the term *entrepreneur* a fourth person, entirely distinct from those just mentioned". The role of this fourth type of agent "is to lease land from the land-owner, hire personal faculties from the labourer, and borrow capital from the capitalist, in order to combine the three productive services in agriculture, industry or trade" (p. 222).

From these role descriptions of the "dramatis personae" (*History*, 554), it is obvious that W-entrepreneurs are at the centre of Walras's model world: Their function is to combine the elements and thereby create an integrated economic system. The individual W-entrepreneur hires the necessary factors of production at given conditions of payment; and he initiates a process of production in which one of the employed workers function as a manager. When the goods have been produced, he sells them at the market price that is prevailing at that time. Then the W-entrepreneur makes up his balance sheet. Depending on the costs of the factors of production and the price of the goods, his bottom line shows either gain or loss. Such positive or negative profits, however, only exist as long as the economic system is in disequilibrium. According to Walras, this disequilibrium is removed by perfect competition among W-entrepreneurs. When this competition has brought the system into equilibrium through a process of trial and error (*tâtonnement*), the W-entrepreneurs "make neither gain nor loss" (*Wesen*, 438).

Walras's equilibrium thus has an apparently paradoxical characteristic: his W-entrepreneurs that are driven by the profit motive obtain no profit whatsoever. Hence, he must conclude that, in equilibrium, the W-entrepreneurs have to obtain their incomes by functioning as landowners, workers, or capitalists. Schumpeter thought "that this entrepreneur is only a fiction" that is created because of the limitations of Economic Statics (*Wesen*, 438). However, he also recognised the possibilities of inserting an alternative entrepreneur, the S-entrepreneur, into a reinterpreted version of the Walrasian system. The background was that he thought that Walras's account for the *tâtonnement* process gives a false picture of the speed of the market process. While Walras saw the groping toward equilibrium as a speedy process without disequilibrium trading, Schumpeter followed the Austrian school by considering it very slow and erratic. It has thus, somewhat provocatively, been remarked that "Walras's *tâtonnement* takes a minute; Menger's *tâtonnement* takes a century!" (Streissler, 1973, 174; emphasis removed). Carl Menger's Austrian successors, especially Friedrich von Wieser, developed this view on the slow process of "imputing" values to intermediate goods and primary production factors, based on the valuations of consumer's goods. It probably is not least the sluggishness and comprehensiveness of Wieser's *tâtonnement* process that Schumpeter refers to when he stated "that L. Walras and [F.] v. Wieser are those authors whom the author [Schumpeter] believes he is closest to" (*Wesen*, ix). While the Walrasian process assumes fixed set of goods and production methods and can be headed by an imaginary auctioneer, Wieser's (1893, 212) slow process not only concerns quantities and prices but also the evolution of the whole economic system:

> "A knowledge of the values of goods, such as has existed in every economy up till now, is ... one of the most valuable of possessions. It is almost as valuable as the possession of the goods themselves, inasmuch as it is the key to their use. The sum of thousands of years of experience concerning the sources of supply of goods, and the suitability or otherwise of the conditions of their production, as well as concerning the amount of demand for them, is represented in the figures of value handed down to us. Were a nation to lose all remembrance of these, it would be an enormous economic misfortune."

Schumpeter's interpretation of the Austrian *tâtonnement* seems to have been that the slow emergence of a consistent price system reflects the adaptation of agents with bounded rationality to the living conditions of a nation. Based on the Statics–Dynamics dichotomy, he saw this as a specification to the "first problem of economics", that is, to find the equilibrium state of the economic system from given parameters. The second great problem of economics, however, was not treated. It concerns the development of a "theory of the change of data [parameters]" (*Entwicklung I*, 464–5). The problem of this theory is how "an economy accomplishes the transition from one level ... to another level" (p. 466). As already mentioned, Schumpeter recognised this problem "when in my beginnings I studied the Walrasian conception and the Walrasian technique" (*S1937*, 165). Although he "as an economist owed more to Walras than to any other influence", he wanted "to construct a theoretical model" that answered "the question how the economic system generates the force which incessantly transforms it" (p. 165). The birthplace of Schumpeter's model can, probably, be found in his account of Walrasian comparative statics (*Wesen*, 441–519).

Schumpeter's name for comparative statics is "the method of variation", that is, the study of the response of the endogenous variables to an exogenous change. He emphasises that this study of the variation of the endogenous variables presupposes that the production functions and consumption functions of the economic system do not change. This means that the response to the exogenous change has to be small and take place quickly. An example of the problems created by a major change is described in terms of the marginal value of money (*Wesen*, 471–4). Since the evaluation of money is the outcome of long experience, a major change means that the economic agents have to restart their " 'learning to count' " (p. 473). This learning is hindered by the fact that the value of money is closely related to the "standard of life" required to be member of a certain economic class. For instance, the consumption functions of the land-owning nobility will only respond gropingly and unwillingly to a downward or upward change in the marginal utility of money (p. 474). This and similar processes are excluded from comparative statics. What is missing is a "theory of the change of data". While "the first problem of economics" concerns a given state of the economy, its second problem is: "How does an economy accomplish the transition from one level ... to another level? This is the question about the essence of economic evolution" (*Entwicklung I*, 465–6; cf. *S1912b*, 94).

Although comparative statics is limited to the "first problem of economics", Schumpeter nevertheless considered it a major contribution to the toolbox of economic analysis. The serious application of this tool requires "higher mathematics", that is, advanced forms of integral and differential calculus. Since the design of *Wesen* excluded the demonstration of the use of these methods, he stated that "I hope soon to have the occasion to add what is missing in this respect" (*Wesen*, 445). This came close to a promise of delivering the book that he later called "The Theoretical Apparatus of Economics, in which I want to improve the quantitative methods of modern economics" (*BL*, 283). *Wesen*, however, concentrates on persuading the reader with only rudimentary mathematical skills that "higher mathematics" is needed. The major example used for this persuasion is the rather equation-dense analysis of the effects of the introduction of a quantity tax on a particular good (*Wesen*, 484–97). Schumpeter applies properties of the functions defined implicitly by the Walrasian system of simultaneous equations to answer the question what happens to the optimal solution because of the imposed tax. The change of the price of the good creates a chain reaction that is decomposed by means of Taylor's formula. The different elements of this decomposition are then studied by integration and differentiation. The conclusion is that the controversy on tax issues emerges because different assumptions are made concerning the form of the functions and because some elements of the decomposition are ignored. A shorter analysis of protective tariffs on imports also reveals that protectionist arguments normally include dynamic effects (pp. 503–9).

Since Schumpeter never developed these sketchy arguments into a systematic use of "higher mathematics", one might suspect that this neglect is based on lacking mathematical skills. This is hardly true. Even thirty years after he had abandoned higher mathematics as his personal research tool, his contributions to *Rudimentary Mathematics for Economics and Statisticians* (*S1947a*, 106–9, 129–33, 159–79) were clearly superior to those of the main author, William Crum. The real reason for abandoning comparative statics and the related types of mathematics is to be found in the seriously limited range of problems that they cover. Of course, if we are willing to renounce on a high level of precision, we may by means of comparative statics "force our way towards the problems of Dynamics" (*Wesen*, 518). How far we can come in this way, however, is an empirical question. Schumpeter clearly thought that the results of comparatives statics with respect to the "problems of Dynamics" would not be satisfactory for him: "We can *never* surge into their core; the great tendencies of evolution [Entwicklungstendenzen] surely go past our systems, plays on other scenes. We perceive their voices only as the rumbling of a distant thunder" (p. 518). His reason was that these "tendencies of evolution" could not be derived from the type of economic agent on which comparative statics is, more or less explicitly, founded: "What a pitiful miserable figure he is, our economic subject who is always looking so anxiously for equilibrium. He has no ambitions and no entrepreneurial spirit; in brief, he is without force and life" (p. 567). It seemed obvious to him that "[e]ven the ordinary process of the economy is full of life and movement and has to be conceptualised in steady evolution [Entwicklung]. However, we stand puzzled towards the phenomenon of evolution and the 'high problems' of economic progress."

These statements are not implying that the search for equilibrium is absent from economic life. Instead, it means that something essential is missing in the Walrasian model. What Schumpeter missed, he most clearly suggested in his second book: "The men who brought forth modern industry were 'all-of-a-piece people' and not pathetic figures who steadily asked whether each effort that they had to perform really gave a sufficient surplus of utility" (*Entwicklung I*, 137). Instead, "[s]uch men create because they cannot do otherwise" (p. 138). Although Schumpeter wanted to develop a model that places such creative entrepreneurs at the centre of the mechanisms of economic evolution, these mechanisms needed the addition of non-innovative agents. The solution was ready at hand. The adaptive response to the deeds of the creative entrepreneurs could be modelled by means of a revised version of the Walrasian system with routine-based behaviour of ordinary agents. The most important of these agents was a reinterpreted version of the Walrasian entrepreneur interpreted as a routine-based manager.

Schumpeter's distinction between innovative and non-innovative agents provided him with the elements of his "theory of the change of data". The data are the parameters that determine the behaviour of non-innovative agents. In contrast, the innovators take these 'parameters' as the starting point for their innovative projects. They can base the evaluation of the profitability of their projects on the Schumpeterian conjecture that the other agents of the economic system do not adapt quickly. This conjecture, which has some similarity with the conjecture that Cournot ascribes to his oligopolists, cannot be based on the extremely flexible behaviour of Walrasian entrepreneurs. It is, instead, based on a revised version of Walras's concept. We are thus facing three different concepts of entrepreneurs:

- The Walrasian entrepreneur adapts promptly to the changed parameters of the economic system and thereby contributes to the equilibration of that system. The formal role of this agent in the general equilibrium system means that although time is not included in the concept, we are actually facing instantaneous reaction.

- The Schumpeterian manager is the Schumpeterian version of the Walrasian entrepreneur. The behaviour of the Schumpeterian manager is based on routines. This means that adaptation takes time and that adaptation to major change is impossible or difficult for this type of agent.

- The Schumpeterian entrepreneur disturbs the equilibrium by buying and using resources to change one of the 'parameters' of the economic system. The financing of the project of making such a change is motivated by an expectation of profit. This expectation is based on the characteristics of the project as well as on the Schumpeterian conjecture of sluggish response from the rest of the economic system.

## 3.6 Conclusion

When *Wesen* had been printed, Schumpeter systematically distributed it to core economists. The most important was Walras. In a couple of letters to him, Schumpeter pointed out that "this is a book of a disciple" and that he wanted to work under the Walrasian "leadership" (*BL*, 43–4). Later, he found time to pay a visit—his first and only—to Walras in Switzerland. However, the elderly Walras thought in terms of a passively adapting economy and did not accept Schumpeter's vision of economic evolution (*S1937*, 166). Since he met similar reactions from practically all the leading economists, he became aware of the hard work in front of him. Nevertheless, Schumpeter always upheld his great respect for Walras.

The reconstruction of the arguments of *Wesen* in the present chapter has demonstrated that the presentation of the complex Schumpeter–Walras relationship as one between a disciple and his master is not totally wrong. Schumpeter was, however, the type of disciple who wants to confront new problems by innovating the teaching of his master radically. The understanding of this point helps to resolve what, probably, is the most difficult of Schumpeter's many paradoxes. For instance, Morishima and Catephores (1988, 42) represent the idea of the Walrasian Schumpeter: while it is "generally believed that Schumpeter's hallmarks were the terms 'entrepreneurs', 'innovation', and 'new productive combination' ", these concepts and the underlying ideas are actually "a direct extension of Walrasian concerns." Schumpeter has also been presented as a "Walrasian Austrian" (Schefold, 1986), while Freeman (1990, 28) has suggested that Schumpeter was a non-Walrasian but bound to Walrasian tools of analysis: "it was Schumpeter's misfortune that he attempted to marry it [the Walrasian equilibrium theory] with his own theory of dynamic destabilizing entrepreneurship". Others see him as an eclectic whose approach involved "brilliant mixture, if not always an internally consistent" blend of "Marxism, Walrasian equilibrium analysis, and German historical scholarship" (Rosenberg, 1986, 209).

All these statements are true—if we do not take them too literally. Schumpeter was inspired by Walras even with respect to Economic Dynamics; his Austrian background had some influence; his Walrasian tools did create problems; and, as we shall see in the next chapter, he did include inspiration from Marxism and the historical school. However, Schumpeter created his evolutionary theory as a new combination of all these, and several other, backgrounds. This evolutionary theory not only consisted of an evolutionary economic dynamics. It also required an evolutionary economic statics that helps the understanding evolutionary dynamics, i.e., an analysis of the conditions under which the evolutionary process comes to a halt. Since these conditions are very strict, the evolutionary statics primarily serves as an analytical tool. However, Schumpeter never specified his evolutionary statics in a satisfactory way. Although the main problem seems to have been that this task is difficult, his respect for Walras induced him to write as if the task had largely been solved by his great master.

The first version of Schumpeter's personal research programme is found in the concluding section of *Wesen* on the "Prospect on Dynamics [Ausblick auf die Dynamik]" (pp. xxxiv, 614–22). Here he stated that the process of "crystallisation" of Economic Statics should be allowed to fulfil its purifying mission. The residual of economic issues that cannot be handled by the emerging "crystal" is a mixed bundle. This bundle includes the problems of economic evolution. The reintegration of these problems would undermine the coherence of Economic Statics and be of little help to the development of analytical tools for Economic Dynamics. Leaving the problems to the members of the historical school was also unsatisfactory. They had done an important job with facts and theories about details but there was no indication that they would turn to formal theorising. Hence the task was left to Schumpeter. The earlier presented quotation from *Wesen* (618–9) is a kind of summary of the guiding principles for his evolutionary research: "it shall always stay our principle to be silent—or ... to delimit ourselves to summaries of facts—about things on which we have nothing *exact* or sufficiently *interesting* to say." Although the ultimate goal was "exact" modelling, there was also a need for summarising the stylised facts of economic evolution and for sketching out theories of the most interesting of these facts.

# 4

# Elitist Dichotomies and General Evolutionary Analysis

We can most easily understand Schumpeter's early move from the Walrasian system to his theory of economic and social evolution within capitalism on the background of elite theory. His knowledge of this kind of theory is beyond doubt. Thus his former student Eduard März (1991, 58) emphasised that "[a]nyone who knew Schumpeter personally can testify that he had a thorough knowledge of theories of the élite of the late nineteenth century (Nietzsche, Pareto, Mosca, Michels, Le Bon) and took pleasure in flirting with such ideas." Actually, März's conclusion is that Schumpeter's "own theory of development was an attempt to give concrete economic substance to the vague theories of his time." This conclusion can also be reached by reading *Entwicklung I*, as when McCrea (1913, 526), in his review of this book, suggested that it "offers a super-man interpretation of economic progress". The theories of the elite were based on the contradistinction between the masses and the elite, and so were many of Schumpeter's early formulations. For him, economic agents can be divided into two groups that are characterised by different levels of competence and will:

- The masses: "Most people tend to their usual daily business and have enough to do at that. Most of the time such people are on slippery ground and the effort to stand straight exhausts their energies and suppresses all appetite for further exploration. ... The daily work keeps them down, organization as well as the influence of their colleagues inflict untearable chains on them. This is the masses." (*S1912a*, 412–13; *Entwicklung I*, 162–3)

- The elite: "A minority of people with a sharper intelligence and with a more agile imagination perceive new combinations. ... Then there is an even smaller minority—and this one acts. ... It is this type that scorns the hedonic equilibrium and faces risk without timidity. He does not consider the implications a failure will inflict upon him, or care whether everyone depending upon him will lose their keep for old age. ... The decisive moment is therefore energy and not merely 'insight'. ... What matters is the disposition to act. It is the ability to subjugate others and to utilize them for his

> purposes, to order and to prevail that leads to successful deeds—even without particularly brilliant intelligence." (*S1912a*, 413–14; *Entwicklung I*, 163–4)

It should be noted that Schumpeter is not assuming that competence and will are characteristics that divide *Homo sapiens* into two distinct and permanent groups. He only assumed the existence of "gradual differences of personalities" (*S1912a*, 434n; *Entwicklung I*, 162n). It is the difficulty of the task of implementing "new combinations" that in any epoch divides the population into two distinct groups. However, the early Schumpeter did not emphasise this point. He simply used his descriptions of the mass and the elite to characterise their respective roles in the process of economic evolution. The mass is the follower of routines and the elite is changing these routines by implementing innovative projects. This elite needs its surplus of mental energy not only because innovation is intrinsically difficult but also because those who "lose their keep for old age" strongly resist innovation.

The emphasis on the innovative deeds of the elite gives Schumpeter's early formulation of his theory of economic evolution many of the characteristics of a hero epic (as it was remarked by Franz Oppenheimer, one of his German critics). At the beginning of the epic, 'progress' has come to a halt because all the economic agents have adapted to a system of routine, but then the first entrepreneur finds an innovative project and enters the stage together with a helpful banker. Their example brings forth a swarm of additional entrepreneurial projects, and a major effect of these projects is a dramatic fight with the routine-loving mass of agents. If the entrepreneurs are successful, some members of the masses of economic agents go bankrupt or lose their jobs while others are forced to adapt. Another effect of entrepreneurial action is that the economic system becomes so disturbed that further innovation exceeds the capabilities of the remaining men of action. Instead, the immediate end of the story is the emergence of a new routine system. Nevertheless, this system sets the stage for a new wave of entrepreneurship in an apparently never-ending epic. This epic of economic evolution can be called Schumpeter's entrepreneurial interpretation of economic history.

## 4.1 The 'lost' chapters of *Entwicklung I* and their translation

*Wesen* had analysed the solution to the great problem of general economic equilibrium by assuming that "[e]veryone adapt to given circumstances as well as possible" (*Entwicklung I*, 465; cf. *S1912b*, 94). In Schumpeter's second book, the task was to complement this solution by taking "the second step towards a total model [Gesamtbild] of the economy". This step concerns "the second great problem of economics" (*Entwick-*

*lung I*, 465–6; cf. *S1912b*, 94) that can be phrased in the following way: "how does an economy carry out the transition from one level—which has be considered to be a point of rest point of equilibrium—to another level? This is the question on the essence of economic evolution [Entwicklung]." The shortest summary of this "essence" of economic evolution is found in the large paper "On the Essence [Wesen] of Economic Crises" (*S1910d*), which was published in 1910. This paper includes accounts for his analysis of economic evolution and its relationship with the Walrasian system. For instance, it announces Schumpeter's evolutionary model in staccato. The basic idea is that "[w]hich ever the economy might be organised, it always has a static equilibrium state. And it must always be directed into new directions by leading personalities" (*S1910d*, 16). The redirection implies "new combinations" and the "function of the entrepreneur lies in the implementation of these new combinations" (p. 15). In contrast to "the accustomed combinations of statics", such new combinations "do not generally become accepted automatically". Instead, "an intelligence and an energy are required which are found only in a minority of economic agents".

This early announcement of Schumpeter's evolutionary theory pointed towards the 550 pages of his second book, which was published when he was 28 years old. The book is best known as *The Theory of Economic Development* from 1934. As already mentioned, *Development* is the somewhat shortened translation of the second edition of *Theorie der wirtschaftlichen Entwicklung* (*Entwicklung II*). He produced this 1926 edition by radically revising and shortening the first edition that came out late in 1911. While *Development* and *Entwicklung II* are so close that they can be considered the same book, the first edition of *Theorie der wirtschaftlichen Entwicklung* should be considered a separate book in order to avoid confusion. *Entwicklung I* is not least characterised by a presentation of the core concepts of entrepreneurs and routine-based agents that differs radically from the way these concepts are presented in later editions. The first edition also includes a never-repeated chapter that summarises the book and adds a sketch of a general theory of social evolution. It is especially in these parts of the first edition of his book that we can study his elitist vision and its relation to his theory of economic and social evolution. Here we find the clearest expressions of Schumpeter's entrepreneurial interpretation of history.

The rediscovery of the early formulations of his ideas has especially focussed on "Schumpeter's 'lost' seventh chapter" (Mathews, 2007). Shionoya (1990, 315) emphasised that this chapter "actually presents ideas for a general theory of social order and social change"; and other researchers have given comments after its recent translation into English (*S1912b*). However, Schumpeter's early ideas can be understood most clearly if we start from the equally "lost" arguments of the second chapter of *Entwick-*

*lung I*. Therefore, it is important that a small part of this chapter has also become available in English language (*S1912a*, 405–14). The detailed contents of both chapters of *Entwicklung I* are recorded in Table 4.1 on the facing page.

*Entwicklung I* and the other version of Schumpeter's book started with the same series of propositions. The starting point is that "[t]he social process is really one indivisible whole" (*Development*, 3; cf. *Entwicklung I*, 1). By reading the whole book, we gradually recognise that this process can best be interpreted as social evolution, but we also recognise that the subject matter is largely economic evolution. Therefore, Schumpeter needed to isolate analytically this integrated part of the indivisible stream of social evolution. This step is described in the following way: "Out of the great stream the classifying hand of the investigator artificially extracts economic facts. The designation of a fact as economic already involves an abstraction". This abstraction is "the first of the many forced upon us by the technical conditions of mentally copying reality." The last three words sounds strange. A look at the German words that were translated in this way reveals that his ambition, of course, was not one of "mentally copying" reality in a naive way but instead the "mental reconstruction of reality [gedanklichen Nachbildung der Wirklichkeit]". The main message, nevertheless, is retained in the English translation: the contribution of any specialised social science is never sufficient for explaining the social process as a whole. Schumpeter wanted to overcome this limitation as far as possible by combining relatively independent studies within several social sciences. This ambition was not explicit in the shortened version of the original text. However, in *Entwicklung I* (p. 1) he, shortly after the quoted sentences, added that only "the last step of our journey will again lead us to the proposition with which we began." This "last step" of returning to the "social process" as an "indivisible whole" was taken in the seventh chapter of *Entwicklung I*. The omission of this sentence was needed in *Development* because the 86 pages of Chapter 7 had simply been removed. By sticking to the introductory proposition, Schumpeter nevertheless emphasised that the revised book was only a torso.

The German title of the last chapter of *Entwicklung I* is "Das Gesamtbild der Volkswirtschaft", and it has been translated as "The Economy as a Whole" (*S1912b*, 93) and as "The View of the Economy as a Whole" (*S1912a*, 414). However, these translations are not entirely satisfactory. Let us first consider the strange word "die Volkswirtschaft". Today, it means both 'economics' and 'the economy', and Schumpeter obviously related to the latter meaning. However, 'das Volk' means 'the people' or 'the nation', so the word 'Volkswirtschaft' points to the economy of the nation—with connotations to the peculiar conception of the nation developed by German romanticism and the historical school. Schumpe-

**Table 4.1.: Lost chapters of *Theorie der wirtschaftlichen Entwicklung* (1912)**

| | |
|---|---|
| Chapter 2. The Fundamental Phenomenon of Economic Evolution | |
| | [1] Preliminary remarks. |
| Stationarity | [2] The stationary economy as a historical phenomenon. [3] Its reason. |
| Behaviour | [4] The two types of human behaviour. [5] Non-hedonic behaviour within the economic areas and its characteristics. [6] Its psychological explanations and its importance for the economy. [7] Answers to criticism and supplements. |
| Innovation | [8] The outward form of evolution. [9] The system of future values. [10] The concept and function of the entrepreneur. [11] The different methods for the implementation of new combinations. [12] The purchasing power of the entrepreneur; the banker. |
| Chapter 7. The Comprehensive Model of the National Economy | |
| | [1] Preliminary remark. |
| Alternatives | [2] The two problems of economic theory. [3] Historical and theoretical problems of evolution. [4] Discussion of the 'environmental theory' of evolution. [5] Discussion of the 'growth theory' of evolution. |
| Economics | [6] Three general propositions on economic evolution. [7] The impact of evolution on individual economic agents. [8] The most important special case. [9] The scheme of the total socio-economic process. [10] Overview of our composition of the process of production and some of its applications. |
| Sociology | [11] On the economic structure of society and the problem of the social party emplacement of the individual. [12] The social atmosphere of the capitalist economy. [13] Analogies to economic evolution in other areas of social life; the social phenomena. |

Comment: While Chapter 7 of *Entwicklung I* was excluded from *Development*, Chapter 2 was radically rewritten and its first half was largely omitted. The start of each section of the book is (normally) indicated by an empty line. Section titles are *only* found in the table of contents of the German editions. The grouping of sections has been added for the purposes of the present book. See also Table 5.1 on page 103. Source: The table translates the detailed contents of Chapters 2 and 7 of *Entwicklung I* (pp. iii, v).

ter did not share this conception, but his chapter demonstrates that he was thinking of a 'national economy' that was related to all other sectors of social life within a national framework. The translation of the word 'das Gesamtbild' also gives problems. We might translate it to 'the total picture', but 'the picture' is not the only possible translation of 'das Bild'. Schumpeter's meaning is rather that of 'das Gedankenbild', that is, 'the mental construct', 'the analytical construct', 'the schema', or 'the model'. This meaning was well known in German epistemology. Thus, Max Weber (1968, 190–1) used "Gedankenbild" in his attempt to resolve the battle of methods by emphasising that the Austrian school was producing "unified *analytical* constructs [einheitlichen *Gedanken*bilde]". The fact that Schumpeter ended his *Entwicklung I* by sketching out an even more comprehensive construct than that of his Austrian teachers can be emphasised by translating the title of his seventh chapter to "The Comprehensive Model of the National Economy".

Chapter 7 of *Entwicklung I* concentrates on summarising and extending Schumpeter's theory, but it includes an exposition and criticism of alternative theories of economic evolution (*Entwicklung I*, 464–88; cf. *S1912b*, 93–107). Thereby, Schumpeter recognised that the historical school and neoclassical economics had already studied economic evolution. Economic historians and descriptive economists had been concerned "with the concrete course of evolution in a particular time and at particular locations", including the change of the characteristics of industries and their rise and decline (p. 467). These studies of evolutionary "episodes" can loosely be integrated in "comprehensive studies"; and here Schumpeter probably not least thought of Sombart's *Modern Capitalism* of 1902. In contrast, economic theorists were concerned with the questions of the abstract mechanisms of evolutionary change: "First, how and by what process do concrete changes occur? Second, is it possible to recognize regularities in the way that everything new arises? And if so, can these regularities be formulated in a general way?" (p. 467). Schumpeter lumped together all general theories except his own under two headings. On the one hand, there is "environmental theory of evolution", according to which, changes in the physical and social conditions determined the evolution of the economic system. On the other hand, the "growth theory of evolution" considers economic development and evolution as caused by the growth of population and physical capital as well as by exogenous factors like inventions, organisational changes, and the development of wants. Schumpeter rejected these types of theory. Concerning the environmental theory, he argued that it is "empty and meaningless, as far as it is true, and wrong, as far as it leads to any conclusions" (p. 471). The growth theory of evolution was dismissed more concretely. Thus, Schumpeter argued that the causal direction was from capitalist evolution to population growth, inventions, and change of wants. For instance, he approvingly quoted Marx's dictum: "Capitalism has stamped populations out of the ground" (p. 478). In contrast, he had to deny Marshall's idea that "the fundamental phenomenon of the economy lies in its steady expansion in well-balanced proportions" (p. 474).

If we consider Chapter 1's model of the circular flow of economic life on the background of the last chapter of *Entwicklung I,* it becomes obvious that it is constructed in a way that excludes the causation of economic evolution by environmental change and simple economic growth. Thereby the circular-flow model sets the stage for the introduction of innovative behaviour as the sole cause of the evolutionary process. This introduction required the application of "Wieser's principle of continuity", according to which, everyone other than the entrepreneur "will cling as tightly as possible to habitual economic methods and only submit to the pressure of circumstances as it becomes necessary. And he will even perform this 'yield' according to the rules provided by experience" (pp. 8–9; partly in *Development,* 8–9).The innovative entrepreneur is seen as a powerful leader who is able to force less able economic agents to adapt to his plans. This enforcement is emphasised by Chapter 7. It gives rise to a process that includes both the creation and destruction of socio-economic positions (pp. 492–511; *S1912b,* 109–20). Since analogous processes can be observed in any sector of social life, it is possible to create a general theory of social evolution (pp. 535–48; *S1912b,* 135–42). However, this conclusion had already been prepared by Chapter 2 of *Entwicklung I.*

The title of the second chapter is "Das Grundphänomen der wirtschaftlichen Entwicklung", and the best translation of this title is "The Fundamental Phenomenon of Economic Evolution". This translation is not the one found in *Development,* where Schumpeter endorsed Redvers Opie's translation of 'Entwicklung' to 'development', but Schumpeter changed his mind immediately after the publication of this book. The use of the term 'evolution' is especially endorsed by its consistent use in its major sequel, *Business Cycles.* His use of the title "The Fundamental Phenomenon of Economic Evolution" does not imply that it is "the phenomenon of evolution" that is "fundamental". Instead, Schumpeter's chapter deals with "the fundamental phenomenon" whose removal would imply that we would also lose the complex "phenomenon of economic evolution". While this basic phenomenon is patently clear in *Entwicklung I,* it became less clear in later versions of the book. Therefore, he explicitly stated that "the ultimate fundamental phenomenon of economic evolution" is found in "the essence of the entrepreneurial function and the behaviour of the economic agents who are its carriers" (*Entwicklung II,* 110–11; confused translation in *Development,* 74). The essence of the function and behaviour of Schumpeterian entrepreneurs is presented very generally in *Entwicklung I.* Actually, much of the argument is so general that it covers the function and behaviour of innovators in any type of social evolution: "the strongest individuals ... will create something new and destroy the old thing, conceive and carry out bold plans, ... [and] subjugate their fellow citizens to their rule" (*S1912a,* 409; *Entwicklung I,* 157). The core concept is "schöpferisches Gestalten". Becker and Knudsen (*S1912a*, 433) translates this term to "creative construction", thereby contrasting it beautifully with the necessary counterpart of "creative destruction", a term that Schumpeter used in *Capitalism*. According to the early Schumpeter, it is "creative construction" that is the intrinsic cause of both economic and social evolution. Its absence would imply an economic and social order that is either stationary or changes because of exogenous disturbances. Since Schumpeter emphasised that "creative construction" takes the form of "new combinations" of existing resources, the term might also be translated to "creative recombination". This translation also points at a possible inspiration. Thus, Wilhelm Wundt, a leading German philosopher and psychologist, had characterised creativity by "creative synthesis" ("schöpferisches Synthese"). However, although Schumpeter demonstrated knowledge of Wundt's work (*Entwicklung I*, 482), he did apply it directly. Instead, he thought that the functional principle of "the carrying through of new combinations" had to be complemented by "the energy principle" (*Entwicklung I*, 170). It is the psychological "energy" of the innovator that allows him "to subjugate others and to utilise them for his purposes" (*S1912a*, 414; *Entwicklung I*, 164).

After the above considerations, we may shortly characterise the contents of Chapter 2 of *Entwicklung I* (see Table 4.1 on page 71). This chapter on "The Fundamental Phenomenon of Economic Evolution" starts by defining "evolution" as including "only such changes in the circular flow of economic life … that the economy generate from itself, … not driven by external impulses" (*Entwicklung I*, 103; cf. *S1912a*, 405). This definition does not necessarily imply that the real phenomena of economic evolution can be described as "evolution" in Schumpeter's sense. On the contrary, he is fully aware that the overwhelming majority of theorists tended to expand their Economic Statics with a type of Economic Dynamics that was either driven by the change of external circumstances or by the gradual growth of the economic system. However, such theories imply that "the phenomenon that we in practice call economic evolution" is not an essential characteristic of the economic system (p. 103). The rest of the book is obviously designed to demonstrate that real economic evolution can largely be modelled as an endogenous economic process that is driven by innovative entrepreneurs. This demonstration sticks to all the assumptions of the model of an equilibrated and stationary "circular flow" but adds "our new agency [Agens]" (p. 105). The characteristics of this agency is dramatised by a discussion of "the stationary economy as a historical phenomenon" (pp. 107–24). Here Schumpeter asks his readers to drop their preconceptions about an economic system that is permeated by the phenomena of economic evolution. What should be surprising is not that, for instance, peasants show resistance to economic change and 'progress' but that this resistance is often overcome. Even "where the peasant comes in direct contact with modern evolution", we see a reaction that "falls totally within the schema of statics". Actually, it is only decades after an effective demand has emerged due to evolution elsewhere that we see reluctant "adaptations to changed circumstances" (p. 112). This and many other examples demonstrate that resistance to change is the norm rather than the exception. The reason for lacking evolution is not that the standard economic agent, the Homo Economicus, does not recognise alternatives to his habitual behaviour and gradual adaptation (p. 117). The reason is rather that he lacks the will and competence to engage in more radical and difficult processes of change.

At this point of the argument, Schumpeter introduces a type of agent that has the necessary will and capability. This is "our man of action", the entrepreneur in the Schumpeterian sense (p. 133). We shall discuss the psychological analysis of this type of agent (pp. 134–56) below. Instead, we immediately turn to the entrepreneurial action, the "carrying out of new combinations of existing economic possibilities". In other words, "the strongest individuals ... will create something new and destroy the old thing" (pp. 157–8; *S1912a*, 409). Since such an entrepreneur starts from a circular flow, he cannot draw upon an "industrial reserve army" in the Marxian sense, nor upon any other unemployed resources. This means that any introduction of the new combination requires "a transfer [Virement] of productive forces" to him from their previous applications (p. 159; *S1912a*, 410). The tool to enforce this transfer is "purchasing power" defined as "the abstract power ... over goods in general" (p. 84). However, since there is no room for the entrepreneurial act in the circular flow, he has to borrow the needed money from sources that are not part of the model (pp. 184–98). This, of course, requires an expected profit from the new combination that is more than sufficient to pay back the loan and the accumulated interest. Therefore, a core aspect of the competence of the entrepreneur is to be able to formulate his plan persuasively in relation to both the present price system and a "system of future values" that, to some extent, is influenced by the implementation of this plan. If he is successful, the rest of the economic agents will be forced to adapt, both when "productive forces" turn from them to the innovative project and when the project starts to deliver its output.

## 4.2 From elite theory to the Schumpeterian dichotomies

In economics, we are accustomed to think in terms of a single type of agent, the optimising Homo Economicus. The different capabilities and energies of individual agents might be included in their endowments, but this is not necessarily the case. Classical economics, especially, was not only characterised by what can be called analytic equalitarianism but also by assumptions on the real homogeneity of human beings. Thus

Adam Smith (1986–99:I, 120) emphasised that the "difference of natural talents in different men is, in reality, much less than we are aware of", and he remarked that the difference "between a philosopher and a common street porter, for example, seems to arise not so much from nature as from habit, custom, and education." This emerging type of human heterogeneity seems to be sufficient to create the processes of change studied by evolutionary economics and evolutionary sociology; but few real results were obtained. The problem seems to have been that human behaviour was considered to be so flexible that it was hard to think in terms of routines and their innovation without the formal tools of evolutionary analysis. In contrast, those who believed in essential human heterogeneity and/or applied analytic anti-equalitarianism were more likely to distinguish between the routine behaviour of the masses and the innovative efforts of the elite. According to Schumpeter, "it should be obvious that analytic equalitarianism is of immense importance, not only for economic sociology ... but also for many problems of economic theory itself. We need only replace it by the opposite assertion of fact to realize that this would change the whole picture of the economic process" (*History*, 122). Actually, Schumpeter applied the "opposite assertion" to a unique degree. He could have phrased his results in terms of different endowments of the economic agents of neoclassical economics, but he did not do so. The reason is probably that modelling by means of a basically homogeneous type of agent did not encourage the mass–elite distinction. Instead, he clearly thought in terms of two types of agent that can be called Homo Economicus and Homo Creativus.

The theories of the elite were part of an antirationalist and antiequalitarian stream of thought that had a significant following in the beginning of the twentieth century (Hughes, 1977). This type of thinking not only included Nietzsche and Pareto but also Henri Bergson's philosophy of "creative evolution". It may be seen as a reaction against predominant tendencies in the social sciences. For instance, Alfred Marshall's (1898, 43) famous statement: "The Mecca of the economist is economic biology rather than in [mechanical] economic dynamics" related to a broadly defined sociology (that also covers economics). Here there had been a large upsurge in evolutionist attempts to explain social change during the last half of the nineteenth century. The main name was the English polymath Herbert Spencer, who combined inspirations from classical economics, biological theories of organismic development, and Comte's positivistic philosophy. The result of this strange mix is a theory of the evolutionary *development* of everything. The general result of "the survival of the fittest" is a movement from primitive homogeneity to complex organisation. The complex organisation is studied in terms of an increasing division and coordination of labour. Actually, Spencer deducted Adam Smith's theory of social evolution from biological considerations; and these considerations influenced him to become a very radical supporter of economic liberalism. However, it was not the policy of Social Darwinism that disturbed Schumpeter. He was rather worried by the underlying vision of automatic "progress". In this respect, he felt closer to the age-old hero theory of social change. Spencer (1896, 351–2) considered "the great-man-theory of social affairs" as suggesting "that Sociology is not possible". He thought that the relevance of this theory is "limited to early societies" (p. 32). Later evolution is not caused by great men. The analysis of later changes should not be made in terms of the "great-man-theory" but in terms of a process of evolution of "new institutions, new activities, new ideas, sentiments, and habits; all of which unobtrusively make their appearance without the thought of any king or legislator."

According to a student of the issue, the "extent to which Spencer's views have influenced modern social thought on the subject of the great man and his environment can hardly be exaggerated" (Hook, 1943, 67n). However, like Schumpeter, his teacher Friedrich von Wieser reacted against this attempt to describe economic and institutional change in terms of spontaneous processes within nameless masses of people. Wieser's reaction against Spencer's belittlement of great men is not discernible in his early works, to which Schumpeter was normally referring; but in the early years of the twentieth century he began his explicit opposition. The results of the new world view were presented in Wieser's (1914) *Social Economics* and in his short articles on "Mass" and "Leadership" (in Elster et al., 1923–29). However, he expressed his critical reaction most clearly—and naively—in an autobiographically oriented paper (Wieser, 1907). Here he told that his world view during his years at grammar school came from Virgil's *Aeneid*, the hero epic on the founding of imperial Rome, which was popular in the Austrian-Hungarian Empire. Wieser saw this epic as suggesting that "new institutions, new activities, new ideas, opinions and customs" originated in the minds of kings and lawmakers (p. 337). Then he met the gradualist evolutionism of Herbert Spencer, who taught that the task was to write "history without names" (p. 339). Spencer's approach persuaded Wieser; and this persuasion must have been further fixed when he fell under the spell of Carl Menger's similar vision. However, Wieser ultimately found his persuasion unsatisfactory. Although Spencer was right that the 'genius' could not do without the mass, he had forgotten that the mass could not do without the innovator: "Without great men, there would be no evolution ... If mankind became reduced to average man, it would be convicted to stagnation" (p. 340). The resistance of the masses against change may seem overwhelming, but still the "innovator enters virgin land; he is the first, the only; he has none beside him and none over him; he stands over everyone else" (p. 343).

Wieser's formulations suggest that it was not least the influence of

Friedrich Nietzsche that turned him away from Spencer—and away from Adam Smith's and Menger's interest in the spontaneous evolution of institutions. In the beginning of the twentieth century, Nietzsche (who died in 1900 after years of mental illness) had become a fashionable and notorious philosopher with works like *Will to Power* and *Thus Spoke Zarathustra*. In contrast to Spencer's anonymous history, Wieser's (1907) regained innovator was like Nietzsche's Zarathustra: "The Superman [Übermensch] only becomes prophet when he speaks to the mass". However, Wieser envisaged a Superman that cared for the masses, so he could not formulate his critique of anonymous history in a fully Nietzschean way. According to the later Nietzsche (1976, 97; emphasis removed), "It stands in my Zarathustra. —and he who wants to be a creator in good and evil has first to be a destroyer and break values. Thus the greatest evil belongs with the greatest good: this, however, is the creative good." His criticism of the mass behaviour is recorded in the earlier essay "On the Uses and Disadvantage of History for Life". Here Nietzsche (1983, 113) suggested that "statistics ... prove how vulgar and nauseatingly uniform the masses are". The facts revealed by statistics "are the effects of inertia, stupidity, mimicry, love and hunger". He did not find such aspects of human behaviour suitable subjects for historical narrative. Nevertheless, "the kind of history at present universally prized is precisely the kind that takes the great mass-drives for the chief and weightiest facts of history and regards great men as being no more than their clearest expression". Nietzsche protested against this "writing of history from the standpoint of the *masses* and seeking to derive the laws that govern it from the needs of these masses, that is to say from the laws which move the lowest mud- and clay-strata of society." Actually, he only thought that the masses deserved notice in three respects: "first as faded copies of great men produced on poor paper with worn-out plates, then as a force of resistance to great men, finally as instruments in the hand of great men; for the rest, let the Devil and statistics take them!" (p. 113).

Here Nietzsche must be referring to Quetelet's pioneering work on statistics with its emphasis on the stable characteristics of Average Man. In contrast, Nietzsche's creative "Devil is the regent of this world and the lord of success and progress" and "the lever of the movements of history" (p. 113). To uphold this creative spirit, he needed a "generation of fighters and dragon-slayers" (p. 121). Each of these Wagnerian Supermen would "rebel against a state of things in which he only repeats what he has heard, learns what is already known, imitates what already exists" (p. 123). This simple negation of anonymous history in terms of "masses" was certainly not enough for Wieser. However, he saw it as reflecting core aspects economic and institutional evolution—and Schumpeter implicitly agreed. Actually, Nietzsche had done a useful job by exaggerating the achievements of supermen; but a full understanding of socio-economic change required an analysis of the interaction between elites and masses—interpreted as "dynamic" and "static" forces. Nietzsche's fantasies of isolated supermen were of little help for such an analysis, so we should not overemphasise his importance for Schumpeter. However, many others expressed similar views in ways that caught broader attention among social scientists.

Among the many authors who, in the years around the beginning of the twentieth century, opposed Spencerian (and Marshallian) influences was the English conservative critic William Mallock. Schumpeter (*History*, 789n) later pointed out that he "was never recognized by the economic profession and seems to be entirely forgotten now" but that he, nevertheless, had "the courage to tell an unpopular truth". Mallock (1901) presented his "truth" in *Aristocracy and Evolution* and in most of his many other literary works. Here he opposed Spencer's "Attempt to Merge the Great Man in the Aggregate" (p. 17). This aggregation means that we ignore that "[e]volution, in fact, is the unintended result of the intentions of great men." More specifically, "the struggle which produces economic progress—and progress of every kind is produced in the same way—is ... a struggle amongst a small section of the community to lead, to direct, to employ, the majority in the best way". On this background, we recognise that "capitalism is an agent of progress" and that "capital itself" is "the result of the previous influence of great men on the industrial actions of the many" by means of "the payment of wages". (p. 158) In his review, Thorstein Veblen (1898a, 430, 435) remarked that the "arguments are presented in a lucid and forcible way" and that "the grotesquely devious ways of its economic argument do not prevent it from being a suggestive contribution to the discussion of cultural development". Since Mallock was too cranky to be cited in scientific publications, we do not know the evaluations of him by Schumpeter and Wieser. Their reactions were probably close to those of Veblen—with the major exception that they considered it worthwhile to apply something like his vision as a starting point for developing the economic analysis of economic change. Their visions, however, differed from that of Mallock—and even more from that of Thomas Carlyle, his predecessor as a conservative critic—by strongly emphasising the unglamorous and unromantic aspects of the great leaders and thus excluding naive "hero worship" (*S1928f*, 65n).

A basic characteristic of Schumpeter's analyses in the whole of *Entwicklung I* is that he performed them in terms of dichotomies that seem to be derived from these types of elite theory. Among the Schumpeterian dichotomies are: the use of fixed combinations and the introduction of new ones; equilibrium and innovation-induced disequilibrium; the "static" and "dynamic" types of income; a rate of interest that is zero or above zero; the innovation-based upswing of the economy and the movement toward equilibrium during the downswing; and the use of similar but looser dichotomies for the studying change in any area of social life. Schumpeter emphasised his use of dichotomies in a specification of his study of economic evolution that he included in both *Entwicklung II* and *Development*. Here he pointed out that

> "our position may be characterised by three corresponding pairs of opposites. First, by the opposition of two real *processes*: the circular flow or the tendency towards equilibrium on the one hand, a change in the channels of economic routine or a spontaneous change in the economic data arising from within the economic system on the other. Secondly, by the opposition of two theoretical *apparatuses*: statics and dynamics. Thirdly, by the opposition of two *types* of conduct, which, following reality, we can picture as two types of individuals: mere managers and entrepreneurs." (*Entwicklung II*, 120–22; cf. *Development*, 82–3)

These three "pairs of opposites" obviously relate to the overall dichotomy between Statics and Dynamics that Schumpeter applied in *Wesen* and *Entwicklung I*. He saw Statics as the study of the routine-based circular flow and the movement towards it through the action of "mere managers". Similarly, Dynamics is the study of the change of the routines of the economic system as initiated by the activity of the Schumpeterian entrepreneur. Actually, Schumpeter initially thought that there is something like a one-to-one relationship between the types of processes, methods, and conducts. However, he focussed heavily on his two types of conduct in the original version of the large Chapter 2 on "The Fundamental Phenomenon of Economic Evolution". In this chapter, he did not only deal with the dichotomy of "mere managers and entrepreneurs". Actually, he contrasted two *general* types of individual whose behaviour determine the evolution not only of economic life but of social life in general. It is, therefore, useful to consider the description of these two types of man in a very broad context.

The elite theoretic views give an impression of forgotten aspects of Schumpeter's vision as well as of the reasons for his characteristic opposition to theories of gradual and automatic evolution. However, although he knew the opinions of the members of our sample of elite theorists, it should be noted that similar visions and oppositions were widespread in the beginning of the twentieth century. Since Schumpeter, furthermore, flirted with elitist terms but hardly described any of the elitist sources of his early work, we cannot make firm conclusions on this matter. In contrast, it is not too difficult to understand how he tried to give precise content to the type of theory under study. Actually, all the above-mentioned dichotomies served this purpose. Of most direct relevance was the contradistinction between two types of conducts and individuals. Since this contradistinction is the major novelty in *Entwicklung I* when compared with *Wesen*, we shall consider its relevance for expressing the elitist world view. When describing his two types, Schumpeter used a cautious double terminology to avoid unwanted connotations and futile controversy (*Entwicklung I*, 127–8):

- The "hedonic-static" type: Considered in terms action (conduct), this type is denoted as both "static" and "hedonic". By characterising the action as "static", Schumpeter apparently emphasised that it refers to a circular flow or the movement to a circular flow. In contrast, "hedonic" means that the action is relating to the goal of utility or pleasure. Considered in terms of individuals, Schumpeter referred to "those from whom we *only* observe hedonic-static action" (*Entwicklung I*, 128).

- The "dynamic-energetic" type: The action of this type is denoted as both "dynamic" and "energetic". A "dynamic" action is one that is related to the disturbance of a given circular flow. An "energetic" action seems to be one that requires a surplus of mental "energy". This energy is needed in "the struggle with those 'constraints' that not everyone can tackle" (*Entwicklung I*, 128). Since all individuals to some extent have to act in the hedonic-static, it is "those who we *also* see acting dynamically-statically" that belong to the "dynamic-energetic" type.

Schumpeter emphasised "that we simply take the term 'energetic action' from everyday language" (*Entwicklung I*, 128). The reason is that he did not want "a dependence on the contents that the word has already found in different areas of knowledge." Therefore, his use of the term should be considered "simply a case of a scientific application of a popular concept". However, the terms "energetic action" or "energetic individuals" are today hardly used widely in everyday language. Furthermore, Schumpeter's account is not very clear. For instance, the degree to which "energetic individuals" are also "non-hedonic individuals" is not clarified convincingly. Therefore, it might be relevant to ignore his warning against looking for its meaning in other areas of knowledge. When he wrote *Entwicklung I*, the most widespread use of the term was in the "energetic theories of culture" that drew inspiration from physics and chemistry, and a leading name was Wilhelm Ostwald. However, Schumpeter probably shared Max Weber's sharp criticism of such theories. It seems more relevant to consider the simpler formulation of a researcher who, together with Ernst Mach, he created the empirio-criticism that preceded logical positivism by considering "science as an economy of intellectual energy" (Pearson, 1957, 221). The researcher is the German psychologist and philosopher Richard Avenarius. Since he developed a very complicated terminology, it is advisable to apply a simplified statement of his analysis of the "natural history of problems" in terms of mental energy (Höffding, 1905, 117–27). Thereby it becomes clear that we are facing an early form of what later has been called "cognitive economics" and the modelling of "bounded rationality".

According to the reconstruction of Avenarius's argument, the situation for an individual is unproblematic as long as its problem-solving "energy" corresponds to the challenge or stimulus [Reizung] from its environment. If we denote the energy of the individual by $E$ and the challenge by the environment by $C$ (both assumed to be measured in units of 'psychic energy'), then the situation is unproblematic as long as $E = C$. A problematic situation emerges if this is not the case. If the individual as not able to—or willing to—mobilise sufficient problem-solving energy ($E < C$), then the environment appears problematic and full of contradictions. Nevertheless, the individual can gradually regain its equilibrium by lowering its level of ambition or by using the solutions of other individuals. Avenarius called the process of moving towards equilibrium "de-problematisation", and this process is both individual and social. If a situation emerges in which equilibrium is obtained for all individuals, then their knowledge may be said to be unproblematic and descriptive, and the system can reproduce itself in a stationary state. According to Avenarius, the adaptive road to equilibrium contrasts with the creative behaviour that emerges because of a surplus of energy ($E > C$). In this case, which became the foundation for Nietzsche's philosophy (Höffding, 1905, 128), the individual will tend to transcend its environment by undertaking challenging enterprises or seeking danger. In this Nietzschean case, the behaviour of the individual becomes "over-energetic" or, with a shorter expression, "energetic". The individual shows a lack of balance, and the disequilibrium will continue until the energy is spent or the environment is transformed to correspond with the ambitious of the individual. Thereby, the equilibrated and unproblematic state became disturbed, and that probably is the reason why Avenarius de-emphasised this type of behaviour.

Even the reconstructed version of Avenarius and Nietzsche's description of human behaviour may appear old-fashioned and naive. Therefore, a translation into, for instance, Herbert Simon's (1982) terms of "bounded rationality" and "satisficing behaviour" would be useful if the terms are related to the "gap between an agent's competence and the difficulty of the decision problem to be solved" (Heiner, 1983, 562). However, we have already a crude but useful tool for understanding important parts of Schumpeter's world of ideas. The starting point is the challenge of operating within a stationary or slightly disturbed economic system. This challenge is limited for two reasons. First, the decisions of production and consumption largely become a matter of routines that can rely in the system of prices. Second, the relevant behaviour routines can largely be borrowed from other economic agents. Therefore, those with only a limited amount of problem-solving energy can reach an unproblematic situation in a direct or indirect way. However, what for the majority of economic agents appears an equilibrated situation is problematic for the type of agent who in the circular flow has a large residual of energy after coping with the immediate problems of his environment. This "energetic" type is the innovative entrepreneur who, for a while, considers the task of following the routines of the equilibrated economic system as unbearable idiotism. He therefore—in contrast to those tied to routine because of limited energy—starts a restless search for opportunities to spend his surplus energy. Since Schumpeter defined economic rationality in relation to the problems of the circular flow, he could not describe the entrepreneur's search for new challenges "by rationalistic routes" (*Entwicklung I*, 116). Instead, he suggested the need of including "Will to Power, joy in effort and similar things". Schumpeter later added "the will to found a private kingdom", "the will to conquer", and "the joy of creating, of getting things done, or simply of exercising one's energy and ingenuity" (*Development*, 93).

In contrast to the Schumpeterian entrepreneurs, the problem-solving energy of the "hedonic or static" individuals corresponds to the challenges posed by their attempts to function within a stationary or slightly disequilibrated economic system. It is in this context that the "behaviour of each economic subject ... can be explained by the aim to achieve the greatest possible satisfaction of needs" (*S1910d*, 411; *Entwicklung I*, 160). This maximisation, thus, takes place within a highly constrained conception of the possibilities. We may even say that the normal economic subjects have adapted their preferences to their possibilities and their possibilities to their preferences:

> "Static action, in itself, need not depend on hedonic [pleasure-seeking] or eudemonic [happiness-seeking] motives just as hedonic or eudemonic motives do not lead to static action by conceptual necessity. In reality, however, static action and hedonic motives coincide de facto. The struggle with the constraints would hardly ever be recommendable from a hedonic standpoint. Furthermore, hedonic motives normally characterise individuals who also have the kind of weakness in making decisions that implies that they stay in the old trajectories." (*Entwicklung I*, 128–9)

## 4.3 The dichotomies of Pareto and Schumpeter

Schumpeter's attempts to develop an evolutionary theory did not solely take place within the confines of even a broadly defined science of economics. On the contrary, when the newspaper of the students of Harvard University in 1944 interviewed Schumpeter (*S1944a*, 1), he declared that his research had "varied but always stayed on the same plane—that of evolving a comprehensive sociology with a single aim." By "sociology", he did not here refer to the activity of professional sociologists but to a theory applicable to social life as a whole as well as to its individual spheres (including economic life). The "single aim" was to analyse the process of social evolution (including economic evolution). Thus, he had wanted to develop a general theory of social evolution that covered economic evolution as a special case. Although the interviewer at Harvard did not raise the issue, a core question for Schumpeter's students was where to find expositions of his General Theory. The main answer would have been that it is present in his accounts for economic evolution. In *Cycles*, he thus pointed to the analogy between economic and political evolution and, in a footnote, added: "The writer believes, although he cannot stay to show, that the theory here expounded is but a special case, adapted to the economic sphere, of a much larger theory which applies to changes in all spheres of social life, science and art included" (*Cycles*, 97n).

While Schumpeter's basic approach to the process of socio-economic evolution had similarities with most available elite theories, it is for modern economists especially interesting to compare with the work of Vilfredo Pareto. When Schumpeter wrote *Wesen* and *Entwicklung I*, Pareto had stopped his teaching at the Swiss University of Lausanne, where he had previously functioned as Walras's successor. Pareto had turned to sociology after finishing economic research with his *Manual of Political Economy*, but he still was internationally renowned as a leading representative of mathematical economics. This was the situation when Schumpeter sent him a copy of *Wesen*. For the want of knowledge of German knowledge, Pareto asked his friend Maffeo Pantaleoni about his opinion. Pantaleoni answered: "The book is rather diffuse, but good. It is useful, very useful for the Germans, but does not contain anything new. He must not be discouraged, however. He does not appear to know your *Manuale* [*Manual*]. But he appears to know Italian because he cites the *Giornale degli Economisti*" (Schneider, 1975, 8). Although these formulations are slightly sceptical, they might have opened for collaboration if Schumpeter had been willing to acknowledge Pareto's leadership. However, Schumpeter insisted on his praise of Walras and on his view that Pareto's economic work is its natural extension (*History*, 860–1). Furthermore, he wanted to include part of what Pareto considered the realm of sociology into economic theory.

Although both Pareto and Schumpeter started from the Walrasian system and both moved beyond it because of their elitist views, their reaction took different forms. While Schumpeter generously acknowledged the importance of Walras's contribution throughout his life, the patrician

Pareto quickly became very hostile to the rationalist, utilitarian, equalitarian, and liberal world view of the petty-bourgeois Walras—and tended even to belittle his contributions to economics (*S1949h*, 119–20). Schumpeter considered such views a minor issue compared with the usefulness of the Walrasian toolbox; but Pareto wanted to make clear that there was a gulf between him and Walras (p. 120). Therefore, he had emphasised his anti-intellectualist, anti-utilitarian, anti-equalitarian views (p. 137) even in his *Manual of Political Economy*—but his treatment of elites is placed outside of the Walrasian core structure. Pareto invoked Darwin's theory of natural selection but he emphasised that selection is not perfect and that it may favour the coexistence of different types of men. Therefore, there is room for human heterogeneity. Although the distribution of physical, moral, and intellectual characteristics is not necessarily well depicted a statistical bell curve, the normal distribution is useful as a first approximation. These human characteristics are related to economic and social inequalities. The economic consequence of their normal distribution, however, is a skew distribution of income with the large majority concentrated near the bottom-level income. Nevertheless, Pareto emphasised that is often also possible separate the population in distinct classes:

> "Just as one distinguishes the rich and the poor in a society even though income increases gradually from the lowest to the highest, one can distinguish the elite in a society, the part which is *aristocratic*, in the etymological sense (*aristos* = best), and a common part. ... If we consider a set of qualities which favor the prosperity of and domination by one class in the society, we have what we will call simply *the elite*." (Pareto, 1972, 90–1)

According to Pareto, the "prosperity and domination" of any social elite is based on the hereditary or acquired "qualities" that are needed from it in a particular society (pp. 90–1). Therefore, "these aristocracies do not endure". Instead, we see that they "are continually being reconstituted; thus we have a phenomenon which could be called the *circulation of the elites*." This circulation cannot be avoided because *all* societies need to be dominated by a selected elite, "even when the regime is seemingly one with the most widespread democracy". The explanation of the "law" of the circulation of elites is rather complex but it is based on restrictions against social mobility into the ruling elite. These restrictions mean that the selection of the "best" does not any longer take place. Instead, the elite in power tends to degenerate with respect to the characteristics required by society. In contrast, a group of people of above-average characteristics from the lower part of the hierarchy gather strength; and ultimately this group replaces the ruling elite.

Pareto felt that is was impossible to develop his elitist vision in relation to economic theory. The problem was not that he felt any need of abandoning what we today would call the General Systems Theory, which he had derived from Walras. The problem was rather that Pareto's economics is based solely on the rational calculations of economic agents that are basically of the same type. In contrast, he conceived sociology as a totally independent complement that opens up for the inclusion of the heterogeneity of instinctive and customary behaviour. He considered this type of behaviour as determining the functioning and change of socio-economic and political institutions. In Pareto's sociological treatise—*Mind and Society*, which was published during World War I—this argument is spelled out at great length. Here we also find a few remarks on two opposing types of elite that can be called "speculators" and "rentiers" (Pareto, 1935, §§ 2233–6). These human types are clearly taken from economic life, where "speculators" (or "entrepreneurs") are oriented towards the discovery of sources of gain, while "rentiers" (or "savers") emphasise the exploitation of given resources. Their long-term coexistence is explained by fact that the one behavioural type creates the conditions of survival of the other type. If the human type of "speculators", for instance, dominates the ruling elite, then its innovativeness creates a lack of stability that is exploited by a competing elite of "rentiers" that emphasises stability. According to Pareto, this is an important part of the history of the circulation of elites that takes place both in aristocratic and in democratic societies.

There is no doubt that Pareto's mode of thinking is rather close to that of Schumpeter. Like Pareto, he emphasised the crucial role of the statistically observable distribution of human characteristics. This distribution determined the leading role of the Schumpeterian entrepreneur. While the "gradual differences of personalities" has no importance in the circular flow, it provides "essential explanatory principles" for the analysis of "the carrying out of new combinations" (*Entwicklung I*, 162n; *S1912a*, 434n). Since most people are simply not able to perform this task, they have to be led by entrepreneurs with powerful personalities. This problem defines Schumpeter's special form of the circulation of elites. The rise of the families of the Schumpeterian entrepreneurs is due to the wealth and power that emerges from successful innovation. Their fall is partly due to the fact that successful innovators and, especially, their families tend to live off past investment—as described in Thomas Mann's *Buddenbrooks: The Decline of a Family*. This behaviour, which Schumpeter later summarised by the proverbial "Three generations from shirtsleeves to shirtsleeves" (*S1927g*, 250), improves the chances for new innovators to move upwards in the social hierarchy. Schumpeter found the same type of movement in all sectors of social life and, like Pareto, tended to analyse it in terms of General Systems Theory.

The comparison with Pareto also serves to clarify other aspects of Schumpeter's work. We have seen that Pareto applied a statistical approach to the study of the characteristics of a population and the consequences of these characteristics. Schumpeter also did this at an early point of time; and the reason, probably, is that he and Pareto used many of the same sources. Modern social statistics emerged from the efforts of Adolphe Quetelet, but his emphasis on Average Man quickly led to statistical studies of the Above-Average Man. Francis Galton (1892)—Darwin's cousin—initiated these efforts through his study of *Hereditary Genius*; and Schumpeter later provocatively stated that his contribution "makes him in my humble opinion one of the three greatest sociologists, the other two being Vico and Marx" (*History*, 791). More specifically, Galton obtained this unique status because he had confronted the issues of human heterogeneity by developing some of the foundations of modern statistics. Galton also helped Karl Pearson—one of Schumpeter's teachers of statistics—to create a research programme for developing more general studies of biological evolution as well as the statistical concepts needed for these studies. Schumpeter had studied their path-breaking statistics that emphasised issues about the social elite, so he recognised the affinity between their work and that of Pareto. Schumpeter probably also recognised the difficulty in operating with a multitude of human characteristics, including the propensities of speculative and rentier-like behaviour; so he chose to concentrate on a single characterisation of human ability. This characteristic was that of leadership abilities. Here "leadership" should be interpreted in the narrow sense of being able to lead others in relation to innovation. Although he initially largely presented this characteristic as a qualitative one that split the population into leaders and non-leaders, it seems clear that it was not a mere afterthought when he turned to a continuous distribution of leadership ability in a large footnote in *Entwicklung II* and *Development*.

Most of Schumpeter's readers had interpreted his distinction between the few entrepreneurs and the many managers as one that grouped individuals into two distinct classes. Therefore, Böhm-Bawerk had criticised Schumpeter for exaggerating the importance of the entrepreneurial class. Furthermore, one of his follow-up papers (*S1914b*) was "charged with introducing an intermediate type ('half-static' businessmen)" (*Development*, 81n). Schumpeter's answer was that it had never been his intention to make a fixed dichotomy. Instead, he was thinking in terms a normal distribution of ability, which may be compared with that of singing ability—ranging from a few non-singers via the great majority of mediocre singers to the top-level Carusos. In Schumpeter's case, the ability was that of performing an innovation, which "presupposes aptitudes differing *in kind* and not only in degree from those of mere rational economic behavior" (p. 81n). With respect to these aptitudes or qualities for innovative efforts, he also could think in terms of a normal distribution of the population. He seems to have applied the custom of calling the lower quartile subnormal, the middle two quartiles normal, and the higher quartile supernormal. The subnormal part "may be so poor in those qualities ... [that] they play a wretched part in the smallest affairs of private and professional life in which this element [innovation] is called for" (p. 82n). Then Schumpeter turned to the normal half og the normal distribution. He emphasised that "[p]ractically all businessmen belong here". The reason is that they "represent a selection" that has been "individually or hereditarily tested." Finally, he came to the part of the population that is "characterised by super-normal qualities of intellect and will." However, this definition of supernormality does not reflect the fact that it is difficulty of the task at hand that determines which percentile of the population that it able to perform it. Therefore, we have to turn to the assumption that there is "a continuous variety of degrees of intensity of 'initiative'. ... Many a one can steer a safe course, where no one has yet been, others follow where first went another; still others only in the crowd, but in this among the first" (p. 82n).

Given any statistical distribution of innovative ability, it is obviously the innovative characteristics of the concrete task that determines the split between the fit and the unfit. Radical innovation is for the innovative Carusos, but they define the task of imitation, which at first has an innovative component. The next task is for the imitative crowd, and then the stage is set for normal businessmen who can function adaptively within a given framework. Finally, there are tasks for those who—like "many of the best clerks" (*Development*, 82n)—can only work within a fully fixed routine system. In his theory of crises, he needed the full distribution of innovative abilities, but to get started, Schumpeter needed to define a task that allowed him to return to the dichotomised simplicity of Pareto. Here he relied on self-sustaining systems of routines and the supposed difficulty in transforming them.

Although Schumpeter largely agreed with Pareto's vision and basic analytical tools, he was sceptical about the invocation of vague and non-observable instincts and of the Paretian analysis at the level of groups. Furthermore, he disagreed that the analysis of the consequences of "non-rational behaviour" and "differences of personalities" should be left to a separate sociological discipline. If economics upheld the assumption that "all economic subjects possess identical vision and energy" (*Entwicklung I*, 162n; *S1912a*, 434n), it would be incapable of treating evolutionary phenomena. However, he saw all the disciplines of the social sciences as dealing with essentially the same mechanisms of evolution. Thus, he had wanted to develop a general theory of social evolution that covered economic evolution as a special case. As he pointed out in 1914, even his theory of "The Wave Movement of Economic Life" should be seen as "the special economic expression of a totally general form of social evolution" (*S1914b*, 301; see also *Cycles*, 97n). This theory depicts social evolution as a sequence of equilibria that are punctuated by revolutions of the routines of social life.

## 4.4 Towards a general theory of social evolution

The main purpose of Schumpeter's criticism of the available theories of economic evolution in the last chapter of *Entwicklung I* was to set the stage for his own theory. He summarised this economic theory in terms of three "general principles" (*Entwicklung I*, pp. v, 490; cf. *S1912b*, 108), which we have already met. First, "evolution is a purely economic phenomenon". Second, "evolution is essentially a disturbance of equilibrium". Third, "[e]conomic evolution is not an organic entity that forms a whole; it rather consists of relatively separate partial evolutions that follow one upon the other." More specifically, "evolution of the economy occurs in a wavelike fashion. Each of these waves has a life of its own." These three principles served the construction of his abstract model of the essence of economic evolution.

Schumpeter's abstract model did not imply that he thought that real evolution could be reduced to purely economic phenomena. The broader issues were especially important in the long-term "up- and downward movements in the life of nations", which he later called Kondratieff waves; but Schumpeter only wanted to "describe an economic process which depends on the particulars of economic life" (*Entwicklung I*, 492). For this process he wanted to avoid the value-laden issue of "[w]hether evolution leads to social welfare or to social misery". He, nevertheless, in a purely descriptive manner emphasised that his process of evolution "necessarily leads to gains in value" but also "[a]lmost always ... leads to losses in value as well" (p. 493). The problem for the individual economic agent "those profits and losses compensate each other only in exceptional cases" (p. 499). The implication is that "a decline in status and class of wide circles goes hand in hand with the upward movement" (p. 503). This "drama" may imply that "entire layers of society lose the ground under their feet", but social degradation is not identical with long-term unemployment. On the contrary, the cyclical character of the evolutionary process means that employment opportunities will reemerge after a few years (pp. 509–10). Nevertheless, the purely economic analysis has pointed beyond economic life to the realm of socio-political conflict on economic policy.

The sociological part of the last chapter of *Entwicklung I* (525–48; cf. *S1912b*, 128–42) starts by considering the hierarchical social structure of capitalist societies. This structure is ultimately determined by the success of his innovative entrepreneur, and in this sense the "entrepreneur not only economically, but also socially has to be at the top of the social pyramid" (*Entwicklung I*, 525). His economic success calls for attention and fascination that ranges into politics and literature (p. 526), but this success also directly influences large social groups. This is obviously true for the social group that consists of the entrepreneurs and their families, but even the social position of other individuals is to a large extent determined by their direct or indirect relationships with successful entrepreneurs. This includes the capitalists that provided the entrepreneurs with their working capital, but even the income of landowners may be influenced by capitalised and distributed entrepreneurial profits. Thus we see that "more or less enduring social positions have arisen from evolution" (p. 531; cf. *S1912a*, 419). Such "positions can only be based on quasi rents, increases in returns due to counter-effects of evolution and interest from realised and invested entrepreneurial profits and savings." It is such income streams that supports a distinguishable layer within the hierarchical social system, "a genuine capitalist class", and within this class emerges "distinctly capitalist interests, life forms, and schools of ideas of the character of traditions" (*S1912a*, 419; *Entwicklung I*, 531).

However, although the capitalist economy as a whole is stable for long periods, "the individual elements of this class are not *very* enduring" (*S1912a*, 419; *Entwicklung I*, 531). The economic reason for this instability of the class positions of individuals and families is that the income derived from innovation is fragile. The next wave of evolutionary change will wipe out apparently safe positions while it at the same time creating new fortunes. This process, which Schumpeter later characterised as "creative destruction", will most directly influence the social group that consists of the entrepreneurs and their families, and the result is a rapid entry to this group as well as a somewhat slower process of return to the lower classes. Furthermore, the capitalist class is characterised by conflicts over the sharing of entrepreneurial profits. Thus, it does not constitute a united front against ordinary workers. On the contrary, the individual entrepreneur not only fights for his share of the profits that he creates. He also needs the collaboration of his workers to increase the size of these profits, and thereby these workers may obtain their share. Thus, although the popular image of the entrepreneur–worker relationship is one of conflict, there is also much room for collaboration. Actually, "the economic conflict of interests between the two parties is not very intense at all". Furthermore, the "entrepreneurs are the best customers of the workers" and a "continuous improvement of the workers' situation is emanating from them". In any case, their conflict of interest "is not larger than the conflict of interest between entrepreneur and capitalist" (*S1912a*, 419; *Entwicklung I*, 532).

The real problem for the capitalist economy is rather related to the reactions against the instability it creates for social positions. Especially those who move downwards in this structure react very strongly (*Entwicklung I*, 534–5; cf. *S1912a*, 421–2). The problem is that the "declassing process is not only an economic, but also a socio-psychic phenomenon". This means that it is not only those who are "being crushed to death, or at least injured" who "react unfavourably to the capitalist evolution [Entwicklung]". The same response is made by "everyone whose economic importance declines in relative terms" and by "everyone whose social position suffers from the appearance of new elements." They all "look critically upon the new men and their actions". These formulations demonstrate that Schumpeter considered "creative destruction" of relative social positions an important issue; and in *Capitalism* he coined this expression to emphasise the core element of his analysis of the decay of capitalist societies. The formulations also seem to conjure Machiavelli's (1998, 21) image of the opposition against the "new prince" or "new man": "there is nothing more difficult to execute, nor more dubious of success, nor more dangerous to administer than to introduce a new order of things; for he who introduces it has all those who profit from the old order as his enemies, and he has only lukewarm allies in all those who might profit from the new."

Schumpeter's gradual expansion from an abstract model of economic evolution to a "Comprehensive Model of the National Economy" created serious methodological problems. His basic problem was whether he could use the Walras-inspired analytical tools for the study of the larger socio-economic issues. This is the core topic of the last section of *Entwicklung I* called "Analogies to economic evolution in other areas of social life; the social phenomena". Here Schumpeter sketched out a general theory of the Statics and Dynamics of social systems and subsystems. Although this theory has similarities with Pareto's systems theory, it emphasised a form of evolution that was basically the same for economic and sociological studies. First, he studied the economic system under the assumption that its behaviour is codetermined by the fixed behaviour other sectors and by the overall environment. Then he studied the influence of a fixed economic system on the behaviour of other sectors. Finally, he turned to the interdependent behaviour of all the sectors of the nation. He called the totality of these sectors "*the social culture* of a people" and he considered it to be "the embodiment of *socio-cultural evolution*" (*Entwicklung I*, 545; cf. *S1912a*, 430).

To handle socio-cultural evolution and its results, he needed to complement explicitly his economic theory with theories of the behaviour of all other sectors. He developed these complementary theories by means of analogy:

> "For the processes of evolution [Entwicklung] portrayed so far, there exist, as previously emphasized in chapter two [on economic evolution in general], remarkable analogies in other sectors of social life. ... [T]hese analogies can contribute to further illuminate our understanding, and to show that existence and activity in these other sec-

tors can be grasped with a perspective parallel to ours. What are these 'other sectors'? Well, for instance, the sectors of politics, art, science, society, or moral opinions, etc." (*Entwicklung I*, 535–6; cf. *S1912a*, 422–3)

Schumpeter developed his analogies both to obtain deeper insight into his evolutionary approach and to demonstrate its general relevance for studying the sectors of social life. Let us, according to the present focus, begin with his types of individuals. His generalisation needed a renaming of these types because the terms "static-hedonic individual" and "entrepreneur" [Unternehmer] were closely related to economic theorising. Therefore, he chose more general terms: "In each sector there are statically disposed individuals and there are leaders [Führer]" (*S1912a*, 428; *Entwicklung I*, 542–3). The former are "characterized by essentially doing what they have learnt, by moving within the received boundaries and by having in a determining way their opinion, dispositions and behaviour influenced by the given data of their sector." In contrast the leaders "are characterized by perceiving new things, by changing the received boundaries to their behaviour and by changing the given data of their sector."

Schumpeter's analogy suggested for him that both possible and fruitful to study the processes of social life in each sector in terms of these generalised types of individuals and behaviour, just as he had already done it for the economic sector. However, the exchange of the term "entrepreneur" by that of "leader" serves to bring out clearly an elitist aspect of his thinking that had largely been implicit in his economic theorising: The "static masses" with relatively little mental energy are the followers of the present, or previous, leaders who have determined their behaviour. Schumpeter thought to have found this way of determining behaviour when dealing with art, science, and political life. He actually emphasised that we "[e]verywhere" find a distinction between "those spirits ... who *create* new 'lines' of art, 'schools', and parties" and "those spirits who *are created* by the 'lines' of art, 'schools', and parties" (*S1912a*, 428; *Entwicklung I*, 543). The latter type of individuals are those "adopting, accepting and adjusting to a given situation of material and ideal nature"; the former type are those who are "changing the given situation itself." Since Schumpeter had first-hand knowledge about scientific evolution, it was natural for him to take the history of science as his major example (*S1912a*, 428–9; *Entwicklung I*, 543–4). Here it is especially clear that the "bare new thought alone is not enough and is never carried out 'by itself' ". A new thought is not readily "taken into serious consideration and accepted in free decision by those involved". Instead, the "usual process is rather that the new thought is taken up by a powerful personality and is implemented by his influence". This leader is "here like everywhere else" characterised by "the energy of action and not the energy of thought". Without this energy of action the new thought would hardly be perceived as something "one must take into consideration, acknowledge, adapt to".

Schumpeter's analysis of functioning of leaders and their, direct or indirect, followers depended on definitions of a static system of behaviour in each sector of social life and of the mechanisms of change. These definitions, however, was readily at hand because his approach to the equilibration of the economic system and to its evolution was very general. The starting point was the analysis of the state of each sector as the result of the influences from its environment that were assumed to be fixed, and the next step was to introduce the general evolutionary mechanism that involved leadership. Thus he had to face "two different problems" when studying the economic sector and all other sectors of social life (*Entwicklung I*, 540; cf. *S1912a*, 426). The first problem is that of explaining a stationary state of the sector. The other problem is that of explaining the "mechanism of evolution" that changes that state. The first "problem to specify how the matters of each of the sectors shape under given circumstances, that is, in which way a certain environment necessitates a certain shape of the sector. The other problem is that of the mechanism of evolution [Entwicklung], as we can say for short." The combined solution of these two problems "completes the task of the mental reconstruction of real life."

Schumpeter's statement that his "mental reconstruction"—the problem that he had stated in the first paragraph of *Entwicklung I*—was finished with the analysis of the state and the evolution of each individual sector is case of exaggeration, however. What is missing is the study of the mutual determination of the states of the sectors and of the parallel processes of evolution within these sectors. He was fully aware of this fact. Therefore, he continued by pointing out that "the next step in providing insight is to substitute 'understanding based on causal chains' by the element of 'general interdependence' " between the sectors (*S1912a*, 427; *Entwicklung I*, 541). It was through the analysis of this interdependence that he could obtain an "understanding of the holistic state [Gesamtzustand] of social life as the result of the holistic state at the point in time preceding the one under consideration". However, this understanding would only cover how "interventions from without and changes in the natural data" causes "the social life of a people". As we already know, Schumpeter considered this exogenous determination as radically incomplete for the understanding of the evolution of social life. He, however, had only specified his additional cause of evolution within the context of individual sectors. If each of these leadership-based processes took place in complete isolation, then it would be difficult to explain "the organic unity of culture". However, the processes in different sectors are closely connected. He had already emphasised that "economic evolution [Entwicklung] also entails social changes of a non-economic type". Now he emphasised that this is "just one instance of a general phenomenon":

> "Success in any sector initially affects social organisation by elevating the position of the successful 'leader' and by affording him, to a greater or lesser extent, a social power base. Success in any sector influences the social values in general, what is drawing attention, what is being considered good or desirable. And in this way achievements in any one sector of social activity will end up having a shaping effect across all sectors of social life. ... We thereby untie things from rigid causal chains and give them back their life. And in this holistic understanding [Gesamtauffassung] of cultural evolution [Entwicklung] the economy has its particular place." (*S1912a*, 431, translation modified; *Entwicklung I*, 547).

Schumpeter's suggestion of bringing "social life" back to the analysis of social phenomena through an encompassing ('holistic') analysis of cultural evolution was not supported by any extended argument. It served instead as a programmatic statement that was placed just before the concluding paragraph of *Entwicklung I.* The placement clearly indicated a reference back to the very first sentences of Schumpeter's book. The starting point had been that while "[t]he social process is really one indivisible whole", the scientific procedure requires that the investigator "artificially extracts" facts on the analytically defined sectors of the overall process ( *Entwicklung I*, 1). Thus "mental reconstruction" of social evolution is radically imperfect. Schumpeter wanted to overcome this imperfection as far as possible by combining relatively independent studies of the major sectors of the overall social evolution. He, in Chapter 1, hinted at this ambition by pointing out that "the last step of our journey will again lead us to the proposition with which we began" (p. 1). We have above considered his sketchy treatment of the problem of the "social process" as an "indivisible whole" in Chapter 7 of *Entwicklung I.* Here Schumpeter did not solve his initial problem; but he formulated a research programme that consisted of two parts. The first part of his programme was to study the evolutionary process within the other major sectors of social life in the same way as he had studied evolution within the economic sector. Each study of an individual sector would reduce the influence of all the other sectors to the formal role of parameters. The second part of the research programme was to allow for the co-evolution of the different sectors. Presumably, he would start by studying the co-evolution between two sectors while treating the influence of other sectors parametrically. However, he ultimately wanted to treat the simultaneous co-evolution of all sectors. Actually, he in the above quotation presented his solution to that problem. Here the overall evolutionary coordination emerged from the influence of successful leadership in one sector on behaviour in other sectors.

As we shall see in Chapter 7 of the present book, Schumpeter had reached a much more sceptical view on the co-evolution between the sectors of capitalist societies when he returned to his general theory of social evolution in *Capitalism* from 1942. In the meantime, he had published several important papers that clearly related to his general theory. Thus he, during and after World War I, used his theory to confront the great questions of the war and the subsequent peace. "The Sociology of Imperialisms" (*S1919b*) denied that the nature of capitalism points to the type of imperialism that was at the root of the War. "The Crisis of the Tax State" (*S1918b*) accepted that the state was based on an exploitation of the capitalist process but denied that this fact would necessarily lead to a collapse of capitalist economic evolution. "Socialist Possibilities Today" (*S1920b*) served to give a realistic evaluation of the socialist proposals for the solution of the postwar problems. It emphasised that the whole point of "socialism" in Marx's sense is to slow down economic evolution and concentrate on other sectors of social evolution. However, postwar reconstruction suggested the use of the mechanisms of capitalist economic evolution—and that Marxian socialism was thus a matter of a relatively distant future. Finally, "Social Classes in an Ethnically Homogeneous Environment" (*S1927g*) developed a theory of the origin of classes based on two cases: the medieval German nobility and the nineteenth-century industrial bourgeoisie. He pointed out that there are significant differences between these cases: "the successful [noble] families established themselves in the positions they had temporarily acquired, as though such a situation must automatically endure ... while position of the industrialists is rapidly dissipated unless it is constantly marked by the same the kind of success that created it" (*S1927g*, 268). Nevertheless, it is obvious that social classes are conglomerates of families that fulfil socially important functions and that class membership is related to individual aptitude for these functions. Furthermore, the class as a whole rise and fall with its particular function. In this respect, the rationalisation and socialisation of the capitalist society is highly problematic for the bourgeois class. This later became a major issue in *Capitalism*, which not least builds on Schumpeter's sociological papers and a modified version of his entrepreneurial interpretation of history (see Chapter 7).

## 4.5 Conclusion

The present chapter has tried to approach the *vision* that seems to have guided Schumpeter's attempt to create a complement to the core neoclassical model. Here the term 'vision' has the meaning that Schumpeter gave it when describing the scientific life projects of other great economists. Although he did not publicly apply the term to describe his own work, its relevance for characterising his research programme is obvious. Just like in his interpretation of the life work of Marx (*History*, 388), Schumpeter obtained "an early conception of all that is fundamental in his scheme of thought". This gave him "a theoretical purpose and a plan that never varied in essentials". Therefore, the scheme of thought was developed and applied with "a large amount of consistency". His name for the economic part of this research programme shifted from "Economic Dynamics" to "the theory of economic evolution" and he hardly had a name for the extension of the programme into evolutionary economic sociology. However, he *did* stick to his "early conception" and tried to develop it "consistently".

It is not easy to describe visions. In their full-fledged form, these visions are only present in the minds of researchers. We only observe the model that is constructed to implement a particular vision, and the arguments used to justify this model may be very different from the underlying vision. However, the features of a specific vision are not beyond all conjecture. Actually, the clues for the detective of the mind are often relatively easy to find. Although the vision may emerge through a flash of insight, its scientific relevance presupposes that it is prepared by hard work on theoretical structures and problems. Furthermore, the vision is often so powerful that it, initially, is expressed in a relatively unguarded manner. Both the preparatory work and the unguarded expression of the vision are especially clear in the case of Schumpeter. The reason is that *Entwicklung I* comes close to a huge research report on his preliminary attempts to implement the Schumpeterian vision. The fact that he published such a work in progress later became so embarrassing for Schumpeter that he, in the second edition, tried to present his results in a more standard manner. Although he did not fully succeed, the core statement of his vision is found in the unedited version of the book that he published at the age of twenty-eight.

Schumpeter's early vision can be described as the entrepreneurial interpretation of history. This name is chosen to emphasise that Schumpeter wanted to match Marx's economic interpretation of history. Both interpretations point toward an analysis of the evolutionary mechanism that produces socio-economic change—and they are most easily applied to capitalism. However, Schumpeter's emphasis on innovative entrepreneurship in the evolution of any sector of social life serves to overcome the idea of automatic evolution that seems implied by Marxian historical materialism. Furthermore, the Schumpeterian emphasis on the relative independence of the evolutionary processes in the different sectors suggests that it is not necessarily the evolution of the economic sector that will ultimately determine the destiny of capitalism. While these specifications characterise Schumpeter's early vision, they do not fully express the vision underlying his later work. Although *Capitalism* continued to emphasise the non-existence of any efficient mechanism of sectoral coevolution, it has apparently come closer to Marx by focussing on the ability of existing organisations to change innovatively in their monopolistic competition. Here firms innovate to obtain extra profits and political parties innovate to gain power. The inclusion of this aspect of organisational behaviour is not really covered by the term 'the entrepreneurial interpretation of history'. Instead, Schumpeter might be seen as having moved to a more loosely defined 'evolutionary interpretation of history'.

Schumpeter probably became embarrassed by the way he had presented his results in *Entwicklung I* for at least two reasons. First, it is a scientific norm that results should be presented in the context of justification rather than in the context of discovery. Although the researcher is allowed to use any means of finding important results, the scientifically relevant task is to present them in a way that allows his colleagues to check their internal consistency and their correspondence with facts. Since the underlying vision is irrelevant for this checking, the exposition of *Entwicklung I* is problematic. Second, although Schumpeter's entrepreneurial interpretation of history was not entirely foreign for Sombart, Max Weber, and some of the neo-Marxists, it was hardly shared by any major economist. This vision interpreted the economic evolution and other types of social evolution as outcomes of the interaction between a creative elite and a routine-following mass of agents. Since it was hardly possible to convince more than a few that this was the right way of seeing economic and social change, the task was to demonstrate that solid results emerged from the Schumpeterian research programme. Other researchers could then try to derive the same results from more generally acknowledged assumptions. Schumpeter later emphasised this view of scientific "progress". Vision is simply a "pre-scientific cognitive act" that gives meaning and motivation for the creation of scientific novelty: "No new departure in any science is possible without it. Through it we acquire new material for our scientific endeavours and something to formulate, to defend, and to attack." Thereby, "the stock of facts and tools grows and rejuvenates itself" (*S1949e*, 286).

# 5

# Evolutionary Dynamics in the Capitalist Economy

The present chapter focusses on the socio-economic and scientific problems that Schumpeter wanted to confront in *Entwicklung I*. The theoretical problems can roughly be specified as concerning the economic functioning of capitalism (see Figure 5.1 on the next page). Although the specification of these theoretical problems of capitalism is crucial for the understanding of his evolutionary research programme, they have not normally been treated explicitly by economists. For instance, Schumpeter (*History*, 552n) later remarked that "while the term Capitalist gained citizenship in the economist's lingo, the term Capitalism was, throughout the nineteenth century, hardly used except by Marxists and writers directly influenced by Marxism." The developments within the economics of the twentieth century have not really changed this situation. The major exception is Schumpeter's own contribution that depicted the major characteristic of capitalism as being its promotion of economic evolution.

The missing treatment of the problems of interest to Schumpeter was remarked by William Baumol's (2002, p. x) in his book on capitalism as an "innovation machine". Baumol emphasised that, when developing the book, he was "unable to find anything that deals *directly* with my subject ... with the exception of very brief discussions by Marx, Engels, and Schumpeter". This comment demonstrates that even Baumol follows the all-dominant practice of focussing on a few of Schumpeter's statements in *Development* and *Capitalism*. In contrast, the present book tries to demonstrate that major parts of these books, and of most of his other works, are closely related to Baumol's topic. Actually, it seems clear that Schumpeter's main ambition was to depict capitalism as an engine of innovation and evolution. He approached this task from several angles, and one of them was income inequality. This angle was emphasised when he, in a letter, commented on *Entwicklung I* thirty year after its publication: "I tried to show that distribution in capitalist society is dominated by the fact that capitalism is an evolutionary process and therefore displays phenomena which we cannot hope to discover in a study of a stationary economy" (*BL*, 333). However, the explicit problems treated in *Entwicklung I* (p. 463) are "[t]he essence of the capitalist economy, the entrepreneurial profit, the interest [on capital], the crises". These are the four major problems that we shall explore in the present chapter.

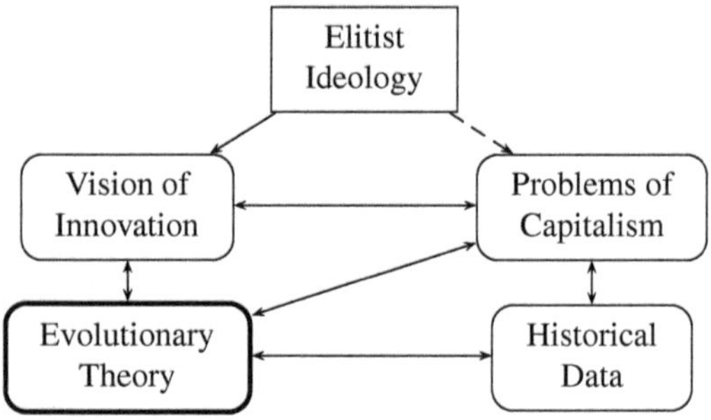


**Figure 5.1.: Schumpeter's basic analytical strategy in *Entwicklung I***

Comment: The vision of innovation as performed by entrepreneurs and enforced on routine-based economic agents has clear connections with an elitist type of "ideology", as it was later called by Schumpeter. He explicitly defined the problems of capitalism and implicitly related to historical evidence. Source: The figure applies Schumpeter's general account for the scientific process (Figure 11.2 on page 360) on his own work.

A simple explanation why Schumpeter wanted to confront the theoretical problems of capitalist economic evolution is that he inherited them from Marx. However, his specification of the problems becomes much clearer when we recognise that, in the beginning of the twentieth century, these problems were at the very centre of a controversy between the Austrian school of neoclassical economics, the German historical school, and some neo-Marxists. In this context, it seems clear that Schumpeter's emphasis on the strict limitations of the tools of neoclassical economic analysis not only reflected the difficulties in expressing his more or less elitist vision of economic evolution. He also emphasised their limitations for the treatment of the theoretical problems under discussion. Actually, he emphasised that the emergence of his evolutionary economics was the outcome of attempts to confront such problems. This background is described in the preface of *Entwicklung I* (p. vii):

> "I took my point of departure in concrete theoretical problems, first, and that was in the year 1905, in the problem of crises. Step by step, I felt driven further into an independent new treatment of still further theoretical problems until it finally became clear for me that I was always engaged in one and the same fundamental idea and that this idea relate to the whole domain of theory and allows moving the boundary stones of theoretical knowledge in the direction of the phenomenon of economic evolution."

## 5.1 Three interpretations of *Entwicklung I*

The dating of the start of Schumpeter's evolutionary research programme to 1905 is of considerable importance. It not only suggests that *Wesen* from 1908 should be considered an integrated part of that programme. The dating also hints at the discussions the Böhm-Bawerk Seminar of 1905 that seems to have served to specify the problems of capitalist evolution that were treated in *Entwicklung I*. In any case, Schumpeter summarised the great theoretical problems that wanted to solve by means of evolutionary theory in the following way: "The essence of the capitalist economy, the entrepreneurial profit, the interest [on capital], the crises; these are the most important individual problems that has concerned us and for which the presented conception was primarily developed" (*Entwicklung I*, 463; cf. *S1912b*, 93). The solution to the problem of interest on capital was first presented in *Wesen* (414–30) and then in Chapter 5 of *Entwicklung I*. Then followed the solution to the problem of crises and business cycles in the 1910 paper and in Chapter 6 of *Entwicklung I*. Finally, Chapters 3 and 4 of that book present the solutions to the problems of defining the capitalist economy and entrepreneurial profit.

To summarise and integrate these results Schumpeter developed the underlying "fundamental idea" into a modelling framework or a "schema in which the facts of evolution [Entwicklung] should be included generally" (*Entwicklung I*, 469; cf. *S1912b*, 96). This "general framework" allowed the "analysis of the purely economic capitalistic society" and it "remained substantially unaltered ever since" (*Development*, ix). However, Schumpeter later separated, for sociological treatment, "what may be called the theory of cultural evolution, which in important points presents striking analogies with the economic theory of this book" (p. xi). In the preface to the first edition of his book, Schumpeter emphasised that it did not contain the final statement of his theory of economic evolution. Instead, the book describe the "essential foundations" for the theory and postponed the task of "full theoretical structure [Lehrgebäude]" (*Entwicklung I*, vii–viii). All he hoped for was that economic theorists would not ignore the "facts and arguments" that were presented "after scrupulous work and from precise knowledge of the state of the discipline" (*Entwicklung I*, viii). The last sentence of the preface reads: "Apart from that I do not wish more than that this work as soon as possible shall be surpassed and forgotten." In other words, Schumpeter wished that research relating to the theory of economic evolution would quickly become part of ordinary scientific practise. Since hardly anyone followed this wish, he had to perform the work himself. The results partly took the form of revised versions of his book. In contrast, the 550 pages of *Entwicklung I* were never reprinted during Schumpeter's lifetime.

There are several ways of summarising the contents of *Entwicklung I* (see Table 5.1 on the facing page). The quickest way is to describe the book as if it was written in three steps. Schumpeter's *first* step was to develop Chapter 1's account for the circular flow model on the background of *Wesen*'s presentation equilibrium economics. His *second* step was, in Chapters 3–6, to use his preliminary definitions of evolutionary statics and evolutionary dynamics to treat four great problems of capitalism that have already been mentioned: (1) the problem of the definition of the capitalist economy, (2) the problem of entrepreneurial profit, (3) the problem of the interest on capital and (4) the problem of endogenous business cycles and crises in the capitalist economy. Schumpeter's *third* step was to specify the fundamental idea that finally became clear to him. This vision and theory of economic and social evolution is specified in Chapters 2 and 7. These essential chapters can be read as the conclusion rather than as the starting point of *Entwicklung I.*

This suggested stepwise reading of *Entwicklung I* is not obvious from the structure of the book. Readers are likely to get immediately attracted to the components of the third step. This strategy is problematic since the contents of Chapters 2 and 7 do not have the same degree of analytical focus as the other chapters. Instead, they are emphasising Schumpeter's vision of two types of behaviour and his evolutionary research programme that not only cover economic evolution but also the evolution of other sectors of social life. These great issues are only partially developed in the rest of the book.

The different reviewers (see Section 5.6) related to different aspects of Schumpeter's multi-purpose book. Some considered the analytical insufficiencies of the theoretical argument of *Entwicklung I,* others its apparently elitist vision, and still others the treatment of the great problems of capitalism. A closer analysis of the book confirms that these three aspects are important for understanding of this book. A further study of his later books and papers demonstrates that they are also important for understanding his work as a whole. Therefore, there is a need of studying carefully both *Entwicklung I* and the subsequent editions of this book in from three different perspectives:

1. The expression of a vision: *Entwicklung I* can be read as implementing Schumpeter's vision of economic evolution as governed by two types of behaviour. The routinised behaviour of the mass of agents secures coordination of a given economic system while the innovative behaviour of the elite promotes the change of that system. It was the extension of this vision to any sector of socio-economic life that suggested his entrepreneurial interpretation of history.

2. The sequel to *Wesen*: *Entwicklung I* can be considered as following the tradition of creating models that allow the systematic treat-

**Table 5.1.:** ***Theorie der wirtschaftlichen Entwicklung*** **(1912)**

| Translated title | Description, pages and comparisons |
|---|---|
| 1. The Circular Flow of Economic Life as Conditioned by Given Circumstances | Neoclassical "Statics" is reinterpreted as covering a routinised economic system whose evolutionary change is solely determined by exogenous factors (pp. 1–102; largely like *Development*, 3–56) |
| 2. The Fundamental Phenomenon of Economic Evolution | Pure economic evolution is described by adding innovative entrepreneurs to the model of the stationary economic system; entrepreneurs are "energetic" while the "mass" of economic agents are followers of routine (pp. 103–98; extracts in *S1912a*; completely rewritten in *Development*, 57–94) |
| 3. Credit and Capital | The "essential" function of the creation of bank credit is to provide the entrepreneur with the buying power ("capital") needed to implement his innovation in the context of the circular flow (pp. 199–277; partly like *Development*, 95–127) |
| 4. Entrepreneurial Profit | Although a skew income distribution is an essential characteristic of any kind of economic evolution, the profit mechanism of capitalism provides an effective means of transforming any given circular flow (pp. 278–323; largely like *Development*, 128–56) |
| 5. Interest on Capital | In the circular flow profit and the rate of interest on capital tend to become zero; thus the incomes of the class of rentiers depend on economic evolution (pp. 324–413; largely like *Development*, 157–211) |
| 6. The Essence of Economic Crises | The process of economic evolution necessarily takes the form of a succession of prosperities and depressions; economic crises might emerge during this process (pp. 414–62; included in *S1910d*; completely rewritten in *Development*, 212–55) |
| 7. The Comprehensive Model of the National Economy | Theories of harmonious economic growth or exogenously determined evolution are criticised; evolutionary economics gives a better analysis and suggests how to approach evolution in other social sectors (pp. 463–548; translated in *S1912b* and partly in *S1912a*; omitted from *Development*) |

Comment: Since *Entwicklung I* is only available in German, the table presents its structure and main arguments. The table also provides information on partial translations of *Entwicklung I* and on the relationship between its contents and that of *Development.* Detailed information on Chapters 2 and 7 of *Entwicklung I* is found in Table 4.1 on page 71.

ment of aspects of economic or social life. Schumpeter largely started from a reinterpreted version of the Walrasian system and then added the elements of economic evolution: the entrepreneur and his profit as well as credit, capital, and interest.

3. The response to the controversy on the great problems of capitalism: *Entwicklung I* seems to present an evolutionary solution to great debates on the definition of the capitalist economy as well as on the explanation of entrepreneurial profit, interest on capital, and economic crises and business cycles.

Since the first two interpretations have been promoted in Chapters 4 and 3, the present chapter provides an attempt to develop the third interpretation. Thus it presents *Entwicklung I* as Schumpeter's attempt to clarify and explain the great problems of capitalism.

## 5.2 Starting at the Böhm-Bawerk Seminar of 1905

Schumpeter's dating of the start of his evolutionary research programme to 1905 hints at a very concrete background for his specification of the problems of capitalist economic evolution. We know that he, in the beginning of 1906, completed his study of economics and law at the University of Vienna. Otto Bauer obtained his degree a couple of weeks before Schumpeter and Ludwig von Mises reached the same goal a week later. All three of them had not only spent the last part of their study period to prepare for the final exams. These brilliant students had also been discussing the basic problems of capitalism, mainly in the abstract terms of economic theories of value. Their discussion took place in connection with the Böhm-Bawerk Seminar in the summer term of 1905.

Böhm-Bawerk was both locally and internationally considered as *the* representative of the Austrian school—and his international fame as an economist was only surpassed by Alfred Marshall. He had obtained this status through his large book on *Capital and Interest* as well as through a very long series of controversies on the topics of this book. His book and the related controversy had special importance since theories of capital and interest were understood as reflecting the essence of capitalism. However, Böhm-Bawerk had lived off past scientific investment while he was serving as the Austrian Minister of Finance. When returning to the University of Vienna, he wanted a review of the criticism against his book and wrote a new version of it. These wishes seem to have set the agenda for his research-oriented seminar that under the general theme of value theory discussed Marx's work and his own theory. His reputation attracted very capable students from both wings of political life (Haberler, 1950, 337–8). While Schumpeter tried to transcend the controversy, the neo-Marxist wing included Otto Bauer (later the Austrian Minister of Foreign Affairs), Rudolf Hilferding (later the German Minister of Finance), and Emil Lederer (later the leading socialist academic of Germany); and the conservative-liberalist wing included Ludwig von Mises (later the leader of the Austrian school in its American exile) as well as Böhm-Bawerk himself.

Böhm-Bawerk's ambition of renewing his *Capital and Interest* as a theoretical system helped to define what Schumpeter called "that activity which will remain unforgettable to all of us—and that series of seminar discussions in the summer semesters" (*S1914c*, 150). The very first of these half-year seminars served to confront Böhm-Bawerk's two-volume book with Marx's three-volume *Capital*. When the third volume of the latter book was published more than 25 years after its first volume, Böhm-Bawerk nearly immediately answered with the booklet "Karl Marx and the Close of His System". Here, he, to the satisfaction of almost all economists and definitely to Schumpeter, demonstrated the impossibility of combining the tools of analysis found in the first and the third volume of *Capital*. The basic problem was that the analytical apparatus that Marx inherited from David Ricardo could not be transformed to the kind of analysis that Marx wanted to perform. Hilferding produced a reply called "Böhm-Bawerk's Criticism of Marx". A glance in this paper shows that even if a defence of Marxian economics had been possible at the level of the basic modelling issues that Böhm-Bawerk had treated, Hilferding was not equipped for such a defence. However, the paper also seems to demonstrate that it is possible to defend Marx's vision and the basic logic of his analyses as much more realistic and fruitful than the rather barren vision and logic found in Böhm-Bawerk's *Capital and Interest*.

The background for Hilferding's response was the neo-Marxist school that had emerged in 1904 with the first volume of *Marx-Studien*. This volume not only contained Hilferding's anticriticism of Böhm-Bawerk's criticism of Marx. It also contained a sketch of a novel scientific methodology of Marxism (by Max Adler) and a pioneering work on the sociology of law (by Karl Renner, later the Prime Minister and President of Austria). In the following year, Otto Bauer had published an account for the Marxist theory of crises. Thereby, these researchers created "the Neo-Marxist school which flourished in the two first decades of this [twentieth] century. Vienna was its center; Otto Bauer, Rudolf Hilferding, Max Adler was its leaders" (*Capitalism*, 49; see also Bottomore and Goode, 1978). On this background, Bauer and Hilferding were well prepared for the discussions with the person they considered the leading representative of neoclassical economics. According to Ludwig von Mises (1978, 24), the "discussion between Bauer and Böhm—the other participants remained in the background—occupied the whole winter [should probably be: summer] semester". In this discussion, "Bauer's brilliant intellect was luminously displayed, and he showed himself a worthy opponent of the great master whose critique had mortally wounded Marxist economics". His ambition was "from a Marxist standpoint to dismember subjective value theory."

Since Schumpeter was not a Marxist, one might have expected that he would side with Böhm-Bawerk and Mises. This was not the case, however. Haberler (1950, 338) found out that "in the heated debates ... Schumpeter attracted general attention through his cool, scientific detachment". Although Böhm-Bawerk "at once recognized his talents", his detachment as well as the "seemingly playful manner in which he took part in the discussion" were "mistaken by many for a lack of seriousness or an artificial mannerism". However, his way of contributing was rather the result of a rejection of the premises of the controversy. In another context, Schumpeter's presented his own approach to controversy:

> "Good sense lies in the saying: to understand all is to forgive all. Still more to the point: who understands all sees that there is nothing to forgive. And that certainly holds true of Science. ... Like many contemporary colleagues, I have come to the conviction that nearly every 'school' ['Richtung'] and every individual author are right in their propositions: *as* they are meant, *from the standpoint of the purposes* for which they are intended, most propositions *true* ... We want to *understand* and not to fight down; *learn*, not criticise; *analyse* and work out the correct in each proposition, not merely accept of reject." (*Wesen*, v–vi)

This approach to scientific controversies explains Schumpeter's "cool, scientific detachment" with respect to both neoclassical economics and Marxism. He had no ambition of simply defending neoclassical economics in the form that it had taken in the hands of the Austrian school. Instead, his task was to work out the "correct" kernel of the theoretical structures of economics. He presented the result of this study in *Wesen*. Here he had cut the Gordic knot of the wide-ranging syntheses presented in works like Böhm-Bawerk's *Capital and Interest* and Marshall's *Principles of Economics*. He had also suggested that neoclassical economics should concentrate on Economic Statics in its Walrasian form. Similarly, he did not want "merely accept of reject" Marxism but to understand and learn from this stream of thought. For this purpose, Böhm-Bawerk's criticism of Marx's analytical difficulties and errors was of little help. On the contrary, the task was to remove the formally erroneous part of Marxian analysis and to focus on the scientifically important contribution of *Capital*: the unique treatment of the problems of capitalist evolution. As demonstrated in *Wesen*, it was exactly these problems that Schumpeter wanted to remove from neoclassical economics. Here, concepts like profit, capital and interest had suffered from becoming part of Economic Statics—and its extension into a non-evolutionary theory of economic growth. Furthermore, the concept of endogenous crises could not be included in this type of thinking. The consequence of the inherent limitations of neoclassical economics was that the study of capitalism as an evolving system driven by the prospects of profit and interest on capital seemed to be left over to the younger members of the historical school and to the neo-Marxist school. The challenge from the combined efforts of these schools was especially clear in the German-speaking countries in the beginning of the twentieth century. At that time, new life was brought to the historical school by the Marx-inspired work of Sombart, Spiethoff, and Max Weber. These important researchers brought new life to Schmoller's historical school by confronting great problems of capitalism—like the essence of its historical emergence, the long trends of the capitalist economy, the occurrence of crises, and the overall structure of capitalist society. However, Schumpeter thought that their contributions were seriously constrained by unwillingness to produce the analytical tools for handling these extremely complex problems. This unwillingness was not present among Marxist theorists. They wanted to move from Marx's *Capital* to the theoretical problems raised by modern capitalism in general and the Austrian-Hungarian Empire in particular. They analysed the emergence of large-scale enterprises, relations between industrial and financial capital, business cycles, imperialist tendencies, nations and multi-culturalism, and methodological issues. Schumpeter's complex relationship to this type of work has been emphasised by Tom Bottomore (1992). In should, however, be noted that Schumpeter thought that the burst neo-Marxist creativity was seriously flawed by adherence to the Marxian type of grand synthesis between economics, sociology, history, and policy analysis. The alternative, which was chosen explicitly by Schumpeter, was to cut the Gordic knot of synthesis and try to establish more limited and sober forms of coordination of knowledge (*Capitalism*, Ch. 4).

Although Schumpeter wanted "to *understand* and not to fight down" and to "*learn*, not criticise" the works of the younger members of the historical school and of neo-Marxist scholars, they did not teach him the conclusions. Instead, they provided scientific problems for his train of thought. This train of thought led far beyond its starting points, because it was supported by his modified tools of neoclassical analysis and by his selective use of the data provided by the historical school.

## 5.3 Theories of interest and of capitalism

The most systematic and harshest evaluation of *Entwicklung I* was produced by Böhm-Bawerk (1913). Actually, this criticism was put forward in no less than 60 pages of the scientific journal of the Austrian economists. The book was criticised partly for serious logical errors, partly for not reflecting reality. Schumpeter's 40-page answer (*S1913a*) was not able to satisfy Böhm-Bawerk, who in his "Concluding Remarks" pointed out that it does not help that a wrong theory has an outstanding advocate. He simply could not accept the idea of the innovative entrepreneur as the source of profit and interest. However, it is through Schumpeter's conception of entrepreneurial profit that his otherwise mysterious theory treatment of the problem of interest on capital becomes clear (see Chapters 3 and 5 of *Entwicklung I*). His model simply assumes that there is no credit for consumptive purposes and no credit for the maintenance of established firms. Instead, the innovative entrepreneurs provide the sole demand for credit. This credit is used for establishing their innovation-based firms and it is gradually paid back when these firms begin to function. The supply of money for these innovative projects is provided by the creation of credit by capitalist bankers. Thereby, they obtain a positive rate of interest. In contrast, the rate of interest is zero in the circular flow, that is, the version of the Schumpeterian model that does not include innovative entrepreneurs and thus also excludes the services of capitalist bankers.

It was especially this analysis of the rate of interest that motivated Böhm-Bawerk to launch his attack on *Entwicklung I*. The reason was probably that he considered a positive rate of interest as the pillar of a sound national economy. Nevertheless, as Shabbir Khan (1957, Ch. 4) has emphasised, it is hardly possible to understand major parts of Schumpeter's account for his theory without taking into account his critical interpretation of his great teacher. In the obituary paper from 1914, Schumpeter observed that although Böhm-Bawerk had been a student of the leaders of the older historical school in Germany, he quickly grasped the central message of Carl Menger: the theoretical economics of consumption and production is ultimately based on decisions made in terms of marginal utility. He also, according to Schumpeter, decided to develop "an all-embracing theory of the economic process" that was based on "the one problem whose solution seemed to him to be still missing. This was the problem of interest, of the net return to capital, the most difficult and the most important in economics" (*S1914c*, 147). The importance of this problem is that our understanding of and attitude to "the nature and meaning of capitalism depend on our view of the meaning and function of profit and interest". Just as in the case of Marx, the theory of interest and profit is scientific core from which "everything else follows more of less conclusively" (p. 147).

The scientific core of Böhm-Bawerk's *Capital and Interest* is based on the fact that production takes time. Hence the output that is presently consumed by workers and landlords is determined by past decisions of capitalists. Such a capitalist is most easily understood as "an industrialist working with his own capital" (*S1914c*, 164). In sharp contrast to Menger, he states that "capital is 'nothing but the total of intermediate products which are generated in the various stages of roundabout production' ". Consequently, "capitalism" is simply defined as an economic system of "roundabout production", that is, production that takes time. For instance, a socialist economy would, by Böhm-Bawerk, be classified as a "capitalist" economy with essentially the same problems as market economies (pp. 165–6). Even if we stick only to this definition, it is clear that Schumpeter agreed with those neo-Marxists and those members of the historical school who judged *Capital and Interest* like they would have judged a rewrite of Shakespeare's *Hamlet* in which the Danish Prince played hardly any role. Such a rewrite would have no fascinating driving force and no dramatic conclusion. Similarly, Böhm-Bawerk could tell no other story than that of the gradual lengthening of the multi-stage transformation of raw materials and labour into consumption goods. He even excluded the entrepreneurial efforts needed for the introduction of his roundabout methods, because he largely considered them automatic. Furthermore, no economic crises had to be included into the story, because he considered them purely accidental phenomena. It was simply the more and more prudent saving behaviour of the capitalists that was the force behind the ever-increasing capital–labour ratio and the ever-increasing consumption per capita in the Böhm-Bawerkian system. Prudent saving was considered the undramatic essence, or "spirit", of capitalism.

While Schumpeter in *Entwicklung I* totally accepted the pivotal role of the theory of profit and interest and made his own attempts to determine the common problems of all economic system, he rejected Böhm-Bawerk's reduction of capital and capitalism to a technical issue that largely concerned "roundabout production". However, this theory can partly be expressed in terms of the time-consuming innovations that are needed for the change of the degree of "roundaboutness" in a capitalist economy in the sense of Marx. This influenced Schumpeter's reading of Böhm-Bawerkian specification of his explanation of the observed positive rate of interest on capital by means of the phenomenon of "roundabout" production in market-based capitalism. According to *Capital and Interest*, the positive rate of interest is the logical result of "three reasons" that, according to him, have to be generally accepted. The first two reasons, which Schumpeter considered inessential for his own purposes, concern the supply of savings that is made because of a premium (or agio) for postponed consumption. The famous "third reason" concerns the demand for working capital. The size of this demand is determined by the method of production measured by the time needed for production.

Since more time-consuming methods of production have a higher productivity, those capitalists willing to wait longer are able to pay a higher premium for loans (*S1914c,* 177). If the rate of interest approaches zero, then an unlimited expansion of working capital becomes profitable.

However, this expansion implies a scarcity of goods available for present consumption and this scarcity leads to an increased rate of interest. Hence this interest is "the brake, or governor, which prevents individuals from exceeding the economically admissible lengthening of the period of production, and enforces provision of present wants" (p. 182). The rate of interest can also be used as an indicator of "the cultural level of a nation": a higher cultural level means a higher willingness to wait and a lower demanded rate of interest. On the supply side, the low interest both means long periods of production and low yields from further lengthening, because of decreasing returns to production time (p. 182). Thus, the willingness to supply working capital at relatively low rates of interest is the precondition for economic progress; and differences with respect to the subjective comparison of present and future consumption determine the fate of both social classes and whole countries.

Although Schumpeter in *Entwicklung I* essentially accepted Böhm-Bawerk's premium theory of interest, he did not accept sole focus on "roundaboutness" and the related idea of decreasing returns to innovative activity in a mature economy. Furthermore, he inverted the direction of causality: it is not saving that promotes innovation, but innovation that creates the phenomenon that is normally called saving. Finally, Schumpeter saw the essential phenomenon of interest as related to the finance of innovation and with no relation to the ordinary demand for working capital. However, even in this interpretation, the rate of interest paid by entrepreneurial project is crucial. He later remarked the similarity between Böhm-Bawerk's account for interest and that of Marx: "Being a general time discount that applies to the returns from productive services of all kinds, interest as it were preys upon them all". This gives the theory an "affinity ... with the exploitation theory of Marx" and it might "have been presented as a novel exploitation theory" (*History*, 931–2).

Schumpeter even thought that Böhm-Bawerk's concrete model can most easily be expressed in Marxian terms. Marx decomposed the capital needed for production into "constant capital" (the value of physical capital and intermediate goods) and "variable capital" (the value needed for hiring labour); and he considered the ratio between constant capital and total capital, the "organic composition of capital", as a crucial variable for his analyses in *Capital*. Similarly, Böhm-Bawerk's "performance centered in a highly simplified picture of the manner in which, given a certain supply of labor and of subsistence, interest and wage rates are simultaneously determined and in turn determine the organic composition of capital" (*History*, 927). Just as in the case of Marx, Böhm-Bawerk's formal model did not fully cover the underlying vision because it only solved a problem that concerned a stationary economy. Although Böhm-Bawerk in this respect "surpassed the Marxist schema by treating a problem not explicitly treated by Marx" (p. 928n), his performance was not without flaws. According to Schumpeter (p. 847n), even Carl Menger "severely condemned" Böhm-Bawerk's theory of capital and interest (probably especially his productivity theory): "In his somewhat grandiloquent style he told me once: 'The time will come when people will realize that Böhm-Bawerk's theory is one of the greatest errors ever committed.' " However, Schumpeter did not want to condemn but to rescue important insights by emphasising that the Böhm-Bawerkian theory was a contribution to "dynamics" rather than to "statics" (*Wesen*, 408–13, 428).

Schumpeter's presentation of his emerging alternative theory of interest should logically have been postponed to Chapter 5 of *Entwicklung I*. However, he considered it so important that he included the "Prolegomena to a dynamic theory of interest" in *Wesen* (414–30). Here he suggested that his theory might partly be expressed in terms of Böhm-Bawerk's concept of the productivity of roundabout methods of production. The main point is that the use of roundabout methods does not generally and permanently create an ability to pay a premium to the capitalist savers. Instead, it is the investment of entrepreneurs in innovations with longer production periods that provides the profit that can be used to pay interest. They do so by solving an inter-temporal problem. The presently available "world of goods" is governed by a system of prices that has no room for the innovative project. However, if some of these goods are brought into a "new combination", then they can be used to produce a future output that can pay a premium when measured in terms of the present price system. The condition is that—although the present has no room for it—credit money is created for the innovative project. With this money the entrepreneur detracts some of the present goods for his innovative purpose. Thus, money is a real force in the economic system and not just a "veil" as thought by Böhm-Bawerk and many other contemporary economists. Schumpeter abandoned the Böhm-Bawerkian theory and sided with Menger's monetary theory of capital that better reflected the real power of money to promote economic evolution (*Wesen*, 427n; *Entwicklung I*, 251–4).

This account for Schumpeter's evolutionary theory of interest was expanded in Chapters 3 and 5 of *Entwicklung I*. The essence of capitalism is *defined* as the creation of credit for innovative projects. The starting point is that there is no free money in the circular flow. However, additional money can, to some extent, be created by standard banking procedures. This credit expansion is built on the expectation that the supported projects in due time are able to pay a monetary surplus from which interest can be extracted. In this perspective, "capital" is the purchasing power supplied to the entrepreneurs, "a process, a method, to carry out new combinations" (*Entwicklung I*, 271). It is a source of income whose value is determined in the Schumpeterian market for money (or "capital"). Here we "see how the credit need of industry expresses itself and how the banking world shortly supports and encourages it, soon tries to brake, and soon after denies it any further satisfaction" (p. 275). Personal influence supplements the multiple rates of interest as regulators of this market that ranges from short-term credits to long-term loans and the buying of shares. Such is "the headquarters of the capitalist system" with its "trading in credit" for the purpose of "financing evolution" (*Entwicklung I*, 276–7; cf. *Development*, 126–7). The conspicuous "fortunes" made in this market are derived from activity of the Schumpeterian entrepreneurs, and these fortunes tend to disappear.

It was on this background that Schumpeter developed his theory of interest. The first part of the argument is that this rate of interest is zero in the circular flow. This process is characterised by the routinised reproduction of existing goods and the corresponding circulation of money. Since accidents and personal life cycles are excluded as inessential, there is neither demand for nor supply of loans. Since there is no market for the *ownership* of land, the fact that a zero rate of interest would increase its price to infinity does not matter. However, the need for a money market emerges as soon as we introduce innovative entrepreneurs into the model. If they can obtain a surplus from commanding purchasing power for implementing their projects, they are willing to pay a positive rate of interest. The actual rate of interest is determined by the supply to the money market. The higher the interest, the more difficult is entrepreneurial task. Therefore, interest is "a brake … on evolution, a kind of 'tax on entrepreneurial profit' " (*Entwicklung I*, 413; cf. *Development*, 210).

Although Schumpeter probably took Böhm-Bawerk's theory of interest as the starting point and tried to provide a novel interpretation that avoided the logical errors of the Böhm-Bawerkian system, their conclusions were radically different. Schumpeter based his argument on heretic theories of the role of money and of the transformational nature of economic evolution. Therefore, he emphasised that it is not the savers who create investment, but innovative investment that supplies the possibility of creating fortunes. Furthermore, the fortunes that provide the upper strata of society with their core incomes are conspicuously fragile since they vanish as soon as the underlying innovations have been brought into common practice. Finally, the basic conflicts in capitalist economies are not just between workers, landlords, and capitalists. There are also basic conflicts over the rate of interest between credit-providing capitalists and innovative entrepreneurs as well as a struggle for economic life of the presently innovative entrepreneurs against the managers and former entrepreneurs who base their activities on the innovations of the past. These conclusions were part of a larger endeavour to turn the Böhm-Bawerkian system upside down by reversing the status of consumption and production as well as of the cautious saver and the creative entrepreneur. This inversion provided a fresh look on the problems of real capitalism, including the conflicts over distribution and the phenomenon of economic crises. More specifically, while Böhm-Bawerk, in a non-historical manner, studied the issues of present and future consumption, Schumpeter chose the viewpoint of production or, rather, the viewpoint of the entrepreneur who created radically new lines of production in a historically given economic system. Thereby, he was approaching core problems of Marxian analysis.

## 5.4 The evolutionary function of business cycles

Since Schumpeter stated that his evolutionary research started in 1905 with "the problem of crises", it might appear paradoxical that *Entwicklung I* begins with an extensive account for his famous model of "the circular flow of economic life". The name of the concept of the circular flow is apparently borrowed from his teacher Eugen von Philippovich (*Entwicklung I*, 1n; *Development*, 3n). The model is about a state of the economic system in which the means of production have "static uses" that only change due to "circumstances" external to this system. The response to exogenous change is a *tâtonnement* process of convergence to a new stationary state; and similar processes emerge as responses to endogenous change. Although Schumpeter did not change his model when revising his book, he did change his argument about its place within the structure of the evolutionary theory. In *Entwicklung I* (viii), Schumpeter states that this chapter "introduces the reader to those theoretical conceptions with which all the rest works". In *Entwicklung II* (xii), he instead suggests that the chapter is of no interest for those trained in economic theory, except for a few sentences needed for the theory of interest. This suggestion implies that Schumpeter's theory of economic evolution is a simple extension of the concerns of economic theorists—and we already know that this is not true. Therefore, it is wise to stick to his original emphasis on the crucial role of the first chapter. Especially, it should be noted that Schumpeter excludes his innovative entrepreneur from Chapter 1. Even the Walrasian entrepreneur is only included in the form of what he later called a "mere manager" (*Development*, 83; cf. *Entwicklung I*, 69). Together with the rest of the economic agents, this manager is engaged in an economic process that is groping for a state in which their routine-based behaviours are mutually consistent (pp. 11–14, 38–40). This state, which might today be called a complex Nash Equilibrium of routine-based behaviour, is "the circular flow of economic life".

However, the addition of the innovative entrepreneur and the capitalist banker to the circular flow creates a model that serves the development of a macroscopic account for the evolutionary process in a capitalist system. This account is presented in Chapter 6 as the solution to the problem of economic crises. Schumpeter considers these crises as reflecting an evolutionary process that continuously moves from one equilibrium state to the next. The driver of this process is groups of entrepreneurial projects that provides "waves of evolution [Wellen der Entwicklung]" (*Entwicklung I*, 435; cf. *S1910d*, 31). The reason for these groups of projects is double. On the one hand, the success of the first innovative project in the money market paves the way for the next, and the successful completion of a project in an even stronger signal to the imitators. On the other hand, the macroeconomic effects of a large group of projects will at some point become detrimental for further innovative activity. Furthermore, when the impulse of the wave starts to vanish, an enormous pressure of adaptation is put on the part of the economy characterised by old routines; and this pressure lead to panics and crises .

Through these analytical moves, Schumpeter clearly departed from the analysis of Walras. Assuming that he had the problem of crises in mind from 1905 and onward, he must have been provoked by reading Walras's *Elements of Pure Economics*—especially by its Part VII. At that point of his book, Walras has already moved through a series of approximations to economic reality. Thereby, "the system of the economic universe reveals itself, at last, in all its grandeur and complexity: a complexity at once vast and simple, which, for sheer beauty, resembles the astronomic universe" (Walras, 1954, 374). However, he continues by sketching a couple of further approximations to reality under the heading "Conditions and Consequences of Economic Progress". He first deals with the reproduction of equilibrium as a stationary process—like the closed input-output system that reproduces year after year in Quesnay's *Tableau Économique*. This process is described in the following way (p. 378):

> "We shall suppose the basic data [parameters] of the economic problem ... to remain fixed, so as to give us something in economics analogous to what is called a *stable system* in mechanics. Moreover, we shall assume not only that the preliminary phase of groping has been completed with equilibrium established *in principle*, but also that the phase of static equilibrium has actually commenced, so that equilibrium is established *in fact*."

Here Walras has moved from the abstract study of equilibrium to a stationary equilibrium. This is a routine-based equilibrium into which the Schumpeterian entrepreneur can be introduced. However, Walras does not consider this possibility. Instead, he simply assumes the exogenous change of the "basic data", i.e., the parameters, of his model of the economic system. Hence, when he turns to the step of passing "from the static to the dynamic state", he supposes "the annual production and consumption, which we had hitherto represented as a constant magnitude, change from instant to instant along with the basic data of the problem" (Walras, 1954, 380). As examples of these data he mentions "the initial quantities possessed, the utilities of goods and services, the technical coefficients, the excess of income over consumption, the working capital requirements, etc." The change of these data means that the economic system is "perpetually tending towards equilibrium without ever actually attaining it, because the market has no other way of approaching equilibrium except by groping, and, before the goal is reached, it has to renew its efforts and start over again". Therefore, "the market is like a lake agitated by the wind, where the water is incessantly seeking its level without ever reaching it" (p. 380).

Schumpeter obviously did not subscribe to this analysis of the emergence of disequilibrium due to exogenous parameter change and the re-establishment of equilibrium by endogenous forces. He hardly found better the improved Walrasian image of an agitated lake that strived towards a gradually increasing level because of exogenous scientific progress and endogenous savings. While this image is implicitly suggested by the extended analysis of the last edition of *Elements*, we shall stick to the earlier version of the analogy of the disequilibrated lake. This analogy was developed when Walras (1954, 381) turned from "wind" to "storm". He did so in relation to the problem of crises: "just as a lake is, at times, stirred to its very depths by a storm, so also the market is sometimes thrown into violent confusion by *crises*, which are sudden and general disturbances of equilibrium." Here it is the "storm" that, as an exogenous force, creates major oscillations around the equilibrium level of the "lake". Walras specified the economic waves in terms of the movement of a single price. The upswing of the wave implies that "a selling price will remain for long periods of time above cost of production and continue to rise in spite of increases in output". The crisis implies that "a fall in price, following upon this rise, will suddenly bring the selling price below cost of production and force entrepreneurs to reverse their production policies."

Schumpeter was challenged by the Walrasian analogy and wanted to replace it by "the perennial gale of creative destruction" (*Capitalism*, 84). While he largely accepted Walras's story of the movement from disequilibrium back to equilibrium, he rejected the story of the creation of disequilibrium because of the influence of exogenous forces. This rejection did not imply that he excluded the creation of waves of economic activity by exogenous factors. It instead implied that he found exogenous factors radically insufficient for his explanation of long-term evolution. Instead, Schumpeter's probably used his emerging theory to reinterpret Walras's statements as reflecting the more or less endogenous activity of his innovative entrepreneurs. If the reactions to entrepreneurial projects are slow, then their selling prices remain above costs for a relatively long period; and the average prices can increase because of the additional demand for resources created by a group of such projects. However, when this type of demand disappears, the average prices will fall below the costs of at least those following old routines, and their economic responses are either adaptation or bankruptcy. Schumpeter used the analogy of a lake, or an ocean, to describe his problem. For instance, he stated that "the waves of the economy do not always return to the same level as the waves of the ocean do, they indeed oscillate around a level but not always around the same one". Therefore, we need an answer to the second great problem of economics: the movement from one level to another one. This is "the essence of economic evolution" (*Entwicklung I*, 465–6).

Through the introduction of the innovative entrepreneur as a disturber of equilibrium, Schumpeter clearly departed from neoclassical economics in general and particularly from Böhm-Bawerk. Considered as a response to Marx, a major limitation of Böhm-Bawerk's theory of capitalism is that it totally ignores the possibility of endogenous economic crises. Otto Bauer cannot have failed to point out that Böhm-Bawerk thereby revealed that this macroeconomic phenomenon, which to some extent was covered by Marx, seemed to be beyond the reach of the Austrian school. However, in the beginning of the twentieth century, it became more and more difficult to explain away crises as merely accidental. Actually, Clément Juglar had already, in 1860, used time-series data to demonstrate that crises represent a cyclical phenomenon that occurred roughly every ten years. This defined an agenda of detailed research for Arthur Spiethoff, the most thorough member of the emerging new leadership of the historical school. The results of this partly Marx-inspired research meant that, "with the possible exception of Marx, Spiethoff was the first to recognize explicitly that cycles are not merely a non-essential concomitant of capitalist evolution but the essential form of capitalist life" (*History*, 1127). Schumpeter sharpened his provocative conclusion that cycles are the necessary form of capitalist *progress*; and he wanted to formulate this insight theoretically.

Although Spiethoff's results were very slowly maturing and published, he allowed himself to publish "Preliminary Remarks to a Theory of Overproduction" in 1902. Here he divided cycles into to phases: a "boom" and a "depression". Spiethoff's (2002, 53–60) boom develops in four stages. First, the "boom starts in particularly promising branches of production" from which capital expects high profitability; "and from there a general upswing develops". Second, the creation of "new production facilities" leads to "a real shortage of goods". Third, the "new businesses" appear as suppliers of goods; and at this stage the "high prices are already at risk". Fourth, we see the "feverishly increased production flinging its products onto the market in the absence of corresponding equivalent consumption." Prices have not been constant during this process but "the mania for keeping them rigid ... usually aggravates overproduction." Spiethoff's (pp. 60–8) depression is caused by psychological and economic reactions to the hopeless state of overproduction. The psychological reaction "is expressed in the fact that the exuberant spirit of enterprise is followed by complete depression." This means that a "widespread distrust in new enterprises develops, and owners of capital frequently prefer to let their resources lie idle". The economic reactions "to production facilities far exceeding long-term requirements" include hoarding of money, bankruptcies, cutbacks, "weeding out defective technological facilities", labour-saving initiatives and harsh times for labour. The reduction of prices and the increase of consumption is much slower but in the end we might see "a certain state of equilibrium". After analysing the interconnected periods of boom and depression, Spiethoff turned to the question of whether this phenomenon will disappear. The "initial problems of the capitalist mode of production" can be seen in the early "orgies of new company formation" like "the South Sea Bubble and the enterprises launched by [John] Law". However, although these days were long over, "a new age of revolutionary inventions providing a constant stream of economico-technological and psychological impetuses" gave him "no way of predicting when crises will come to an end" (p. 75).

Spiethoff's "Preliminary Remarks" were part of an emerging literature that defined the context of Schumpeter's already quoted statement that he took his "point of departure in ... the problem of crises." The fact that he studied the literature on crises and business cycles is revealed by his early reviews of German (*S1906c*) and American (*S1909c*) books. Nevertheless, as he later pointed out, he especially felt in congruence with the work of Spiethoff (*Development*, 214–15). The only difference that Schumpeter emphasised is that Spiethoff failed to acknowledge that "the new does not grow out of the old but appears alongside of it and eliminates it competitively" (p. 216). Schumpeter first presented his vision and analysis of crises as part of economic evolution in his large, and recently translated, paper "On the Nature of Economic Crises". This paper has largely been ignored since most of it was included verbatim in Chapter 6 of *Entwicklung I*. However, this chapter includes a never-repeated and condensed account for the major elements of his general evolutionary analysis (*S1910d*, 5–16, 50) and it, furthermore, was extensively rewritten for *Development*. In any case, we recognise Schumpeter, like Marx and in contrast to most other economists, chose "to look to business cycles for material with which to build a fundamental theory of capitalist reality" (*History*, 1135).

What Schumpeter's paper added to the literature is largely the demonstration that an evolutionary theory of crises follows from his general model of economic evolution. To demonstrate this, the paper had to announce this model in staccato before turning to the problem of crises. This sequence also seems to reflect the underlying research work. Although Schumpeter had started his research work by wondering about the crises phenomenon, it was the problem of interest and profit that led him to his general theory. Since he connected this theory with "the modern theoretical edifice of economics" through a well-defined "chain of considerations", it is important that his specific "crisis theory" follows "as a simple consequence" from the general theory. Furthermore, it was for him noteworthy that this general theory "was not developed with this purpose in mind" (*S1910d*, 5n). However, this statement can also be interpreted as the specification of a problem: Schumpeter's early and quick move into the issues of crises and business cycles continued to trouble him for the rest of his life! More specifically, the paper concluded by proposing that "the fundamental ideas of our argument can be summarised in the following theses". Then he listed nine theses (*S1910d*, 50; see also Andersen, 1994, 40–4). Theses 6–9 concern the application of Schumpeter's basic evolutionary scheme for the study of economic crises while Theses 1–5 summarise his basic evolutionary model in a highly abstract way (see Table 5.2 on the facing page).

The interpretation of Schumpeter's theses is not easy. Thesis 1 seems to provide a summary of *Wesen* by proposing that is possible to use the Statics–Dynamics dichotomy to decompose economic processes into two classes. However, we should note that both classes are defined for the purposes of evolutionary economics. Thus, the movement toward a stationary state is understood in terms of Schumpeter's modification of Walras's equilibrium-seeking dynamics. Thesis 2 emphasises that it is not the exogenous change of the economic structure but only what Schumpeter considered endogenous mutation that is included in the core version of his evolutionary economics. Thesis 3 adds that the phase of evolution basically has to be considered as the disturbance of a general "static" equilibrium in which evolution has come to a halt. This mutative phase was in 1910 called "Entwicklung". As we have seen, the disturbance may be specified as a challenge to the apparently given parameters of the economic system. Thesis 4 tells that the innovative disturbance ("Entwicklung") provokes the response of "the static masses" in the direction of a new equilibrium. In the original formulation, Schumpeter used the German word "Statisierung" to denote his selection-oriented version of the *tâtonnement* process. Thesis 5 describes the equilibrating process as the adaptation to the new system of marginal value and price. It emphasises that the adaptive response by the "static masses" is not an easy one. Instead, the Schumpeterian *tâtonnement* implies that at least some of the old economic positions lead to bankruptcy rather than becoming adapted to the new parameters of behaviour. Elsewhere in the paper, Schumpeter stated that it is not only the evolution itself, but also the "spasms of the collapse" after a crises that create "untenable situations" that can "be transferred only step by step—*par tâtonnement*—into an equilibrium state" (*Entwicklung I*, 454).

As we have already seen in Section 3.5, the five theses were not de-

**Table 5.2.: Schumpeter's early theses on economic evolution and crises**

| Label | Thesis |
|---|---|
| 1. Process types | "The economic processes fall into two separate and, in practice, clearly distinguishable categories: static and dynamic." |
| 2. Evolution | "The dynamic category constitutes the pure economic development [Entwicklung as the emergence of novelty], that is, those changes in the appearance of the economy that develop out of the economy itself." |
| 3. Disturbance | "Economic development [Entwicklung as the emergence of novelty] is essentially a disturbance of the static equilibrium of the economy." |
| 4. Equilibration | "This disturbance causes a reaction in the static masses of the economy; namely, a movement towards a new equilibrium state." |
| 5. Reorganisation | "The process of convergence to the static state [Statisierung] necessarily creates an end to each specific phase of development and causes a reorganisation of the value and price system of the economy and a general 'liquidation'." |
| 6. Cycles | "These statements [Theses 1–5] explain the phenomenon, which is popularly characterised as the change between prosperity and depression." |
| 7. Crises | "During the process of convergence to a static state and, especially, during the time of its inception, collapses can easily develop, which we term economic crises κατ' ἐξοχήν [*par excellence*] and which render the process 'abnormal'." |
| 8. Coincidences | "The economy—and, indeed, this includes the static economy—is also exposed to coincidental disturbances, which, if they are sufficiently significant, can cause such crises." |
| 9. Unimportance | "But these [exogenously determined] crises present no problem, they can indeed be effortlessly understood." |

Comment: Schumpeter theses are only found in his paper on economic evolution and crises (*S1910d*). Theses 1–5 present Schumpeter's general scheme for analysing economic evolution in terms of the routines of the circular flow and their change. Here the term 'Entwicklung' could have been translated by 'mutation' since it only covers the emergence of novelty; the full evolutionary process also includes the process of 'Statisierung' (which includes selection). Theses 6–9 propose that the general scheme can be used to explain why endogenous economic evolution takes a cyclical form. Source: *S1910d* (50).

signed for the study of economic crises. The rest of the theses serve to claim a place for Schumpeter's general theory of economic evolution in the literature on the problem of crises. Let us start with Theses 8–9. They point out that economic crises may have exogenous causes (like bad harvests or the ending of war activities). According to Schumpeter, the economic consequences of such phenomena can easily be handled by means of Economic Statics. However, the basic issue is whether or not the economic system has an intrinsic tendency to produce recurrent crises. With respect to the capitalist economy, the answer is yes; but the argument is not easy to find in Theses 6–7. Instead, these theses simply postulate that Schumpeter's model can be used for the explanation of business cycles and crises to the extent they are generated from within the economic system. The immediate reason for the difficulty of understanding these theses is that the whole paper had dealt with the question, so Schumpeter could summarise his results briefly. Elsewhere, he stated that the first entrepreneur paves the way for the next, and so on. Thereby, their innovative activity may obtain macroeconomic importance. Furthermore, the step in the overall process of economic evolution is necessarily brought to an end by the forces that promote equilibrium:

> "The counter-movements do not merely obstruct evolution [Entwicklung], they put an end to *this* evolution. A great many values are annihilated; the fundamental conditions and presuppositions of the plans of the leading men in the economy are changed. The economic system needs rallying before it can go forward again; its value system needs reorganising. And the evolution that starts again is a new one, not simply the continuation of the old." (*Entwicklung I*, 415; cf. *S1910d*, 17)

Thus, Schumpeter emphasised that the complicated process of returning to equilibrium from a highly disequilibrated state puts a brake on further innovation. As a result, the innovative demand that had created the prosperity disappears; and, at that point, the economic system can easily move into a full-fledged depression. However, the economic system will ultimately reach a new equilibrium from which a new round of innovation and adaptation can begin.

It is clear from even a very quick inspection the paper (and Chapter 6 of *Entwicklung I*) that these claims on cycles and crises are *only* based on theoretical deductions from the general evolutionary theory. However, the literature in which this theory wanted a place was heavily empirically oriented. Therefore, Theses 6–7 cannot be considered real conclusions, but rather a research programme. The core tasks of this programme are to demonstrate that there is a real tendency toward a stationary economy and that this tendency punctuates any phase of economic evolution. Schumpeter often ignored these tasks. Even in *Development* (83n), he stated that the "stationary economy is ... an incontrovertible fact, apart from the fact ... that there is a tendency towards a stationary state in every period of depression." Although these statements are hardly incorrect, they do not substantiate the claim of a simple coupling between his theoretical waves of evolution and the facts of business cycles. However, Schumpeter probably thought that Spiethoff's data supported his claim. He furthermore, in 1914, tried to place conception of capitalist cycles in an empirical context. This was done in a paper called "The Wave Movement of Economic Life" (*S1914b*) from 1914. This paper was published in *Archiv für Sozialwissenschaft*, the journal that Weber and Sombart had started ten years before.

Schumpeter's 1914 paper presents six stylised facts that separate prosperity and depression: (1) raising versus falling prices; (2) raising versus falling interests, wages, and rents; (3) large versus low (or negative) entrepreneurial profits; (4) increasing versus decreasing output and employment; (5) deterioration versus improvement of the relationship between the banks' obligations and their reserves; and (6) intensive versus vanishing creation of new firms (*S1914b*, 281–2). These stylised facts are presented as summarising much empirical evidence, including Wesley Mitchell's *Business Cycles* from 1913. However, hardly anyone had emphasised the creation of innovative firms. Schumpeter did so with the following chain of causation of prosperity (p. 287): We start with the emergence of entrepreneurs with plans about new enterprises; they obtain newly created credit from banks; the use of the additional money leads to higher prices of productive services and then to higher prices of consumption goods; these changes leads to the creation of more new firms and to the expansion of old ones; hence, both production and employment raise; and the different types of profit generally become large. After depression has set in, the chain of reaction is similar (p. 288): entrepreneurial investment vanishes, and thus demand for credit expansion; then we see a fall in prices; and so on. Schumpeter's paper, however, concentrates on the question why a depression is necessary when a steady expansion seems possible. The complex answer, which is a kind of prologue to Schumpeter's later work, includes reasons for the emergence of entrepreneurial projects as a mass phenomenon as well as for the subsequent slow and painful reorganisation of the economic system (pp. 289–301). The most surprising aspect of this account is that it does not include any specification of the statistical data needed for proving that "The Wave Movement of Economic Life" is the necessary form of economic evolution under capitalist conditions. However, Schumpeter's paper for *Archiv für Sozialwissenschaft* demonstrates that he considered his theory of economic evolution not only to be a contribution to theoretical economics in the narrow sense but also to the research programme of the historical school and to the great debates of the Böhm-Bawerk Seminar.

## 5.5 The "spirit of capitalism" and the system of concepts

Each of the great problems that Schumpeter treated in *Entwicklung I* provided him with a 'window' through which he could look upon the evolutionary mechanisms of the capitalist economy. His study of the interest on capital (see Section 5.3) had tried to demonstrate that capital accumulation and the great industrial fortunes are not the cause, but the result of innovative entrepreneurship. His study of business cycles and economic crises was made in a similar vein (see Section 5.4). He tried to argue that business cycles reflect the shifting between periods that are dominated by the immediate effects of innovative entrepreneurship and periods in which the economic system is undergoing an adaptation to the innovative activity of the immediately preceding period. Although these two 'windows' for looking at the inner mechanisms of capitalism provided him with many insights, there is no doubt that his study of the problem entrepreneurial profit provided him with the most direct view of this mechanism. The moving force of Schumpeter's model of the capitalist economy is obviously the innovative entrepreneur; and the pivot of the efforts of this entrepreneur is profit. Therefore, the analysis of the problem of entrepreneurial profit in Chapter 4 of *Entwicklung I* is crucial even for the Schumpeterian treatment of the great problem of defining the essence of capitalism. Thus we are approaching the very core of the early Schumpeter's entrepreneurial interpretation of economic history. The task of the present section is to demonstrate that this entrepreneurial interpretation is not least an indirect response to Marx's work. More specifically, Schumpeter seems to have responded to the Marx-related discussion on "the spirit of capitalism" by Werner Sombart and Max Weber in the beginning of the twentieth century.

While both the original title of Chapter 4 and its title in *Development* is "Entrepreneurial Profit", this chapter has in *Entwicklung II* (207) the heading "Entrepreneurial Profit or Surplus Value". This title serves to emphasise the basic point in the whole book, but also an ambiguity. Actually, it is "profit *and* interest" that "can be described as surplus value in the Marxian sense" (*Development*, 143n). Furthermore, this "surplus" is not conserved as in classical economics, but present only as long as the entrepreneurial project is facing a system of prices that has not reached a new equilibrium. The gross profits from a set of innovative projects, nevertheless, induce an "exploitation of the leaders": the bankers obtain their share through the rate of interest while other shares are captured by land owners and workers with strong positions (*Entwicklung I*, 302, 264, 397, 413). Hence, the temporary "surplus" created by the entrepreneurs is even relevant for the economy as a whole (p. 297). However, analytical difficulties hindered Schumpeter in developing this core message of his book. Instead, he started from the circular flow in which the rate of profit is zero. The innovative project is based on the expectation of a surplus when the needed inputs are bought at the prices determined by the model. The implementation of the project by means of credit changes these prices; but the sluggishness of the response may allow the expectation to become a temporary reality. This reality is related to individual projects that have "exceptionally different returns" (*Entwicklung I*, 321). In this world, Marx's emphasis on the idea of the "tendency of equalisation of entrepreneurial profits" is misleading. This "equalisation" only emerges when the introduction of new combinations has ceased and the streams of profits disappears temporarily.

Schumpeter nevertheless recognised that, before his efforts to change the situation, "the only larger attempt concerning the problems of evolution [Entwicklung] is that of Karl Marx" (*Entwicklung I*, 98). This reference does not concern the economic interpretation of history but the Marxian "attempt to treat economic evolution itself by means of economic theory". This treatment is designed explain "the development [Entfaltung] of economic life and not just its circular flow". However, "even though the tone of evolution breathes and the aspect of statics is de-emphasised", Marx sticks to the statical foundations of classical economics. Since Schumpeter agreed with Böhm-Bawerk that these foundations were faulty and since evolutionary analysis, in any case, had to break away from "Economic Statics", it was Marx's sketchy evolutionary story rather than his specific economic analysis that interested Schumpeter.

Although Marx's story is badly represented by his application of the analytical tools of classical economics, it is clear throughout his writings. The story is basically the same as the one told by Malthus's formula of population growth. In Darwin's hands, this formula of a population's potentially geometrical increase became, in a multi-population setting, "a force like hundred thousand wedges trying to force every kind of adapted structure into the gaps of the oeconomy of nature, or rather forming gaps by thrusting out weaker ones" (quoted by Ospovat, 1995, 60). Actually, it is only the fact that a single species cannot adapt to all niches in the ecology of nature that hinders it in obtaining a perfect monopolistic position. Marx found his Malthusian formula for economic life under capitalism by rethinking Aristotle's sketchy remarks on economic issues in terms of a new notation (Meikle, 1995, 52). The great Greek philosopher argued that the introduction of money is perfectly legitimate since it does not change the consumptive purpose of the exchange but simply makes it more efficient. However, the fact that money ($M$) can be stored and accumulated allows for an illegitimate inversion of means and ends: selling of commodities ($C$) in order to buy (and consume) becomes buying in order to sell. This inversion from $C$-$M$-$C$ to $M$-$C$-$M'$ implies a hunt for a monetary surplus ($\Delta M = M' - M$) that can be reused to obtain more money in exponential progression. Although this ambition cannot be fulfilled, it results in an unlimited process that is described as disturbing and potentially destroying the Greek city states. According to Marx, the real revolutionary force of monetary accumulation was only seen much later when the hunt for money had subsumed production under its purposes. Here we have two types of "commodities". The capitalist buys productive services ($C$) that after a period of production provides saleable goods ($C'$). Thus, the general formula of an individual capital is $M$-$C \cdots C'$-$M'$, where $\cdots$ represents the time needed for production and sales. The only meaning of this process is to maximise the individual surplus ($\Delta M$); and this can not least by done by means of innovative activity. Thus, the Marxian formula reflects a vision the evolutionary force under capitalism that is hardly less powerful that Darwin's vision of the "hundred thousand wedges". It will force all industrial capitalists to *try* to expand their capitals according to their categorical imperative: "Accumulate, accumulate! That is Moses and the prophets!" (Marx, 1990–92:I, 742).

This vision is not only applied throughout *Capital* but also, " 'with a miracle of compression' ", in the *Communist Manifesto* (*S1949f*, 294). Here we, according to Schumpeter, find "the social vision that [his] work has to implement". This vision "implied a program of research" that not only covered economic sociology but also economics (p. 305). The economic elements are based on "a panegyric upon bourgeois achievement that has no equal in economic literature". This achievement is founded in "the *creative* role of the bourgeois class that the majority of the most 'bourgeois' economists so persistently overlooked". The creative role can be defined as "a 'constant revolutionizing of production', creation that spells obsolescence and consequent destruction of any industrial structure that exists at any moment". Therefore, "capitalism is a process, stationary capitalism would be a *contradictio in adjecto* [contradiction in terms]" (p. 302). The capitalist process "tends to evolve the giant concern" and it is marred with " 'the epidemic of overproduction' ". Both centralisation and crises are signs of the ultimate limits of the capitalist economy that are also indicated its creation of a "proletariat" (pp. 303–4).

Schumpeter's evaluation of Marx's analytical implementation of this vision (*Capitalism*, 43–4) seems modelled after the discussions at the Böhm-Bawerk Seminar. "In the court that sits on theoretical technique, the verdict must be adverse." Marx was using an inadequate and obsolete analytical apparatus and made mistakes that influenced essential conclusions. The "qualification of the verdict" is that "his critics were far from being always right" and that he had made quite a number of significant analytical contributions on business cycles and multi-sectoral modelling. "But the court of appeal—even though still confined to theoretical matters—might feel inclined to reverse the verdict altogether." The argument for this reversal is that Marx "was the first economist of top rank to see and to teach systematically how economic theory may be turned into historical analysis and how the historical narrative may be turned into *histoire raisonnée*". Such a reasoned history is "still the economic theory of the future" (pp. 43–4). Marx's idea of an *histoire raisonnée* is not directly represented by the enormous historical and statistical material that, in *Capital*, he used to illustrate and clarify his theoretical argument. Instead, "a reasoned (= conceptually clarified) history … of the economic process" (*Cycles*, 220) implies an inversion of the roles of theory and empirical evidence in the treatment of long-term economic evolution. The resulting treatises would not only be much less elegant than their theoretical counterparts. They would also serve as important challenges for the existing theoretical toolbox and suggest the invention of new tools.

The university researcher who, in Schumpeter's youth, most clearly adopted Marx's vision and brought the related concept of capitalism into academic focus was Werner Sombart. The starting point is found in his evaluation of Marx's "economic system". In this evaluation from 1894, Sombart dismissed Marx's formal economics. He instead emphasised that Marx had performed an important exercise of "pure logic" that should not be confused with empirical reality. While Sombart's paper was condemned by Böhm-Bawerk (1949, 102–18) as a lame replacement of the erroneous Marxian economics with loose speculation, it was praised by Friedrich Engels in the last piece he ever wrote. However, Engels (in Marx, 1990–92:III, 1033) also pointed out that "insufficient regard is paid to the fact that what is involved is not only a logical process but a historical one". This hint was sufficient for Sombart, the most brilliant student of the leader of the historical school (Schmoller), to engage intensively in an exercise that combined Marx's evolutionary "logic" with historical fact. The result was published in 1902 in two large volumes on *Modern Capitalism*, of which the latter volume was called *Theorie der kapitalistischen Entwicklung*. Sombart's book presented an idiosyncratic mix of theoretical sketches and historical facts that, according to Schumpeter, "even out-Schmollered Schmoller". Although this book "shocked professional historians by its often unsubstantial brilliance … it was in a sense a peak achievement of the historical school, and highly stimulating even in its errors" (*History*, 816–17n). It also secured him a place in the new triumvirate of leaders of the historical school, the others being Spiethoff and Max Weber. As noted in previously, Schumpeter wondered whether "the area of 'Dynamics' belongs to them" (*Wesen*, 18). Although Schumpeter understood and respected the development of "individual hypotheses" by Sombart and the other members of the historical school, he had come to see himself primarily as an exact theorist. Therefore, his immediate goal was to handle economic evolution by developing an abstract theoretical system. Nevertheless, his development of this system seems to have received a stimulus from Sombart's transformed version of Marx's research programme and from some of its concrete propositions.

The core of Sombart's "reasoned history" concerned the movement from the craft system, which formed a relatively independent part of Medieval feudalism, to the capitalist system. This movement became a big problem since he described the craft system without internal mechanisms that could promote a major change. Instead, he described the change in terms of the emergence of a new type of man. He gradually came to specify this type, the modern-style bourgeois or entrepreneur, as representing a form of human motivation that combined a "trinity" of motivations in order to succeed in capitalist competition: the motivations of the conqueror, the organiser, and the trader. The material background for this version of the Nietzschean Superman was provided by profit opportunities, new types of organisation, and the rationalising methods of double-entry bookkeeping. However, as soon as these preconditions were available, it was not simple accumulation but this type of dominating individuals that provided the driving force of the system. In *Modern Capitalism*, Sombart (1902:I, 378) characterised this driving force as the "spirit of capitalism", and he developed the concept in later works.

Max Weber gave a somewhat different explanation of the driving force of the capitalist system. In a famous essay from 1904–05, which later was published as an English book with the title *Protestant Ethics and the Spirit of Capitalism,* he suggested that that "spirit of capitalism" is based in the ethos of capital accumulation provided by Calvinist Protestantism. Schumpeter did not accept this explanation. He recognised that Sombart's process started in the Catholic Italy of the Renaissance. This was also the birthplace of double-entry bookkeeping and of relatively modern types of banking and credit. So it was here that he would look for the preconditions for the emergence of the driving force of the capitalist system. Furthermore, he recognised that Weber's (and Sombart's) reasoning in terms of dichotomies hindered the analysis of social evolution. According to his later formulations, there "[t]here was no such thing as a New Spirit of Capitalism in the sense that people would have had to acquire a new way of thinking in order to be able to transform the feudal world into a wholly different capitalist one" (*History,* 80). The reason is that "the feudal ages contained all the germs of the society of the capitalist age". In this setting, the "capitalist methods" and the "capitalist 'spirit' " emerged in a stepwise manner (p. 81). Thus, the strong form of the problem of the "spirit of capitalism" represents one of the "Spurious Problems, that is to say, of those problems that the analyst himself creates by his own method of procedure" (p. 80n). In Weber's case, the method was a form of reasoning in terms of "Ideal Types" that was not able to cope with the evolutionary transformation of one "Real Type" into another. We may add that, in this respect, he repeated the methodological error of the biologists that had tried to understand species in terms of Plato's "Ideas" and, therefore, had to assume an exogenous force (interpreted as Divine intervention) when new species emerged.

The fact that Schumpeter found a "fundamental methodological error" (*History*, p. 81n) in the main argument of the *Protestant Ethics* did not mean that that this book was without interest for him. For instance, Weber (1958, 66–9; cf. MacDonald, 1965, 375–8) included a case story that was easily handled by Schumpeter's scheme of economic evolution. This case, the early evolution of the textile industry, must immediately have caught Schumpeter's attention because of his background in an old textile family. Weber started from the routines of traditional economic life at a time when hand-loom weavers in the countryside worked for urban textile merchants ("putter-outs"). The "peasants came with their cloth" and they "received the customary price for it" (p. 66). Similarly, the customers "also came to him, ... or, long before delivery, placed orders which were probably in turn passed on to the peasants". In this state of the economy, moderate earnings allowed the putter-out to "lead a respectable life". On this background, Weber introduced a radical disturbance: "some young man from one of the putting-out families went out into the country, carefully chose weavers for his employ, greatly increased the rigour of his supervision of their work, and thus turned them from peasants into labourers" (p. 67). The innovative act that "suddenly destroyed" the state of "leisureliness" should be considered a deed rather than an automatic event. Weber's argument is that the "revolution" was not brought about by the "stream of money invested in the industry", but by "the new spirit, the spirit of modern capitalism", Furthermore, "the first innovator" had to overcome a "flood of mistrust, sometimes hatred, about all moral indignation" (pp. 68–9). This opposition to the "entrepreneur of this new type" was well founded: "those who would not follow suit had to go out of business. The idyllic state collapsed under the pressure of a bitter competitive struggle, respectable fortunes were made, and not lent out at interest, but always reinvested in the business" (p. 68).

We may conclude that Weber related to both "Economic Statics" and "Economic Dynamics" in a way that fitted well into the evolutionary model that Schumpeter developed. The innovator took his starting point in an equilibrated economic system, the innovation was difficult because of opposition, and the result was the destruction of the old system and the emergence of a new one. However, while Schumpeter saw the emergence of a new system of routines as something that was destroyed by the next round of innovation, Weber was, in his *Protestant Ethics*, satisfied with a *single* occurrence of creative destruction. The reason is that he only wanted to characterise the *emergence* of the general "spirit of capitalism". Thereby, he, however, came to glorify routine behaviour within the capitalist system—especially the reinvestment of profits in existing enterprises. In contrast, Schumpeter thought that routine behaviour destroyed the possibility of accumulation. Furthermore, he considered routines as the build-up of a resistance against the next innovative deed. Finally, Schumpeter wanted to explain the consequences of the innovative deeds for the waves of reconstruction of the economic system. It was this conception of the economic system as a whole that Weber was unable to develop. In this respect, Sombart performed somewhat better—but he was still far from producing a general economic model that included the "spirit of capitalism".

Schumpeter's development of a model that included his own version of "spirit of capitalism" provided him with an alternative to both Weber and Marx. This alternative is very radical because the economic system repeats an extended version of the Weber story each time it has entered a state of equilibrium. We may also say that Marx's "primitive accumulation" has to be repeated each time an equilibrium is reached. This equilibrium might be described as a Marx-like process of simple reproduction in which there are no free resources and no monetary surpluses. The question is how the innovative entrepreneur can conquer the resources that are needed for his project. The answer is found in the fact that the money made available to him will be used to create a temporary stream of monetary surplus. The analysis of this rather uncertain stream is much more difficult than the analysis of the outcome of the simple Marxian process of production ($M$-$C \cdots C'$-$M'$). Nevertheless, it is on the expectation of a share of a positive $\Delta M$ that the capitalist banker is persuaded to create credit money for the entrepreneur. However, neither innovative entrepreneurs nor capitalist bankers are present when the economic system again reaches a state of equilibrium. Instead, they have retreated as landowners, managers, or workers. Hence, in the abstract Schumpeterian model of economic evolution, it is the *whole* of the bourgeois class that has to be recreated each time the system sets off into a wave of economic evolution. Therefore, Schumpeter's version of the categorical imperative of capitalism is: Innovate, innovate! That is Moses and the prophets! This imperative cannot be followed by established firms that, in the Schumpeterian model, normally have to be satisfied with a stationary state and ultimately disappear when their economic profits becomes negative. The imperative is only followed by a new combination of innovative entrepreneurs and the creators of credit for them. Schumpeter thought that it is only here that Marx's (1990–92:III) analysis of the conflict between "finance capital" and "industrial capital" is of relevance. This conflict not only raises the question about what is left of the innovative surplus for the innovative entrepreneur, but also how much the underlying entrepreneurial activity is controlled by the financing agents. In contrast to Hilferding (1981), who was already outlining his *Finance Capital* during the time of the Böhm-Bawerk Seminar, Schumpeter thought that the direct influence of banking institutions on entrepreneurial projects was very limited (as he later emphasised in *Cycles*, 349).

The structure and contents of *Entwicklung I* (see Table 5.1 on page 103) demonstrates that Schumpeter wanted to achieve what neither Marx nor Weber and Sombart had accomplished. Like Marx, he recognised that he had to start from a well-defined model of an economic system in its steady state. Since this model could not be provided by the defunct classical economics, he turned to the Walrasian system. However, he was fully aware that Walras's model had not been developed as the starting point for the study of economic evolution. Therefore, he reinterpreted the model as reflecting the routinised behaviour of ordinary economic agents. He also carefully removed any potential for endogenous evolution. Thereby he clearly departed from Marx who had seen the accumulation of money capital in relation to incumbent firms as the ultimate cause of evolution. In contrast, Chapters 1 and 2 of *Entwicklung I* seems to describe incumbent firms as representing a conservative force that have to be overcome by the entry of innovative entrepreneurs and their new firms. The consequence is that the capitalistic features of the economic system are unstable. Actually, these features tend to disappear as the system approaches the circular flow so that they have to be recreated periodically. The elements of this recreation are presented in Chapters 3–6, which describe capitalism as an engine of economic evolution that is based on the finance of innovation by credit creation. The role of the capitalist within this engine is best described as a banker who specialises in giving credit to innovative entrepreneurs. However, it is the demand for money by the innovative entrepreneur that is the real driver of the capitalist process of evolution. Furthermore, the fluctuations of innovative activity produce business cycles as a basic characteristic of the capitalist economy.

Schumpeter's account for this theory of economic evolution applied standard concepts of economics like profit, credit, capital and business cycles, but these concepts were interpreted within a framework of routine and innovation. The list of definitions recorded in Table 5.3 on page 131 serves to emphasise this point. The starting point is the circular flow in which everyone applies routinised "production combinations". On this background, "economic evolution" is defined as the endogenous change of the routines by the introduction of innovations, called "new combinations". The innovative projects are called "enterprises" and the agents who carries them out are the Schumpeterian "entrepreneurs". The carrying out of new combinations creates a temporary "surplus value" that is mainly shared by the entrepreneurs and the suppliers of the necessary "credit", the capitalist bankers. From another viewpoint, the money supplied by the bankers are the "capital" of the model, and the owners of this capital can be denoted "capitalists". However, seen from the viewpoint of the circular flow, there is no unused money in the system. Therefore, the purchasing power for the entrepreneurs has to imply the creation of credit money. This is seen as the defining characteristic of the "capitalist economy".

The starting point of the chapter on entrepreneurial profit is an example from the textile industry. This example is not the reformation of the putter-out system but the introduction of a new type of machinery, the power-loom (*Entwicklung I*, 280–4; *Development*, 129–32). Böhm-Bawerk had treated this example in the first volume of his *Capital and Interest*. He concluded that free competition secures that the replacement of labour by the new machine only produces a short-term profit that cannot serve as a real source of interest. Schumpeter presented a somewhat different interpretation in terms of a drama. First, we see a textile industry that has reached a stationary equilibrium based on handlooms. Second, an entrepreneur who has the ability of "overcoming all the innumerable difficulties" (p. 129) engages in implementing the powerloom innovation. The money for this time-consuming project is lent by a bank (a capitalist) on the expectation of a surplus of revenue over costs (including interest). The availability of this surplus can be argued in present prices; but for large projects, the entrepreneur has to correct for the influence of his project on future prices. Third, the successful completion of the innovative project means that "the spell is broken and that new businesses are constantly arising under the impulse of the alluring profit" (p. 131). The increasing competition enforces a process of reorganisation during which non-adapting firms disappear and unemployment may arise. Fourth, we see a new equilibrium emerge based on powerloom technology. In this new routine system no profits are made and no payment of interest is due to entrepreneurial initiative.

One might argue that Schumpeter's version of the powerloom example is nearly identical to that of Böhm-Bawerk. However, the focus on the long-term outcome is replaced by a study of the mechanisms of change with special emphasis on their driving force. Thereby, it becomes clear that the innovative entrepreneur and his immediate followers do obtain a temporary surplus. This surplus is due to the fact that they have "carried out new combinations" of pre-existing resources by means of borrowed money (that in older days were provided by themselves). Therefore, the surplus is a premium for their innovative will and ability. Their abilities not only include those needed for the overcoming on constraints. They also include thinking in terms of a present system of routine-based prices and a future system of prices that includes their own project but still not have reached a new equilibrium. This way of thinking presupposes that the old system of prices has a significant degree of inertia. Schumpeter explained this inertia is the result of custom: the prices have been learnt through long experience and are not easily changed in a consistent manner. Given inertia, economic surplus can be derived from innovative projects and shared between the entrepreneur (as profit) and the capitalist

**Table 5.3.: Hierarchy of concepts relating to entrepreneurs and capitalism**

| Concept | Definition |
|---|---|
| Circular Flow | A "mental picture" in which "the economic system will not change capriciously on its own initiative" (*Development*, 5, 8–9; *Entwicklung I*, 4, 8–9) |
| Production combinations | "[C]ombinations of productive forces" whose "results are products" (*Development*, 15; *Entwicklung I*, 23) |
| Economic evolution | "[S]uch changes of economic life as are not forced upon it from without but arise by its own initiative, from within" (*Development*, 63; *Entwicklung I*, 103) |
| New combination | An unused production combination that is beyond the horizon of the agents of the circular flow but nevertheless advantageous (*Entwicklung I*, 168–9; cf. *Entwicklung I*, 65–6) |
| Enterprise | "The carrying out of new combinations" (*Development*, 74; *Entwicklung I*, 177) |
| Entrepreneur | The "man of action … who carries out new combinations" (*Entwicklung I*, 172; cf. *Development*, 74) |
| Entrepreneurial profit | The entrepreneur's part of the "surplus over costs" arising from "new combinations"; the "surplus value in development" is partly distributed to other agents (*Development*, 128–9, 143; *Entwicklung I*, 278–9, 297) |
| Credit | "[E]ssentially the creation of purchasing power for the purpose of transferring it to the entrepreneur, … the method by which development is carried out in a system with private property and division of labor" (*Development*, 107; *Entwicklung I*, 214) |
| Capital | "[T]hat sum of means of payment which is available at any moment for transference to entrepreneurs" (*Development*, 122, emphasis removed; cf. *Entwicklung I*, 236) |
| Capitalist economy | "That form of economic organisation in which the goods needed for new production are withdrawn from their settled place in the circular flow by the intervention of purchasing power created *ad hoc*" (*Development*, 116; cf. *Entwicklung I*, 226) |

Comment: The fact that Schumpeter's concepts relating to entrepreneurs and capitalism can be organised in a hierarchy demonstrates that they are closely connected. If we read the above list backwards, it becomes clear that a major goal of the Schumpeterian concepts is to define the capitalist economy in terms of innovative entrepreneurship and credit creation.

(as interest). Schumpeter added: "Only this surplus ... can be described as surplus in the Marxian sense. No other surplus exists" (*Development*, 143n; cf. *Entwicklung I*, 297). Thus he, in contrast to Böhm-Bawerk and partly to Marx, emphasised that there is no permanent and secure income for entrepreneurs and capitalists. These incomes derive from an innovative surplus that is repeatedly recreated and destroyed. Thereby, both parties depend on economic evolution: "Without evolution there is no entrepreneurial profit, without entrepreneurial profit no evolution" (*Entwicklung I*, 322; *Entwicklung II*, 236; cf. *Development*, 154). Schumpeter added that evolution and profit are also the precondition for "the great social phenomenon" of the "accumulation of wealth".

## 5.6 Conclusion

Schumpeter had in *Wesen* not only emphasised the fundamental distinction between equilibrium economics and evolutionary economics. He had also begun his exploration of the latter branch of economics. He presented the preliminary of his evolutionary economics in *Entwicklung I*. In the preface to this book, he concluded that he had not provided a full theoretical structure but rather the "essential foundations" for such a structure. In this connection, he stated that he had two wishes. On the one hand, he wished that economic theorists would not ignore the "facts and arguments" that he had presented (*Entwicklung I*, viii). On the other hand, he wished that his book "as soon as possible shall be surpassed and forgotten." In other words, Schumpeter wished that research relating to the theory of economic evolution would quickly become part of a tradition of evolutionary economic research. Since hardly anyone followed this wish, he had to perform the work himself. The results partly took the form of revised versions of his book. In contrast, the 550 pages of *Entwicklung I* were never reprinted during Schumpeter's lifetime.

The basic response to Schumpeter's book could have been predicted. The proposal for a new field of theoretical economics and for related work in sociology was too radical and too underdeveloped to establish a research tradition. Even the reviews were hardly surprising. Let us consider a few examples. Roswell McCrea (1913, 526) suggested that Schumpeter "offers a super-man interpretation of economic progress, in main outline quite analogous to the sociological system of Gabriel Tarde." A similar interpretation led Franz Oppenheimer (1916, 211, 222) to consider the book as "hardly more than an economic novel ... much more a hero epic than sober science!" The most systematic and harshest evaluation was, however, produced by the teacher who had conducted the seminar at which *Entwicklung I* was probably born. Böhm-Bawerk (1913, 2) thought that Schumpeter had become "intoxicated by a brilliant idea" and that he had not put this idea under "a sufficiently sober and cautious cross questioning". The idea was that 99 per cent of the entrepreneurs do not deserve their name. Thereby, he ignored the significant "intermediate zone" between "the purest statics" and "the creative dynamics" (p. 4). However, Böhm-Bawerk only mentioned such issues in passing. Instead, he reserved his great argumentative skills on the systematic destruction of Schumpeter's theory of interest as a solely dynamic phenomenon.

*Entwicklung I* also received more positive reviews. Benjamin Anderson (1915, 660) hoped that "Professor Schumpeter—whose command of English style is faultless—will give us an English edition of this interesting and important book." John B. Clark (1912, 875) considered the book "a study of capital, interest, and profits, with an incidental inquiry into the causes of commercial crises." He emphasised that "[a]ll these phenomena it connects with economic evolution, and it studies them as they appear in a world of change and progress." Given Clark's own ambitions, it is not surprising that he thought that *Entwicklung I* "makes an important addition to the limited amount of scientific literature which deals consciously and systematically with what is commonly termed 'Economic Dynamics'." He especially emphasised that "Schumpeter's work discusses commercial crises and treats them very properly as dynamic phenomena—the outcome of a certain unbalances and uneven process." With respect to crises and some of the other topics of the book, Clark suggested that "the reader will find himself in a world of reality abounding in critical issues on which the work sheds welcome light" (p. 875). The book, however, was verbosely written. This is the reason why Navratil (1913, 444) suggested that a shortening to half of the original size would make the book much easier to understand. When Schumpeter largely had followed this suggestion in the revised German edition from 1926, Oscar Morgenstern (1927, 282) produced a review in which he concluded that "this is one of the most stimulating and fascinating books that has been written on economic theory. It is, since it gives the first elaborate *dynamic economics* in the proper sense, very revolutionary". The revolutionary nature of Schumpeter's work meant that even 12 years after its first edition had come out, Morgenstern suggested that "[i]t seems safe to say that the real influence of this treatise, whatever may remain of the substance of the theories involved, has just begun".

Morgenstern, who is known for the help he provided John von Neumann in developing classical game theory, was right that the influence of *Entwicklung I* and *Entwicklung II* had "just begun". However, he failed to predict that Schumpeter during the rest of his life would not achieve significant help for the development of his evolutionary theory. Schumpeter's starting point had, probably, been the problem of explaining economic crises and business cycles. The apparent straightforwardness of his answer seems to have seduced him to consider his theory of economic evolution as primarily being an account for the intrinsic business cycles of the capitalist economy. It was only later that he recognised the enormous difficulties of using his analytical apparatus for that purpose. Actually, the task presupposed solutions to all the other scientific problems that he was facing. The most important of these problems was the basic analysis of the mechanisms of evolution under capitalist conditions. Schumpeter had probably little doubt that the core mechanism of innovation is driven by the entrepreneur, who can thus be considered the "spirit of capitalism". The combination of this mechanism of innovation with the mechanism of adaptation presupposed a basic modification of Walras's analytical tools. Instead, *Entwicklung I* provided a quick addition of the innovative entrepreneur to a routinised circular flow; and it continued by redefining accordingly the concepts of profit, credit, capital, and the capitalist economy. However, the too-quick integration of these concepts into his basic model of economic evolution gave serious problems for Schumpeter's subsequent work in relation to *Development, Cycles,* and *Capitalism.*

# Part II

# The Evolutionary Trilogy

# 6

# Approaching the Evolutionary Trilogy

According to the Schumpeter's theory of scientific creativity, the initial "period of sacred fertility" produces the fundamental vision and the fundamental results that are "subsequently worked out" (*S1921*, 87). He also suggested that "the roots of great scientific achievements, especially those of a theoretical nature, predominantly lie in the twenties of their author's life" (*DBA*, 16–17; omitted in *S1914c*). In Schumpeter's case, the especially fertile years covered the period 1903–13. During these years, he had not only produced *Wesen* and *Entwicklung I* but also developed many related ideas. The subsequent development of his early results primarily took place after he had resumed academic life in 1925. In the intermediate years, he had been involved in politics and private banking—including his short and rather unsuccessful service as the Austrian Minister of Finance (see Appendix A). Now he established himself as an internationally recognised professor of economics—first at the University of Bonn in Germany from 1925 to 1932; and then at Harvard University from 1932 until his death in 1950. At the University Bonn he radically revised *Theorie der wirtschaftlichen Entwicklung* and prepared to extend its analyses of the evolutionary functions of money and business cycles. Nevertheless, the three books from which his evolutionary economics and evolutionary sociology are today primarily known were published during his Harvard period. These books are *The Theory of Economic Development* from 1934; *Business Cycles* from 1939; and *Capitalism, Socialism and Democracy* from 1942. It is these books that are at the focus of the present chapter; and each of the next three chapters is dedicated to one of the books.

## 6.1 The evolutionary trilogy and its name

If we for a moment postpone the problem whether *Development* and *Entwicklung I* can be understood as a single book, Schumpeter can be said to have produced six major books. Three of them were produced before World War I. In this early period, Schumpeter not only produced *Wesen* and *Entwicklung I* but also *Doctrine* on the history of economics. The three new books produced during the Harvard period are *Cycles*, *Capitalism*, and *History*. On this background, Shionoya (1997, 16, 23) suggested a distinction between the "early trilogy" and the "later trilogy". However, this distinction does not suit the present purposes for reasons that are developed in the following text. Nevertheless, it is convenient to state three of the reasons immediately. First, Shionoya (1990) has himself been effective in demonstrating that *Entwicklung I* and *Development* come close to being two different books; and the present book tries to emphasise this point. Second, although *Doctrine* has been published as an independent book in English, it is really a huge entry commissioned by Max Weber for an extensive handbook of economics in the broad sense. Furthermore, it can, for most purposes, be considered as having been replaced by *History*. Third, *History* is so different from the other books that it is best treated separately as part of a postscript to the whole of Schumpeter's work. For these and other reasons the present book does not apply the partitioning in an "early trilogy" and a "later trilogy".

An alternative interpretation of the basic architecture of Schumpeter's work was presented in Section 1.2. This architecture consists of three parts. The first part can be described as programmatic and consists of *Wesen* and *Entwicklung I*. This part suggests a division of labour between equilibrium economics and evolutionary economics. It also developed the latter branch of economics and suggests complementing it with evolutionary sociology. The second part of Schumpeter's work develops his early evolutionary results and suggestions. It can be called the evolutionary trilogy and consists of *Development, Cycles,* and *Capitalism*. The third part consists of works in progress. This part should not be considered as covering the whole range of planned books and unfinished manuscripts that emerged during Schumpeter's academic life. Instead, the third part of the architecture consists of central plans and manuscripts that emerged during and after his production of the second part. The most conspicuous manuscript is that of *History,* which was published posthumously. However, as we shall see in Part III of the present book, less developed plans and manuscripts are also important from the viewpoint of the development of his evolutionary economics and for understanding its relations with equilibrium economics and other social sciences.

Before we interpret Schumpeter's evolutionary trilogy as a whole, it is convenient to characterise the three books from the present perspective.

1. *The Theory of Economic Development* has 255 pages. *Development* provides the core statement of Schumpeter's analysis of economic evolution. The book is not a simple translation of *Theorie der wirtschaftlichen Entwicklung* from the beginning of 1912 (or rather the end of 1911), which covered 548 pages. *Entwicklung I*'s confusing presentation of the elements of his evolutionary economics was improved in the later and much shorter editions of the book (*Entwicklung II* and *Development*). Schumpeter focussed on the economic function of innovative entrepreneurs and removed much of the presentation of their personalities. Furthermore, he totally removed the

idea of social innovators and the related idea of how to analyse the evolution of other sectors of social life.

2. *Business Cycles* has 1095 pages. *Cycles*—with the subtitle *A Theoretical, Historical, and Statistical Analysis of the Capitalist Process*—complements *Development* in three ways: (1) it restates and develops the basic theory of economic evolution and the closely related theory of waves of evolution; (2) it uses these theories to analyse historical and statistical evidence on the history of capitalist economic evolution; and (3) it presents itself as a contribution to the literature on business cycles. The complex structure and the unfinished nature of *Cycles* meant that it largely failed to convince its readers. Even the abridged and much more readable edition by Rendigs Fels (*Cycles Abr.*) has not hitherto changed the destiny of Schumpeter's most ambitious book.

3. *Capitalism, Socialism and Democracy* has 431 pages. *Capitalism* consists of a series of essays. The core of *Capitalism* complements *Development* and *Cycles* in at least ways. First, the book analyses the evolution of the institutional features of capitalist economy with special emphasis of the increased capabilities of firms. Second, it presents a modified model of economic evolution in which it is the oligopolistic competition between established firms that drives the evolutionary process. Third, it suggests two models of the evolutionary process within democratic political systems. Fourth, it provides elements of an analysis of the co-evolutionary interaction between the economic system and the socio-political system.

Although it is from these three books that modern economists and social scientists largely derive their knowledge of Schumpeter's evolutionary contributions to the social sciences, very few consider them as forming an evolutionary trilogy that to some extent has to be studied as a whole. Actually, even each individual book is seldom studied as a whole. Instead, an industrial economist is likely to look up what Schumpeter on a few pages of *Capitalism* wrote on the so-called "Schumpeterian hypothesis" of the relationship between firm size and innovation. Similarly, a political scientist is likely to consult a couple of *Capitalism*'s chapters to understand the Schumpeterian theory of "competitive elitist democracy". However, while such strategies are quick, they hinder the understanding of Schumpeter's actual contributions on these individual topics and even more his potential importance for the general development of evolutionary economics and evolutionary political science. The combination of *Development, Cycles,* and *Capitalism* under the heading 'The Evolutionary Trilogy' serves to emphasise these points. The word 'trilogy' is normally used to denote a group of three related literary works.

The relatedness of *Development, Cycles,* and *Capitalism* can partly be characterised by their common study of evolutionary processes. Their relatedness can also be described in terms of a sequential movement. We move from the simple analysis of economic evolution of *Development* to the very complex analysis of *Cycles*; and we move from the analysis economic evolution of these two books to the integrated treatment of economic evolution and socio-political evolution in *Capitalism*. These types of relatedness reflect the way in which Schumpeter produced the books and provided pointers between them. The forward-pointing hints or references become especially clear if we start with *Entwicklung I*. This book demonstrates that *Cycles*'s topic of the waves of economic evolution was at the centre of his early research and that the encompassing evolutionary analysis of *Capitalism* was prepared methodologically in the last chapter of *Entwicklung I*. On this background, it is not difficult to recognise that even the shortened and streamlined *Development* is pointing toward the two other parts of the trilogy. Furthermore, half of the contents of *Cycles* is an evolutionary history of capitalist *economic* evolution that suggests the urgent need of the broader analysis of *Capitalism*.

Rendigs Fels (1964, viii) seems to have been the first to describe *Development, Cycles,* and *Capitalism* as a trilogy. He also tried to specify their relatedness by presenting the tree books as "the trilogy setting forth 'the Schumpeterian System' ". Thereby he pointed at *The Schumpeterian System* by Schumpeter's former students Richard Clemence and Francis Doody (1950). Their ambition was to demonstrate that "[t]he Schumpeterian System is an imposing analytical machine" and that it can be defended against "a set of 'standard criticisms' " (pp. 5–6). Schumpeter did not consider the problem in this way. He did not see himself as a system builder but rather as a scientific innovator who was inducing criticism and further work. One way of emphasising the openness and unfinishedness of his contributions would be to describe the totality of the evolutionary trilogy as a work in progress. This term is presently interpreted as including works that point beyond themselves and towards the efforts of other researchers. However, we shall in the present book reserve the term 'Schumpeter's works in progress' to the narrower set of works covered by Part III.

The suggestion that Schumpeter's evolutionary trilogy as a whole can be considered work in progress serves to emphasise that this idea is not intended to become part of a "mythology of coherence" (Quentin Skinner; quoted by Emmett, 2003, 528). Neither these three books nor his work as a whole developed in a preplanned manner. Instead, it evolved in multiple directions. In the case of the evolutionary trilogy, these directions were to some extent suggested by the two programmatic works of his youth. However, it is hardly possible to interpret all his projects in this way. He apparently told one of his friends that his plans at the beginning of his Harvard period "encompassed 17 books, including 2 novels" (Harris, 1951a, 97). Another friend pointed out that when Schumpeter in the early 1940s had finished *Cycles* and *Capitalism*, "he had planned not only the *History*, but a treatise on money, a general theory, further sociological work, a theory of estetics and several novels to express those of his ideas that could not be fitted into a scientific mould" (Smithies, 1950, 633). However, we—especially for the evolutionary trilogy—have to recognise a large degree of actual coherence. This coherence is, to a large extent, captured by the Schumpeterian proposition of the importance of the yearly years as a period that "creates what is subsequently worked out" (*S1921*, 87). The resulting degree of coherence is important for the open-ended task of making rational reconstructions of his work.

An important implication of the idea of the evolutionary trilogy is that it suggests that Schumpeter's least readable and least successful book cannot be passed over in merciful silence. Although many economists know some of the core ideas and powerful formulations of *Development* and *Capitalism*, they also know that *Cycles* is generally considered a failure. This impression of failure cannot be removed totally by the present evolutionary interpretation of Schumpeter's work. It is, however, possible to rescue many of its insights and results for further work. This attempt is in accordance with Schumpeter's suggestion that we "should look upon the book merely as something to shoot at and to start from—as a motivated program for further research" (*Cycles*, v). The easiest way of starting this exploitation of *Cycles* is to consider whether the book could have had a better title than *Business Cycles: A Theoretical, Historical and Statistical Analysis of the Capitalist Process*. As we shall see, the contents of the book are much better reflected by a title like 'Theory, History, and Statistics of the Waves of Capitalist Economic Evolution'. In the same vein, we could replace *The Theory of Economic Development* by 'A Theory of the Mechanisms of Economic Evolution'. Finally, *Capitalism, Socialism and Democracy* could be thought of as 'Capitalist Economic Evolution and Its Socio-Political Limits'. While the titles that Schumpeter actually chose do not draw attention to the core contents of his programme for further evolutionary research, the alternative titles seem to give the shortest possible abstract of this programme.

## 6.2 The fields of evolutionary analysis

Schumpeter made his first presentation of his evolutionary economics in *Wesen*. Here he presented it as a necessary complement to equilibrium economics. His exposition of the double programme for the science of economics was made by means of the Statics–Dynamics dichotomy (see Section 3.4). Although he took this dichotomy from John B. Clark, he seems to have interpreted it by means of Stuart Mill's *System of Logic*.

Schumpeter later suggested that this book's treatment of the social sciences "is of first-rank importance for us" (*History*, 451). In this part of Mill's (1868:II, 513) book, we find the formulations on "Social Statics" and "Social Dynamics":

> "The Empirical Laws of Society are of two kinds; some are uniformities of coexistence, some of succession. According as the science is occupied in ascertaining and verifying the former sort of uniformities or the latter, M. Comte gives it the title of Social Statics, or of Social Dynamics; conformably to the distinction in mechanics between the conditions of equilibrium and those of movement; or in biology, between the laws of organization and those of life. The first branch of the science ascertains the conditions of stability in the social union: the second, the laws of progress. Social Dynamics is the theory of Society considered in a state of progressive movement; while Social Statics is the theory of the consensus ... existing among the different parts of the social organism[.]"

Schumpeter was clearly thinking of this statement when he in *History* (452) emphasised the distinction "between the problems of the effects that follow from a given cause under given conditions and the problems of the 'laws' that determine those social conditions themselves". By making and developing this distinction, "Mill unfolded a program that harnessed the purest of pure theory and the most concrete of institutional research into a peaceful co-operation and this without emasculating either" (p. 452, emphasis removed). Schumpeter also suggested that "[t]he distinction that had to be fought for, sixty years later". Since the *System of Logic* was originally published in 1843, the fight took place around 1903. Although Schumpeter might have referred to several contributions to this fight, he must have included his own efforts. Actually, he turned to theoretical economics in 1903 and quickly made Mill's distinction into the cornerstone of his research programme. However, Schumpeter rejected Mill's definition of "Social Dynamics" as the study of "the laws of progress". He did not subscribe to the theory of lawful and automatic progress that had developed within classical economics. Furthermore, he did not believe that the outcome of the process of evolution could be described by the term 'progress'. This application of this term involves a highly controversial value judgement—which he, furthermore, found unwarranted. Therefore, he tried to avoid the term or to use it as a quote. Nevertheless, Schumpeter's early formulations of his research programme can immediately be related to Mill's formulations. The Schumpeterian version of Economic Statics (equilibrium economics) is an instance of Mill's Social Statics. It thus solely engaged in studying the laws of coexistence. Schumpeter's Economic Dynamics (evolutionary economics) is an instance of Mill's Social Dynamics. It is thus engaged in studying the "laws" of "progress".

Although Schumpeter's evolutionary economics primarily can be considered as belonging to Social Dynamics, it also includes an evolutionary version of Social Statics. This type of Social Statics considers an economic system that works according to the laws of coexistence as the product of the adaptive part of the evolutionary process. It is only if evolution has produced such a system that we can apply the tools of static analysis—including comparative static analysis—with some degree of confidence. Schumpeter emphasised this point:

> "The assumption that conduct is prompt and rational is in all cases a fiction. But it proves to be sufficiently near to reality, if things have time to hammer logic into men. Where this has happened, and within the limits in which it has happened, one may rest content with this fiction and build theories upon it. ... But this holds good only where precedents without number have formed conduct through decades and, in fundamentals, through hundreds and thousands of years, and have eliminated unadapted behavior. Outside of these limits our fiction loses its closeness to reality." (*Development*, 80)

Schumpeter concluded this description of the limited range of Social Statics by emphasising that comparative statics will only work for the analysis of "small variations at the margins, such that every individual can accomplish by adapting himself to changes in his economic environment, without materially deviating from familiar lines" (*Development*, 81). He also pointed out that "[s]mall disturbances" may "in time add up to great amounts" (p. 81n). This fact could be used to develop a theory of economic evolution that is very different from that of Schumpeter, and this theory of "progress" had actually been developed in by the tradition of informal evolutionary analysis from Adam Smith via Mill to Marshall. However, Schumpeter did not belong to this tradition. Instead, he applied a radical version of the Statics–Dynamics dichotomy. According to him, Statics concerns "the circular flow of economic life as conditioned by given circumstances". This circular flow is essentially conservative; and it is the task of static evolutionary economics to study this conservative system. In contrast, his dynamic evolutionary economics includes innovations that are, by definition, unprecedented and whose logic has to be hammered into routine behaviour by means of a difficult process of adaptation. It is his assumption of a conservative circular flow that is changed by non-incremental innovations that allowed Schumpeter to stick to the sharp distinction between Social Statics and Social Dynamics.

Although Schumpeter's evolutionary economics includes the fields of dynamics evolutionary economics and static evolutionary economics, he never made a systematic distinction between the latter field and the corresponding field of static equilibrium economics. One of his reasons might have been that he wanted to uphold a bridge between the two basic branches of economic analysis. However, this praiseworthy purpose also served to constrain his development of the field of static evolutionary economics. Actually, many of his readers got the impression that it had largely been developed by Walras. Furthermore, his formulations often suggest that he thought in the same way. Nevertheless, the present book emphasises static evolutionary economics as a separate field of analysis (see Figure 6.1 on the next page). The purpose is not to ignore what Schumpeter actually said. Instead, the separation serves to focus attention on the ambiguities of his core concept of economic equilibrium that is the result of the mixing of static evolutionary economics with static equilibrium economics. As emphasised by the above quotation, the ambiguities of his concept of the circular flow of economic life can hardly be overcome unless it is considered part of his static evolutionary economics. Another set of ambiguities becomes clear if we make the distinction between evolutionary economics and evolutionary sociology. Although Schumpeter did make the distinction (phrased in different words), he does not seem to have developed it systematically. One of the reasons seem to be that while the distinction between equilibrium economics and equilibrium sociology can be made easily, the borderline between economics and sociology becomes blurred when we turn to evolutionary analysis. Unfortunately, this important issue cannot be covered systematically by the present book.

## 6.3 The evolutionary mechanisms of the capitalist engine

The study of Schumpeter's evolutionary trilogy as a more or less integrated whole is difficult. Although it largely concerns the evolutionary part of his research programme, it cannot be fully decomposed into static evolutionary economics, dynamic evolutionary economics, and evolutionary sociology. The task of reading the trilogy involves many additional difficulties. Some of the difficulties are due to his verbose style of writing as well as his frequent use of metaphor and ambiguity—often with unintended side-effects. Other difficulties are due to the fact that *Development, Cycles,* and *Capitalism* do not apply a consistent terminology. Further difficulties are connected to the fact that his evolutionary models are underspecified so that the consequences of his shifting assumptions are difficult to detect. Nevertheless, the primary difficulties emerge from the facts that evolutionary processes are notoriously difficult to conceptualise and that any conceptualisation of them tends to clash with accustomed modes of thinking in economics and other social sciences. These primary difficulties have to be confronted before we turn to the secondary

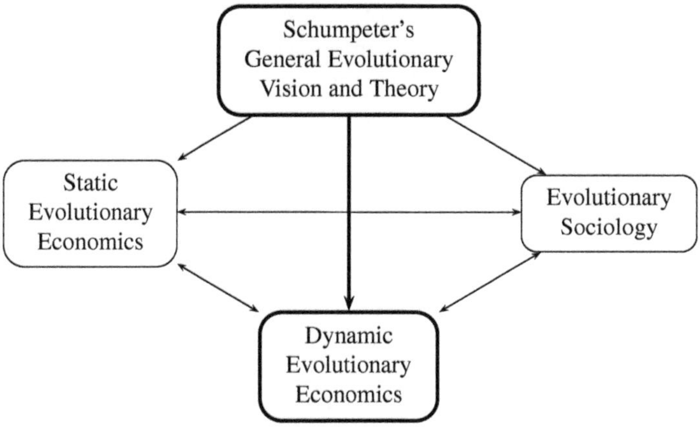


**Figure 6.1.: Schumpeter's evolutionary pivot and his fields of analysis**

Comment: Much of Schumpeter's later work consisted in elaboration his early vision and analysis of economic evolution. This work did not only concern dynamic evolutionary economics, but also static evolutionary economics, that is, the study of the functioning of states in which evolution has come to a halt and the comparison between such states. His work also included evolutionary sociology, which is even more complex than evolutionary economics. The present figure is largely a truncated version of Figure 2.2 on page 36.

ones. This is the reason why many evolutionary researchers develop a strong interest in methodological issues. Although this interest can produce important results, it can also produce sterile battles of methods. The present book follows Schumpeter's advice to reduce explicit metatheoretical discussion to a minimum. However, the troublesome history of evolutionary economics suggests that we cannot do without some efforts in the field of evolutionary economic methodology. This field might even exploit some of the ways of overcoming the difficulties of evolutionary analysis that have been proposed by biologists and philosophers. The 'translation' of these contributions, of course, has to be complemented with an exploration of the many additional methodological problems that emerge when we analyse the interaction of economic agents with myopic experiences and expectations.

For the purposes of clarifying the analytical strategy of the present book, the contributions by the philosopher Jon Elster seem sufficient. Although he has produced several relevant books, reference shall be made only to Elster's (1983; 1998) somewhat dated book on *Explaining Technical Change* and to his short paper called "A Plea for Mechanisms". Elster focussed on the structure of scientific explanations. Since the rational-choice explanation implies that evolutionary processes do not exist and since more realistic types of intentional explanation are too complex to cover here, we shall immediately move to causal explanation. This type of explanation starts from a phenomenon that should be explained (the explanandum, $E$). We explain the phenomenon causally by listing the conditions ($C_1, C_2, \ldots, C_n$) that produces the phenomenon. The standard form of causal explanation starts by referring to a scientific law: 'if $C_1, C_2, \ldots, C_n$, then always $E$'. This type of explanation ends with the statement: 'since $C_1, C_2, \ldots, C_n$ are true, we have explained $E$'. However, evolutionary explanation cannot normally use this explanatory strategy because evolutionary processes are too complex and because they involve essential elements of indeterminateness. Therefore, we would end up in an analytical quagmire with pseudo-laws like 'if $C_1, C_2, \ldots, C_n$, then sometimes $E$' or 'if $C_1, C_2, \ldots, C_n$, then sometimes $E_1$ or $E_2$ or $\ldots$ $E_m$'. Although explanation by law is the goal, evolutionary explanation is in practice constrained to the more modest explanation by evolutionary mechanisms. One of the consequences is that evolutionary explanations cannot be used to predict concrete outcomes of the evolutionary process. While a scientific law of the form 'if $C_1, C_2, \ldots, C_n$, then always $E$' can not only be used to explain existing phenomena but also to predict future ones, evolutionary explanations are primarily suited for the treatment of existing phenomena. If they are nevertheless used for prediction, the statements concern classes of outcomes rather than individual outcomes.

Although Schumpeter did not engage in this type of abstract methodological discourse, he seems to have recognised the need of approaching evolutionary processes by means of a specification of the underlying evolutionary mechanisms as well as the resulting need of limiting prediction to a very cautious format. For instance, he *did* use evolutionary mechanisms to explain the fact that railways at the end of the nineteenth century dominated long-distance land transportation while mail coaches had been driven to extinction. In contrast, he *did not* try to predict the next revolution of the routines of transportation. The crucial issue is how he specified the mechanisms of economic evolution (and of the evolution of other sectors of social life). Here he put overwhelming emphasis on the mechanism of innovation. According to Elster (1983, 112), Schumpeter's "key explanatory idea is that of the *entrepreneur*—a unique historical figure, of supernormal energy and will. Rather than gloss over the creative and unpredictable aspects of innovation, he made these into the cornerstone of his theory." He even used this idea as the primary explanation of waveform economic evolution and de-emphasised other aspects of business cycles. On the one hand, he was "concerned with the innovations that provide the irregular series of shocks" (p. 113). On the other hand, "Schumpeter was comparatively uninterested in the mechanisms of diffusion and propagation that determine the precise shape of the fluctuations". This emphasis on the mechanism of innovation did not imply that he ignored the mechanism of adaptation. However, the result of his priority is that his exposition has appeared unsatisfactory for most of his readers. Actually, the mechanism of adaptation is much easier to understand than the mechanism of innovation. Furthermore, the mechanism of innovation can only be presented clearly on the background of an exposition of the functioning and the results of the mechanism of adaptation.

Since his readers were not used to think in terms of evolutionary mechanisms, Schumpeter used his great writing skills in an attempt to force them to *see* the evolutionary process that he tried to depict. He thus coined and increasingly used the expression of "the capitalist engine". He especially used this expression throughout Part II of *Capitalism*, but it is also found elsewhere (see *S1927a*, 183; *Cycles*, 111, 118, 155, 700, 702; *S1950b*, 419, 423n). The image of the capitalist engine is a rhetorical device. Although the processes of capitalist economic evolution are obviously not the same as those of the engine of an automobile, the engine metaphor serves to prepare the reader for the real topic of Schumpeter's exposition. Thereby, the expression 'the capitalist engine' serves as a pedagogical metaphor to transfer given ideas rather than as a heuristics metaphor that produces new ideas (see Klamor and Leonard, 1994, 45–8). Like several of the other expressions he coined (not least 'creative destruction'), it has a powerful influence on the reader's imagination. The metaphor of the capitalist engine depicts the economic system of capitalism as a strange engine. The fuel of this internal-combustion engine is provided by Schumpeterian entrepreneurs. Their innovations "locate the ignition of the process" (*Cycles*, 102; *Cycles Abr.*, 76), and this process includes responses in the form of imitation, adaptation, and destruction. The result of the working of the "capitalist engine" is "economic evolution".

Schumpeter not only used an engine as a metaphor for economic evolution but also in relation to scientific activity and other areas of social life. However, while Schumpeter's metaphor of the capitalist engine concerned a socio-economic process, his metaphor of the "analytic engine" or "analytic machine" (*S1927i*, 23, 27) concerned the individual process of producing scientific results. This expression seems to relate back to Alfred Marshall's use of the metaphor and maybe even to the Analytical Engine, that is, Charles Babbage's planned predecessor of modern computers. Schumpeter might have started to think of science as an evolutionary process produced by the scientific engine (as he did in *History*; see Chapter 11); and the result of the working of this scientific engine in the improvement of the analytic engine available for individual scientists. However, there were probably more important reasons for his use of the expression 'the capitalist engine'. In any case, his use of the engine metaphor in *Capitalism* spread from economic evolution to political evolution and aspects of general social evolution. Thus he spoke of "the political engine" (see Section 7.4) and "the rationalist engine" (see Section 7.5). Nevertheless, it is the capitalist engine that serves to express his central concern. It is also this engine that is at the centre of attention in the present chapter and the next two chapters of the present book. Therefore, it is helpful to relate parts of Schumpeter's evolutionary theory to his powerful metaphor.

The type of evolution produced by Schumpeter's capitalist engine is not adequately covered by terms like 'growth' or 'progress'—since these terms ignore the painful transformation of the economic system. It is even less acceptable to depict the functioning of the capitalist engine as if it was simply 'designed' to provide the idle running of the stationary circular flow. The predominant emphasis on this idle running motivates studies of "how capitalism administers given structures". It also turns attention toward the fine-tuning of the capitalist economy according to the welfare criterion of Pareto efficiency. Against these predominant perspectives for the study of capitalism, Schumpeter used his idea of "creative destruction". As we have seen in the above quotation, he argued that "the relevant problem is how it [capitalism] creates and destroys" what is normally conceived as "given structures". We have to recognise that we are facing a "process of industrial 'mutation'—if I may use that biological term—which incessantly revolutionizes the economic structure *from within*, incessantly destroying the old one, incessantly creating a new one" (p. 83).

Although the informal writing style of *Capitalism* allowed the use of the word "incessantly", Schumpeter emphasised in a footnote that "[t]hose revolutions are not strictly incessant". Instead, "they occur in discrete rushes which are separated from each other by spans of comparative quiet" (*Capitalism*, 83n). Thus, the working of the capitalist engine is either characterised by "revolution" or by the "absorption of the results of revolution". He considered this the essence of "what are known as business cycles". They are thus, in their primary form, an expression of the working of the engine of capitalism. We have already met the analytical scheme that covers this two-stage working of the engine on page 12 and in Table 5.2 on page 119. It can be summarised in the following way:

(1) A circular-flow system governed by the adapted routine behaviour of the mass of agents
(2) is challenged by the innovative behaviour of Schumpeterian entrepreneurs,
(3) but sooner or later the selective adaptation of mass behaviour will establish a *new* system of routines in which the S-entrepreneurs and their innovations have been absorbed;
(4) and then the story starts once more.

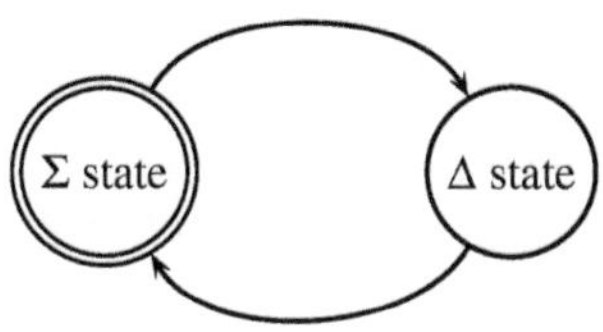


**Figure 6.2.: The abstract capitalist engine with a two-stroke cycle**

Comment: The mechanism of innovation (the $\delta$ mechanism) moves the economic system from a stationary ($\Sigma$) state to its maximally disequilibrated ($\Delta$) state. The mechanism of adaptation (the $\sigma$ mechanism) moves the system back to a $\Sigma$ state. In this state, previous innovations have been absorbed in an equilibrated system of economic routines.

The "incessant" repetition of this process demonstrated for Schumpeter that the "capitalist process, not by coincidence, but by virtue of its mechanism, progressively raises the standards of life of the masses" (*Capitalism*, 68). Since he did not believe in any well-defined measurement of "the standards of life", he turned to the common sense of that term. Thus he pointed out that the "capitalist achievement does not typically consist in providing more silk stockings for queens but in bringing them within the reach of factory girls in return for steadily decreasing amounts of effort" (p. 67). This was the result of the painful workings of the capitalist engine by means of a mechanism of innovation and a mechanism of adaptation. The integrated reading of the evolutionary trilogy suggests that he mainly conjured the image of a two-stroke engine in which these two mechanisms work sequentially (see Figure 6.2). The cycle of this engine starts from a stationary circular flow. The fuel for the propulsive stroke is provided by innovations while no further energy is needed for the reactive stroke that brings the engine back to an adapted circular flow.

Figure 6.2 applies a strange notation. The stationary state of the economic system is called the $\Sigma$ state while the maximally disequilibrated state of the economic system is called the $\Delta$ state. Furthermore the mechanism of innovation that moves the system from a $\Sigma$ state to a $\Delta$ state is called the $\delta$ mechanism; and the mechanism of adaptation that moves the system from a $\Delta$ state to a renewed $\Sigma$ state is called the $\sigma$ mechanism. Finally, the $\Sigma$ states are characterised by a double circle while the $\Delta$ state are characterised by a single circle. This pictorial part of the notation simple emphasises that the evolutionary process can stop in any $\Sigma$ state while a $\Delta$ state cannot serve as the terminal state of the process. The use of small Greek letters to denote the evolutionary mechanisms serves several purposes. One of them is to emphasise that the mechanisms can be specified in a number of different ways. For instance, the $\delta$ mechanism does not have to be specified by the Schumpeterian mechanism of economic innovation. It can also be used to denote the mechanism of innovation in other social sectors and for replacing Schumpeter's mechanism by other mechanisms with a similar function. The use of large Greek letters to denote the states of the evolutionary process serves similar purposes. Furthermore, by avoiding any immediate reference to Schumpeter's definitions of these states, it becomes easy to recognise that these definitions are not fully adequate.

Figure 6.2's specification of the two-stroke cycle in terms of two mechanisms emphasises that Schumpeter was focussing on the transformation of the economic system as a whole. This becomes especially clear when we try to use his metaphor to characterise the waveform evolution that he thought is underlying business cycles. The propulsive stroke can be identified as an innovation-induced 'upswing' and the reactive stroke as 'downswing" that performs the process of adaptation, which has become necessary because of the innovations of the upswing. The idle state of the circular flow can be called the $\Sigma$ state of the capitalist engine while the maximally disequilibrated state produced by the upswing is the $\Delta$ state. It is not difficult to recognise that a capitalist society whose economic life is dominated by the workings of such an engine is characterised by serious socio-political conflict. Most of its members would prefer that the engine stayed near the $\Delta$ state or that the difficulties of the selective and adaptive return to the $\Sigma$ state could be avoided. However, attempts to fulfil these preferences would hinder the working of the capitalist engine and the result would be that the system does not move through a series of different $\Sigma$ states but stays in a once-and-for-all given $\Sigma$ state.

The fact that the two-stroke engine is a metaphor for Schumpeter's evolutionary interpretation of history can best be recognised by adding time to its sequence of states. This brings out the historical dynamics of the capitalist engine (see Figure 6.3 on the facing page). Economic history is basically described as a series of stationary circular flows, which we can denote $\Sigma_0, \Sigma_1, \ldots$. These states are qualitatively different and cannot be satisfactorily compared by any measure of economic growth. Nevertheless, the two evolutionary mechanisms that bring the capitalist engine from one idle state to the next one have characteristics that produce 'progress' in a very loose sense. The mechanism of innovation (the $\delta$ mechanism) presupposes that entrepreneurial projects have characteristics that in some sense are better than some of the routines of the previous circular flow. Furthermore, the mechanism of adaptation (the $\sigma$ mechanism) also secures some degree of localised 'progress'. However, the question whether the result of overall change is an improvement is an empirical one that does not allow any easy social consensus or any clear-cut welfare theory in the style of Pareto. Nevertheless, it is, without any

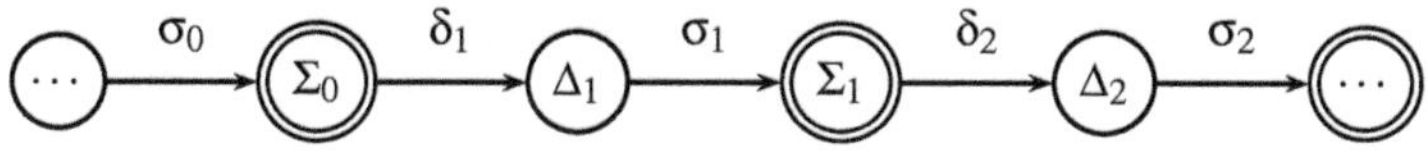


**Figure 6.3.: The historical dynamics of the capitalist engine**

Comment: The long-term functioning of capitalist engine produces a stylised version of economic history. The movement from $\Sigma_0$ to $\Sigma_1$ and to later states represents the evolution of the routine-based economic system. It cannot be described as unconditional "progress" since the evolutionary process involves both creation and destruction. The great question is whether the two mechanisms of evolution—the innovative $\delta$ mechanism and the adaptive $\sigma$ mechanism—change over time, i.e. whether $\sigma_0 \neq \sigma_1 \neq \sigma_2$ and $\delta_1 \neq \delta_2$.

value judgement, possible to describe the efficiency with which the two mechanisms are working. The mechanism of innovation is influenced by the financial and social conditions for the Schumpeterian entrepreneurs; and the mechanism of adaptation is influenced by the combined functioning of the markets for products, labour, and finance.

A related question is whether and how the two evolutionary mechanisms are changing over time. This question has often been addressed in terms of economic policy, but Schumpeter's question was rather whether the mechanisms of the capitalist engine are changing by themselves. A similar question on the evolution of the evolutionary mechanisms has been addressed in evolutionary biology under the heading "the major transitions in evolution" (Maynard Smith and Szathmáry, 1997). In *Capitalism*, Schumpeter actually suggested that such a transition had taken place within capitalist economic evolution. His semi-formal analysis of economic evolution had hitherto assumed that innovations are carried out by creating of new firms and that the resulting profits are competed away by "perfect competition" in a very loose sense. Now he emphasised the historical transition from "the capitalism of perfect competition" to "big-business capitalism" (p. 107). Furthermore, this transition did not lead to the result expected by most economists. On the contrary, they had to recognise the "shocking" possibility "that big business may have had more to do with creating that standard of life than with keeping it down" (p. 82). Actually, he argued that the innovative activities of large firms imply a speeding up of evolutionary change.

The emergence of a new mechanism of innovation suggests the existence of two brands of the capitalist engine. The innovative mechanism of Mark I is based on the establishment of new firms by innovative entrepreneurs. In contrast, Mark II is based on the innovative activities of incumbent firms in their oligopolistic competition. Schumpeter seems to have been aware of the existence of the capitalist engine Mark II from the very beginning of his academic work. Nevertheless, he nearly exclusively developed his analytical work in terms of Mark I. One of the reasons seems to be that it is very difficult, and maybe impossible, to define a plausible Σ state for the Mark II engine. Without such a state he probably felt that he would have to discard the type of evolutionary theorising that is found in *Development* and *Cycles.* Instead of a two-stroke engine, he would have to think in terms of mechanisms of innovation and adaptation that incessantly work in parallel. This possibility suggests a much looser metaphor of the capitalist engine.

The distinction between the two versions of the capitalist engine is crucial for the rest of the present book. Therefore, it is important to recognise their basic characteristics:

- The capitalist engine Mark I. The driving force of this engine is the innovative projects of Schumpeterian entrepreneurs. These projects imply the establishment of new firms by means of borrowed money. The pioneering S-entrepreneurs operate on the background of an equilibrated system of economic routine. Their immediate followers as well as the economic activities of economic agents who directly or indirectly support them bring the system to its maximally disequilibrated state. Here innovation has come to a halt. The system is brought back to its equilibrium state by a competitive process of adaptation. The result is that the routine system has evolved.

- The capitalist engine Mark II. The driving force of this engine is the innovative projects of established firms. The firms are engaged in a process of monopolistic competition that focusses on new products, new processes, and new forms of organisations. These projects might require borrowed money by they can also be financed by the profits of the past. The degree to which this Schumpeterian competition implies a two-stage process is unclear. The easiest interpretation is that the mechanism of innovation and the mechanism of adaptation work in parallel. Furthermore, the adaptation to the innovations can, to a significant extent, take place within established firms. These features suggest a significant increase in speed of the evolution of the routine system.

In *Capitalism,* Schumpeter did not mention the Mark I model. Instead, he simply jumped to Mark II and left to his readers to recognise that it is radically different from the Mark I model of *Development* and *Cycles.* One of the implications of basing *Capitalism* on Mark II with its simultaneous processes of innovation and adaptation is that Schumpeter could avoid dealing with the complexities of waveform economic evolution. Instead, he could concentrate on the basic features of any model of the capitalist engine. These features are the mechanism of innovation and the mechanism of adaptation. Nevertheless, he hardly abandoned the idea of a two-stoke engine in which these mechanisms works sequentially. The reason is probably that it is easier to analyse the consequences of process of economic evolution in the Mark I model.

The problem of Schumpeter's two models of the capitalist engine is, as already emphasised, not the only one that confronts any attempt to study his evolutionary trilogy as a whole. Actually, we have to confront so many difficulties that a rational reconstruction that ignores the fact that the trilogy consists of three relatively independent books does not seem warranted at the present state of development of the Schumpeter literature. This is the reason why each of the following three chapters concentrates on a single book. If the traditional treatment of the books had been followed, these chapters would have presented them in chronological order. Thus we would have moved from *Development* via *Cycles* to *Capitalism*. However, the present book presents these books in reverse order. A reading that moves from *Capitalism* via *Cycles* to *Development* has several advantages. For instance, we start with Schumpeter's most popular book. Furthermore, we consider his historical and theoretical accounts for the waves of economic evolution before arriving at the most abstract part of the trilogy. Another advantage of the reverse exposition of the books of the evolutionary trilogy is that the comprehensive nature of the sketchy evolutionary modelling in *Capitalism* can help to clarify the rather narrow assumptions on which Schumpeter constructed the evolutionary models of *Cycles* and *Development*. Nevertheless, the major advantage of reading the evolutionary trilogy backwards is that is that we thereby clearly recognise that Schumpeter's evolutionary economics is an open-ended research programme and that he left its results in unfinished shape.

# 7

# The Capitalist Engine and Socio-Political Evolution

Schumpeter's *Capitalism, Socialism and Democracy* was published in 1942—after he had published *Business Cycles* in 1939. These books contrast sharply. While *Capitalism* can be read fairly easily by interested laymen, even experts find that *Cycles* is hardly readable in its totality. *Cycles* represents Schumpeter's research efforts during the 1930s to integrate theory, history, and statistics in order to explain business cycles as reflecting the process of capitalist economic evolution. Its architecture can be compared with that of a Gothic cathedral: out of an initially simple structure emerges an enormous degree of complexity. In comparison, *Capitalism*'s treatment of the topic of the destiny of capitalism has a functionalist architecture.

The preparation of the relatively straightforward treatment of the problems of capitalism started in parallel with the production of *Cycles* and it was completed between 1939 and 1942. Therefore, *Capitalism* has sometimes been interpreted as the outcome of Schumpeter's need to relax. This interpretation explains the writing style and the provocative formulations that have served to make *Capitalism* his most-read book. It also explains why Schumpeter defensively told Paul Samuelson (2003, 465) that he had produced an "off-the-cuff pot-boiler"—that is, a rough sketch that is written to make money. The famous economist Lionel Robbins drove the idea of relaxation to the extreme by suggesting, in a conversation, that the book provides "supremely intelligent after-dinner talk" (quoted by Elster, 1983, 112). Nevertheless, Schumpeter's *Capitalism* does not reflect quickly generated ideas like those that are expressed after a good conference dinner. Instead, he had for more than thirty years been thinking about the problem of the destiny of capitalism as well as of his way of approaching this problem analytically. Furthermore, he wanted to express his results about the transformation of the capitalist society into something else as clearly as possible. We should exploit these facts through a serious study of *Capitalism*.

The early Schumpeter had sketched out a general theory of economic and social evolution in the last chapter of *Entwicklung I* (see Section 4.4). This approach to intra-sectoral evolution and inter-sectoral co-evolution can be considered as expressing his entrepreneurial interpretation of history. In contrast to the rigid causality of Marx's economic interpretation of history, Schumpeter's interpretation suggests that overall "socio-cultural evolution" (*Entwicklung I*, 545) is the outcome of relatively independent evolutionary processes within the individual sectors. This relative interdependence helps to explain why *Capitalism*'s clarification of the great problems related to the destiny of capitalism is constructed as a series of loosely interconnected essays. The essayistic style is thus not only a sign of relaxation after hard work. It also reflected the difficulty of describing in an integrated way the major institutional changes that had occurred in capitalist societies in the first forty years of the twentieth century. Schumpeter wanted to be explicit about two related processes of institutional change. On the one hand, he wanted to analyse the relatively independent institutional evolution *within* the economic sector that had turned classical capitalism into something like big-business capitalism. On the other hand, he wanted to confront analytically the deep-going institutional change that resulted from the co-evolution between the economic sector, the political sector, and other sectors of social life. His analyses can be seen as combining his special type of evolutionary economics with what may be called evolutionary political science and evolutionary sociology. The present chapter covers his contributions to all these emerging disciplines, but the focus is on Schumpeter's evolutionary economics.

## 7.1 Two ways of reading *Capitalism*

The remarks above suggest that there are two main ways of reading *Capitalism*. The first way is to focus on its structure and overall argument. According to this reading, the evolutionary theorising found in the book is only relevant to the extent that it promotes the argument on the destiny of capitalism. This reading is well represented by the title *Capitalism, Socialism and Democracy*. The second way of reading this book is to focus directly on its evolutionary theorising. According to this reading, *Capitalism* is applying, modifying, and extending the theorising found in Schumpeter's previous works. This reading can be supported if we largely ignore the title he chose for his book and instead think of it in terms of 'Capitalist Economic Evolution and Its Socio-Political Limits'.

The reading of *Capitalism, Socialism and Democracy* that emphasises its title is well supported by the structure of the book. *Capitalism* is organised as five relatively independent parts on (I) Marx, (II) capitalism, (III) socialism, (IV) democracy, and (V) socialist parties (see Table 7.1 on page 158). The introductory part consists of an extensive account of Marx's work, and it serves to prepare Schumpeter's readers to think about the evolution and the destiny of capitalist society in the most general terms. Since we have already considered core elements of this analysis, we can move directly to Schumpeter's response to the Marxian challenge. This response starts from two questions. The question for Part II is "Can capitalism survive?" and the answer is "no". The question for Parts III–IV is "Can socialism function?" and the answer is "yes". Finally, Part V gives a historical account for the evolution of socialist parties, which can be ignored since it contributes little to the overall argument.

Schumpeter's presentation of his analysis was made in a way that served to force his readers—critics and defenders of capitalism alike—to rethink their own preconceptions. The way of doing so was to show, in a more or less ironic manner (see Muller, 1999), that their preconceptions were more or less unfounded. First, since his answers to the basic questions were the same as those of the neo-Marxists, Schumpeter could expect them to follow his argument and find out that their reasons for these answers were highly problematic and that they would rather achieve dystopia than utopia. Second, Schumpeter addressed his irony to Keynes and the American Keynesians. They had expanded the experiences of Britain and of the depression and weak recovery of the 1930s into a general theory of "vanishing investment opportunity"; and they seemed happy to administer a stagnating capitalist economy. Schumpeter suggested that such an economy might be even less efficient than the stagnating economy that would ultimately be the result of socialist management. Third, his "prognosis" of the death of capitalism and the feasibility of socialism could provoke the defenders of capitalist society to rethink their naïve preconceptions of societal stability. This aspect of Schumpeter's implicit message had been ignored by many of his readers. In the preface to the second edition of *Capitalism* (*S1947b*), he commented on "the charge of 'defeatism' " against his overall argument. He pointed out that the real danger for those defending capitalism is "escapism" rather than confronting the challenges to the system. He also pointed out that "this is not a political book" that has to confront this type of issues. Nevertheless, he wanted to force all his readers to think creatively about the great problems of capitalism rather than to follow accustomed modes of thought.

Schumpeter's provocative conclusion that capitalism cannot survive is, of course, based on the analyses presented in *Capitalism*. Therefore, it serves to draw attention to the structure of Schumpeter's argument on the destiny of capitalism, which is presented in Part II of *Capitalism*. The argument begins by stating the conclusion and the main thesis:

> "Can capitalism survive? No, I do not think it can. ... The thesis I shall endeavor to establish is [1] that the actual and prospective performance of the capitalist system is such as to negative [negate] the idea of its breaking down under the weight of economic failure, but [2] that its very success undermines the social institutions which protect it, and 'inevitably' creates conditions in which it will not be able to live and which strongly point to socialism as the heir apparent." (*Capitalism*, 61)

**Table 7.1.: *Capitalism, Socialism and Democracy* (1942; 1950)**



Comment: The standard interpretation of *Capitalism* as a set of essays on the destiny of the capitalist system is clearly supported by its table of contents. However, the book can also be interpreted as providing an extension of Schumpeter's evolutionary economics and sketches of his evolutionary sociology and evolutionary political science. Part II contains famous analyses of the dynamics of big-business capitalism and the increasing resistance against it due to socio-political evolution. The potential result of these processes are discussed in Part III, while Part IV contains Schumpeter's theory of democracy.

Schumpeter's thesis is thus that it is not the capitalist economy but the system of socio-political institutions that determines the destiny of capitalism. His arguments for the two propositions of his thesis are not equally developed. Since he had spent most of his research efforts on the first proposition, his arguments for the economic proposition are more convincing than his arguments for the institutional proposition. However, the interpretation of *Capitalism* as a book on the destiny of capitalism has to put the main emphasis on the relatively weakly defended proposition that the economic success of capitalism undermines its necessary institutional setting. It also has to explore in detail his even weaker defence for the feasibility of socialism and for potential compatibility with democracy. Nevertheless, if this reading is performed carefully, it clearly demonstrates that Schumpeter argued that socialism would lead to dystopia rather than to utopia. This reading also demonstrates that he argued that a new round of capitalist economic evolution perform much better in terms of raising standards of living than had been the case during the 1930s.

The alternative way of reading *Capitalism* is to focus on its applications, modifications, and extensions of the evolutionary theorising that is found in Schumpeter's previous works. This reading suggests that the two mentioned propositions relate, respectively, to economic evolution and to socio-political evolution. With respect to economic evolution, the proposition is that "the capitalist engine will—or would if allowed to do so—work on in the near future, say for another forty years, about as successfully as it did in the past" (*Capitalism*, 110). With respect to socio-political evolution, the contrasting proposition is that the capitalist system not only transforms "its own institutional framework but it also creates the conditions for another" (p. 162). To develop the arguments for these propositions, Schumpeter was forced to sketch extensions of his previous analyses of the mechanisms of economic and socio-political evolution. He even provided a sketch of the mechanisms of political evolution in democratic societies. This sketch is of major importance for the arguments on institutional change within capitalist societies in Part II of *Capitalism*. Actually, major elements of the treatment of democracy in Chapters 21 and 22 of Part IV (see Table 7.1 on the facing page) clearly belongs to the treatment of capitalist societies. The present account focusses on these parts of *Capitalism* while the rest of the book is largely ignored.

It seems likely that Schumpeter wanted *Capitalism* to be read not only as an essay on the destiny of capitalism but also as a defence for and development of his type of evolutionary analysis. Although this wish might have influenced his demonstrations that this type of analysis can be extended into sociology and political science, it will suffice to give an example of his promotion of evolutionary analysis within economics. We have in Chapter 3 seen that he considered economics at dividable into equilibrium economics and evolutionary economics. Although Schumpeter had tried to develop the latter branch, economics had during the 1930s become dominated to an even larger degree than before by equilibrium and quasi-equilibrium modes of thinking. Now he criticised this predominant style of economic analysis for studying "competition within a rigid pattern of invariant conditions, methods of production and forms of industrial organisation" (*Capitalism*, 84). This predominant analysis made clear that

> "the problem that is usually being visualized is how capitalism administers existing structures, whereas the relevant problem is how it creates and destroys them. As long as this is not recognized, the investigator does a meaningless job. As soon as it is recognized, his outlook on capitalist practice and its social results changes considerably." (*Capitalism*, 84)

The type of capitalist practices that Schumpeter especially focussed on in *Capitalism* are "monopolistic practices". This fact helps to explain Joan Robinson's reaction. It was probably on the background of her own pioneering work on the theory of imperfect competition that Robinson (1943, 382) remarked that Schumpeter's argument on monopolistic practices "blows like a gale through the dreary pedentary [pedantry] of static analysis". However, she also remarked that the effect of the book is increased by the fact that it "is arranged on the plan of a detective story" in which "none of the obvious suspects are guilty" of killing capitalism (p. 382). This arrangement, as well as the richness of theoretical accounts, means that "the reader is swept along by the freshness, the dash, the impetuosity of Professor Schumpeter's stream of argument." Robinson concluded that "no matter whether it convinces or not, this book is worth the whole parrot-house of contemporary orthodoxies, right, left, or centre" (p. 383). Nevertheless, the later discussion of *Capitalism* has been dominated by an irresistible urge to compare Schumpeter's "prediction" with later events. This urge even dominated a volume celebrating *Capitalism* with 40 years of hindsight and with contributions by, for instance, Paul Samuelson, Gottfried Haberler, Arthur Smithies, and Robert Heilbroner (in Heertje, 1981). As Herbert Gintis (1983, 85) remarked in his review of that book, this kind of assessment is a major error: "We have long abandoned the attempt to assess Marx' social theory on the basis of his 'predictions', and would never conceive of applying such an approach to Weber, Ricardo, Smith, or Marshall. Why then apply such a criterion to Schumpeter?" Furthermore, "Schumpeter can be taken seriously only within the context of a commitment to the development of a science of social change"; but most of his economic and sociological commentators have not even recognised "that there *is* such a science". In contrast, Schumpeter assumed that it is possible to detect the basic mechanisms and tendencies that explain the transformation of capitalist societies.

The analytical strategy applied in the present chapter is to study the evolutionary mechanisms on which Schumpeter based this story of the destiny of capitalist society. This strategy is in some ways similar to the one Schumpeter applied in his extensive account for Marx's work. Here Schumpeter emphasised that he did not consider Marxian thinking as an indivisible whole. On the contrary, he made a decomposition of Marx's work into that of an economist, a sociologist, a teacher and a prophet (*Capitalism*, Part I). If we make a similar decomposition of Schumpeter's contribution, and if we ignore his roles as teacher and prophet, we arrive at a picture of a researcher who is more sober and operational than it appears from a first reading. Furthermore, the present analytical strategy also implies that we should resist the temptation of integrating the Schumpeterian contributions to evolutionary economics and evolutionary sociology. Even in this respect we follow Schumpeter who emphasised that "Marx's system illustrates well that, though synthesis may mean new light, it also means new fetters" (*Capitalism*, 45). Synthesis means "coordination of the methods and results of different lines of advance" and it might easily lead to "worse economics and worse sociology" (p. 46). Schumpeter's solution was to cut the Gordian knot of Marxian synthesis and study its individual components separately. His own study of the evolution of capitalist society tried to avoid producing a similar Gordian knot. We have to do similarly. The reason is that although Schumpeter made careful analytical separations, he also sketched a reasoned history of capitalist society based on his general theory of social evolution. Such an effort is necessarily of a synthetic nature. This synthesis has to be de-emphasised in order to move attention to his analysis of individual mechanisms of capitalist evolution.

## 7.2 Mark II of the capitalist engine and its implications

Schumpeter's analysis of how the capitalist economic system "creates and destroys" existing structures represents an important example of what William Baumol (1959, Part I) called "magnificent dynamics". This type of dynamic analysis is obviously far removed from the formal analysis of the change of non-evolving economies that Baumol tried to promote through his own book on *Economic Dynamics*. However, he later approached the problems of economic transformation in a book on capitalism as an "innovation machine". Baumol (2002, p. x) emphasised that, when developing this book, he was "unable to find anything that deals *directly* with my subject" in Schumpeter's work "with the exception of very brief discussions". This comment demonstrates that even Baumol follows the all-dominant practice of focussing on a few of Schumpeter's statements in *Capitalism*. In contrast, the present book tries to demonstrate that major parts of this book, and of several of his other works, are closely related to Baumol's topic. Actually, it seems clear that Schumpeter's main ambitions were to depict capitalism as an engine of innovation and evolution as well as to study the prospects for this engine under the conditions created by the evolution within other social sectors.

*Capitalism*'s analysis of prospects for the future working of capitalist economic evolution can be phrased in terms of the metaphor of the capitalist engine (see Section 6.3). For this analysis, Schumpeter needed a model of the capitalist engine that included the important feature of brakes. It is not difficult to put two brakes on the capitalist engine. The primary brake influences the mechanism of innovation. The secondary brake influences the mechanism of adaptation. The primary brake influences the motivation, expectations, and practical possibilities of the Schumpeterian innovators. The secondary brake influences the response to the challenge provided by innovation. The function of this secondary brake is to reduce the pressure on the firms that are threatened by innovative change. However, this pressure can also be reduced by using the brake on the mechanism of innovation. The inducement to use the two brakes of the capitalist engine is suggested by Schumpeter's emphasis on "the process of creative destruction which we have taken to be the essence of capitalism" (p. 104n). The economic evolution of the routine system consists of a series of routinised equilibria and innovative disturbances that challenges given routines. Thus it is clear that Schumpeter was thinking in terms of two related concepts: "creative destruction" is the selecting out of firms or their routines by the pressure from an innovation and "the process of creative destruction" is a combination of this kind of selection and the innovative activities that drives the process. However, he *never* used the term 'creative destruction' in isolation. Instead, he used the term "the process of creative destruction" (*Capitalism*, 81, 83, 87, 89, 90, 91, 95, 96, 101, 104, 105, 194) and, as a synonym, "the perennial gale of creative destruction" (pp. 84, 87, 88, 90).

It was through his vision of "the process of creative destruction" that Schumpeter effectively pushed aside standard ideas about economic change and 'progress'. Economic evolution is not a simple growth process in which all sectors of economic life expand in a balanced way. Neither is it a process that can be described by the standard model of perfect competition. It is instead characterised by the creation of novelty and the destruction of old products and processes. Furthermore, many of the existing firms and other organisations do not smoothly upgrade their competencies and switch their areas of specialisation. They instead often perish in the evolutionary process. Finally, employees who lose their jobs are often facing great stress and significant welfare losses that seem more obvious than the long-term advantages of capitalist evolution. However, by using "the process of creative destruction" as a synonym for "capitalist evolution", Schumpeter emphasised that the benefits of this evolution cannot be obtained without losses and social inequalities. This provocative proposition seems to have been the starting point for his analysis of socio-economic evolution in the capitalist society. Since reactions against "creative destruction" cannot be avoided, capitalism is characterised by an essential conflict between innovative initiatives and old routines as well as the derived conflict between the defenders of the capitalist process and those who want less stressful conditions.

The above account for the capitalist engine and its problems in functioning within a socio-political context that put increasing emphasis on predictability and social security is not simply a summary of major arguments in *Capitalism*. Instead, it is based on the reading of this book as part of Schumpeter's evolutionary trilogy. The major novelty in *Capitalism* is the development of the argument that the long-term socio-political trend in capitalist societies points toward the increased use of the brakes of the capitalist engine. Schumpeter suggested that only short-term benefits can be obtained in this way. According to him, even the Keynesian-type interference with the capitalist engine could have the long-term consequence of bringing it to a full stop. More radical interventions of the socialist type would produce the same result without providing an alternative engine of economic change that would function in the long run. Schumpeter wanted both the critics and the defenders of the capitalist engine to understand the consequences of hindering its working. However, his use of metaphors to depict this machine of evolutionary change serves to emphasise why he described *Capitalism* as a "pot-boiler". Nevertheless, his best-seller was not only intended for the mass market. It also served to present the integrated evolutionary analysis of economic and social affairs as precisely as he was able to do. Thereby he seems to have been calling for further research.

It was probably Schumpeter's writing of the historical parts of *Cycles* that convinced him of the necessity of treating the Mark II model of the capitalist engine. He had tried to summarise the history of 300 years of capitalist evolution in terms of Mark I; but he had to recognise that incumbent firms are at least the major source of incremental 'innovations'. Since these changes can be described as an adaptive response to given circumstances, they cannot be called Schumpeterian innovation in the narrow sense. However, during the twentieth century, the established firms had increasingly become engaged in a broad range of innovative activity. Thereby they had emerged as the major driver of economic evolution. He considered his model of entry-based innovation as representing "Competitive Capitalism" of the past; now he also had to recognise the emergence, especially from the 1890s, of "Trustified Capitalism". He emphasised that "[e]conomic evolution or 'progress' would differ substantially from the picture we are about to draw, if that form of organization prevailed throughout the economic organism" (*Cycles*, 96; *Cycles Abr.*, 71). At that point of the argument, he emphasised that "giant concerns" do not "dominate the picture in any country." Furthermore, "[e]ven in the world of giant firms, ... [i]nnovations still emerge primarily with the 'young' ones, and the 'old' ones display as a rule symptoms of what is euphemistically called conservatism" (*Cycles*, 97; *Cycles Abr.*, 71). Although he hardly gave up the latter view, he at the end of *Cycles* (1044; *Cycles Abr.*, 417) stated that "economic 'progress' in this country [the USA] is largely the result of work done within a number of concerns at no time much greater than 300 or 400" and that "any serious threat to the functioning of these will spread paralysis in the economic organism". Here his emphasis has moved from innovative entry to innovation within dominant incumbent firms. This shift was radicalised and completed in *Capitalism*.

In his most popular book, Schumpeter did not state explicitly that he moved from the Mark I model of the capitalist engine to his Mark II model. Therefore, it is not surprising that quite some confusion emerged among his readers. This confusion could have been avoided if it had been recognised that he had been interested in both models at least since the Böhm-Bawerk Seminar of 1905. At that time, he was confronted with Marx's theories of the concentration and centralisation of capital. This theory was largely based on the competition between incumbent firms. Although it was not stated in such terms, it is obviously the evolutionary process "which compels capitalists to accumulate irrespective of what they feel about it" (*Capitalism*, 30). This imperative is due to the fact that "the profit of each plant is incessantly being threatened by actual or potential competition from new commodities or methods" so that it will "sooner or later turn into a loss" (p. 33). To avoid this threat, profits are used for the expansion and improvement of the plant and the firm. This leads to "the tendency of the capitalist process to increase the size both of individual plants and of units of control." Marx's very early "prediction of the advent of big business" (p. 34) was not guarded by a careful analysis of the underlying process. Furthermore, Schumpeter was not convinced by the related analysis of "finance capital" by Hilferding (1981), because it overemphasised the function of banks and underestimated the instability of modern capitalism. However, the emergence of "big business" was becoming a major issue in the early years of the twentieth century, and this influenced both the early Schumpeter and the later Schumpeter. Therefore, it is not correct to consider the Schumpeterian idea of competition between established firms as having emerged around 1940.

Schumpeter's analysis of the monopolistic aspect of economic evolution and its tendency to grow in importance has three major elements. First, the temporary existence of monopolistic positions is an essential element in any capitalist economic evolution since the entrepreneurs are motivated by the profits that can only be reaped during the period in which their innovations give them some monopoly power. Second, the process of capitalist evolution includes organisational innovation that in more and more industries imply that big business becomes predominant. The large corporations sustain their competitiveness by a stream of innovations of different types. These innovations are largely produced in-house through a series of innovation projects by departments of research and experimental development. Third, the use of R&D departments exemplify that the emergence of large corporations has to a large extent solved the difficult problem of finance of innovation. This problem had been a primary one for the S-entrepreneurs of Mark I. The reason is that they need external finance from banks, and these banks not only perceive a conflict of interest with respect to the sharing of the gross profit from the innovative project. They also have to recognise that the entrepreneurs are normally better informed about the innovative projects than themselves. This problem has not been treated systematically before the emergence of the modern economic theory of asymmetric information (by Joseph Stiglitz and others); but even while Schumpeter was developing his Mark II model, it was clear that the (partial) finance of innovation by the internal resources of large corporations provided a solution to the information problem and the problem of conflict of interest. If we add that even a corporation with an apparent monopoly is normally facing a dynamic form of "workable competition", we have to rethink our conception of the relationship between innovative change and the strategy of "the large-scale establishment or unit of control":

> "What we have got to accept is that it [the large firm] has come to be the most powerful engine of ... progress and in particular of the long-run expansion of total output not only in spite of, but to a considerable extent through, this strategy which looks so restrictive ... In this respect, perfect competition is not only impossible but inferior, and [it] has no title to being set up as a model of ideal efficiency." (*Capitalism*, 106)

This proposition means that "monopoly" is not a brake on the evolutionary process but instead its promoter. Schumpeter emphasised four arguments against the application of the perfect competition assumption of equilibrium economics for the analysis of economic evolution. First, "perfectly free entry into a *new* field may make it impossible to enter it at all" (*Capitalism*, 104–5). The reason is that the perfect promptness of entry would destroy the possibility of profit. The alternative is not classical monopoly. Instead, the realistic picture suggests that the first innovative entrant is threatened by further entry in a way that seriously limits its possibility "of practices that aim, through restricting output, at conserving established positions and at maximizing the profits accruing from them" (p. 87). Second, the standard criticisms against "rigid prices" does not hold. Actually, "perfect and instantaneous flexibility may even produce functionless catastrophes" (p. 105). More sticky prices serve to avoid wild fluctuations, while repeated innovations explain why "there are no major instances of long-run rigidity of prices" (p. 93). Third, the perfectly competitive model secures that profits vanish instantaneously and thus it leaves no real room for "business strategy". However, "in the process of capitalist evolution these profits acquire new organic functions" (p. 105). Actually, the instantaneous vanishing of profits would stop capitalist evolution. Fourth, "a perfectly competitive economy is comparatively free from waste", but this result is obtained by "excluding the most characteristic features of capitalist reality". It is not obvious that perfect competition produces less "waste" than "the process of creative destruction". On the one hand, much of what looks like "waste" is an essential part of the evolutionary process. For instance, what looks like "excess capacity" might be the capacity that is needed at the next peak of the business cycle. On the other hand, we see major "wastes of opportunities" when industries that comes close to "the perfectly competitive arrangement" are studied from an evolutionary viewpoint (*Capitalism,* 106).

In *Capitalism,* Schumpeter largely studied "the process of creative destruction" and the reactions against it in relation to the oligopolistic practices of large firms. In this context, he emphasised that "[e]very piece of business strategy acquires its true significance only against the background of that [evolutionary] process and within the situation created by it" (*Capitalism,* 83–4). Therefore, business strategy "must be seen in its role in the perennial gale of creative destruction; it cannot be understood irrespective of it or, in fact, on the hypothesis that there is a perennial lull." These statements are apparently contradicted by Schumpeter's sociological proposition that capitalist economic evolution gradually produces a system that is less creative and less destructive. He had already formulated this proposition in his paper in the *Economic Journal* of 1928. Here he suggested that innovation "becomes 'automatised', increasingly impersonal and decreasingly a matter of leadership and individual initiative" (*S1928f,* 71). The reason is that while "[i]nnovation in competitive capitalism is typically embodied in the foundation of new firms ... this is different in 'trustified' capitalism" (p. 70). Here innovation is typically performed "within the big units now existing, largely independent of individual persons." In this setting, innovation "tends to be carried out as a matter of course on the advice of specialists" and "by taking a long-term view". However, it became clear in *Capitalism* that this transformation of the functioning of the capitalist economy was the outcome of "the process of creative destruction". Big-business capitalism was the result of organisational innovations that largely competed away the simple evolutionary mechanisms of competitive capitalism.

The shift toward this idea of the obsolescence of individual entrepreneurship had been prepared by remarks in *Entwicklung II* that are included in *Development*. In an important passage, Schumpeter tried to deduce the conclusion that "the importance of the entrepreneur must diminish just as the importance of the military commander has already diminished" (*Development*, 86). He arrived at this conclusion by a logical syllogism about the obsolescence of the entrepreneurial function. The first premise is that "[t]he more accurately ... we learn to know the natural and social world, the more perfect our control of facts becomes" and "the more the significance of this [entrepreneurial] function decreases". The second premise is that "time and progressive rationalisation" actually create a situation "within which things can be simply calculated, and indeed quickly and reliably calculated". Therefore, the conclusion is that "the importance of the entrepreneur must diminish".

If we ignore the context of Schumpeter's syllogism, then it has rightly been criticised by Richard Langlois (2007, Ch. 2). In his section on "The Schumpeterian Tension", Langlois considers the syllogism as representing "a strange commingling of an empiricist and a rationalist theory of economic knowledge." The empiricist theory of entrepreneurship emphasised "the impossibility of surveying exhaustively all the effects and counter-effects of the projected enterprise" (*Development*, 85). The rationalist theory of a transformed society suggests that this impossibility can be overcome. However, the rationalist theory cannot hold for a society that is characterised by an open-ended evolutionary process. Therefore, the second premise is either false or is based on a transformation towards a society characterised by a circular flow or harmonious growth. It is only in these unlikely cases that realistically produced knowledge and fully rational calculation would reach the same result. The context of Schumpeter's syllogism demonstrates that he must have been thinking of such cases. He was arguing that innovative leadership is needed because of the nature of the task of performing something novel, the psychological requirements needed for fulfilling this task, and the fact that novelty is met by hostile reactions from the social environment. In the context of this argument, he had to recognise three reasons why "there would be no special function of leadership as distinguished from routine work" (*Development*, 87–8). Innovative leadership would *not* be needed (1) "[i]f social life had in all respects the relative immutability of, for example, the astronomical world" or (2) if "mutability were yet incapable of being influenced by human action" or (3) if the type of action needed for mutation were "equally open to everyone" (p. 87). Since Schumpeter did not believe in any of these conditions is fulfilled by the capitalist economy, which he in *Capitalism* characterised by "the process of creative destruction", his arguments about the vanishing of the entrepreneurial function have to be studied very carefully.

Schumpeter's section on "The Obsolescence of the Entrepreneurial Function" (*Capitalism*, 131–4) starts by mentioning "the possibility that the wants of humanity might some day be so completely satisfied that little motive would be left to push productive effort still further ahead" (p. 131). The result would be a "more or less stationary state" in which there "would be nothing left for entrepreneurs to do". Instead, "the management of industry and trade would become a matter of current administration". In such a society, capitalism "would become atrophic", "[h]uman energy would turn away from business", and "[s]ocialism of a very sober type would almost automatically come into being." Since "[f]or the calculable future this vision is of no importance", we have to consider a possibility that would affect "capitalist society nearly as much as the cessation of economic progress would." This is the possibility that "[p]rogress itself may be mechanized as well as the management of the stationary economy" (p. 131). Since Schumpeter followed Max Weber in assuming that such a mechanisation and rationalisation of economic change actually took place, he could formulate his view about the future of capitalism in terms of a new syllogism (p. 134). The first premise is that "if capitalist evolution—'progress'—either ceases or becomes completely automatic, the economic basis of the industrial bourgeoisie will be reduced eventually to wages such as are paid for current administrative work". The second premise is that "capitalist enterprise, by its very achievements, tends to automatize progress". The conclusion is that capitalism "tends to make itself superfluous—to break into pieces under the pressure of its own success." Since Schumpeter thought that the core element of this paradoxical type of success is organisational innovations that automatise progress, he could state his conclusion in provocative form: "[t]he true pacemakers of socialism were not the intellectuals or agitators who preached it but the Vanderbilts, Carnegies and Rockefellers" (*Capitalism*, 134).

The most controversial part of Schumpeter's syllogism is probably the second premise that "capitalist enterprise ... tends to automatize progress". Actually, this statement seems to have given Burton Klein (1977, 133) the "impression that there were two Schumpeters: Schumpeter the revolter against determinism, and Schumpeter the determinist". This impression, however, depends on an implausible interpretation of Schumpeter's proposition. Another understanding of this proposition can be obtained by studying the meanings of the terms "enterprise", "progress", and "automatize". 'Capitalist enterprise' cannot be taken in the sense of founding innovation-based firms since he also includes the innovative activities of established firms. 'Progress' is for Schumpeter a synonym for an evolutionary process in which both technology and preferences change in a largely unforeseeable manner. This means that statements about 'progress' do not imply full foresight and they do not have an easy interpretation in terms of social welfare. 'Automatize' relates to such an open-ended evolutionary process. Therefore, the automatising of 'progress' cannot mean that that its precise outcomes become predictable.

What become 'automatic' are, instead, the innovative activities of large firms: they tend to use a fixed proportion of their revenue for innovation and hire employees for performing this function. The result is that "capitalist enterprise ... tends to automatize progress" (*Capitalism*, 134) in a sense that is not found in Schumpeter's basic model of economic evolution. While this model depicted capitalist economic evolution as the outcome of the will of the S-entrepreneurs, the new model depicts economic evolution as the 'automatic' result of the oligopolistic competition of large firms. The evolution within the population of large firms has created research departments and marketing departments within what appears to be "perfectly bureaucratized giant industrial units" (p. 134). These large units "largely create what they exploit ... in the process of creative destruction" (p. 101). Since the increasing importance of the Mark II model does not represent a new brake on the engine of capitalism, we even for this model have to think in terms of two exogenous brakes: the brake on the mechanism of innovation and the brake on the mechanism of adaptation. Nevertheless, Schumpeter emphasised that the endogenous transformation of the capitalist engine from something close to Mark I to something increasingly dominated by Mark II increased the motivation to use the brakes as well as the possibilities of doing so.

## 7.3 Emergence of the capitalist engine and the tax state

Schumpeter's abstract analyses of the capitalist engine and its brakes were meant to complement his studies of the history of the capitalist economy and its relationship with the state. Although *Capitalism* is mainly concerned with the extrapolation of this history beyond 1942, it also gives hints about the beginning of the history of capitalism and the related state. If we add *Cycles* as well as several of Schumpeter's sociological papers (see *ESC*), we find a whole sketch of the history of capitalism from its beginning to the early 1940s. Thereby, we recognise that he was approaching the ideas of stages of capitalist evolution that had been proposed by Sombart, Spiethoff and other members of the German historical school. Even more than them, Schumpeter emphasised the evolution of the inner organisation of the capitalist economy as well as of the co-evolution between economic life and socio-political life under capitalism. This led to historical expositions that are partly covered below, partly in Section 8.2. *Capitalism*'s theme of the exogenous use of the brakes of the capitalist engine suggests that we should emphasise his analyses of the evolution of the socio-political system. Nevertheless, it is useful to start with a few remarks on his analysis of the emergence of the capitalist engine.

In contrast to Max Weber, Schumpeter did not believe in explaining economic behaviour of entrepreneurs and capitalists by "Puritan attitudes" and the related "abstaining from hedonist enjoyment of one's profits" (*Capitalism*, 30). He wanted to explain the long-term evolution of the capitalist system in terms of an endogenous, rather than an exogenous, "social psychology". This "psychology" can be derived from his already mentioned definition of "capitalism" as based on the carrying out of innovations "by means of borrowed money, which in general, though not by logical necessity, implies credit creation" (*Cycles*, 223; *Cycles Abr.*, 179–80). This definition implies that "we shall date capitalism as far back as the element of credit creation." Although Schumpeter did not develop the consequences in this connection, it is clear that a basic point was that credit freed the innovator from feudal constraints. These constraints meant that the social costs of the innovative enterprise were considered when it was decided whether or not the enterprise should be allowed. Since it was those who faced these costs who normally were dominant, innovations in feudal agriculture and handicraft were seldom accepted. However, the emergence of banks and other means of finance gave the possibility of ignoring the complex social consequences of the innovative project and concentrate on its isolated profitability.

The search for the emergence of credit creation brings us further back than the birth of Max Weber's "Protestant ethic". In the case of "Southern Europe this would carry us to the close of the twelfth century and the beginning of the thirtieth century" (*Cycles*, 224; *Cycles Abr.*, 181). The conflict of interest between banks and the S-entrepreneurs and their firms served to enforce increasingly strict calculations, and these calculations were much eased by the emergence of double-entry bookkeeping during the fifteenth century. This "element has been stressed, and *more suo* [in his usual manner] overstressed by Sombart" (*Capitalism*, 123n). However, this formalised type of bookkeeping was just "the last step on a long and tortuous road." The evolutionary process that led to credit creation and double-entry bookkeeping demonstrates that "there is no need to speak, as Sombart and others did, of a new 'spirit' (*Geist*) having come about somewhere in the stretch between 1400 and 1600" (*Cycles*, 228). Actually, such ways of speaking reflect non-evolutionary modes of thinking. What appears as "a big innovation hardly ever springs out of the current events as Athene did from the head of Zeus" (p. 227). Even such innovations have to be treated in accordance with "the principle of continuity". Schumpeter could have illustrated this principle of evolutionary thinking by the evolution from an antelope to a giraffe through incremental steps that never allow a shift in terminology. The ordinary form of classificatory theorising, "typological thinking", is defenceless against the problem of continuous transformation that can only be resolved by evolutionary analysis, "population thinking" (Mayr, 1976). Sombart and Max Weber ignored the evolutionary principle and imposed logical distinctions of typological thinking on the gradual transformation of reality

(*Cycles*, 228–30). Therefore, they saw this reality as making system-level jumps. Schumpeter, in contrast, accepted to find the primitive and impure forms of the logic of enterprise and its finance "in a small capitalist milieu which is surrounded by a noncapitalist world" (p. 225).

Schumpeter's analysis of the history of capitalism also included what he, in an early work, called "the tax state" (*S1918b*). This term emphasises that he approached the relationship between the political system and the economic system in terms of the mechanism of taxation needed by the state to finance its expenditures. His study of this mechanism was motivated by the fiscal difficulties of the state around the end of the First World War. At that time, he tried to confront the great question whether a state that is built on taxing the capitalist economy is sustainable in the long run. He answered this question in a booklet that in English translation is called "The Crisis of the Tax State". Here he related to the idea of the neo-Marxist Rudolf Goldscheid (1919) that the budget of the state is nothing but "the skeleton of the state stripped of all misleading ideologies" (quoted by *S1918b*, 100, emphasis removed). The translation of the paper in which Schumpeter implemented this idea was made by his students Wolfgang Stolper and Richard Musgrave. However, its analyses of the government's taxes and expenditures are very different from the economic theory of public finance that Musgrave was developing. The analysis of the history and structure of the tax state led to a mix of economics, sociology, and political science that is even more encompassing than the mix we find in the public choice theory of James M. Buchanan.

Schumpeter's starting point was that the "public finances are one of the best starting points for an investigation of society, especially though not exclusively its political life" (*S1918b*, 101). The "fiscal history" covers the "spirit of a people, its cultural level, its social structure, the deeds its policy may prepare— ... stripped of all phrases". In this perspective, we may in the public finances recognise "the thunder of world history more clearly than anywhere else" (p. 101). Under the conditions of massive war, the state budget obviously had a huge deficit but he argued that the capitalist tax state could survive for the time being (*S1918b*, 128, 130). Although "the state lives as an economic parasite", it can survive as long as it "withdraw from the private economy only as much as is consistent with the continued existence of this individual interest in every particular socio-psychological situation" (p. 112).

The "parasitic" origin of the tribute that the state extracts from the economy becomes clear from another paper that Schumpeter wrote at the end of World War I. In the paper on "The Sociology of Imperialisms" (*S1919b*), he emphasised the irrationality of imperialist wars. In the old warrior societies, the state emerged as a war machine; and this machine was, in principle, engaged in limitless conquest (pp. 156–79). Each conquest was expected to bring tributes from the newly suppressed populations, and these tributes were used to finance further expansion. This form of expansion for its own sake was inherited by the absolutistic monarchies of Europe (pp. 179–87); but the economic tributes tended to be much smaller than the costs of conquering foreign land, of defending it against rival imperialists powers, and of extracting foreign tribute. The huge financial needs of European states meant that they became increasingly dependent upon monetary taxation of the capitalist sectors of their economies, and this dependence pointed in a new direction (pp. 187–214). The reason is that the basic capitalist process is disturbed and not promoted by imperialist endeavours. Instead, capitalism points toward free trade and international alliances between firms. Although World War I seemed to contradict this proposition, Schumpeter argued that it is only when nationalistic protectionism becomes prevalent that some dominating firms combine with old-fashioned and expansionistic parts of the state apparatus to an imperialistic force. Such an imperialism is "atavistic in character" (p. 188), and it does not change the basic tendency of capitalist society. Instead, anti-imperialism must be expected "whenever capitalism penetrates the economy and, through the economy, the mind of modern nations" (p. 191). This penetration partly works through "democracy—in its 'bourgeois' sense" (p. 192).

When the imperialist model of public finance had lost its efficiency, it became clearer that the state depended on the efficiency of its fiscal apparatus as well as on the wealth of its own citizens and their willingness to pay taxes. The "huge expansion through the centuries" of the tax state meant that the "limits of the tax state" were often forgotten (*S1918b*, 117). Although there were ultimate limits of the size of a capitalist state, its concrete limits could be moved by innovating the routines of the fiscal apparatus and by changing the antitax routines of the citizens of the state. These government strategies help to explain the transformation of the European states from obvious "parasites" that extracted the classical forms of tributes from the private sector to symbiotic institutions that, more or less, were accepted by their citizens. Since the collection of tributes by a foreign state was costly and gave poor results, the further evolution was driven by "fiscal necessities". Such necessities explain "the friendliness toward the peasants ... of the princes" as well as their "assumption of the role of representatives of the interest of the 'country as a whole' " (p. 137n). The emerging capitalist economy provided a much improved source of government finance, and this gave new contents to the population-friendly strategy of the rulers. The competition with rulers of other states implied an endless pressure for expanding expenditure that had to be financed. Since mercantilist policies were not very efficient in expanding the tax base, the state largely relied on its ability to increase the willingness of wealthy individuals to pay taxes in exchange of an increasing number of benefits from government. One of these benefits was the right to vote for an advisory parliament, as it was still the case in the Austrian-Hungarian Empire when Schumpeter was young.

The further development of the state can be described in three ways. First, "the actual master of the state was usually the prince from whose hands the modern democracy of the Continent received the state or is about to receive it" (*S1918b*, 111). Second, the later evolution of the state implies that "one could say more frequently of the bureaucracy that it was the state". Finally, "the state could penetrate so deeply into the consciousness of the people" that it "may perhaps continue to exist as a mere habit of thought of its citizens." Schumpeter did not analyse these characteristics of modern states in "The Crisis of the Tax State". Instead, he emphasised that any type of tax state "has its definitive limits" and these limits "are, of course, not conceptually definable limits of its field of social action, but limits to its fiscal potential." Such limits vary according to the wealth of different countries and the nature of this wealth. There are thus large differences "between new, active, and growing wealth and old wealth, between entrepreneurial and rentier states." Nevertheless, we see a general tendency towards a "bourgeois tax state" that "confronts the private economies with relatively few means" and this state "remains something peripheral, something alien to the proper purpose of the private economy, even something hostile, in any case something derived" (pp. 111–12). The "meaning of the organization of the tax state lies in the autonomy of the private economy and of private life and would be lost when the state can no longer respect this economy" (p. 139n). When this borderline is approached, "an absurd waste of energy" is needed to "enforce the tax laws, tax inquisition becomes more and more intrusive, tax chicanery more and more unbearable". The question is whether the new rulers of the tax states can avoid such problems.

Although *Capitalism* does not use the term "the tax state", we find little evidence that Schumpeter changed his analysis of this type of state. Thus we read that the "outstanding feature of commercial society is the division between the private and the public sphere" (*Capitalism*, 197). This feature is outstanding because "in commercial society there is a private sphere which contains so much more than either feudal or socialist society allocates to it." Actually, the two spheres or sectors are "to a great extent manned by different people" and they are "organized as well as run on different and often conflicting principles, productive of different and often incompatible standards." These different standards imply a "friction" that "developed into antagonism in consequence of the wars of conquest waged upon the bourgeois domain with ever-increasing success by the men of the public sphere." We may also speak of "government interference"; and this interference has tended to "hamper and paralyze the private engine of production." Thus Schumpeter thought that a "considerable part of the total work done by lawyers goes into the struggle of business with the state and its organs" (p. 198).

## 7.4 Democratic political evolution: Mark I and Mark II

Since it is primarily the politically governed sector of society that can create and destroy the institutional framework needed for the capitalist engine, the history of this economic engine can fairly easily be combined with the history of the tax state. However, *Capitalism* contains a relatively independent analysis of political evolution. Actually, Schumpeter used this book to sketch two models of democratic political evolution. He defined democracy as a political system characterised by competition for power. This definition has caught much attention in political science since Anthony Downs (1957, 29n) in his *An Economic Theory of Democracy* claimed that "Schumpeter's profound analysis of democracy forms the inspiration and foundation for our whole thesis, and our debt and gratitude to him are great indeed." However, the whole "inspiration and foundation" for both Downs and most other political scientists seem to derive from only a couple of chapters of *Capitalism* (pp. 250–83). These chapters present a theory of the democratic process that can hardly be understood fully in isolation from Schumpeter's general evolutionary theory. Therefore, it is not surprising that recent contributions to the debate (Brooker, 2005; Medearis, 2001) have tried to put Schumpeter's narrow definition of democracy into a broader context that includes innovative leadership as well as the socio-political processes of democratisation and socialisation. The debate, nevertheless, fails to recognise that Schumpeter considered the role of the state and the related political system as undergoing an evolutionary process that had many connections to economic evolution and socio-cultural evolution.

We have already considered Schumpeter's theory of the emergence and functioning of the tax state. He emphasised the connections between the tax state and the capitalist economy. The analysis of democracy in *Capitalism* is more abstract. It focusses on the mechanisms of democracy; and it is developed in relative isolation from the other parts of the book. The simplest explanation of this isolated treatment of democracy is found in the structure of the book (see Table 7.1 on page 158). Schumpeter developed his analysis of the political system of democracy in relation to his treatment of the problem of socialism (Part IV of *Capitalism*) and not in relation to the treatment of the social evolution that undermines capitalism (Part II). It is thus the architecture of *Capitalism* that excludes a systematic analysis of the role of democratic political evolution in the transformation of capitalism. Nevertheless, the book seems to suggest that it is the co-evolution between the political sector and the economic sector that ultimately determines the destiny of the capitalist system.

Schumpeter had already presented his idea of developing a theory of political evolution in *Entwicklung I.* Here he had proposed that such a theory could start from the analogy with economic evolution (see Section 4.4). Just like the economic system is analysed in terms of a stationary state that evolves because of the activity of the innovative entrepreneurs, the routine-based activities of the political system evolves because of the occasional emergence of innovative political leadership. Even in the political sector, we have to discern between "those spirits who *create* ... new parties" and "those spirits who *are created* by parties". It is "through the armour of the school or the party" that a "new tendency ... can demand attention, discussion and finally victory" (*S1912a*, 428, 430; *Entwicklung I*, 543, 545). The argument of *Capitalism* demanded that this idea was developed into a model of the mechanism of political evolution. Schumpeter's mature analysis of "the working of the political engine" (*Capitalism*, 273) concentrated on democratic systems. Such systems can be defined in a variety of ways, but his definition concentrated on the democratic mechanism. He suggested that we should take the view that

> "the role of the people is to produce a government ... And we define: the democratic method is that institutional arrangement for arriving at political decisions in which individuals acquire the power to decide by means of a competitive struggle for the people's vote." (*Capitalism*, 269)

Although Schumpeter largely used his definition of democracy to discuss whether this political system will function under the conditions of socialism, his model of democracy is general. His first modelling considerations are found in the paper on "The Meaning of Rationality in the Social Sciences". Here he emphasised that the "old theory of democracy as formulated in the seventeenth and eighteenth centuries *presupposes* degrees of awareness of one's interests, clearness of ends, rationality in the perception and use of means and, most important of all, accessibility to rational argument which are altogether unrealistic" (*S1940b*, 328). To overcome this problem, he wanted to develop a "reformed theory of democracy". This theory "would have to drop, not wholly but to a considerable extent, the hypothesis of conscious rationality". Thereby, it would "arrive at a totally different picture of democratic processes, particularly as regards the role to be assigned to rational arguments" (p. 328). Schumpeter implemented this research programme in Part IV of *Capitalism*. The basic assumption is that democracy "is a political *method*, that is to say, a certain type of institutional arrangement for arriving at political—legislative and administrative—decisions" (*Capitalism*, 242). It is "a mere method that can be discussed rationally like a steam engine or a disinfectant" (p. 266).

Schumpeter's way of thinking about democracy directly opposed "the classical doctrine of democracy", which made its definition in what he considered mysterious terms. The classical thinkers assumed that "the democratic method is that institutional arrangement for arriving at political decisions which realizes the common good by making the people itself decide issues through the election of individuals" (*Capitalism*, 250). These individuals "assemble in order to carry out" the will of the people. Even in the most developed utilitarian approach, of John Stuart Mill and others, it was assumed "that there exists a Common Good, the obvious beacon light of policy". If this common good exists, then "there is also a Common Will of the people". The reason is that the people, "by the force of rational argument", can recognise the common good. Schumpeter believed as little of utilitarian stories within political science as within economics (pp. 251–2): the common good of the people cannot be defined; the notion cannot cover the details of policy-making; individuals and, especially, groups of individuals have insufficient capabilities with respect to rational decision-making. We have already seen that Schumpeter, together with Pareto, emphasised "the importance of the extra-rational and irrational element in our behaviour" (p. 256). With respect to the analysis of democracy, he added problems detected by social psychology. When compared with individual decision-making, "[e]very parliament, every committee, every council of war" displays "a reduced sense of responsibility, a lower level of energy and greater sensitiveness to non-logical influences" (p. 257). This points to "the vital fact of leadership" (p. 270), especially the individual leadership to cope with novel situations or to innovate existing political routine. Furthermore, just like consumers are "amenable to the influence of advertising and other methods of persuasion" (p. 257), so are the voters to an even larger degree influenced by those who compete for political leadership. Therefore, the "Common Will" is really "Manufactured Will"; and this type of will of the people "enters on the ground floor" of the model of democracy (p. 270).

Schumpeter's own model of democracy was intended to analyse "some of the more important features of the structure and working of the political engine in democratic countries". In this engine, "the primary function of the elector's vote is to produce government", and this includes the "acceptance of leadership" (*Capitalism*, 273). However, the driving force of the political engine is the "competitive struggle for the people's vote" (p. 269). It is through this competitive struggle that new issues are introduced to the political agenda. If a political leadership is ignoring an important issue, the opposition is likely to recognise this in order to increase its competitiveness. However, "there are cases in which the political engine fails to absorb certain issues either because the high commands of the government's and the opposition's forces do not appreciate their political values or because these values are in fact doubtful" (p. 281). Such issues might be the entry tickets for outsiders who want to join the competitive struggle. However, it is clear that "concept of competition for leadership ... presents similar difficulties as the concept of competition in the economic sphere" (p. 271). The problem is that "competition is never completely lacking, but hardly ever perfect." Even in political life, we find "cases that are strikingly analogous to the economic phenomena we label 'unfair' and 'fraudulent' competition or restraint against competition" (p. 271). Actually, such phenomena might increase because the evolution of the political engine.

Schumpeter did not explicitly explore the process of political evolution that takes place within his model of democracy. Instead, he seems to present two versions of this model. He started from a political system that seems similar to his Mark I model of economic evolution. This Mark I model of democracy is exemplified in terms of the functioning of the English Parliament in the nineteenth century (pp. 274–82). Here groupings of members (the early form of parties) would try to sustain themselves in a routine-like manner. However, an innovative aspirant for leadership, like Disraeli or Gladstone, would sometimes take over such a group, or try to create a new group. By emphasising new issues, this leader might become prime minister, but he would sooner or later be challenged by other innovative parliamentarians. This system was not stable since the competitive struggle also brought organisational innovations that gradually led to the emergence of modern political parties. Although this evolution was especially clear in the United States, we, in general, recognise the emergence of a system that Schumpeter seems to describe as a caricature of his Mark II model of economic evolution.

In Schumpeter's Mark II model of democracy the "party is a group whose members propose to act in concert in the competitive struggle for political power" (p. 283). Such a party needs a significant amount of routine work. This means that the "psycho-technics of party management, slogans and marching tunes, are not accessories. They are of the essence of politics." The party also needs to define its characteristics in terms of "principles or planks". These characteristics are "as important for its success as the brands of goods a department store sells are ... important for its success" This, however, does not make it "impossible for different parties to adopt exactly or almost exactly the same program" (p. 283). This sceptical description of the activities of established political parties could easily have been extended to their use of 'innovation' in their oligopolistic struggle for power. Thereby, it would become clear that the quantity of innovative activity might be larger in this system than in more loosely organised competitions for power. However, Schumpeter clearly felt more at home in the Mark I model of democracy, and he did not explore the full implications of his Mark II model. These implications can be worked out by studying other parts of *Capitalism*. Here we recognise that oligopolistic competition does not necessarily mean that the selected political leadership effectively solves the problems of the capitalist society. The introduction of these problems on the political agenda is neither secured by this form of competition nor by the related influencing of the preferences of voters. Although the parties have to recognise that new entrants might take up these problems, they can for a long time serve as filters against real innovation. They might also select against types of politicians that are best suited for solving the long-term problems. Finally, this selection is also dependent on the supply of politicians. If the prospect of working for a long time within the constraints of party routine scares away those individuals that are most capable of performing political leadership, then the democratic political engine will work less efficiently with respect to problem solving.

For those who think in terms of democracy as a whole, it is not easy to recognise the evolutionary nature of Schumpeter's two models of the democratic process. What evolves is not immediately this political system, but the more or less routinised behaviour of those people who specialise in politics. The Mark I model is based on the competition between loosely organised parties. This means that the routines of the mass of parliamentarians can to some extent be influenced individually by the political innovator. In contrast, the highly organised parties of Mark II represent strong filters against this type of innovative change. Instead, they build up internal resources for creating innovative responses in their oligopolistic struggle for the people's vote. The question is whether this shift has effects that are analogous to the effects of shift from competitive capitalism to monopolistic capitalism. Schumpeter's answer is not stated explicitly. However, he hardly believed that even his very cautious concept of "economic progress" had any counterpart in political evolution. Furthermore, he was very sceptical about the innovative potential of modern mass politics. Finally, he probably considered the major innovative potential of Democracy Mark II as relating to the undermining of the capitalist economic system. Such conclusions might not create much enthusiasm for establishing an Evolutionary Political Science. Nevertheless, clear thinking in this field seems even more needed than with respect to economic evolution, and this thinking might not arrive at Schumpeter's conclusions.

Schumpeter's models of the democratic process took for granted the institutional framework that allows political competition. However, his historical approach to political evolution did not allow this assumption. Since democracy had emerged through small steps from a non-democratic system, it could also vanish in the same way. Actually, "the modern democracy rose along with capitalism, and in causal connection with it" (*Capitalism*, 296). A core question is, thus, whether democracy will disappear together with the transformation of capitalism into something else. He might have considered the gradual transformation of Democracy Mark I into Democracy Mark II as part of such a process.

The expansion of voting rights to the whole population had been connected to types of political behaviour that, at least for him, became increasingly unpredictable. This might be one of the reasons he abstained from discussing the future of democracy. Instead, he emphasised that the democratic process builds on rather strict conditions. These conditions are especially violated in the case of war. Actually, "democracies of all types recognize with practical unanimity that there are situations in which it is reasonable to abandon competitive and to adopt monopolistic leadership" (p. 296). The major reason is that democratic leadership is seriously constrained by concerns about upholding its popularity. In this respect, "the prime minister in a democracy" can be compared "to a general so fully occupied with making sure that his army will accept his orders that he must leave strategy to take care of itself" (p. 287). However, if the monopoly created in emergency situations "is not limited as to time", it transforms into "dictatorship in the present-day sense" (*Capitalism*, 296).This condition is only one of several that are necessary for upholding democracy, and Schumpeter emphasised five further conditions.

Schumpeter's first condition builds on the fact that the democratic process requires a substantial number of "individuals of adequate ability and moral character" (*Capitalism*, 290). The reason is that "the competitive struggle for responsible office is ... wasteful of personnel and energy." The danger is that although these individuals are available, "the democratic process may easily create conditions in the political sector that ... repel most of the men who can make a success in anything else." The second condition for success of democracy is "that the effective range of political decision should not be extended too far" (p. 291). Even an "all-powerful parliament must impose limits on itself" and to "accept the specialist's advice". The third condition is that democracy is provided with "the services of a well-trained bureaucracy of good standing and tradition, endowed with a strong sense of duty and a no less strong *esprit de corps* [team spirit]" (p. 293). The fourth condition is "democratic self-control" (p. 294). This self-control implies that both electorates and politicians have "an intellectual and moral level high enough to be proof against the offerings of the crook and the crank". Furthermore, the politicians should not "embarrass the government each time they could do so." Finally, the fifth condition is that there exists "a large measure of tolerance for difference of opinion" (*Capitalism*, 295). This condition includes genuine respect for other opinions and "willingness to subordinate one's own opinions." Therefore, "democratic government will work to full advantage only if all the interests that matter are practically unanimous".

Schumpeter's five conditions for the sustainability of democracy might be considered much too strict. Actually, we recognise that all these conditions have, at some points of time, been violated by some of the political systems that are democratic according to his criterion of competition for leadership. Therefore, we might dismiss the strict conditions as reflecting his personal scepticism about the efficiency of the democratic method. However, it seems more helpful to consider his worries as relating to a process of socio-political evolution in which "a century is a 'short run' " (*Capitalism*, 163). This very long-term perspective calls for clear thinking about the ways in which democracy can be sustained under conditions that are obviously increasingly removed from the classical form of the capitalist tax state.

## 7.5 The endless economic frontier and the sociological trend

Schumpeter's analysis of the mechanisms of political evolution represents an interesting application of his general evolutionary theory that might have relevance for political scientists and sociologists. However, in relation to the major argument of *Capitalism*, the focus is not on democracy as such but on the degree to which democratic evolution transmits the positive and negative social reactions to economic evolution considered as a process of creative destruction. In other words, the question is whether the socio-political system is increasingly likely to use the brakes of the capitalist engine. If the social reactions become increasingly hostile to creative destruction and if the democratic system transforms these reactions into intervention, then the capitalist engine can slow down or stop. Before turning to this possibility, we have to consider the question whether this engine slows down or stops because of its own working.

We have already met two brakes of the capitalist engine. The first influences the mechanism of innovation and the second influences the mechanism of adaptation. The extension of equilibrium economics into the realm of economic evolution had suggested "monopoly" as a third brake. Section 7.2 described Schumpeter's rejection of this interpretation. Instead, he implicitly argued that the transition from Mark I to Mark II resulted in an increase of the speed of economic evolution. Nevertheless, there might still be an additional brake. If the set of opportunities for innovative investment is finite, then these opportunities will ultimately be exhausted. Since this assumption was common in the economic debates of the 1930s, Schumpeter had to demonstrate that this version of the third brake has no relevance "for the next forty years" (*Capitalism*, 111). This is what he tried to establish in his criticism of the theory of "vanishing investment opportunity". Thereby he tried to demonstrate that only two brakes were left. These were the brakes that could be manipulated by the socio-political system.

Schumpeter emphasised the issue of the potential stoppage of the capitalist engine because he, in contrast to leading Keynesians, did not believe the capitalist economy could "settle down to a stationary state without being vitally affected" (*Cycles*, 1033; *Cycles Abr.*, 405). This statement was emphasised because his evolutionary theory suggested that "stabilized capitalism is a contradiction in terms" and that the survival of the capitalist system is only possible in an "atmosphere of industrial revolutions". He summarised the background for this conclusion in the following way: "without innovations, no entrepreneurs; without entrepreneurial achievement, no capitalist returns and no capitalist propulsion". Therefore, he was worried about "the theory vanishing investment opportunity" (exemplified by Hansen, 1938). Even if *Capitalism* had replaced individual entrepreneurs by established firms, he had not changed opinion about the impossibility of stagnating capitalism.

The idea that "industrial revolutions" was a phenomenon of the past had been seen before in the history of economic thought. Thus Schumpeter thought that "John Stuart Mill compromised himself by holding in 1870 that the possibilities of capitalist enterprise were substantially exhausted" (*Cycles*, 1037; *Cycles Abr.*, 410). Schumpeter did not want to join a similar bandwagon of pessimism. Therefore, his task "reduces to substituting for unconvincing reasons why investment opportunity should be vanishing, a more convincing one". What he considered "unconvincing reasons" were those given by his fellow economists—not least the Keynesians—who ascribed to the theory of vanishing investment opportunity. This theory suggested that the great industrial breakthroughs were over and that the capitalist economy was missing this crucial stimulus to private investment spending in face of an increasing propensity to save. In contrast, Schumpeter gave strong arguments why capitalist economic evolution is exploiting an endless frontier of opportunities. He found two convincing reasons for the temporary capitalist stagnation that actually took place. First, the economic conditions of the 1930s were characterised by a Kondratieff downswing with low innovative activity. Second, the capitalist system produces not only economic evolution but also a process of socio-political evolution that leads to the increasing use of the brakes of the capitalist engine. The actual stagnation was not least the result of the application of these brakes. The fiscal policies advocated by the Keynesians tended to perpetuate this stagnation by creating a "capitalism in the oxygen tent—kept alive by artificial devises and paralyzed in all those functions that produced the successes of the past" (*S1943*, 185).

Schumpeter's argument against the theory of vanishing investment opportunity had the time perspective of forty years. He squarely rejected the problem of saturation of demand as relevant and this meant that the worries about the decline in the birth rate of the population were of little importance. However, he had to admit that "the unique opportunity for investment" due to the opening up of new lands "cannot recur" (*Capitalism*, 115). Nevertheless, this does not mean "that the 'closing of the frontier' will cause a vacuum". The major reason is that the "[t]echnological possibilities are an uncharted sea". We have to be cautions when arguing about this sea of possibilities. Especially, "[f]rom the fact that some of them has been exploited before others, it cannot be inferred that the former were more productive than the latter" (p. 118). On the contrary, Schumpeter's basic vision of the *economic* prospect of capitalism can be formulated in terms of the metaphor of "the endless frontier". He came close to using this metaphor to express his ideas, but it was actually phrased by Vannevar Bush (1945) to characterise the conditions of science. His phrase relates to facts of, or the myths about, the early economic history of the United States as characterised by a moving "frontier" between unexplored possibilities and well-established activities. However, while the geographical frontier in the end was closed, no similar closure needs to occur in scientific and economic evolution.

The main mover of Schumpeter's endless frontier is the S-entrepreneur in the narrowest sense. For instance, the early pioneers of "railroadization" had to overcome the hindrance of preconceived ideas of the carrying capacity of their innovation in a world characterised by mail coaches: "We may smile now at the opinion of the age that railroads were being overdone, seeing how small a part of what we now know had to be done was accomplished then. But certainly they were in advance of what was then required" (*S1934a*, 111). Thus, the representatives of the established world of economic routine were right, since the pioneers were overdoing things when judged by old standards: "Of course they were, for they themselves created the economic world, which was to provide the demand for their services and which never could have developed without them." When their "economic world" turned into routine, there were new S-entrepreneurs to move the frontier of the economic system. The question, however, is whether this repeated renewal of the potentially endless frontier of economic life would at some point come to an irrevocable stop. One part of his answer is that the ultimate stoppage of evolutionary renewal is unlikely to be caused by exhaustion of innovative opportunities. The other part is that the emergence of a capitalism based on large corporations promotes rather than hinders evolutionary renewal.

The existence of an endless frontier in the foreseeable future meant that Schumpeter could concentrate on the brakes of the capitalist engine that could be manipulated by the socio-political system. His proposition was that these brakes would be used increasingly because of the basic trend of socio-political evolution. Even the Mark II version of the capitalist engine is characterised by the process of creative destruction that produces uncertainty and insecurity for firms and households. However, Schumpeter argued that uncertainty and insecurity were becoming increasingly opposed. The very achievements of the capitalist engine meant that firms and households had more to lose than before; and they had a weaker social network to rely on if their established positions were lost. If we add the short-term nature of mass behaviour, the conclusion is that an increasing opposition against creative destruction was emerging. This suggested the use of the brake on the mechanism of adaptation. As Schumpeter had tried to demonstrate, the use of this brake would also influence the mechanism of innovation. Furthermore, the search for certainty and security could also lead to the direct use of the brake on this mechanism. In contrast, the part of the population that gained directly from innovative activity had decreased because of the transition from Mark I to Mark II. Therefore, there was an increasing pressure on the state to control and delimit capitalist economic evolution. The functioning of the democratic system according to its own Mark II model serves to complete this abstract account of Schumpeter's considerations. In the competitive struggle for the people's vote, those parties who promoted certainty and security were likely to win.

The analysis in *Capitalism* is much more complex than suggested above. However, this analysis does suggest a co-evolution between economic life and socio-political life. Thereby, it serves to implement the ideas that Schumpeter sketched on the last page of *Cycles*. Here he suggested that "the sociological drift cannot be expected to change" (*Cycles*, 1050; *Cycles Abr.*, 423). For the narrow purpose of studying business cycles, this "sociological drift" had to be treated as representing "factors external to our process" of capitalist economic evolution. Nevertheless, "in a wider sense those factors and the mentality and moral code behind them are not external to the process of economic evolution but part of it, a part as essential ... as any 'objective' shrinkage of investment opportunity could be" (p. 1050n). The intimate relationship between economic evolution and socio-political evolution suggested that it would be radically wrong to say that because "'only' politics and humors being the matter, pristine vigor of the capitalist process could easily be restored at the next swing of the electoral pendulum." Actually, he emphasised that "capitalism produces by its mere working a social atmosphere ... that is hostile to it, and this atmosphere, in turn, produces policies which do not allow it to function" (*Cycles*, 1038). It was largely these suggestions that Schumpeter developed in *Capitalism*.

At the most abstract level, *Capitalism* tries to demonstrate that the hostile "social atmosphere" against the capitalist process was largely the result of a routinised form of "rationality" that was produced by the capitalist engine. He told, "[i]n desperate brevity", this story in the chapter on "The Civilization of Capitalism" (*Capitalism*, 121–30). Actually, he did not even define the concept of civilization but only remarked that he turned "to the cultural complement of the capitalist economy ... and to the mentality that is characteristic of capitalist society and in particular of the bourgeois class" (*Capitalism*, 121). Elsewhere he, however, defined "the concept of civilization" in terms of "a system of beliefs, a schema of values, and attitude to life, a state of the arts, and so on" (*S1948d*, 429). His major point was that the capitalist process had induced a "civilization that was essentially rationalist and utilitarian. It was not favorable to national glory, victory, and so on. That civilization required rationalist credentials for everything it was doing" (*S1941b*, 340). In *Capitalism* (124) he emphasised that "the evolution of economic rationality" produced a "type of logic or attitude or method". This logic can be traced back to "rational cost-profit calculations" and made a crucial step forward through the emergence of "double-entry bookkeeping" (p. 123). From here "the rationalist engine" made a "conqueror's career of subjugating—rationalizing—man's tools and philosophies, his medical practice, his picture of the cosmos, his outlook of life, everything in fact including his concepts of beauty and justice and his spiritual ambitions" (pp. 124–5). Schumpeter's acceptance of the revolutionary role of "rationality" seems to be based on a concept of routinised rationality. This concept seems related to Max Weber's idea of the spread of rationality that he even had emphasised as the common theme of the huge handbook of social economics that he edited. In the preface to the volume in which Schumpeter's *Doctrine* appeared, Weber in 1914 wrote that "the point of departure" of the whole handbook "is the view that the development [Entfaltung] of the economy must be understood as part of the general rationalisation of life" (Bücher et al., 1914, vii). Although Schumpeter's evolutionary approach normally made him very cautious of extrapolating observable trends, he nevertheless made heavy use of the trend of "rationalization" in *Capitalism*.

Schumpeter's description the spread of "rationality" from the core of the capitalist economy to the whole of capitalist society might look like that of a success story. Nevertheless, Schumpeter did not believe that "the rationalist engine" was sufficient in economic and social affairs. He had made that clear in a paper on "The Meaning of Rationality in the Social Sciences" (*S1940b*), which he wrote for a book that the sociologist Talcott Parsons planned but never published. Schumpeter emphasised that it is a problem that we, as researchers, assume that the studied social actors are "subjectively rational" with respect to the means of achieving their goals (see Table 7.2 on the facing page). Subjective rationality implies that the agents have used given goals to reach the best of the possible solution. However, since we do not know the goals of the agents and the information available to them, it is too easy to think that their decisions are subjectively rational. If we add full information, this solution is also objectively rational. We have to recognise that it is often the scientific observer who transforms the assumed rationality and information of the agents of an abstract model to the observed actors. The realistic search for the subjective rationality of actual agents will rather demonstrate "the narrow limits of its range" (*S1940b*, 330). Its range is

**Table 7.2.: Schumpeter's analysis of rationality in the social sciences**

| | | |
|---|---|---|
| I. Rationality of the Observer | A. Rationality of Method | |
| | B. Rationalism | |
| II. Rationality in the Observed | A. Conscious or Subjective Rationality | $\alpha$. Rationality of Ends |
| | | $\beta$. Rationality of Means |
| | B. **Subconscious or Objective Rationality** | |

Comment: Schumpeter's "observer" is the social scientist who studies the role of "rationality" among "observed" human agents. The core form of rationality is "objective rationality". It is the result of following rules of thumb that have proven their adequacy in practice. It seems to be the outcome of an evolutionary process. Source: *S1940b* (323).

actually "immeasurably smaller than our predecessors in economic and political theory thought". It is much better to think in terms of the actor's "subconscious rationality" (also called "objective rationality"). This type of rationality implies the application of rules of thumb that produce the result that would have been reached by subjective rationality—like in Herbert Simon's (1982) models of bounded rationality. Since these rules are the outcome of processes of social selection and social imitation, they are not necessarily adequate, however. This type of rationality is looking backwards at established experience even when solving new problems. Therefore, it is especially inadequate for innovative activity. Here creativity based on more or less irrational emotions also play an important role (*S1947c*).

Although the different versions of the concept rationality are confusing, they all emphasise the removal of emotions from social affairs. This removal played an important role for Schumpeter's suggestion that the brakes of the capitalist engine would be used and that the capitalist society would be transformed into something else. Although the hostility against the process of creative destruction was partly based on emotions, Schumpeter suggested that the defenders of capitalism could not respond. His reason is that the "rationalist and unheroic" bourgeois "can only use rationalist and unheroic means to defend his position or to bend the nation to his will" (*Capitalism*, 137). These means might appear fully sufficient since capitalism "has plenty of utilitarian credentials to present" (p. 144). However, "[p]olitical criticism cannot be met effectively by rational argument" since it "can never reach the extra-rational driving power that lurks behind it." The problem is that "[c]apitalist rationality does not get away from sub- and super-rational impulses" and that its removal of irrational traditions merely means that such impulses get out of control. Such impulses are, according to Schumpeter, necessary when we are looking for "a prime mover of group action". Furthermore, "any pro-capitalist argument must rest on long-term considerations", while it for the majority "is the short-term view that counts" (*Capitalism*, 144–5). This short-termism is actually "perfectly rational" when evaluated on the basis of "individualist utilitarianism". Since a society cannot be built on such a shaky foundation, it has to create an "*emotional* attachment to the social order—i.e., the very thing capitalism is constitutionally unable to produce".

Whether this result of Schumpeter's macro-sociological analysis is true is obviously of much interest to those who read *Capitalism* as a grand essay on capitalism, socialism and democracy. In the present context, the question is rather whether the result is based on his attempt to complement evolutionary economics with an evolutionary sociology. This might actually be the case. However, Schumpeter, at best, presented his evolutionary sociology in a very rudimentary form. In this form, it is difficult to present the mechanisms that might change the supposed trend of rationalisation. It is also difficult to include factors—like nationalism—that emotionally gather a population in defence of its given economic and social system. Therefore, Schumpeter's book should not be read for its "predictions" but for its challenges to the thinking and research about the relationship between the capitalist engine and socio-political evolution.

## 7.6 Conclusion

The standard way of reading *Capitalism, Socialism and Democracy* is reflected by its title. This reading focusses on Schumpeter's arguments to the degree that they promote his story about the destiny of capitalism. The present chapter reflected another way of reading his book by focussing directly on its evolutionary theorising and the related historical material. Core elements of *Capitalism* were interpreted as containing the application, modification, and extension of his previous studies of the evolution of the capitalist economic system and of the capitalist society in general. According to this interpretation, Schumpeter's book could also have been called 'Capitalist Economic Evolution and Its Socio-Political Limits' or 'The Capitalist Engine and the Increased Use of Its Brakes'. The interpretation behind such alternative titles is based on the idea of treating *Capitalism* as part of Schumpeter's evolutionary trilogy. In relation to *Development* and *Cycles, Capitalism* can be seen as providing the most mature expositions of Schumpeter's evolutionary economics and his of general theory of social evolution. Since the book, furthermore, has been studied more than *Development* and *Cycles,* it can serve as the starting point for the study of these other parts of the evolutionary trilogy.

The chapter did not try to cover Schumpeter's magnificent dynamics of the capitalist society as an integrated whole. Instead the overall evolutionary dynamics were decomposed into a set of relatively independent processes—each of which is governed by its own mechanisms of innovation and adaptive selection. He used these mechanisms to study the evolution within the political and social sectors of capitalist societies. With respect to the political sector, his results not only included an account of the historical evolution of the tax state, but also a sketch of a theory of the evolution within democratic political systems. Actually, he seems to have come close to proposing an evolutionary political science. His analysis of the socio-cultural sector was more encompassing, but also more difficult to handle. Although his type of macro-sociology seems closely connected to his entrepreneurial and evolutionary interpretation of history, he did not develop this research programme explicitly. The easiest way of reaching a more operational definition of his macro-sociological and micro-sociological studies of socio-cultural evolution is probably to compare them with the contribution of Max Weber. In this context, Schumpeter's type of evolutionary sociology can be characterised by its emphasis on the evolutionary processes that are created through the interplay between routinised behaviour and innovative behaviour.

The processes of political and socio-cultural evolution have obvious connections with the working of the capitalist engine of economic evolution. While the political response to the capitalist engine can support its process of creative destruction, Schumpeter considered other responses much more likely. They include the provision of compensations for the social costs of the capitalist engine, attempts to put a brake on it, and more radical forms of getting rid of creative destruction in its capitalist form. Nevertheless, the analysis of the independent functioning of the capitalist engine was at the centre of his research efforts. It is the parts of *Capitalism* that are devoted to this analysis that are most crucial for the understanding of his evolutionary trilogy as a whole. In this context, the major contribution *Capitalism* is its sketchy development of what has been described as the Mark II model of the capitalist engine. The driver of this version of capitalist economic evolution is the oligopolistic competition between established firms. We shall return to this type of model in Chapter 9. However, in relation to the rest of the evolutionary trilogy, this Mark II model largely serves to emphasise that Schumpeter here applied his very different Mark I model of the capitalist engine. According to this model, innovations "entail construction of New Plant"; they are "embodied in a New Firm founded for the purpose"; and "innovations are always associated with the rise to leadership of New Men" (*Cycles*, 93–96; *Cycles Abr.*, 68–70).

# 8

# Waveform Economic Evolution and Business Cycles

Although Schumpeter made several attempts of elaborating and applying the basic model of the capitalist engine that he had presented in *Entwicklung I* and *Development*, the results of his major effort are found in *Business Cycles*. This book has roots back to his student years. At that time, members of the historical school (like Sombart and Spiethoff) had begun to focus on the problem of crises and cycles. Furthermore, this problem seems to have been a hot issue in Vienna among the participants in the seminars of the economic theoretician Böhm-Bawerk and the statistician and economic historian Inama-Sternegg. It is thus hardly surprising that Schumpeter put it at the top of his research agenda from the very beginning (see Section 5.4). However, he thereby not only set himself apart from the older generation of neoclassical economists who tended to ignore crises and cycles. It later became clear that his evolutionary theorising about business cycles was also radically different from the studies of business cycles that boomed in the 1920s and 1930s. The most obvious difference is that while nearly everybody else considered business cycles pathological, Schumpeter considered the phenomenon as part of the healthy process of economic evolution under capitalist conditions. Actually, his two-stroke model of the capitalist engine (see Figure 6.2 on page 149) suggested that waves are essential for capitalist economic evolution. Metaphorically, they should rather be considered expressions of the beat of the heart of economic life than as inflamed tonsils that can be removed by surgery:

> "Analyzing business cycles means neither more nor less than analyzing the economic process of the capitalist era. … Cycles are not, like tonsils, separable things that might be treated by themselves, but are, like the beat of the heart, of the essence of the organism that displays them." (*Cycles*, v)

This conception suggested a research strategy that is similar to the one that Schumpeter found in Marx's work: "to look to business cycles for material with which to build the fundamental theory of capitalist reality" (*History*, 1135). Schumpeter's specific connection between business cycles and capitalist evolution is simple. Capitalism is seen as an engine of a form of economic evolution that does not take place gradually and continuously but rather in waves. Thus, his task was to connect the waves of evolution with the business cycles that are observable in the capitalist economy. In Chapter 6 of *Entwicklung I*, which he rewrote completely in 1926, he had already tried to clarify this connection. In the 1930s he wanted to do so in a more systematic fashion and in close relation to empirical evidence. The result was *Business Cycles: A Theoretical, Historical, and Statistical Analysis of the Capitalist Process*.

Even a quick inspection of *Business Cycles* demonstrates that Schumpeter had made his book appear unconvincing by presenting it as giving a general explanation of business cycles. In practice, he put the emphasis on irregular waves of economic evolution and the related macroeconomic phenomena. This emphasis could have been reflected in the title of the book. He could, thus, have called it 'Innovation-induced Business Cycles', 'Evolutionary Business Cycles', or 'The Waves of Capitalist Economic Evolution'. By adding the main point of the subtitle, he could have arrived at a title like 'Theory, History, and Statistics of the Waves of Capitalist Economic Evolution'. This alternative title reflects the present interpretation of Schumpeter's most ambitious and least successful book. Although *Cycles*'s unfinished combination of theoretical, historical, and statistical analyses helps to explain its lack of success, the book's main message is that we have to combine theory and fact.

## 8.1 The complex contents of *Cycles*

According to the present interpretation, Schumpeter planned *Business Cycles* to form the centre of his evolutionary trilogy as well as his most important scientific work. Although *Capitalism* became his most-cited book and *Development* has emerged as Schumpeter's *magnum opus*, it is suggested that his core ambition was to make a theoretical, historical, and statistical analysis of the capitalist engine. This interpretation does not necessarily contradict John Kenneth Galbraith's verdict that *Business Cycles* "is now only a scholastic oddity" (quoted by Swedberg, 1991, 128). However, this verdict seems to be based on a lack of interest in Schumpeter's evolutionary analysis. As long as we interpret Schumpeter's book as representing one of the many early theories of business cycles, it largely must be considered a failure. Another interpretation of the book emphasises its overwhelming amount of historical and statistical material. This interpretation suggests that Schumpeter tried to overcome the dichotomies of the battle of methods by combining theoretical, historical and statistical analysis. This is probably the main reason why Erich Streissler (1994, 37) has suggested that "*Business Cycles* may, in fact, be considered the last (and one of the most important) monographs of the Younger German Historical School". However, this characterisation of the book is at best a half truth and it does not emphasise his evolutionary analysis. In any case, we seem to be facing three different interpretations of the basic ambition of *Cycles*:

1. Explaining the phenomenon of business cycles: *Cycles* should be considered a contribution to business cycle theory and evaluated as such.

2. Resolving the battle of methods: *Cycles*'s combination of theoretical, historical, and statistical analysis should be interpreted as the implementation of the bridge between neoclassical economics and the historical school that Schumpeter envisaged in his youth.

3. Extending and "verifying" the basic theory of capitalist economic evolution: *Cycles* provides an extension of all parts of *Development.* It presents Schumpeter's most mature treatment of his basic model of capitalist economic evolution and of the ways it can be related to empirical evidence. Furthermore, it provides many refinements of his theories of innovation and of the reactions to innovation.

For those who have read the previous chapters of the present book, it should come as no surprise that the third interpretation is the one that will dominate the present account and that it is argued that the two other interpretations give little meaning except in the light of evolutionary analysis. This interpretation is supported by Schumpeter's formulations in the papers in which he announced his research project on evolutionary business cycles. The goals of the project that led to *Cycles* are most clearly specified in the preface to the German reprint of *Entwicklung II*, which Schumpeter prepared at the end of 1934 (as well as in *S1933a*). At that time, he—unrealistically—thought that he would publish his next book within a year, and this expectation influenced the German preface. Actually, Shionoya (2004, 132), who translated most of the preface to English, considered it a "prolegomenon to *Business Cycles*". In the preface, Schumpeter pointed out that although he did not consider "the text" of *Entwicklung II* (and *Development*) "as altogether final", he had found no time to improve it (*S1935f*, xiii). Instead, he had "in the last years" been engaged in developing complementary analyses. The background was that "anybody who reads this book will soon understand that not only its last chapter but actually each page deal with the problem of business cycles ... [because] the cyclical changing of business situation is the way of life of capitalism" (p. xiii). He had used this feature when treating the "statistical and historical material". Actually, "my theoretical schema or model has become a research instrument, and I had rich opportunity to experience how it functions as an Ariadne's thread." The fact that the

Schumpeterian model functioned as a red thread through the maze of data on capitalist dynamics served as its partial "verification", but the empirical studies also suggested important revisions of the model.

At the time when Schumpeter completed his preface to the reprint of *Entwicklung II,* he also wrote an elegant paper on "The Analysis of Economic Change". This paper proposed "a research program" with two goals. The first was "the goal of establishing the validity of the schema of innovation" (*S1935d,* 149). The second goal was that of "showing how innovation produces, together with its monetary complement, the particular kind of waves inherent to the economic life of the capitalist society and paralleled by similar phenomena in other fields of human activity." While the treatment of "other fields of human activity" was largely postponed (to *Capitalism,* where waves were de-emphasised), the paper suggested that the analysis of business cycles presupposes the coordination "of the historical, statistical and analytical modes of approach" (*S1935d,* 134). However, each of the approaches are "thwarted by that reluctance to coöperation incident to the differences in training, tastes and horizons of the individual workers". To overcome this problem, Schumpeter suggested an extension of the econometric alliance between theorists and statisticians to include historians in a way that would have pleased Gustav von Schmoller. This extended alliance should, in principle, study the phenomenon of business cycles as caused by both "outside factors" and by "the working of the capitalist machine" (p. 136).

Schumpeter described the movement of the capitalist machine as driven by discrete innovations. He chose the example of railway innovation. Although the passenger wagons of the first railways were built like series of main coaches, the discreteness of the innovation was clear: "Add as many mail coaches as you please, you will never get a railroad by so doing" (*S1935d,* 138). Furthermore, "the carrying into effect of innovations ... *cluster* at certain times" because of "various kinds of social resistance to something that is fundamentally new and untried" (p. 141). Thus the "first success will always produce a cluster" that disturbs the functioning of the economic system, but the effects of such a cluster depend on the characteristics of the innovation: "The railroadization ... of a country, for instance, may take between one-half and the whole of a century and involve fundamental transformations of its economic and cultural patterns, ... while other innovations or groups of innovations may arise and disappear within very few years" (p. 143). This difference suggests "that there are *many* cycles rolling on simultaneously", but three types of cycle seems sufficient for practical analysis: the Kondratieff "long wave" of more than fifty years, the Juglar cycle of about ten years, and the Kitchin cycle of a little more than three years (*S1935d,* 143–6). Schumpeter's research programme suggested that these simultaneous cycles should be mapped statistically and historically, and that they should be explained in terms of underlying innovations. In this respect, he had a certain preference for the long waves: "the waves of the *historical* stream of evolution find more and more satisfactory interpretation in the waves of the *theoretical* model of evolution, the longer the waves are" (*S1935f*, xviii).

These formulations were written early in 1935—and it took longer than expected to turn the research programme into a book. In a letter to Mitchell a couple of years later, he reflected on the underlying difficulties of accounting for waveform capitalist evolution. He strongly rejected the idea "that our phenomena are simple and can directly be handled by simple methods either theoretical or statistical" (*BL*, 301). Instead, the core task was the "immensely laborious analysis of every historical pattern". Unfortunately, he could only "scratch the surface of that herculean task"; and he sometimes became "quite melancholy to think that what I have really to say will have to remain unsaid forever". This is obviously not a state of mind that promotes the coherence of *A Theoretical, Historical, and Statistical Analysis of the Capitalist Process*. The consequence was that when *Business Cycles* finally came out in the autumn of 1939, it lacked the extensive coordination of theoretical, historical, and statistical analyses that Schumpeter had promised in his programmatic papers. Instead, his three types of analysis were presented with little mediation (see Table 8.1 on the following page).

The 160 pages of the theoretical part of *Cycles* start with a restatement of Schumpeter's basic model of the capitalist engine. This restatement actually covers nearly 100 pages. It is mainly characterised by a modification of his previous terminology (see Table 9.2 on page 247) but it also includes some theoretical modifications. The change of terminology and the modification of theory serve the presentation of the Schumpeterian theory of the waveform contours of economic evolution. According to this theory, the evolutionary process can be divided into innovation-induced waves that begin and end in "neighborhoods of equilibrium". However, some neighbourhoods of evolutionary equilibria are closer to equilibrium than others. Weak matches are found at the beginnings of individual Kitchin cycles and Juglar cycles, while the closest match to evolutionary equilibria is found at the beginning of the Kondratieff waves—that is, only once every fifty or sixty years.

Schumpeter did not jump directly to his model of multiple and simultaneous waves of evolution. His strategy was instead to extend his basic model of capitalist economic evolution into a very different type of model in a stepwise manner (*Cycles*, 130–92; *Cycles Abr.*, 105–62). This was done according to what Pareto (1972, 9) had called "the method of successive approximations". This method of "approaching reality" suggests that we make theories that "generally become more and more complex". Schumpeter approached the phenomenon of evolutionary business cy-

**Table 8.1.: *Business Cycles* (Cycles, 1939; Cycles Abr., 1964)**

| | *Cycl.* | *Abr.* |
|---|---|---|
| **Theoretical analysis** | | |
| 2. Equilibrium and the theoretical norm of economic quantities | 30 | 8 |
| 3. How the economic system generates economic evolution | 72 | 46 |
| 4. The contours of economic evolution | 130 | 105 |
| **Historical analysis** | | |
| 6. Historical outlines I: Introduction; 1787–1842 | 220 | 177 |
| 7. Historical outlines II: 1843–1913 | 303 | 201 |
| 14. 1919–1929 | 692 | 277 |
| 15. The world crisis and after | 906 | 330 |
| **Statistical analysis** | | |
| 1. Introductory | 3 | 1 |
| 5. Time series and their normal | 193 | 163 |
| 8. The price level | 449 | – |
| 9. Physical quantities. Employment | 483 | – |
| 10. Prices and quantities of individual commodities | 520 | – |
| 11. Expenditures, wages, customers' balances | 544 | – |
| 12. The rate of interest | 602 | – |
| 13. The central market and the stock exchange | 639 | – |
| Appendix: Description of the statistical material | 1051 | – |

Comment: *Business Cycles* has the subtitle *A Theoretical, Historical, and Statistical Analysis of the Capitalist Process*. This subtitle clearly reflects the contents of Schumpeter's complex book. The abridged edition (*Cycles Abr.*, 1964) excludes practically the whole of the statistical analysis as well as the parts of the historical analysis that do not concern the United States.

cles in four steps. We may say that he presented what can be considered his 'zeroth' approximation under the heading "How the Economic System Generates Evolution" in Chapter 3 of *Cycles*. This approximation (see Figure 6.2 on page 149) had actually already been developed in *Entwicklung I* and *Development*. Chapter 4 of *Cycles*—on "The Contours of Economic Evolution"—presents three further approximations to what Schumpeter considered the reality of waveform economic evolution. The full sequence of approximations is:

0. The zeroth approximation is simply Schumpeter's basic model of economic evolution. This model does not fully explain why economic evolution is a waveform process. This zeroth approximation is the topic of Chapter 9 of the present book.

1. The "first approximation", which he also called the "Pure Model", adds assumptions that produce a cyclical behaviour of the zeroth model of economic evolution. Especially, the "swarming" of entrepreneurs is needed to produce macroscopic waves. This first approximation is the topic of Section 8.4 of the present chapter.

2. The "second approximation" introduces the indirect effects of a "cluster" of innovative projects. These macroeconomic effects produce the "Secondary Wave". Other strict assumptions of the "Pure Model" are also replaced by more realistic ones. This second approximation is the topic of Section 8.5.

3. The "third approximation" consists in the introduction of "many simultaneous cycles"—especially Kondratieff waves and Juglar cycles. This third approximation is the topic of Section 8.3.

Through these four steps in the 160 pages on theory, Schumpeter had prepared himself for the statistical and historical parts if *Cycles* (see Table 8.1). The statistical part covers 320 pages and has several purposes, including the provision of institutional facts. However, the most obvious purpose is to help decomposing 150 years of capitalist history into nearly three Kondratieff waves that each consists in six Juglar cycles; and these Juglar cycles are, in turn, decomposed into three Kitchin cycles. The huge historical part of about 590 pages presents the "Historical Outlines" of capitalist economic evolution from 1787 to 1939. It provides historical detail on three Kondratieff waves and about seventeen Juglar cycles; to a lesser extent it also tries to cover some of the more than fifty Kitchin cycles that were expected by Schumpeter. The main emphasis is put on the analysis of the period 1920–1939, which in 360 pages is interpreted as being dominated by the innovation-poor second half of a Kondratieff wave. Nevertheless, the present book concentrates on the account for capitalist economic evolution of the nineteenth century (see Section 8.2).

Schumpeter's most ambitious book did not enhance his scientific fame. Three of the reasons why this book received little attention have often been repeated. They are that *Cycles* suffered from a bad title, bad timing, and bad editing. We have already considered the problem that the title *Business Cycles* does not cover the contents of the book. The next problem is that *Cycles* was published in 1939 when Keynesian economics was still at the centre of attention and when the exploitation of the given economic system for war purposes emerged as the core issue. Finally, a good editor should have emphasised that the 1100 pages of the book do not provide the impression of a coherent scientific monograph. Instead, such an editor should have pressed Schumpeter to focus his book on its strengths: the clarification of his idea that economic evolution largely takes place in waves and his historical evidence about the waveform history of the capitalist economic system. However, the book shows no sign of such an editorial effort, and this effort instead had to be performed twenty-five years later by Rendigs Fels. By focussing on Schumpeter's evolutionary theory and his account for the economic history of the USA, Fels produced a fairly condensed and readable abbreviation (*Cycles Abr.*). However, Fels could only abbreviate and not make changes. Especially, he could not replace the title *Business Cycles* by a more adequate one. In any case, Fell's (1964, viii–ix) argument that the failure of *Cycles* is largely due to a bad title as well as bad timing and bad editing is a very partial one. The basic reason for the failure of *Cycles* is that Schumpeter had set himself the ambitious goal of analysing the historical working of the capitalist engine and that he failed to demonstrate how ambitious readers could continue his movement toward that goal.

Schumpeter's basic proposition in *Cycles* is that the speed of capitalist economic evolution varies over time. This proposition allowed him to decompose the waveform history of economic evolution into distinct periods. He defined the starting point of each period as a situation in which evolution has nearly reached a complete stop. He called this situation a "neighborhood of equilibrium" and assumed that he could operate with "units separated from each other by neighborhoods of equilibrium" (*Cycles*, 138; *Cycles Abr.*, 114). Since this use of the term equilibrium differed radically from its use in equilibrium economics, it created much confusion. If Schumpeter had provided a formal account for his evolutionary process, this confusion might have easily been overcome. However, the available analytical tools did not allow him to specify precisely what he meant by "equilibrium" as the stoppage of evolution. Instead, he tended to mix two relatively independent propositions: (1) that capitalist economic evolution occurs in waves and (2) that the waves of evolution explain the observable phenomenon of business cycles. He also failed to confront directly alternative propositions that could have been derived from the contemporary literature on business cycles. For instance, he could have considered business cycles as a relatively independent macroeconomic phenomenon. This phenomenon involves upswings and downswings in investment that might promote waves of capitalist economic evolution—rather than vice versa. Furthermore, business cycles might occur in situations with few possibilities for innovation and selective adaptation. Such relatively pure business cycles might actually be more severe that those related to economic evolution. These issues would probably have interested Schumpeter's audience more than his primary propositions. Nevertheless, *Cycles* contains only scattered remarks that point to the study of business cycles as a phenomenon that is not immediately coupled with economic evolution.

Schumpeter's difficulties in *Cycles* can be understood by pondering on his general analytical strategy; and an attempt to reconstruct this strategy is depicted by Figure 8.1 on the next page. This figure suggests that his starting point was a vision of capitalist economic evolution that emphasised innovation: the great innovators define the parameters to which imitators and mere managers adapt. The implementation of this vision into a model of economic evolution pointed at a solution to a central problem of capitalism: the problem of crises and business cycles. Schumpeter's solution was that business cycles are basically the form in which capitalist "progress" takes place and that crises are epiphenomena. As he had suggested in *Development* (223–36), the example of the first innovator creates a "swarm" of innovative activity, and this "swarm" causes a period of "prosperity". When the "swarm" dissolves, the result is "depression" (or, in the terminology of *Cycles*, "recession"). It is during this "recession" that the routines of the economic system adapt to the conditions defined by the innovative activity. According to Schumpeter's analytical strategy, this model of the waves of evolution had to be related to empirical analyses based on historical and statistical methods. Such analyses would not only check the validity of the model, but also suggest how the model could be refined. Furthermore, the model-related work with historical and statistical evidence would suggest improvements of the empirical methods. Finally, the refinement of the model would require the invention of new elements of the conceptual framework and new mathematical tools. We may thus consider Schumpeter's analytical ambitions as consisting of two parts. The first ambition was to contribute to the understanding of the problems of capitalism by developing his evolutionary theory and by using it for historical studies. The second ambition was to develop the analytical tools needed for fulfilling his first ambition. Although the results of following the first ambition could in itself contain major scientific contributions, his lasting influence on the science of economics would depend on his success with respect to the extension of the toolbox available for the analysis of economic evolution.

Schumpeter's failure to fulfil his ambitions is hardly surprising. Actually, they seem more suited for an army of scientists than for an individual researcher. Such an individual should be expected to focus his efforts on a limited field in order to produce convincing results; but *Business Cycles* shows few signs of such a focus. He apparently personally wanted to do *everything* suggested by his ambitious analytical strategy, and therefore many of his readers have concluded that he achieved next to nothing. The question is, of course, whether Schumpeter's ambition can be fulfilled in its whole scale and scope—even with the improved analytical tools that are available today. The answer seems to be in the negative. Even if it showed up to be possible to develop a satisfactory model of evolutionary business cycles according to Schumpeter's specifications, this model would hardly be able to explain more than a relatively small part of the empirical phenomenon of business cycles. Furthermore, such

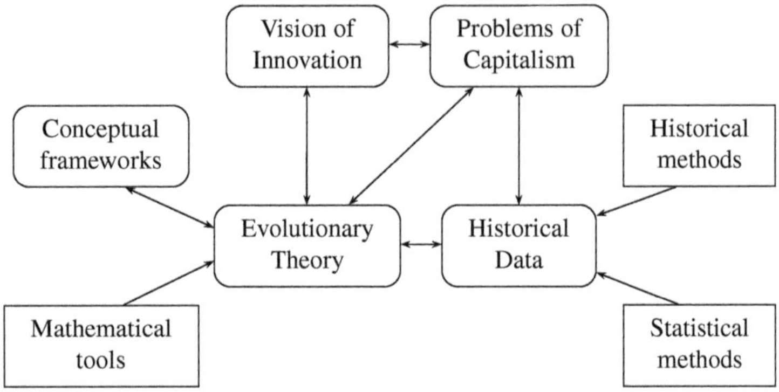


**Figure 8.1.: Schumpeter's complex analytical strategy in *Cycles***

Comment: The figure presents Schumpeter's actual practice rather than his ambitions of studying the waves of capitalist economic evolution. Although he wanted to develop new mathematical tools as well as new statistical and historical methods, he in practice had to apply the given tools. Furthermore, his attempt to modify conceptual frameworks was not an unconditional success. In general, the available analytical tools seriously constrained his study of the waves of economic evolution. This fact was not emphasised by Figure 5.1 on page 100.

an effort would distract attention from the independent task of exploiting the insights of *Cycles* and overcoming its deficiencies for the purpose of understanding the many-sided evolutionary processes within capitalism. By moving attention from cycles to evolution, we, for example, recognise that support for expressing clearly the insights of the historical parts of the book comes from another kind of statistics than that of the aggregate time series. In general, we should confront the "glaring lacunae [gaps]" and the "unfulfilled desiderata [desired things]" that were emphasised by Schumpeter in the preface of *Cycles* (p. v). Any attempt to specify the gaps and desired things of *Cycles* demonstrate that we are facing a work in progress rather than a finished treatise on the capitalist engine of evolution. Nevertheless, we may also appreciate the improvements that he actually achieved with respect to the terminological, conceptual, and modelling treatment of this engine as well as his historical accounts for the functioning of capitalist economic evolution.

## 8.2 Towards a reasoned history of the capitalist process

Schumpeter's ambition in *Cycles* was to provide some sort of "verification" of his basic scheme of economic evolution by demonstrating that it can be used for developing an analysis of the history of waveform economic evolution. This ambition is reflected in the structure of the book.

The theoretical part provides the approximations needed for handling economic history and the statistical part serves to decompose this history. Then the task of the historical part is to demonstrate the actual working of the mechanisms of the capitalist engine. There is, however, no need of taking this structure of *Cycles* too seriously. Actually, the understanding of Schumpeter's evolutionary theory seems better served by starting from the historical account and considering his theory as designed to cover history in a highly stylised manner.

The number of pages of *Cycles* that are dedicated to theoretical analysis is small when compared with the space given to historical analysis. The difference in writing style is even more conspicuous since the theoretical arguments are highly structured while the historical analysis is loose and discursive. This feature of the 590 pages of the historical chapters is probably sufficient for most readers to ignore them. Schumpeter, nevertheless, considered them to contain the most important contribution of his book: his attempt to provide a "reasoned history" of the "capitalist process". This type of history represented the "Historical Approach to the Problems of the Cyclical Process of Evolution"; and the "importance of such an approach has to be emphasized from the outset" (*Cycles*, 220; *Cycles Abr.*, 177). Schumpeter's historical approach had roots in his early research. In a letter that Schumpeter wrote in 1941 (*BL*, 333), he told that his early work developed in three stages. He had initially been following "sociological and historical interests". Then he decided to "become an economic theorist" and, after some years, he "summed up" the results of his efforts in *Wesen*. Finally he turned to "that theory of economic evolution into which personal observation, historical studies, and theoretical work enter in proportions which it is difficult to define". The research project that led to *Cycles* may be described as resuming the early combination of observation, historical study, and theorising of economic evolution. Since the project was conceived as a major scientific effort, Schumpeter could not discuss its results in terms of his "personal observation"—and his related vision of capitalist reality. However, he could try to mobilise the personal experiences of his readers. He did so early in *Cycles* by making the statement that

> "nothing can be more plain or even more trite common sense than the proposition that innovation, as conceived by us, is at the center of practically all the phenomena, difficulties, and problems of economic life in capitalist society and that they, as well as the extreme sensitiveness of capitalism to disturbance, would be absent if productive resources flowed—either in unvarying or continuously increasing quantities—every year through substantially the same channels toward substantially the same goals, or were prevented from doing so only by external influences." (*Cycles*, 87; *Cycles Abr.*, 62)

This is a provocative statement. On the one hand, Schumpeter made the radical proposition that his interpretation of innovation explains "practically all" interesting features of capitalism. On the other hand, he claimed that this proposition is "trite common sense". One might argue ironically that "nothing" is *less* plain for most economists than the latter claim. It presupposes what needs to be demonstrated by means of a theory based on his very special concept of innovation. Nevertheless, there is no doubt that Schumpeter had come to think in this way and that he wanted his readers to do the same before they worked their way through his three successive approximations to a model of the innovation-based capitalist evolution. He realised in advance how "difficult it may turn out to be to develop that simple idea so far as to fit it for the task of coping with all the complex patterns with which it will have to be confronted" (*Cycles*, 87; *Cycles Abr.*, 62). He also realised how "completely it may lose its simplicity on the way before us". Nevertheless, he emphasised that "it should never be forgotten that at the outset all we need to say to anyone who doubts is: Look around you!"

After having related his basic model of evolutionary business cycles to Juglar cycles and Kondratieff waves, Schumpeter tried to enforce his readers to "look around" by providing plentiful historical evidence from Great Britain, the United States and Germany. Thereby, he wanted to "verify" his model of evolutionary business cycles by demonstrating that it provided the tools for writing a "reasoned history" of capitalist economic evolution:

> "Since what we are trying to understand is economic change in historic time, there is little exaggeration in saying that the ultimate goal is simply a reasoned (= conceptually clarified) history, not of crises only, nor of cycles or waves, but of the economic process in all its aspects and bearings to which theory merely supplies some tools and schemata, and statistics merely part of the material. It is obvious that only detailed historic knowledge can definitively answer most of the questions of individual causation and mechanism and that without it the study of time series must remain inconclusive, and theoretical analysis empty." (*Cycles*, 220; *Cycles Abr.*, 177)

Such histories that included both cycles and economic evolution had to some extent been written, and Schumpeter had especially enjoyed Spiethoff's (1953) condensed account. However, the theory of *evolutionary* business cycles was not well supported by the available literature; and no reasoned history of this type had been produced. The problem was especially that Schumpeter's theory required researchers to move "far beyond mere description of spectacular breakdowns, on the one hand, and of the behavior of aggregative quantities, on the other". The task that confronted him was "the formidable one of describing in detail the industrial processes" that are ultimately explaining both crises and the movement of the aggregates (*Cycles*, 221; *Cycles Abr.*, 178). Since the general literature "did not touch upon the essential thing at all", he wanted to exploit "the innumerable monographs of individual industries" as well as "on individual concerns and entrepreneurs". Thereby, he defined a "vast program" that included the "[c]oordination and systematization" of monographic work. Thereby, he hoped to "go some way toward filling the bloodless theoretical schemata and statistical contour lines with fact and toward making our meaning clearer and more vivid".

When writing his history of evolutionary business cycles since the late eighteenth century, Schumpeter made ample use of the assumption that "every cycle is a historical individual" (*Cycles*, 156; *Cycles Abr.*, 132). By this expression he emphasised that individual cycles are not fully comparable since each of them is carried by a different set of innovations that have different effects on the economic system. He largely concentrated his account of two types of cycle: Kondratieff waves (50–60 years) and Juglar cycles (nearly 10 years). In contrast, the short Kitchin cycles of about 40 months play no important role in the argument, since they have very little to do with the evolutionary process. This argument is instead influenced by the fact that it was possible for Schumpeter to decompose each the three Kondratieff waves that he observed into six Juglar cycles (see Figure 8.2 on the following page). For instance, he described the Kondratieff wave that was influenced by the introduction of railways in the nineteenth century as characterised by an upswing and a downswing. The Kondratieff upswing was characterised by the social and financial difficulties in introducing the railway innovation and the related innovations. The upswing could not be performed in a single step but required three Juglar cycles that were more or less carried by these innovations and the related "manias" of railway-related investment. The Kondratieff downswing also required three Juglar cycles before the economic system was brought to a neighbourhood of evolutionary stability.

Schumpeter generalised his empirical finding to the general assumption that every larger cycle consists of an integral number of smaller cycles. This assumption allowed a top-down approach in which main sections of his book could cover Kondratieff waves while its subsections could deal with Juglar cycles. However, he did not follow this approach systematically. He only focussed on the underlying Juglar cycles of the Kondratieff downswing of the 1920s and 1930s. The rest of his exposition of Kondratieff waves is primarily decomposed according to country and areas of industrial activity as well as to macroscopic phenomena. This strategy of exposition reveals how fond Schumpeter was of thinking in terms of Kondratieff waves. These long waves seemed particularly suited for discussing "the particular innovations that carry a given cycle, the actual structure of the industrial organism that responds to them, and

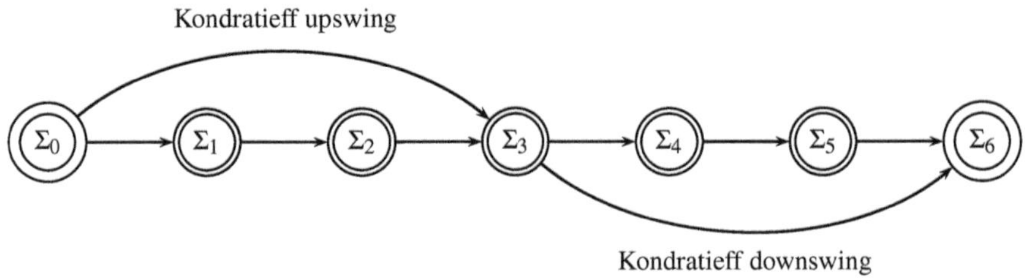


**Figure 8.2.: A Kondratieff wave that consists of six Juglars cycles**

Comment: From the viewpoint of the Kondratieff wave, only $\Sigma_0$ and $\Sigma_6$ can be described as something like equilibrium states. However, the movement between these states is performed by six Juglar cycles. Each of them moves the system to a relatively equilibrated state. For instance, the effect of the first Juglar is to move the system from $\Sigma_0$ to $\Sigma_1$.

the financial conditions and habits prevailing in the business community in each case" (*Cycles*, 143; *Cycles Abr.*, 119). The systematic analysis of such processes of evolutionary transformation of the "industrial organism" requires a huge amount of data. Although Schumpeter's account largely started with the "Long Wave from 1787 to 1842", he found reasons to believe that this Kondratieff was "not the first of its kind" (*Cycles*, 252; *Cycles Abr.*, 182). This suggestion is important since the materials that he used only covered less than three Kondratieffs and since "no phenomenon of an essentially historic nature can be expected to reveal itself unless it is studied over a long interval" (*Cycles*, 220; *Cycles Abr.*, 177). Actually, he suggested that an "intensive study of the process in the last quarter of the seventeenth and in the eighteenth century is hence a most urgent task, for ... a period of 250 years may be called the minimum of existence of a student of business cycles".

It has already been mentioned (in Section 7.3) that Schumpeter would "date capitalism as far back as the element of credit creation" (*Cycles*, 224–5; partly in *Cycles Abr.*, 180). In the case of Southern Europe, this type of dating points at the end the twelfth and the beginning of the thirteenth century. However, he did not expect to find anything like macroscopic business cycles at the beginning of capitalism. His reason was that "the impact of innovation will evidently be felt differently in a small capitalist milieu which is surrounded by a noncapitalist world of much greater quantitative importance". Instead he would look for the emergence of cycles in "the 300 years preceding the epoch usually studied for the purpose of business-cycle analysis" (*Cycles*, 231–52). This period is shortly described in Schumpeter's contribution on capitalism to the *Encyclopaedia Britannica*. While "early capitalism" had no precise starting point, it was transformed into "mercantilist capitalism" through a "symbiosis" with "the rising national states" (*S1946b*, 189–92). This symbiosis lasted from the sixteenth century to the end of the eighteenth century. It is thus in the period of "mercantilist capitalism" that Schumpeter would look for the first signs of evolutionary business cycles; but he was hardly able to present precise results of this search.

Let us instead return to the three Kondratieffs that Schumpeter argued he was able to discern: "the Kondratieff of the Industrial Revolution", "the bourgeois Kondratieff", and "the Neo-Mercantilist Kondratieff" (*BL*, 322). Since he considered the macroscopic units Kondratieffs and their phases as defining the rough contours of the process of evolution, their naming and dating are important. The names and dates are conveniently summarised in a letter that he wrote to Kuznets (*BL*, 321–2); and they are reproduced in Kuznets's (1940, 261n) review of *Cycles*. However, he also emphasised the need of additional information on the three Kondratieffs:

1. The Kondratieff of the Industrial Revolution covers 1787–1842 (*Cycles*, 252–302; *Cycles Abr.*, 182–200). It is decomposed into prosperity 1787–1800, recession 1801–13, depression 1814–27, and recovery 1828–42. Especially in England, it was dominated by cotton textiles, "the evolution of which, together with all the effects on and reactions from the rest of the economic system, affords as clear an illustration of our cyclical process as we can ever hope to find". However, canal construction was also important, and economic life was influenced by the Napoleonic Wars.

2. The bourgeois Kondratieff spans the period 1843–97 (*Cycles*, 303–97; *Cycles Abr.*, 201–46). Its phases are prosperity 1843–57, recession 1858–69, depression 1870–84/85, and recovery 1886–97. The recession period serves to demonstrate the special meaning Schumpeter gave to that term. According to him 'recession' has no welfare connotations but simply represents the more routinised spread of the carrying innovations. These innovations suggests that the whole Kondratieff should be called the age of "the railroadization of the world" or, somewhat broader, the age of "railroads, steel, and steam". However, Schumpeter primarily named it "the bourgeois Kondratieff". Thereby, he emphasised that it was the period in which not only the economic system but also social life in general was dominated by competitive capitalism.

3. The Neo-Mercantilist Kondratieff started in 1898 and was not over in 1939 (*Cycles*, 397–448, 692–1050; *Cycles Abr.*, 246–423). Its phases were prosperity 1898–1911, recession 1912–24/25, depression 1926–38, and recovery 1939–?. From the viewpoint of innovation, its "ignition" was "due to electricity" and it may be called "the Kondratieff of electricity, chemistry, and motors". Nevertheless, apart from "electrification" and "automobilization", an important feature

of the age was innovations that led to "trustification". Furthermore, Schumpeter chose to label it "the Neomercantilist Kondratieff" because "the social atmosphere characteristically changed about the late nineties". This change included a "new spirit of fiscal and social legislation" as well as phenomena like "protection and the increase in expenditure on armaments". Some of these phenomena seemed to "grew out of the very logic of capitalist evolution". Nevertheless, he largely postponed this issue to *Capitalism* and he considered World War I an obvious example of an "external factor".

Since the Second Kondratieff and its "railroadization of the world" seems to have influenced the emergence and development of Schumpeter's evolutionary model, it is hardly surprising that the analysis of the period from 1843 to 1897 includes some of his best performances. In contrast, his handling of Third Kondratieff is more confusing. The weaker fit of his model and the much richer materials available combined to make the analysis next to impossible. Furthermore, while he apparently had planned to analyse the relatively pure process of economic evolution, it became increasingly clear that he could not ignore the changing institutional framework. While his model largely assumed a given capitalist framework, the historical data reflected an evolution of the institutions of capitalism. For instance, he tried to analyse economic evolution within capitalism in terms of firms that were largely unchanged after their creation by entrepreneurs (the capitalist engine Mark I), although he was clearly aware that large firms with long-term innovative capabilities had emerged (the capitalist engine Mark II). More importantly, the socio-political environment of the capitalist economy was also changing. This becomes especially clear at the end of *Cycles*. Here we recognise that he had come to see socio-political change as the main explanation for the recession of 1937. By developing this explanation (*Cycles*, 1011–1050; *Cycles Abr.*, 384–423) he clearly went beyond the limits that he had set for himself at the beginning of *Cycles*. The main issue at stake was the theory of economic "maturity" and long-term "stagnation". Keynes had hinted at this theory in the *General Theory*, and it was being developed by his Harvard colleague Alvin Hansen. The only way of reinterpreting and criticising this theory in Schumpeter's evolutionary framework was to turn to the "inhibitions" to entrepreneurship caused by public resentment and long-term institutional change. As we have seen, this issue had engaged him extensively in the development of *Capitalism*.

Since we have already considered Schumpeter's theorising about institutional change, it is presently most enlightening to concentrate on a more pure case of capitalist economic evolution by returning to the age of the Second Kondratieff. This age was characterised by "the railroadization of the world". Schumpeter emphasised that during the initial phase of railroadization each individual railway line should be considered "an innovation within our meaning of the term" while sectional or national systems "constitute innovations of a higher order" (*Cycles*, 304; *Cycles Abr.*, 202). The different types of railway-related innovations are all characterised by their "comparatively long periods of gestation". Further characteristics are "the quantitative importance of the expenditure involved, the consequent dislocation of all the data of economic life, the new investment opportunities and the new possibilities that are created for further innovation, and the (cyclical) disturbances in turn caused by these". Schumpeter not only exploited these characteristics in his analysis of the Second Kondratieff, but even when he dealt with parts of the previous and subsequent one (especially *Cycles*, 277–8, 291–2, 325–51, 402–8). In combination, these analyses appear to constitute the skeleton of an account for railroadization as the core example of his evolutionary analysis of the capitalist economy. This account includes the issue of "railway manias" and the general influence of railway construction on the Juglars. However, the main emphasis seems to be put on broader issues. Thus Schumpeter starts his analysis of the Second Kondratieff by stating that "nobody could fail to associate it [the whole period 1843–97] with what we call the railroadization of the world" (*Cycles*, 303; partly in *Cycles Abr.*, 201). Railroadization in that period was "though not the whole of it, yet its outstanding feature." This statement "particularly applies to this country [the US], the Western and Middle Western parts of which were, economically speaking, created by the railroad." With respect to England and Germany, "the importance of their own railroads was absolutely and relatively much smaller". Therefore, the "statement should be modified to read that railroad development in the world was the outstanding feature that dominated economic activity in those countries also." In England, this dominance was maintained by exports to and financing of the railroadization of other countries.

The ambition of the following summary of Schumpeter's attempt to write a reasoned history of the age of railroadization is not to defend his proposition that the railway innovation significantly helped to produce the business cycles of that age. The present task is rather to demonstrate that Schumpeter's attempt to produce a reasoned history forced him to develop and exemplify his concept of innovation. By following Schumpeter's account, we recognise that the railway innovation was a product of the First Kondratieff. The use of railways with locomotives started in British coal mines in the beginning of the nineteenth century and the modern type of railway services with its combined transportation of goods and passengers emerged at the end of the 1820s. This form of railway operation builds on the gradual development of railway technology through a large number of inventions—like it had been emphasised generally in Colum Gilfillan's (1935) *The Sociology of Invention*. However, Schumpeter could accept this theory of invention "and yet accept another point of view for our purposes" (*Cycles*, 85n; cf. 227–8n; *BL*, 265). According to this viewpoint, the significant effects on the economic system of the "varying periods of incubation of inventions" began to take place suddenly in the form of innovations. The evolution of the routines of the economic system was started suddenly by concrete railway projects—especially by the Liverpool–Manchester railway that was completed in 1830. However, this project cannot be conceived as *the* innovation and the rest of the railway projects as examples of mere diffusion or imitation. According to Schumpeter, each of the early railway project was an "innovation" for which the S-entrepreneur with difficulty obtained credit and overcame other hindrances. However, these hindrances gradually became more moderate and this decreased the required level of entrepreneurial ability. By dating the start of the Second Kondratieff to 1843, Schumpeter ensured that it included the English "railway mania" that lasted to 1848. However, it was after this year that we see large-scale railway construction in the United States. The accumulated results of railway construction in the US and Great Britain is recorded in Figure 8.3 on the next page. This figure also includes the kilometres of railway line open in Austria-Hungary. This country is not only an example of a laggard that used British know-how and imported British products for railway construction. It is also the country in which Schumpeter originally met the effects of the railway innovation.

Let us concentrate on the data for Great Britain. While the length of the British railway line had been 160 km in 1830, it reached nearly 10,000 km in 1850. The logarithmic form of Figure 8.3 serves to demonstrate that the high growth rate of early years of British railway construction was lowered radically in the first half of the 1840. In the early years of the 1830s, the railway projects took the form of relatively small and isolated efforts. Each of these projects was a difficult innovation; and the aggregate effects of their implementation were modest. Actually, a decade of rapid expansion ended with a significant slowdown. However, "although all the essentials of railroad enterprise—types of entrepreneurs and methods of financing included—stand out fully fledged in the thirties", the "great development that within a few years created almost the whole skeleton of the English railroad system was the work of the forties" (*Cycles*, 278). Two things were becoming clear in the early years of the 1840s. First, railway technology had matured and old and new suppliers of finance and specialised inputs became eager to contribute to new projects. Second, the individual lines were becoming part of a national system of railways to which new lines could connect and thereby provide improved services for their potential customers. These emerging characteristics meant that the innovative aspects of railway projects were to some extent turned into simpler issues of imitation. Thereby, it became much easier to function as a railway entrepreneur, and this was the start

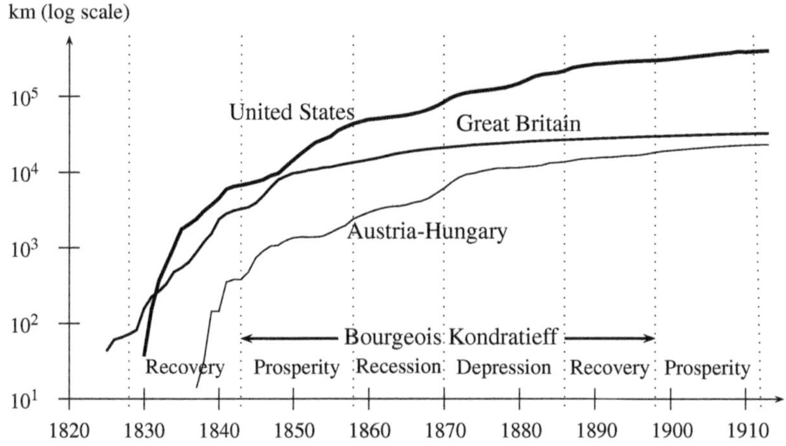


**Figure 8.3.: The railways and the Bourgeois Kondratieff**

Comment: The figure combines two types of information: (1) Total kilometres of railway line open in Great Britain, United States and Austria-Hungary, 1825–1913. (2) Schumpeter's dating of the phases of Kondratieff waves. The dotted vertical lines represent starting years of phases. For instance, the prosperity of the "Bourgeois Kondratieff" starts in 1843. Sources: Railway data are from Brian Mitchell (1981, 609–11; 1983, 656–8; 1988, 588); they have been used by Andersen (2002, 51). Schumpeter's dating is from *BL* (322).

of an explosive development. This explosion (the "railway mania") was sufficient to induce macroeconomic effects—both directly and because of the substantial amount of related investment in existing firms and within new industries and new railway towns. Later railway projects became a matter of routine and thereby totally lost their character of innovations. In any case, "the heroic age of railroad innovation that revolutionized the economic system was entirely over by about 1860". After that year, "English railroad development ... was a consequence of growth in our sense ... responding at every step to existing conditions, rather than an active factor of evolution" (*Cycles*, 342).

However, the growth rate of the British railways had already become very modest in the 1850. After that time, parts of the British economic and financial system served the railroadization of the world, including Austria-Hungary. The railway development of this country not only lagged after the leaders, it also created serious trouble that influenced the socio-political environment in which the early Schumpeter and his teachers worked. Actually, it was here that the financial crash which symbolised the start of the depression period of the Second Kondratieff broke out in 1873. The years before that crisis are "known as the 'promoter's time' (*Gründerzeit*)" (*Cycles*, 362). During this period, "[e]nterprise, spreading from the railroad business and allied lines, extended light-heartedly to everything imaginable, both methods and schemes being clearly fraudulent in many cases" (pp. 362–3). The period was also characterised by "[m]ushroom banks—many little better than bucket shops". After a period of increasingly speculative financial activities, stock market prices began to fall and "[t]he 'crisis' broke in Vienna on May 8, 1873, in a most dramatic way". This event was a major factor of economic and political life of Austria-Hungary, and it was still referred to when Schumpeter was a student. Furthermore, the crisis spread to other countries. For instance, there was "great epidemic of financial and industrial bankruptcies" in Germany, but here it created "much less panic."

The economic importance of such a crisis should not be overestimated. Actually, Schumpeter did not consider crisis "part of the logic of our process", but instead one of the "surface phenomena of the Secondary Wave". With respect to Germany, he characterised the basic issue that "[p]rices high enough for the new firms in an industry were much too low for the old ones. This spelled bad business, loss, death for large strata of the economic structure". Since the "possibilities for advance were obviously exhausted", the whole economic "milieu had to adapt itself the best it could" (*Cycles*, 363–5). Even in the United States, "the larger process—mainly associated with railroad construction—within which the events of 1870 to 1873 constitute a step, had so revolutionized the economic system that liquidation, absorption, adaptation ... was an unusually long and painful affair" (*Cycles*, 338–9; *Cycles Abr.*, 228). However, this depression and the subsequent recovery were not without innovations in the wider sense. Actually,

> "Kondratieff downgrades and revivals precisely display a wide variety of induced or complementing innovations which develop and carry to their limits possibilities opened up before, of which railroad building was but the most important. Accordingly, railroad construction, increasingly settling into a predetermined framework and exploiting pre-existing investment opportunities, became during the period under discussion much more (though not yet entirely) a function of railroad business and, hence, of the rest of the business organism than it had been before, and the relation became substantially one of mutual dependence." (*Cycles*, 339; *Cycles Abr.*, 229)

Schumpeter thus suggested a movement from the asymmetric impulse and response of the innovation-based periods via "mutual dependence" to the circular flow of economic life. In the strict version of the latter state, evolution has come to a halt. However, such a definition of the situation cannot explain the start of the next Kondratieff wave. Therefore, he had, in his analysis of the depression and recession of the Second Kondratieff, to look for the emergence of the carriers of the Third Kondratieff.Since he already knew these carriers, the task was not difficult. Thus, he in the data on the late 1870s, could recognise the beginnings of the "use of petroleum for other purposes that lighting". This innovation was, "like electricity, a 'carrying' innovation of the next Kondratieff, and was in the incubating stage during the second" (*Cycles*, 384; *Cycles Abr.*, 233). The innovations related to the use of electricity emerged similarly. The subsequent "electrical developments that we observe in the later nineties, spreading their effects over the industrial field" actually seem "in themselves to have been sufficient to produce what we call a Kondratieff prosperity and to impress a dominant contour line on the subsequent business situations" (*Cycles*, 411; *Cycles Abr.*, 258).

The questions of the existence and explanation of Kondratieff waves are today controversial, and Schumpeter's answers are outdated. For instance, a great collector of the longest possible economic time series, Angus Maddison, rejected the existence of Kondratieff waves and wanted instead to use his data as evidence of differently defined "phases of capitalist development". In this perspective, Maddison (1982, 78) pointed out that "Schumpeter's cycle analysis runs to 1,050 pages and is highly discursive. Judged on its statistical evidence alone, it would have been long discredited." He emphasised, instead, that the book's "power lies in the imaginative theory he supplies to explain long waves and the highly illuminating commentary on many aspects of German, British, and American economic history" in terms of "upsurge in innovation and entrepreneurial dynamism" as well as "creative destruction, during which old products, firms, and entrepreneurs were eliminated and new products were conceived". Attempts to rescue Schumpeter's basic idea of Kondratieff waves are found in theories of the sequence of economic transformations due to "general purpose technologies" (Lipsey et al., 2005) and "techno-economic paradigms". The latter concept was expanded in the recent account for the history of capitalism in terms of Kondratieffs by Christopher Freeman and Francisco Louçã (2001). However, a researcher like Carlota Perez (2007; 2002, 23n, 11) has suggested that it is better to drop the concept of Kondratieff waves and instead to speak of "great surges of development" with an added phase of the microscopic start. For instance, she would date the start of the "surge", based on steam and railways, to 1829, when the steam engine for the Liverpool–Manchester railway was successfully tested. In contrast, dating according to the start of the related Kondratieff wave ranges from 1843, used by Schumpeter, to Freeman and Louçã's more standard year of 1848.

## 8.3 The Kondratieffs and Juglars of the third approximation

Schumpeter's final step to develop theoretical support for his historical studies, which were characterised in the previous section, was to formulate his "third approximation" to the historical and statistical complexities of real evolutionary business cycles (*Cycles*, 161–74; *Cycles Abr.*, 137–50). The previous "approximations" had assumed that "there is in our material a single sequence of cycles, each of which is of the same type as all its predecessors and successors." The last version of his model changes this assumption by including "many simultaneous cycles". The reasons for this extension are both empirical and theoretical. The empirical reason is that "the impression we derive from any graph of economic time series [does not] lend support to a single-cycle hypothesis". The theoretical reason is that Schumpeter's general theory and his two first approximations provide "no reason why the cyclical process of evolution should give rise to just one wavelike movement." Since innovations are of many types, it seems more reasonable "to expect that is will set into motion an indefinite number of wavelike fluctuations, which will role on simultaneously and interfere with one another in the process" (*Cycles*, 161–2; *Cycles Abr.*, 137–8).

Schumpeter specified his theoretical reasons for expecting more than one "wavelike movement" in terms of three alternative cases (*Cycles*, 166–9; *Cycles Abr.*, 141–4). In the first case, innovations have different "periods of gestation and absorption", which will generally correspond to their relative importance. This means that innovations with short periods between initiation and routinisation can function "on the back of the wave created by ... innovations of relatively long span". The reason for this kind of interference is that when the "wave of a long span is in its prosperity phase, it will be easier for smaller waves ... to rise". In contrast, "in the depression phase of the underlying wave it may be impossible for them to rise visibly at all". However, if the smaller waves emerge during the depression of the larger wave, their small-scale prosperities soften this depression while their depressions serve to intensify the longer-term depression. The second case does not assume that innovations are independent: "Major innovations hardly ever emerge in their final form or cover in one throw the whole field that will ultimately be their own." As a result, we see that when "some innovation has been successfully carried into effect, the next wave is much more likely to start in the same or a neighboring field than anywhere else". For instance, "[o]ne railroad or a few lines may be all, or more than all, that can be successfully built in any given environment at a time. Reaction and absorption may have to follow before a new wave of railroad construction becomes possible." Thus we see that "innovation is carried out in steps each of which constitutes a cycle." Furthermore, although we may think in terms of an overall cycle of "railroadization", this cycle consists of smaller cycles and "has no existence of its own." The third case is more complex. It suggests "a cluster of cycles of various span" that are not only "superimposed on each other" but also produces a cyclical "change in the economic and social structure of society". The main exemplification again comes from "railroadization":

> "Expenditure on, and the opening of, a new line has some immediate effects on business in general, on competing means of transport, and on the relative position of centers of production. It requires more time to bring into use the opportunities of production newly created by the railroad and to annihilate others. And it takes still longer for population to shift, new cities to develop, other cities to decay, and, generally, the new face of the country to take shape that is adapted to the environment as altered by the railroadization." (*Cycles*, 168; *Cycles Abr.*, 143)

This example obviously applies Schumpeter's model of waveform economic evolution in a way that opens up for a multiplicity of cycles of different spans. The overall wave of railroadization is not an example of ordinary business cycles, but of "what it has become usual to call the Long Wave". This wave is completely different from his two other cases of multiple forms of waves. The long wave provides a story about an epoch and concerns "all industrial and commercial processes in that epoch". The example of railroadization also demonstrates that we are facing "a real phenomenon and not merely the statistical effect of a sequence of real phenomena". However, he did not give up the other cases. Instead, he upheld three alternative theories of waveform economic evolution, probably because he considered them complementary. Nevertheless, he did "make a *decision*" about the number and types of cycles that he wanted to include in his analysis. This decision was made because "it would be highly inconvenient ... to attempt to work with an indefinite number of cycles or classes of cycles." He had attempted to work with five cycles, but "came to the conclusion that the improvement in the picture would not warrant the increase in cumbersomeness." His conclusion was "to content ourselves, for the rough purposes of this volume [*Business Cycles*], with three classes of cycles, to which we shall refer simply as Kondratieffs, Juglars and Kitchins" (*Cycles*, 169; *Cycles Abr.*, 144). The present account focusses on the Kondratieff waves and Juglar cycles. In contrast, the Kitchin cycles of about 40 months are largely ignored. Schumpeter added them to cover a type of oscillation that has more to do with the change of inventories than with economic evolution.

By working in terms of three simultaneous cycles, Schumpeter claimed that "we fall in with the general tendency in the study of business cycles". To demonstrate this claim, he provided a brief history of this field of study (*Cycles*, 162–5; *Cycles Abr.*, 138–41; cf. *History*, 1122–35). It started with primitive theorising about "crises" and the great advance "is primarily associated with the name of Clément Juglar". This researcher had "a clear perception of how theory, statistics, and history ought to cooperate in our field." He also changed the field by focussing on "the mechanism of alternating prosperities and liquidations" and suggested that the latter reflects the "reaction of the economic system to the events of the former." He, and his followers, did so by studying "a single wavelike movement". One of these followers was the early Schumpeter "who, when starting work on the business cycle nearly 30 years ago, also accepted the single-cycle hypothesis". However, the researchers faced "irregularities which now crowded upon them." Their sluggish response was to exchange "the problem of *the* wave" with "the problem of identifying and, if possible, isolating the many waves and of studying their interference". A major contributor had been Arthur Spiethoff (see Section 5.4), but Schumpeter emphasised the more operational contributions of Joseph Kitchin and Nicolai Kondratieff. The former had (together with Crum) discovered that in addition to the ordinary Juglar cycle of 9–10 years, the time series data also reveal a shorter cycle of about 40 months. In contrast, Kondratieff (1935) had, in a paper originally published in Schumpeter's *Archiv für Sozialwissenschaft*, described statistically the "Long Wave" of between 50 and 60 years and presented it as a "characteristic of the capitalist process". Although Schumpeter was fully aware of other cyclical patterns with economic causes as well as the cyclical effects of "external factors" such as climate and wars (see *Cycles*, 174–192; *Cycles Abr.*, 150–62), he decided to focus on the Kondratieffs, Juglars, and Kitchins.

In accordance with the tradition of business cycle studies, Schumpeter started with the "analysis of time series" that cover as long periods as possible—in principle from the late eighteenth century to 1939. The time series of his many wave indicators were largely collected and graphed by his research assistants, but he had to discuss their relevance for the understanding of his theory of the capitalist process. This treatment fills many of the pages *Cycles* (Chs 5, 8–13; *Cycles Abr.* only includes parts of Ch. 5), but they mainly showed up to have a very limited purpose: "from our standpoint the first and foremost task of time-series analysis" is to locate "the intervals on our graphs which correspond to neighborhoods of equilibrium" (*Cycles*, 206; *Cycles Abr.*, 171). Ever since Schumpeter's first formulation of a set of theses on economic evolution and business cycles (see Table 5.2 on page 119), he had been convinced that a cycle begins and ends in a state of economic "equilibrium". This cycle basically has two phases—in *Cycles* called "prosperity" and "recession". Prosperity is the movement away from the initial equilibrium caused by entrepreneurial activity. Recession is the movement to a new equilibrium in which the results of the innovations of the S-entrepreneurs have been absorbed in what Schumpeter called "the economic organism" (*Cycles*, 68; *Cycles Abr.*, 42). Thus, he did not only consider prosperity and recession as labels for periods that can be determined by analysing time series of economic variables, but also as names for two basic mechanisms of the capitalist engine. However, he could only make a quite limited—and from an evolutionary point of view rather arbitrary—use of the available statistical data for the description of these mechanisms. Instead, he followed Ragnar Frisch's idea of defining the relatively equilibrated situation at the start of a cyclical unit of economic evolution by means of the inflection point of a smoothed version of an oscillating time series (*Cycles*, 208–11). Given such determination of the historical timing of the cycles, he could proceed to a verbal interpretation of the underlying evolutionary processes and their apparent influence on the aggregates.

Since Schumpeter considered the statistical time series as reflecting "economic growth and the cyclical process of evolution as distorted by the influence of external factors" (*Cycles*, 193; *Cycles Abr.*, 163), the very limited use of them were hardly satisfactory from the viewpoint of his analysis. Actually, he lacked indicators that directly reflected economic evolution. More specially, he lacked evolutionarily relevant time series of microdata and methods for linking them with the macrodata that he primarily relied upon. In this respect, today's situation is obviously very different—as demonstrated by reviews of the use of longitudinal microdata for productivity studies (Bartelsman and Doms, 2000). Although Schumpeter had some time series for individual commodities, even these data were so aggregative that they hardly gave any indication of the underlying evolutionary processes. Therefore, it may seem surprising that he did not use the emerging data on the fate of individual industries and the relationship of this fate to the cycles and "secular trends" of the American economy. Although these data had been collected and analysed by Simon Kuznets (1930) and Arthur Burns (1934), it is hardly an explanation that they, in contrast to Schumpeter, belonged to the institutionalist camp of economists. A better explanation is that Kuznets and Burns used their data to develop a loosely defined "theory of retardation" of the growth of the economic system and that they applied methods that suited this purpose (Metcalfe, 2003). In contrast, Schumpeter's evolutionary theory concentrated on the swarms of industrial renewal and growth while the subsequent phase of retardation must have seemed obvious to him. Furthermore, the analysis in terms of industries as defined by statistical offices is definitely not the best way of detecting waves of economic evolution. Nevertheless, he seems to have missed an opportunity for describing parts of his process statistically. The consequence of Schumpeter's frustrated efforts of linking his time series with the details of his process of evolution is partly that the statistical chapters of *Cycles* are strangely uninspired, and partly that he tried to draw conclusions from aggregate data that really cannot be drawn without complementary microdata.

Schumpeter discussed the expected macroscopic effects of his three simultaneous waves by means of the *only* analytically motivated drawings found in *Cycles* (213–14; *Cycles Abr.*, 175–6). A simplified version of the

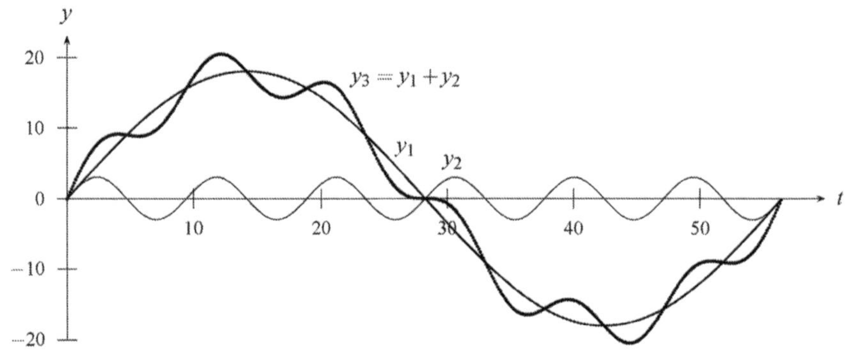


**Figure 8.4.: Schumpeter's didactic example of his third approximation: two simultaneous sine waves and their aggregate**

Comment: Schumpeter interpreted $y_1$ as the Kondratieff wave and $y_2$ as Juglar cycles. His even shorter Kitchin cycles have been excluded. $y_3$ is the aggregate effect of the two included oscillations. His original figure was, of course, not intended to be realistic. Actually, his underlying idea is probably better represented by Figure 8.2 on page 202. Source: The figure was drawn after the mathematical specifications in *Cycles* (p. 1051).

basic drawing is reproduced in Figure 8.4. Here a Kondratieff and six Juglars are represented in their simplest possible form—as sine waves. Furthermore, the two types of waves are assumed to interact in an additive way. Thus the $y_1$ of the Kondratieff is added to the $y_2$ of the Juglars to produce the aggregate result $y_3$. One of Schumpeter's points was that the "extreme regularity" of the components results in "so very irregular-looking a composite" (*Cycles*, 212; *Cycles Abr.*, 174). Actually, the composite that Schumpeter commented upon is more irregular than that of Figure 8.4. The reason is that it was not only produced by a Kondratieff and six Juglars, but also by 18 Kitchins. He was obviously happy with this ingenious construct, but his use of Kitchins in his statistical and historical analysis is very limited. Since these short cycles furthermore show up to be unexplainable by means of his evolutionary theory, he would have served his purposes better by trying to remove them from his time series. In any case, the waves of the drawing were characterised by perfect regularity and Schumpeter had to emphasise that "of course, we do not, as a matter of principle, postulate either internal regularity or sine form" (*Cycles*, 212; *Cycles Abr.*, 174).

Since students often emphasise graphical elements of books more than warranted, it is important to study the totally unrealistic assumptions underlying Schumpeter's chart of the three sine curves (*Cycles*, 1051). These assumptions can be related to the famous mathematical constant

$\pi \approx 3.14$. Let us call the length of the sine wave the 'period' and its maximum displacement around the mean the 'amplitude'. Then function $a \sin(bt)$ has an amplitude of $a$ while $b$ is the number of radians per unit of time (Schumpeter used degrees per unit of time). If $a = b = 1$, we have an ordinary sine wave with an amplitude of 1 and a period of $2\pi \approx 6.3$ years, which is not the characteristic of any typical business cycle. However, if $a = 1$ and $b = 2$, we get something like Kitchin cycles with period of $\pi \approx 3.1$ years. Juglar cycles might be considered as having triple amplitude but also a period that is three times as long. Thus $a = 3$ and $b = 2/3$, which gives a period of $3\pi \approx 9.4$ years. Kondratieff cycles are represented by $a = 18$ and $b = 2/18$, which gives a period of $18\pi \approx 56.5$ years. This result is extremely elegant. Kitchin suggested a wave close to $\pi$, Juglar one that encloses $3\pi$, and Kondratieff one that encloses $18\pi$. Furthermore, Schumpeter had found that "it is possible to count off, historically as well as statistically, six Juglars to a Kondratieff and three Kitchins to a Juglar—not as an average but in every individual case" (*Cycles*, 173–4; *Cycles Abr.*, 149). Now, he, by the simplest possible means, produced this result. Since he apparently used the result reversely as a guide for his statistical and historical analysis, we recognise why his "schemata of three interacting cycles … seemed less commanded by the facts of economic history than by Schumpeter's fascination with the mysteries of harmonic analysis" (Samuelson, 1981a, 3). This is the background for Samuelson's comment that the business-cycle project "began to smack of Pythagorean moonshine" (quoted by McCraw, 2007, 253). Such moonshine has sometimes led to scientific progress, like when Kepler's Laws of planetary movement emerged from the search for an application of what has been called the Platonic, or Pythagorean, solids (Koestler, 1989). With respect to evolutionary business cycles, however, "the logical expectation from the fundamental idea would be irregularity" (*Cycles*, 174; *Cycles Abr.*, 149). The elegant construct leads away from this insight.

Schumpeter, of course, was very cautious about the sine-formed version of his three-cycle scheme. He presented it as "an illustration of all the boldest assumptions which it is possible, and to some extent permissible to make in order to simplify description and to construct an ideal schema with which to compare our observations" (*Cycles*, 212; *Cycles Abr.*, 173–4). One of the assumptions that he did defend was that of an integral number of smaller cyclical units within the unit of a larger cycle: "Each Kondratieff should contain an integral number of Juglars and each Juglar an integral number of Kitchins" (*Cycles*, 172; *Cycles Abr.*, 147). His warrant was given by "the nature of the circumstances which give rise to multiplicity." Even if the innovations of a shorter wave "are entirely independent of the innovations which carry the longer wave", the longer wave defined the conditions for innovative activity related to the shorter wave. These conditions are different in Schumpeter's three cases of multiple cycles, and the rather shaky reasons for an integral number of smaller cyclical units within the larger cycles have to some extent been given. If the innovations carrying the larger cycle are independent from the smaller innovations, the larger cycle will nevertheless define the conditions for their launching. These conditions are best when the larger one has reached an equilibrium state—like in Figure 8.4, where both the Kondratieff and its first Juglar start at the same point of time. If the larger wave simply consists of a series of similar innovations, like railway projects, it will obviously end when this type of innovation and the copying of it have lost its potential. If we are facing a Kondratieff that restructures the whole of the economic system, then its cyclical potential ends when the system has adapted to, for instance, "railroadization".

The interaction between Juglars and Kondratieffs suggested to Schumpeter that his original assumption that a smaller wave starts in a "neighbourhood of equilibrium" cannot be upheld. Actually, full evolutionary equilibrium is not even theoretically possible except in relation to Kondratieffs. Since such equilibrium cannot be reached when this wave turns from recession to depression, it can only exist at the beginning of the Kondratieff—that is, once every 50 to 60 years! Furthermore, even this equilibrium is hardly a strict one since the innovation that carry a Kondratieff typically emerges in the downswing of the previous Kondratieff and has effects of some influence in the upswing of the next Kondratieff. However, Schumpeter only emphasised that "shorter waves must in most cases rise from a situation which is not a neighborhood of equilibrium but disturbed by the effects of the longer waves". Therefore, he had to "modify our previous assumption that the process of innovation starts from such neighborhoods only, as well as our concept of neighborhood of equilibrium itself" (*Cycles*, 173; *Cycles Abr.*, 148). He was not too clear about the needed changes. Nevertheless, with respect to smaller cycles, the alternative it to assume that businessmen "act as a rule on the assumption that the conditions of a phase of longer cycles as if these conditions were permanent." This is obviously a weak substitute for the strong concept of equilibrium and the stationary circular flow that Schumpeter started from. Apparently, the logic of the scheme of multiple and simultaneous cycles of evolution forced him to make this substitution before he had time to think out the issue in terms of the underlying evolutionary process. He had found out that this process really consists of sub-processes that work in parallel, but he had no clear concept of the way in which these processes are coordinated on a short-term basis. Part of the solution seems to be available today in the form of replicator dynamics that are based on the distance of evolutionarily relevant characteristics of firms from the statistical mean of these characteristics (see Appendix D.2).

Although Schumpeter appears to have come close to giving up his concept of equilibrium, even in its weak form of a neighbourhood, he nevertheless upheld his idea that the economic system can be close to and far from a relatively coordinated state. Therefore, the stylised dynamics depicted by Figure 8.4 still had meaning for him. This meaning is related to the qualitative difference between the state of the cyclical movements at extremal points and at their mean. However, since he did not assume sine waves, he preferred Frisch's idea of defining the relatively coordinated situations in terms of inflection points. These points, are of course, also present in the abstract model of sine waves. They become clearly visible if we take the derivative of the composite wave (*Cycles*, 214, 1051; *Cycles Abr.*, 176). In terms of this derivative, the speed of change is at its maximum at the start and the end of a Juglar as it is defined by Schumpeter. Thus we can measure waves from peak to peak, and in the case of the first Juglar these peaks are at $t = 0$ and $t = 9.4$. Furthermore, we recognise the influence of the Kondratieff on the economic performance during the Juglars. While we, during the first Juglar, only very shortly experience negative rates of change, the change rates are negative during most of the third and the fourth Juglar. However, the whole construction builds on the additivity of the sine waves: we simply add the effects of the Juglars to that of the Kondratieff. This is not one of the assumptions that Schumpeter wanted to uphold. For his "rough purposes" he, instead, assumed "that they are logarithmically additive". In any case, this modification did not influence a feature that can most easily be grasped by considering the effect of the depression phase of fifth Juglar in Figure 8.4. The feature is "that the coincidence at any time of corresponding phases of all three cycles [including Kitchins] will always produce phenomena of unusual intensity". His example was that the "three deepest and longest 'depressions' within the epoch covered by our material—1825–1830, 1873–1878, and 1929–1934—all display that characteristic" (*Cycles*, 173; *Cycles Abr.*, 149).

## 8.4 The pure model of the first approximation

Although Schumpeter's reasoned history of capitalist economic evolution made heavy use of his version of Kondratieff waves and Juglar cycles (see Section 8.2), most of his analysis seems made by means of simpler versions of his evolutionary theory. Furthermore, his third approximation is not fully understandable except on the background on his simpler models. Therefore, it is time to consider his "first approximation" (in the present section) and his "second approximation" (in the next section).

Although Schumpeter's first approximation to his model of evolutionary business cycles is apparently not the one he applied in most of *Cycles*, it has a very prominent place in the exposition. Actually, his treatment of the first approximation in the beginning of Chapter 4 of *Cycles* (130–

45; *Cycles Abr.*, 105–21; cf. Oakley, 1990, Ch. 8) is prefaced by nearly a hundred pages on what may be considered the closely related 'zeroth approximation'. The reason for this emphasis is obviously that Schumpeter felt at home when dealing with his basic model of the capitalist engine—while the subsequent approximations to reality forced him into foreign ground. The assumptions with which Schumpeter felt at ease were a starting point in a stationary circular flow characterised by a "neighborhood of equilibrium" and perfect competition; a subsequent disturbance caused by innovations that are carried out by new men with new plant in new firms; and a return to a new equilibrium caused by competitive forces. A cyclical unit thus "consists of two distinct phases, during the first of which the system, under the impulse of entrepreneurial activity, draws away from an equilibrium position, and during the second of which it draws toward another equilibrium position" (*Cycles*, 138; *Cycles Abr.*, 114).

Schumpeter's two-phase model of evolutionary business cycles is depicted by Figure 8.5 on the facing page. This figure is designed to emphasise that we are facing the logical reconstruction of what he considered to be the facts of the evolutionary history of the capitalist economy. The starting point is Circular Flow #1. Its transformation into Circular Flow #2 requires an innovative prosperity and an adaptive recession. Prosperity #2 starts at time $t_1$ and ends at $t_2$. The subsequent Recession #2 begins when the finishing of the innovative projects causes both a monetary deflation and a destructive competition between new and old firms. This recession covers the period from $t_2$ to $t_3$. The result is Circular Flow #2, where all firms have adapted to the new situation or exited. Sooner or later, the new innovative Prosperity #3 will emerge. Although Schumpeter did not specify the duration of Circular Flow #2 ($t_4 - t_3$), Figure 8.5 on the next page is exaggerated. Actually, the duration of circular flows might approach zero. In any case, the duration of the whole of Schumpeter's cyclical unit of evolutionary transformation is best measured by the distance between the start of one prosperity and the start of the next, that is, by $t_4 - t_1$.

Schumpeter emphasised that "there is nothing in the working of our model to point to periodicity in the cyclical process of economic evolution if that term is taken to mean a constant period" (*Cycles*, 143; *Cycles Abr.*, 119). The durations of prosperities and recessions "will depend on the nature of the particular innovations that carry a given cycle, the actual structure of the industrial organism that responds to them, and the financial conditions and habits prevailing in the business community in each case". It might be added that the malfunctioning of the banking system may cause "crises", often around the top of the wave; and these crises will not only influence the form of the cycle but also the duration of the recession phase. Furthermore, random events will probably influ-

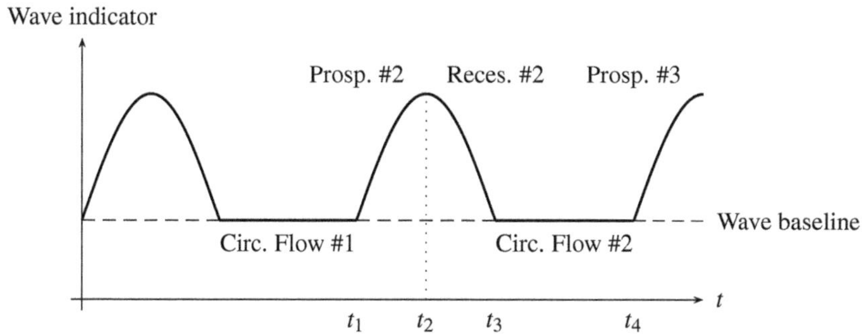


**Figure 8.5.: The first approximation: two-phase waves of evolution**

Comment: The historical sequence of a wave starts from Circular Flow #1. Through the Prosperity #2 and the Recession #2, the Circular Flow #2 is created. The wave indicator is *not* physical production, which moves counter-cyclically. Furthermore, other standard aggregates hardly serve as convincing indicators. Instead the indicator might perhaps be thought of as the statistical variance of evolutionarily relevant behaviour (see Appendix D.2). Source: The figure is inspired by Marschak (1940, 890) and Oakley (1990, 171).

ence the duration of circular flows. In any case, Schumpeter emphasised that "it seems entirely unjustified to deny the existence of a phenomenon because it fails to conform to certain arbitrary standards of regularity" (*Cycles*, 143; *Cycles Abr.*, 119).

Schumpeter did not provide anything like Figure 8.5 in *Cycles*. Therefore, he was not forced to specify what is measured on the vertical axis. Instead of defining a clear-cut wave indicator, he produced an unstructured list of 41 such indicator variables (*Cycles*, 15–17). He probably considered the issue an empirical one, but it is also of theoretical importance. For instance, the wave indicator of the figure cannot be interpreted as some index of physical output. The reason for this is that Schumpeter argued strongly that "the picture of the working of our model ... does not give to prosperity and recession, relative to each other, the welfare connotations which public opinion attaches to them" (*Cycles*, 142; *Cycles Abr.*, 118). On the contrary, it suggests that output moves in a countercyclical manner. The explanation is that resources are withdrawn from the purposes of consumption during prosperity, while the availability of increased quantities of consumption goods is a characteristic of Schumpeterian recession. The figure could, of course, be redrawn according to this pattern; but then it would not fit Schumpeter's more complex models in which output to some extent behaves according to "public opinion". A stronger candidate for a wave indicator variable would be the price level. Under the model's assumption of full employment, this level must increase during prosperity when S-entrepreneurs bid up prices by means of newly created credit and the price level decreases during the harsh competitive conditions during recession. However, Schumpeter abstained from giving priority to this indicator because he did not believe in a meaningful index of the price level. Instead, he chose to apply many indicators in order to follow "the pulse of economic life as a whole". This means that he could pick "series of price levels or of physical volume of production, or ... series of interest rates, clearing-debits, unemployment, pig-iron production, ... or the sum total of deposits" (*S1935d*, 146). One possibility that Schumpeter did not even consider is to define an indicator of economic evolution that does not necessarily correspond to the indicators of ordinary business cycles. Such an indicator might be provided by measuring the movement of the mean and the variance of routine-based characteristics of the behaviour of firms. Then the variance of evolutionarily relevant behaviour is increased by prosperity, decreased by recession, and reaches a minimum in the circular flow. Schumpeter had obviously given up the search for such a statistical indicator and turned to economic history to describe the direct pulse of the evolutionary process. This turn to qualitative data reflects the fact that the plentiful quantitative data only cover ordinary business cycles and does not measure directly the waves of evolution. However, this difference excludes a systematic analysis of the interplay between standard business cycles and economic evolution. We shall return to this issue in Appendix D.2.

Schumpeter's two-stroke model of the capitalist engine suggested that the waves of evolution are nearly identical to the waves of profit-based business activity—just like he had done in this youth. Nevertheless, the later Schumpeter made important progress with respect to the specification of and arguments for this proposition. His improvements in *Development* concerned both the prosperity phase and the recession phase. With respect to the former, the analytical improvements partly consisted in a careful description of the real and monetary consequences of the activity of the S-entrepreneurs. However, Schumpeter also had to explain the increase of the variance of evolutionarily relevant economic behaviour. To increase variance significantly, the activity of a single S-entrepreneur is not sufficient. Instead, he needed "the swarm-like appearance of entrepreneurs" as the "cause of periods of boom" (*Development*, 214). Emil Lederer suggested, by interpreting a survey be Adolph Lowe (1925), that Schumpeter's theory of prosperity is "not satisfactory because it does not try to explain why the entrepreneurs appear periodically in swarms as it were, what the conditions are under which they can appear and whether they will always appear and why, if the conditions are favorable for them" (quoted in *Development*, 214). Thereby Lederer seems to have asked for a deterministic model of swarming. This was not provided by Schumpeter. His argument was instead that if and when the first entrepreneur has demonstrated the possibility of an innovation, then a swarm of related innovators and imitators is likely to emerge. Furthermore, he argued that the best conditions for the emergence of the first entrepreneur were those of the equilibrium state.

Although Schumpeter's arguments on swarming are not fully satisfactory (*Development*, 223–36), he did formulate swarming as a testable hypothesis. As emphasised by Silverberg and Verspagen (2003), the null hypothesis of no swarming is a Poisson process in which the stochastic waiting time for an innovation is constant (or, for instance, exponentially decreasing). Actually, Schumpeter related to this test in the beginning of his answer to Lederer: "new combinations are not, as one would expect according to general principles of probability, evenly distributed through time—in such a way that equal intervals of time could be chosen, in each of which the carrying out of one new combination would fall—but appear, if at all, discontinuously in groups or swarms" (*Development*, 223, emphasis removed). Schumpeter also emphasised that the "swarm-like appearance of entrepreneurs, which is the only cause of the boom, has a qualitatively different effect upon the economic system from that of a continuous appearance evenly distributed in time ... [I]t does not ... mean a continuous, and even imperceptible, disturbance of the equilibrium position but a jerky disturbance, a disturbance of a different order of magnitude" (p. 231). His question and answer was then: "Why do entrepreneurs appear, not continuously, ... but in clusters? Exclusively because the appearance of one or a few entrepreneurs facilitates the appearance of others, and these the appearance of more, in ever increasing numbers" (*Development*, 228, emphasis removed).

Although Schumpeter in *Cycles* (100; *Cycles Abr.*, 75) only used the term "cluster", the image of a "swarm" of S-entrepreneurs is apt. Like swarms of bees gather around queens, swarms of innovating and imitating S-entrepreneurs emerge in the fields of activity created by great innovators ('kings'). This swarming depends not least on three processes. First, it becomes gradually easier to recognise and exploit the profit opportunities indicated by a swarm of S-entrepreneurs. One reason is that it is easier to detect a big swarm than an entrepreneurial 'king' or a small swarm. Another reason is that the pioneers have made things easier for their followers: "whenever a new production function has been set up successfully and the trade beholds the new thing done and its major problems solved, it becomes much easier to do the same thing and even to improve upon it" (*Cycles*, 100; *Cycles Abr.*, 75). In other words, the experiences from the first projects can be transferred to next ones—both directly and through the suppliers of equipment and consultancy. Furthermore, it is easier to obtain credit when the banks know the basic characteristics of the innovation—as demonstrated by the railway projects of the nineteenth century. Second, swarming depends of the distribution of abilities of would-be S-entrepreneurs. We can, in principle, rank these potential entrepreneurs according to their abilities to carry through S-enterprise (see Section 4.3). Only very few of them can overcome the enormous difficulties related to the carrying through of innovations in areas where the process of swarming has not yet begun. There are more, but still relatively few, who can follow immediately after the pioneers, since their task has both elements of pure innovation and imitative activity. Even more have the ability of imitating the first group of imitating innovators, and so on. Thereby, the swarm of S-entrepreneurs grows faster and faster. Third, swarming can be described as a process that spreads from the point of "ignition" in the network of economic routine to still more distant parts of the network. This idea implies a testable hypothesis: "that innovations are not at any time distributed over the whole system at random, but tend to concentrate in certain sectors and their surroundings". The implication of this concentration is that "[p]rogress—in industrial as well as in any other sector of social and cultural life—not only proceeds by jerks and rushes but also as one-sided rushes productive of consequences other than those which would ensue in the case of coordinated rushes" (*Cycles*, 100–2; *Cycles Abr.*, 75–6). For instance, the early railway projects did not only suggest "rushes" of other railway projects but also innovations within supporting industries, and these innovations suggested further renewal in a widening circle of industries. This means that we see a whole "cluster" of more or-less-related S-enterprises and that capitalist "evolution is lopsided, discontinuous, disharmonious by nature".

The above analysis of the process of swarming does not fully cover *Development*'s response to the mentioned criticism by Lowe and Lederer. They had claimed that Schumpeter did not specify the conditions under which entrepreneurial swarming is likely to occur. He gave two variants of an answer. The first answer deals with the reasons why swarming will come to an end. One reason is that the impulse of swarming tends to disappear when "so many new enterprises spring up that they would produce, when in full swing, a quantity of product which, through the fall in prices and rise in costs[,] … would eliminate entrepreneurial profit" (*Development*, 235). Another reason is that "the action of the group of entrepreneurs has in the meanwhile altered the data of the system, upset its equilibrium". This disequilibrium "makes accurate calculation impossible in general, but especially for the planning of new enterprises". It represents " the characteristic uncertainty which results from the new creations of the boom". Yet another reason is "acts of intervention from without" like "a conscious pull on the reins by the central bank" (p. 236). The closely related second answer is that the best conditions for starting the process of swarming are found in a stationary circular flow. The idea is that the decision-making of the pioneering S-entrepreneurs is influenced by their evaluation of the "general business situation" is "normal or above of below normal". This situation is "normal if it allows all [incumbent] firms not working under advantages or disadvantages particular to them" to earn enough "to go on without either increasing or decreasing their investment" (*Cycles*, 4). This situation provides would-be S-entrepreneurs with favourable conditions for calculating the profitability of their S-enterprises. Among the favourable conditions are that prices of the needed inputs are well known and that expectations about revenues are supported by the "normal" quantities and prices of competing products. Although innovative projects are characterised by irremovable uncertainty, the equilibrated economy gives the best possible conditions for profitability calculations and the related persuasion of bankers to provide credit. Furthermore, the equilibrated economic system is characterised by the lowest possible rate of interest and the best possibility for credit expansion. The assumptions of the circular-flow model imply that the low rate of interest is not exploited for the expansion of incumbent firms. Instead, it benefits would-be S-entrepreneurs. When a few pioneers have persuaded the banks that it pays to finance their innovative projects, substantial credits can relatively quickly be mobilised at a relatively low rate of interest for further S-enterprise. It is the combined expectations of the entrepreneurial world and the banking world that creates the process of swarming—and thus prosperity.

Schumpeter interpreted Juglar's statement that "the only cause of depression is prosperity" as meaning "that depression [read: recession] is nothing more than the economic system's reaction to the boom ... so that its explanation is also rooted in the explanation of the boom" (*Development*, 224). The basic explanations of recession are that the swarm of S-entrepreneursdissolves and that a crowding out of old routines takes place. Let us shortly summarise the reasons for the narrow limits of the absolute size of a swarm of innovative projects. First, the resources of the economic system can only support a limited number of S-enterprises at any point of time. Second, even when the implementation of innovative projects has become easier, there are still only a limited number of persons whose abilities allow the needed effort. Third, the disturbance of the routine system caused by the swarm of projects will become so large that it is impossible to evaluate the profitability of further innovative projects. These constraints mean that the swarm of S-entrepreneurs will ultimately dissolve.

The other explanation of recession is that the "economic system needs rallying before it can go forward again; its value system needs reorganising" (*Development*, 217). This reorganisation implies "the diffusion of the achievements of the boom over the whole economic system through the mechanism of the struggle for equilibrium" (p. 251). The complex mechanism is based on the competitive struggle between firms based on new and old routines. The core of this mechanism is that S-enterprises are turned into established firms that for a while produce output at a gross profit that is partly used to repay entrepreneurial loans. This implies the destruction of credit—because banks have no incentive to create new credits. The resulting deflation combines with the output that flows from the new firms to sharpen the business conditions for the old firms that are facing the threat of creative destruction. More neutrally, we may speak of a reduction of evolutionarily relevant variance that is produced by the "adaptation to the new things created"; and this adaptation includes the "elimination of what is incapable of adaptation, resorption of the results of innovation into the system, reorganization of economic life so as to make it conform to the data altered by [innovative] enterprise, remodelling of the system of values, liquidation of indebtedness." (*Cycles*, 137; *Cycles Abr.*, 113). It is especially the terms of "elimination" and "liquidation" that serve to characterise the predominant psychological reactions to recession. Nevertheless, we are facing a downswing that is largely experienced at luxury class. Although temporary technological unemployment cannot be avoided, average real incomes increase. This increase is, according to Schumpeter, ultimately caused by the S-enterprises of the prosperity period. Their direct effect on output is that resources that during the upswing were used for innovative investment are now being turned into consumption goods. The indirect effects of innovation are more complex. On the one hand, the new firms create a competitive pressure on old firms. This pressure forces them to adapt by increasing their productivity or to exit and leave room for the expansion of more productive firms. On the other hand, the stoppage of innovative investment implies a deterioration of the conditions for all kinds of economic activity. The reaction of firms with sufficient resources is to reorganise production. In general, we recognise that "recession, besides being a time of harvesting the results of preceding innovation, is also a time of harvesting its indirect effects" (*Cycles*, 143; *Cycles Abr.*, 119). Schumpeter emphasised that at least five phenomena characterise recession in this sense:

> "[1] new methods are being copied and improved; [2] adaptation to them or to the impact of the new commodities consists in part in 'induced innovations'; [3] some industries expand into new investment opportunities created by the achievements of entrepreneurs; [4] others respond by rationalization of their technological and commercial processes under pressure; [5] much dead wood disappears." (*Cycles*, 143; *Cycles Abr.*, 119)

These five characteristics of recession serve to emphasise that it is a complex competitive process that changes the routines of the economic agents. Some of these changes can be characterised as the simple routinisation and diffusion of the innovations of the prosperity period. However, the new routines are not only being copied but also improved. Furthermore, the routines of old firms are also being changed. The core issue is captured by "the concept of Induced Innovation" that Schumpeter used "to denote those additional improvements which present themselves in the process of copying the first innovators in a field and of adaptation by existing firms to their doings" (*Cycles*, 101n). This concept is distinct from that of innovation in the narrow sense. It represents "adaptive response" and is thus not relating to the efforts of the S-entrepreneur. Nevertheless, "induced innovations" seem to explain a substantial part of the potential and actual expansion of output. However, although Schumpeter had a vivid vision of this aspect of the process of economic evolution, he was not able to model it formally.

Although the effects of the repayment of loans by the resulting new firms may still be insignificant, their output represents a competitive challenge to some of the old firms. Thus the realistic version of what he in *Capitalism* called the process of creative destruction starts before the harsh conditions of recession have emerged and before full knowledge can be assumed about the characteristics of the subsequent circular flow. In addition, bad timing of some of the later S-enterprises may imply that they are destroyed during recession—rather than the established firms that they challenge. The general rule, that also covers other issues, is that "[f]or success in capitalist society it is not sufficient to be right *in abstracto*; one must also be right at given dates. In this lies one of the difficulties of remedial policies" (*Cycles*, 412; cf. *Cycles Abr.*, 259). The timing can be defined in relation to the process of swarming. An otherwise solid innovative project that arrives late in this process is facing a radically increased danger of failure. This danger helps to explain the dissolution of the swarm, and it could also have been used as the starting point for discussing the "errors" that are a crucial part of the process of economic evolution. For instance, Fritz Machlup had, in Schumpeter's course on business cycles, pointed out that "entrepreneurial risk of failure is at a minimum in equilibrium and slowly rises as prosperity develops" while "[e]ntrepreneurial activity stops at a point when that risk is a maximum" (*Cycles*, 136n). Although Schumpeter could easily have integrated this mechanism into his first approximation to evolutionary business cycles, he largely abstained from doing so. His reason seems to have been that he wanted to focus attention on a much more serious source of "error" that is at the centre of his second approximation to what he considered the reality of the waves of economic evolution.

## 8.5 The second approximation with the secondary wave

We have seen above how Schumpeter extended the zeroth approximation to his model of evolutionary business cycles into his first approximation, but he needed further approximations to analyse the complex reality of the movement of the capitalist economy. We have already considered the third approximation that introduced the famous analysis of Kondratieffs, Juglars and Kitchins (see Section 8.3). However, the analysis of these cycles presupposes the development of his model of the second approximation. This model introduces the "Secondary Wave" and changes the strict assumptions of the "Pure Model" of the first approximation into more realistic ones. In this respect, he could draw upon the works of other economists. In his review of Pigou's *Industrial Fluctuations* (*S1927i*, 34–5), Schumpeter had thus, with respect to the "secondary wave", stated that in order "to put the necessary flesh on the bare bones of our argument" he "ought to superimpose on it the whole of Professor Pigou's mighty structure". This help was needed because "the phenomena of the secondary wave may be and generally are quantitatively more important than those of the primary wave" (*Cycles*, 146; *Cycles Abr.*, 122). Furthermore, since these secondary phenomena cover "a much wider surface" of the economic system, "they are also much easier to observe". In contrast, "it may be difficult, especially if the innovations are individually small, to find the torch responsible for the conflagration." Although Schumpeter was mainly interested in the innovative beginnings of the intense and uncontrolled fire of the economic boom, he nevertheless wanted his model to account for this booming inferno as well as for the subsequent freezing during depression.

Schumpeter's second approximation can most easily be grasped by thinking of a capitalist engine that, instead of two phases (upswing and downswing), operates with four phases of its cyclical movement. The four phases are prosperity, recession, depression, and recovery. Thus we are moving from the metaphor of a two-stroke engine to that of a four-stroke engine (see Figure 8.6 on the facing page). This metaphor is mainly introduced to illustrate the complexities that emerged when Schumpeter tried to expand his analysis—and it does not fully reflect his analysis of the four-phase cycle in *Cycles* (see Section 8.5). Nevertheless, some considerations on the working of the four-stroke engine might serve to prepare a fuller analysis. The essential feature the four-stroke engine is a doubling of the propulsive stroke of the two-stroke engine. While this stroke previously only covered the immediate effects of innovative projects, a secondary stroke is added *after* the primary stroke (in contrast to *Cycles* in which these strokes take place in parallel). The primary propulsive stroke produces the movement from a $\Sigma$ state to a disequilibrated $\Delta_\alpha$ state. The secondary stroke is a macroeconomic boom that is due to the optimistic expectations created by the primary stroke. It moves the engine from a $\Delta_\alpha$ state to an even more disequilibrated $\Delta_\beta$ state. This degree of disequilibrium emerges because the secondary prosperity produces investments that are not sustainable in an equilibrated system.

The combined effect of the two types of prosperity is an economic sys-

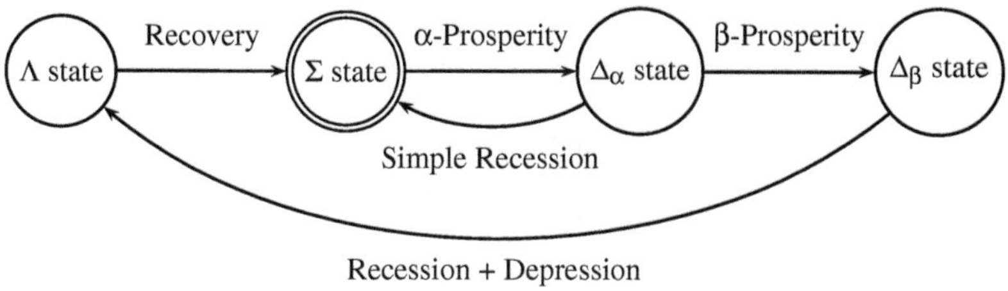


**Figure 8.6.: An interpretation of the capitalist engine as producer of four-phase waves**

Comment: Schumpeter's four-stroke engine of capitalism starts from a $\Sigma$ state—just like his two-stroke engine (see Figure 6.2 on page 149). However, while the simpler engine only includes the disequilibrated state directly induced by innovation (presently called $\Delta_\alpha$), the extended engine adds the even more disequilibrated state produced by "secondary effects" ($\Delta_\beta$). As a result, the engine has to pass the depressed state with low activity (the $\Lambda$ state) before returning to a neighbourhood of equilibrium. The distinction between two $\Delta$ states as well as the combined arrow of "Recession + Depression" are controversial in relation to the text of *Cycles.*

tem that is not only disturbed in the sense that some old firms have to adapt to, or be replaced by, the new firms; even old firms that are not threatened by innovative competition have expanded to a degree that is not sustainable during the subsequent circular flow. This means that the ordinary evolutionary mechanism of adaptation will not normally be sufficient to restore systemic equilibrium. When the boom of $\beta$ prosperity breaks down, the resulting combination of a recession and a depression "will generally, although not necessarily, outrun (as a rule, also miss) the neighborhood of equilibrium toward which it was heading and enter upon a new phase". This phase is characterised by "Abnormal Liquidation" that implies "shrinkage of operations ... below their equilibrium amounts" (*Cycles*, 149; *Cycles Abr.*, 125). Let us call the resulting state of low-activity depression the $\Lambda$ state. The movement of the economic system back to its equilibrated $\Sigma$ state requires an additional phase of recovery.

Since the described four-stroke engine comes close to a caricature of Schumpeter's second approximation, we have to develop this model more carefully. His starting point was a stationary circular flow and its disturbance by a swarm of S-entrepreneurs. What is added in the second approximation is that the prosperity created by the innovative projects induces an expansion of many of the already established firms. A simple example of such an expansion is the effects of a "new factory in a village". This factory "means better business for local grocers, who will accordingly place bigger orders with wholesalers, who in turn will do the same with manufacturers, and these will expand production or try to do so, and so on" (*Cycles*, 145; *Cycles Abr.*, 121). This expansion is made "on the assumption that that the rates of change they observe will continue indefinitely" and the consequence will be "losses as soon as facts fail to verify that assumption." This example can be expanded to many other activities, including "[s]peculation in the narrower sense", but they all "*presuppose* an actual or expected rise in prices to become possible". Since these prices are not sustainable during the subsequent circular flow, the "secondary wave" necessarily represents "cyclical clusters of errors, excesses of optimism and pessimism and the like". In contrast, although the "primary process" of swarming might involve errors, it can "produce ups and downs and ... losses without any errors" (*Cycles*, 146; *Cycles Abr.*, 122). Therefore, he only considered errors seriously in relation to the secondary wave. These errors are closely related to the process of formation of expectations; and the theorist's inclusion of this process "changes the whole character of his problem" and makes it "difficult to handle" (*Cycles*, 54; *Cycles Abr.*, 31). He specified the requirement in his review of Keynes's *General Theory* (*S1936c*, 161n). Here he emphasised that an explanation of this type "acquires explanatory value only if we are made to understand *why* people expect *what* they expect. Otherwise expectation is a mere *deus ex machina* that conceals problems instead of solving them'." Since Schumpeter obviously did not want to introduce artificial and improbable modelling features, he made very cautious use of the concept of expectations. However, "if we understand independently how the situations come about in which, for example, windfall gains, rising prices, and so on produce waves of optimism, we are free to use the fact that this optimism will feed upon itself and crystallize so as to become an element of the mechanism of cyclical events and the 'cause' of secondary phenomena" (*Cycles*, 141; 116–17). Together with similar processes of creating pessimism, this crystallisation helps to explain upswings and downturns in the stock market as well as financial booms and crashes. Nevertheless, these explanations were of secondary importance and should not be extended to what Schumpeter considered wrong and misleading theories of "overproduction and underconsumption" or to "the family of monetary theories of business cycles".

As it has been indicated by Figure 8.6, Schumpeter described the second approximation to his final model in terms of a cycle with four phases by adding depression and recovery to the prosperity and recession of the first approximation (*Cycles*, 145–61; *Cycles Abr.*, 121–37; cf. Oakley, 1990, Ch. 9). Nevertheless, the essential feature is an extended concept of prosperity. This concept is different from that of the figure. Instead of letting the secondary prosperity come after the primary prosperity, they work in parallel. In other words, the secondary prosperity is superimposed on the primary prosperity. The point of the figure is that although this superimposition starts at the very beginning of the process of swarming, it is most easily analysed as coming after the initiation of all the projects of the S-entrepreneurs. These projects form what may be called the primary prosperity that brings the economic system to its disequilibrated state while the secondary expansion creates an even more disequilibrated state. Actually, in "the atmosphere of secondary prosperity there will also develop reckless, fraudulent, or otherwise unsuccessful enterprise" (*Cycles*, 148; *Cycles Abr.*, 124). Thus, the combined effect of the primary and secondary expansion is an economic system that is not only disturbed in the sense that some old firms have to adapt to, or be replaced by, the new firms; even old firms that are not threatened by innovative competition have expanded to a degree that is not sustainable during the subsequent circular flow. This means that the already studied mechanism of recession will not normally be sufficient to restore systemic equilibrium. When the boom of the secondary prosperity breaks down, the resulting "process will generally, although not necessarily, outrun (as a rule, also miss) the neighborhood of equilibrium toward which it was heading and enter upon a new phase, absent in our first approximation". This phase is characterised by "Abnormal Liquidation" that implies "shrinkage of operations ... below their equilibrium amounts" (*Cycles*, 149; *Cycles Abr.*, 125). Thus the phase of "depression" may be seen as drawing away from equilibrium through a "Vicious Spiral" in which "pessimistic expectation may for a time acquire a causal role" (*Cycles*, 148; *Cycles Abr.*, 124–5).

Schumpeter's second approximation serves to pinpoint at least three questions. The first question is whether it is possible to decompose the overall period of prosperity into two phases. He seems to have answered in the negative. The most obvious reason is that only four phases could be defined qualitatively in relation to neighbourhoods of equilibrium: While both prosperity and depression bring the system away from equilibrium, recession and recovery move the system towards equilibrium. This neat specification would have been lost if two types of prosperity were introduced. Furthermore, the distinction between primary prosperity and secondary prosperity is made difficult by their simultaneity. Nevertheless, the decomposition of Schumpeter's complex phase of prosperity may ease the understanding of the core elements of his analysis. The second question arising from the formal description of his modelling of the "capitalist engine" is whether recession brings the system back to a neighbourhood of equilibrium immediately before depression moves it away from this equilibrium. As Oakley (1990, 172) has emphasised, Schumpeter's seems to have argued that recession should point to "some sort of 'interim equilibrium' state" although "he knew that the roots of the depression reach back well before" that state. The safest solution is to say that he did not really answer the question and that the function of recession is not well defined in the second approximation to his model of evolutionary business cycles. Actually, the final competition between new and old routines seems, to some extent, included in the function of recovery. The third question is how the redefined equilibrium state of the second approximation compares with that of the first approximation. Schumpeter's clear answer was that they are different. In other words, the two-phase engine produces a result that is different from that of the four-phase engine. The reasons are, on the one hand, that "abnormal liquidation destroys many things which could and would have survived without it". Especially, "it often liquidates and weeds out firms which do not command adequate financial support, however sound their business may be, and it leaves unliquidated concerns which do command such support, although they may never be able to pay their way". On the other hand, "the return of the system from the depressive excursion takes time" during which data change for a variety of reasons so that "what would have been a neighborhood of equilibrium when depression started is no longer one when it is over" (*Cycles*, 149–50; *Cycles Abr.*, 125–6).

The special status of the neighbourhood of equilibrium means that it, in principle, comes close to a halting state of the capitalist engine. This feature was used by Schumpeter to split the cyclical process into theoretical and historical units. In accordance with the methodology of the historical school, he thought that "every cycle is a historical individual and not merely an arbitrary unit created by the observer" (*Cycles*, 156; *Cycles Abr.*, 132). Therefore, "we are not at liberty to count cycles from any phase as we please". In his perspective, "[t]he phenomenon becomes understandable only if we start with the neighborhood of equilibrium preceding prosperity and end up with the neighborhood of equilibrium following revival." This feature of his theory is emphasised in Figure 8.7 on the next page by adding a period of circular flow between each cyclical unit. The sequence of phases is thus circular flow, prosperity, recession, depression, recovery, circular flow, prosperity, ... The difference from the primary wave of the two-phase model derives from the fact that prosperity is not only carried by the primary effects of innovation but also by its the secondary effects and the related occurrence of irresponsible finance. This means (according to the present interpretation) that the breaking down of the prosperity phase is not followed by equilibrium. Instead, the combination of recession and depression brings the system to a very low level of economic activity. From this point, the system has to be brought back to equilibrium before a new cycle can begin.

The Schumpeterian sequence of phases differs from that of most other business cycle theorists. Although Gottfried Haberler (1937, 77–8n) in his survey of the theories remarked that "the idea that the system passes through an equilibrium, or at least approaches a normal position, somewhere between the upper and the lower turning-point seems to have been vaguely envisaged by many writers", he needed to emphasise that

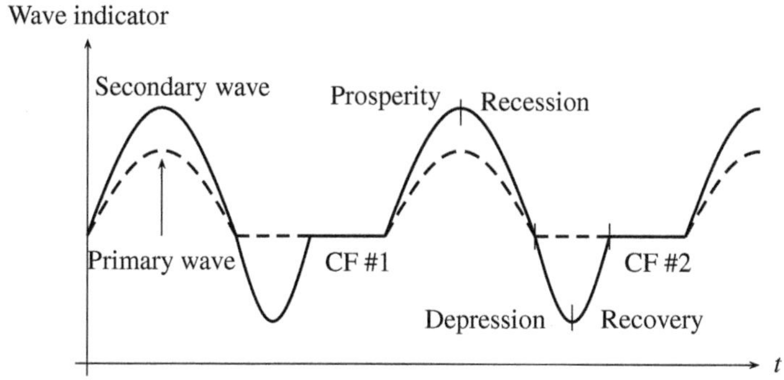


**Figure 8.7.: Schumpeter's second approximation: four-phase waves of innovative investment and erroneous investment**

Comment: The prosperity of Schumpeter's second approximation still starts from a circular flow (= CF); and is still followed by recession. However, the recession turns into depression and has to be followed by recovery before a new circular flow can be reached. When compared with Figure 8.5 on page 219, the secondary wave is superimposed upon the primary wave. If we add Schumpeter's extended assumptions, the duration of the circular flow becomes zero. It might be argued that it is impossible to define the wave indicator and that the figure is thus at best a pedagogical device. Source: The figure is inspired by Marschak (1940, 890) and Oakley (1990, 172–3).

Schumpeter did not present his theory "as an explanation of the lower turning-point, but of the movement of the system away from equilibrium". Most other economists preferred to define the cycle as starting at the lowest point of the depression and thereby to make recovery its first phase. Since Schumpeter's idea of starting from a neighbourhood of equilibrium appeared to be rather fussy while the lowest points of the statistical time series are well defined, Kuznets (1940, 266) pointed out that he actually "scorns the help provided by that statistical characteristic of cycles in time series". Schumpeter's argument, however, was that the "count from trough to trough ... is never theoretically correct" (*Cycles*, 156; *Cycles Abr.*, 132). The majority of business cycles analysts had not understood, or even detected, this evolutionary proposition. Therefore, they had basically thought in terms of a two-phase model in which their version of the prosperity phase includes what Schumpeter called recovery and prosperity. Then their crucial problem was to explain the onset of recovery, and the whole of their "upswing" seemed explainable by the "gradual elimination of the abnormalities then existing—low stocks, unused plant, unemployed labor, idle credit facilities" (*Cycles*, 157; *Cycles Abr.*, 133). Especially, "they find nothing that looks like innovation". If they at all considered this explanation, which was not normally the case, they would—within their framework—have arrived "at the result that innovation has nothing to do with initiating *prosperity*". Schumpeter's reply was that their construction of the cycle easily "drifts into *perpetuum mobile* explanations, particularly of the monetary type" (*Cycles*, 157; *Cycles Abr.*, 133). This type of machine "lacks motive power" and is instead driven by a "*deus ex machina*" (*Cycles*, 139–40; *Cycles Abr.*, 116). Schumpeter's rejection of models of perpetual motion may sound as if he accused their constructors of denying the laws of thermodynamics. However, he rather shared the criticism of most evolutionary thinkers against model systems in which motion is conserved forever. He developed this criticism of models of "Waves of Adaptation or Oscillations" (*Cycles*, 183; *Cycles Abr.*, 157). He considered such oscillations as explained by the fact that "[e]quilibrium has to be found, as Walras put it, *par tâtonnement*". This suggested damped oscillations; and with respect to models without this feature, Schumpeter thought they presupposed the existence of perpetual motion machines. Damping is not found in Michal Kalecki's model of cycles of investment, which Schumpeter considered at some length. However, "since it is not intended to be another *perpetuum mobile* theory of business cycles but the presentation of a piece of the mechanism, we can not only enjoy its simplicity, but also use it to demonstrate the possibility of a distinct type of oscillation" (*Cycles*, 189). Nevertheless, he did not want to contribute to the analysis of this kind of oscillation.

The major reason why Schumpeter upheld the idea of the circular flow as the starting point for a cyclical unit and abstained from exploiting the emerging macroeconomic analysis that ranged from Paul Samuelson's simple cycle model to the Marx-inspired model by Kalecki was that these modelling traditions assumed that it is the direct interaction of aggregate variables that produce business cycles. In contrast, Schumpeter's aggregates were averages that moved because of the existence of underlying heterogeneity of behaviour. Actually, he was a "staunch opponent of aggregation" (Fels, 1964, ix). His very first papers had been on statistical measurement, and one of them included the warning: "Beware of Statistical Aggregates" (*S1905b*, 193; quoted by Haberler, 1950, 338). His analysis of business cycles in *Cycles* served to substantiate this caveat. Here it became clear that evolution means that it is the "disharmonious or one-sided increase and shifts *within* the aggregative quantity that matter. Aggregative analysis, here, as elsewhere, not only does not tell the whole tale but necessarily obliterates the main (and the only interesting) point of the tale" (*Cycles*, 134; *Cycles Abr.*, 110). In Section 8.4, we saw that this "tale" can be expressed in terms of the increased variance of behaviour provided by innovation. With respect to the capitalist engine, we may, with Stanley Metcalfe (1998, 95, 64), say that "innovations provide it with more fuel with which to work" while "evolution consumes its own fuel" because "competition destroys the variety in behaviour on which it depends". Such characteristics of the evolutionary process tend to be ignored when modelling is dominated by aggregate formulations. These formulations also abstract from the evolutionarily irrelevant type of erroneous investment that nevertheless need to be selected away to recreate conditions for a new innovative prosperity.

## 8.6 Extensions of the second approximation

After Schumpeter had formulated the modelling core of his second approximation to the real phenomena evolutionary business cycles, he wanted to complete this model by allowing it to cover "a few other facts" (*Cycles,* 157–61; *Cycles Abr.,* 133–7). These facts suggested modifications of the strict assumptions underlying the "Pure Model". Actually, he thought that the facts contradicted in an essential way the following five assumptions:

1. Each wave of evolution is unaffected by previous evolution.
2. Saving plays no significant role in the evolutionary process.
3. Credit creation is solely made for the finance of innovative projects.
4. Innovation does not create investment opportunities in established industries.
5. The response to innovation takes place under the condition of efficient competition and full employment.

Let us consider Schumpeter's arguments for changing of these assumptions one by one. With respect to the first assumption, he emphasised that "we must drop the assumption, made for convenience of exposition, that our wave is the first of its kind and that ... it is entirely unaffected by the results of previous evolution". Instead, "we must take into account of the fact that each neighborhood of equilibrium contains undigested elements from previous prosperities and depressions" (*Cycles,* 157; *Cycles Abr.,* 133). Here Schumpeter emphasised that some investment industries tend to build and uphold a capacity designed to serve demand during prosperity. This strategy implies that output can expand more readily, and prices rise less, than expected from the first approximation to the model. As demonstrated by the historical parts of *Cycles,* an even more important feature is that a stock of relatively easy innovative projects is often left over from the previous cycle. They quickly contribute to the demand for credit in the beginning of a new prosperity. Actually, the stock of entrepreneurial projects may be sufficient to ignite prosperity so quickly that there is no waiting time between the end of recession and the beginning of the subsequent period of prosperity. These and related "facts" obviously enhances the realism of the model, but Oakley (1990, 206) rightly remarked that "we see here a remarkable array of *disequilibrium* conditions creeping into the 'equilibrium' state".

The next three assumptions of the second approximation were changed by including "saving", system-wide "credit creation", and ubiquitous "investment opportunities" (*Cycles*, 158–60; *Cycles Abr.*, 134–6). Concerning savings, Schumpeter touched upon the possibility of creating an alternative model in which "the financing of innovation [is] by saving, instead of by credit creation". However, he continued to assume that the latter type of finance plays a significant role. In contrast, he recognised the existence of easily available financial resources within business—including innovative firms that instead of repaying their debts uphold a working capital available for investment. Furthermore, he emphasised the fact that banks also finance non-innovative investment during all phases of business cycles. For instance, "the adaptation of old firms and the expansion of some of them into new economic space created by recent innovation will be financed by bank credit". Finally, he suggested that "the effect of innovation in opening new investment opportunities ... cannot be sufficiently emphasised". This point was already touched upon in his analysis of innovative swarming that ends up changing the "economic organism". To do so, this "organism" has to be described in structural terms that allow to follow the effects of the swarm of a broad range of ordinary business investment and on minor innovative activities. However, the description of this structure is not supported by any of Schumpeter's analytical tools. Therefore, one might speculate of the services that might have been provided by marrying Schumpeter's ideas of the network-like investment and innovation activity with the tools of input–output analysis (as developed by DeBresson, 1996). Wassily Leontief (1991) had already, in the late 1920s, sketched out a flexible version of his input–output analysis in a journal co-edited by Schumpeter. However, Schumpeter might have rejected this type of analysis because is always works in terms of the given pattern of interdependence between industrial sectors. In contrast, researchers like Albert Hirschman (1958) developed a theory of "backward and forward linkages" that promote innovation through periods of disequilibrium. Actually, Hirschman seems to have tried to combine Leontief with Schumpeter, and a revised version of his theory might provide some help in developing some of the ideas of both secondary and primary effects of swarming. However, this contribution came too late to influence Schumpeter, and he apparently never considered similar applications of the input–output framework.

The fifth and final change of assumption might be the most important one. Schumpeter changed the assumption of efficient competition and full employment in order to recognise two important stylised facts in the extended version of his second approximation (*Cycles*, 160–1; *Cycles Abr.*, 136–7). The first fact is that of "an imperfectly competitive world" and the second is that of "unemployed resources". These facts are closely related, because an economic system with monopoly power and other imperfections is characterised by underemployed resources—even in what Schumpeter considered to be neighbourhoods of equilibrium. Thus the move to a model of imperfect competition pointed at a world in which "full employment ceases to be a property of equilibrium states and instead indicates—paradoxical though this may sound—disequilibrium of a certain type" (*Cycles*, 161, emphasis removed; *Cycles Abr.*, 137). This consequence of imperfect competition—the existence of what has later been called "natural unemployment" in the equilibrium state—actually serves as a rescue for Schumpeter's evolutionary analysis. This analysis needed to explain the empirical fact that full employment is sometimes reached at the end of the Schumpeterian phase of prosperity rather than at its beginning. Nevertheless, he did not really confront the many problems of oligopolistic competition in *Cycles* but rather in *Capitalism* (see Section 7.2). Here he presented Mark II of the capitalist engine—while both *Cycles* and *Development* had been dominated by Mark I of this engine of capitalist economic evolution.

Since Schumpeter did not implement his change of the five assumptions in a well-defined model, it might appear paradoxical that he actually declared that he would apply his new assumptions. However, the feeling of paradox largely disappears when we turn to the main feature of *Cycles*: its historical account for the history of capitalist economic evolution. This account does not only focus on the broad contours that can be described in terms of Kondratieff waves and Juglar cycles (see Section 8.2). It also tries to cover the details of the working of the capitalist engine. For instance, Schumpeter commented on the effects of George Stephenson's "conspicuous success" in demonstrating the feasibility of the railway for transportation. This success was quickly followed by "all the phenomena of our Secondary Wave"—including what qualifies "for the title of 'bubble speculations'" (*Cycles*, 277). However, it was not the analysis of the early railway age but rather that of the second half of the nineteenth century that really required his modified assumptions. In the United States, this period started with "an atmosphere of boom, unusual expansion of credit and speculation, particularly in land and railroad stock, to all of which Californian gold (since 1850) and the favorable development of foreign trade lent their aid" (*Cycles*, 326; *Cycles Abr.*, 216). The major crash came after the "phenomena of the Secondary Wave were developed to an unusual degree" with "errors and cases of misconduct" (*Cycles*, 336; *Cycles Abr.*, 225). This crash of 1873 was not the outcome of "any 'logic of evolution'" but rather of the logic of the extended understanding of the secondary wave. The same was, according to Schumpeter, the case for the subsequent period of depression.

Schumpeter's remarks on the Second Kondratieff demonstrate that he considered the extended version of his second approximation necessary for handling the early history of business cycles and crises. They also help us to understand why he did not want to make the highly constraining assumptions that are needed to develop the extended second approximation into a precise economic model. For example, he could hardly have used such a model to present the rise and fall of France's Mississippi Scheme and England's South Sea Bubble as "early illustrations of the possibilities and weaknesses of the capitalist machine and of all the rules that apply to its handling" (*Cycles*, 252). Nevertheless, he emphasised that while these lessons are "persistently taught by history", they are also "forgotten each time." Even the solid motivations for government regulation are forgotten. Thus, Alfred Marshall and many others criticised the English Bubble Act of 1719 although "a perfectly good case can be made for it, considering the circumstances of the time and the semicriminal practices that were involved" and although the Act did not prevent "economic evolution" but only "speculative excesses" (p. 248n). Concerning the general lessons, "[n]obody denies that the events 1717–1720 (when in June the flotation of new companies reached its peak) and the bursting of the 'bubbles' looks very much, in major features and in details, like the big crises of 1772" as well as "1825, 1873, 1929" (p. 250). However, what "has been doubted is whether the economic process under this financial surface was also similar". Nevertheless, Schumpeter emphasised that the "mania of 1719–1720 ... was, exactly as were later manias of this kind, induced by a preceding period of innovation which transformed the economic structure and upset the preexisting state of things." As in the later cases, the changes of the "industrial and commercial process" were complemented by "developments in the financial sphere"; this led to a "building boom", to "speculative excesses", and to "the stock exchange crisis".

The person who got most of the blame for the conspicuous failure of France's Mississippi Scheme was John Law. Schumpeter did much to recover Law's reputation. Thus he suggested that "John Law ... is in a class by himself. ... He worked out the economics of his projects with a brilliance and, yes, profundity, which places him in the front ranks of monetary theorists of all time" (*History*, 294–5). Moreover, Law was also one of those innovative entrepreneurs who "add the function of the banker to that of the entrepreneur" (*Cycles*, 251). Schumpeter emphasised that, "for such a banker-entrepreneur, acquisition of the power to create credit will seem much more urgent than any individual one of the innovations which he intends to carry out". Thereby, "the mean becomes more important than the end." Law's "entrepreneurial scheme, which was colonial enterprise in the Mississippi Valley", was "by no means hopeless in itself". The fact that the scheme became based on long-term monopoly rights given by the French government was necessary to reach the goal. However, "Law's great operation" only served to turn France's "national debt into shares of the Mississippi Company"; and this result strangely "kindled the mania by virtue of all sorts of mysterious hopes for unheard-of profits and the general prosperity which it raised" (pp. 251–2). Nevertheless, Schumpeter assumed "that this entrepreneur-banker, dependent as he was ... on the good will of the government and the favor of the Regent, was from the first driven to put his plans into a garb acceptable to them". Under this assumption, "his operations in the field of government finance mean a lapse from the logic of his plan rather than its realization."

Schumpeter's handling of the bubble of 1720 demonstrates that even the extended second approximation is insufficient. One of the problems is that the financial innovations of the time were not primarily used to create credit for the innovative opening of new lands but rather to finance the public deficit that had emerged from war activities. Thus, there was not only a co-evolution between the financial sector and the industrial and commercial sectors. There was also a kind of co-evolution between economic life as a whole and the public sector. However, Schumpeter had explicitly omitted exogenous causation from the analysis of his book: "Among the factors which determine any given business situation there are some which act from within and some which act from without the economic sphere" (*Cycles*, 7; *Cycles Abr.*, 1). The latter factors "must be accepted as data and all we can do about them in economic analysis is to explain their effects on economic life." On institutions, Schumpeter wrote: "we have had examples (changes in tariff policy, taxation, and so on) of what we may term changes in the institutional framework. They may range from fundamental social reconstruction ... down to changes of detail in social behavior or habits" (*Cycles*, 10–11; *Cycles Abr.*, 4). He emphasised that it "is entirely immaterial whether or not such changes are embodied in, or recognized by, legislation. In any case they alter the rules of the economic game and hence the ... systematic relations of the elements which form the economic world." Thus, *Cycles* fully recognised the role of institutions, whether embodied in law or not. Actually, Schumpeter emphasised that "it would be possible to write, without any glaring absurdity, a history of business fluctuations exclusively in terms of external factors" (*Cycles*, 12; *Cycles Abr.*, 5). However, as any analyst, he had to define the boundary of the system under study. In *Cycles*, most of law happened to be outside the analytical boundary; and it was only in the historical parts of the book that he, to some extent, transgressed these boundaries. It was thus not in *Cycles* but in *Capitalism* that he, somewhat more systematically, could analyse the co-evolution between the different sectors of social life (Andersen, 2006).

## 8.7 Conclusion

The present chapter has not followed Schumpeter's way of presenting the complex contents of *Business Cycles* (see Table 8.1 on page 194). Instead, the chapter has considered the historical part of *Cycles* as providing the stylised facts and the sequence of models as an attempt to cover these facts by the available analytical tools. The major conclusion must be that this coverage is incomplete in an essential way. Thus Schumpeter was right when he in the preface pointed at "glaring lacunae" and "unfulfilled desiderata" (*Cycles*, v). The crucial question, however, is whether we should follow what he presented as "a motivated program for further research". This question has been answered in the negative by most economists. Their answer is not least derived from the criticism that emerged immediately after the publication of Schumpeter's most ambitious book in 1939. This criticism was not least provided by Simon Kuznets's famous review of *Cycles*. His conclusion was

> "that the book does not present a fully articulated and tested business-cycle theory; that it does not actually demonstrate the intimate connection between economic evolution and business cycles; that no proper link is established between the theoretical model and statistical procedure; that historical evidence is not used in a fashion that limits sufficiently the area of personal judgment; [and] that the validity of three types of cycles is not established." (Kuznets, 1940, 270)

The present chapter has not tried to suggest that this criticism of *Business Cycles* should be ignored. However, although Kuznets's criticism is largely valid, his evaluation of Schumpeter's book seems based on a too narrow interpretation. Like most of the other reviewers, Kuznets took the book's title too seriously and did not really appreciate it as a general contribution to the analysis of economic evolution under capitalist conditions. A warning against the narrow interpretation of *Cycles* was included in a short comment on Kuznets's review by Mirkowich (1940, 580), where he presented Schumpeter's *Development* and *Cycles* as providing a possible alternative to neoclassical economics and Keynesianism. Mirkowich's comment emphasised that this alternative—or complement—needs separate attention. However, "one of the first conditions would be that his [Schumpeter's] theory of economic development be distinguished from business cycles in their narrower sense, and that all consequences of his concept be accepted, many of which are contrary to the existing structures of equilibrium economics".

The evolutionary interpretation of *Cycles* suggests that it reflects Schumpeter's strategy of looking "to business cycles for material with which to build the fundamental theory of capitalist reality". This material and the related evolutionary theory are very different from what is needed for macroeconomic modelling. Although it might be possible to construct an evolutionary complement to macroeconomics (Foster, 1987; Oakley, 1990), the result of a modern study of *Cycles* might not least be to enforce the specification of core aspects of evolutionary mesoeconomics and evolutionary microeconomics (see Appendix D). In any case, the main conclusion is obvious: complementary research is urgently needed! This conclusion is relevant for many research projects, especially for those dealing with economic evolution. The specific question is whether this complementary research can fruitfully proceed along the lines defined by Schumpeter in *Cycles*. This question, however, cannot be answered in a simple way. The reason is that there is an enormous number of lines to follow beyond a book whose architecture seems to be as least as complex as that of the Gothic cathedrals, which Schumpeter loved to study in his spare time.

There is, however, a predominant architectural structure in *Cycles*: the stepwise extension of Schumpeter's theorising on economic evolution under capitalist conditions. His steps were not confined to the three successive approximations of the complexities of the real capitalist process. Actually, the specifications, modifications and corrections of his analytical tools are, in explicit or implicit form, found throughout his book. The most urgent task seems to be to integrate the best of these extensions into models of economic evolution that are as simple as possible. Unfortunately, the required type of model seems rather far from the basic evolutionary model that he upheld in most of his writings. This has been emphasised by Christopher Freeman (1990, 21). Freeman suggested that basic "criticisms of *Business Cycles*" should not refer to its deficient statistical technique or "to matters of style and presentation". Instead, these criticisms should refer "to fundamental concepts[,] i.e., first to his theory of innovation and second to his theory of equilibrium." It is only through such criticism of Schumpeter's "Pure Model" and through the construction of alternatives that we can rescue the best of his many extensions and sketchy insights from being crowded out by the too easy replication of simplistic versions of his concepts of innovation and evolutionary equilibrium.

# 9

# The Basic Mechanisms of Economic Evolution

After having considered *Capitalism* and *Cycles* in the two previous chapters, we now come to the oldest and most fundamental part of Schumpeter's evolutionary trilogy: *The Theory of Economic Development*. This book is the translation of *Entwicklung II*, that is, the second edition of *Theorie der wirtschaftlichen Entwicklung*. Since we have already studied *Entwicklung I*, the first German edition of the book, in Chapters 4 and 5, it might not be obvious why an extra chapter is needed for treating *Development*. However, the previous chapters did not include a systematic discussion of the basic evolutionary mechanisms that in all editions of the book are at the centre of Schumpeter's attention. The major reasons for this omittance were that *Development* is much more focussed on mechanisms than *Entwicklung I* and that the mechanisms can best be discussed by including the additional treatments of them in *Cycles* and *Capitalism*. These additional treatments are very different. *Cycles* is like *Development* based on the model of economic evolution that has been called Schumpeter's Mark I model. This model of the capitalist engine depicts the mechanism of innovation as based on the creation of new firms,tion because of the conservatism of established firms. In contrast, some of the core arguments of *Capitalism* are based on Schumpeter's Mark II model (see Section 7.2). According to this model, established firms innovate and adapt in oligopolistic competition. Thereby, they become core elements of the mechanism of innovation as well as the mechanism of adaptation. Actually, both mechanisms can, to some extent, be reinterpreted as representing a general mechanism of the adaptation of incumbent firms to the conditions of oligopolistic competition.

Since Schumpeter's Mark II model of the innovative competition between oligopolistic firms is well suited for modern economic analysis, much of the subsequent research has focussed on this model rather than on Mark I. The question is, therefore, whether Schumpeter's presentation of his original model in *Development* is still of any interest. The literature on innovation and entrepreneurship has answered affirmatively since it is in *Development* that we find the clearest formulations of these concepts. In contrast, the answer by most economists seems to be in the negative: there is no need of studying *Development* except for a few of its most conspicuous formulations. Neither of these answers is satisfactory. First, we should not only study *Development* for the sake of its concepts of innovation and entrepreneur but also to understand its account for the mechanism of adaptation and the related concept of equilibrium. Second, Schumpeter probably considered his Mark II model as a complement to Mark I; and he never developed this complement in detail. Therefore, the understanding of his evolutionary economics still requires a study of his Mark I model.

Schumpeter described Mark I of the capitalist engine in *Development* and in the first chapters of *Cycles*. Although these descriptions are closely related, we shall presently focus on the formulations in *Development*. Here the Mark I model is analysed in three steps. The book first introduces a truncated version of an economy from which innovation has been removed; this economy has reached a stationary equilibrium. Then the mechanism of innovation is added and its immediate consequences are studied. Finally, *Development* assumes that innovative activity will sooner or later come to a halt and that the mechanism of adaptation brings the economy back to a renewed stationary equilibrium. Since these three steps are closely related, the isolated analysis of any of them gives little meaning. However, if we start from the argument as a whole, we can decompose it and study its basic elements. This was the approach chosen in a short contribution to the celebration of Schumpeter's 60th birthday in the beginning of 1943. In this contribution, his friend and former teaching assistant Paul Sweezy described the Schumpeterian strategy as having three elements: the isolation of the innovative entrepreneur as the cause of qualitative economic change, the insertion of this agent into a model of an unchanging circular flow, and the study of how the capitalist engine works through the interaction between innovators and the forces of the circular flow. The three elements of this strategy suggested that an evaluation of Schumpeter's theory should answer three questions: "[1] Has he really isolated and abstracted for analysis the *primum mobile* of change? [2] Is the picture of the circular flow fully satisfactory? [3] Is the result of joining the two elements a correct representation of the essentials of capitalist reality?" (Sweezy, 1943, 93).

Although Schumpeter seems to have rejected Sweezy's idea of a "correct representation" of capitalist reality, the three questions are well suited for structuring an evaluation of his basic account for the capitalist engine. The present chapter starts with a short presentation of *Development* as part of Schumpeter's evolutionary trilogy. The following sections consider Sweezy's three questions one by one. Section 9.2 deals with the circular-flow model and the mechanism of adaptation. Then Section 9.3 turns to the question of the innovative entrepreneur as the mover of the capitalist engine. Finally, Section 9.4 studies how the circular flow and the innovative entrepreneur can be combined into a model of the capitalist engine as a whole. It should immediately be noted that Schumpeter was not fully satisfied with the resulting "representation of the essentials of capitalist reality". This is not least demonstrated by his early hints and later development of the alternative Mark II model. The reconstruction of his two alternative models of the capitalist engine is the topic of Section 9.5.

## 9.1 *Development* as part of the evolutionary trilogy

In his preface to *Development* (ix–x), Schumpeter pointed out that "books, like children, become independent beings when once they leave their parents' home. They lead their own lives, while the authors lead their own also." His use of this image is only partially correct. Although it is true that his *magnum opus* had "won its place in the German literature of its time and field", it is not true that he had avoided to "interfere with those who have become strangers of the house." Frank Taussig, the leader of the Department of Economics at Harvard University, had persuaded him to participate in producing the somewhat revised translation into English of *Entwicklung II*; and previously the criticisms, misunderstandings, and his own reflections on *Entwicklung I* had persuaded him to make a radical revision of the first edition of his book. Actually, it is probably because of these interferences that *Development* has survived as a part of the literature that is read today. Nevertheless, his revisions have also created some degree of confusion. The major problems have already been discussed in Sections 4.1 and 5.1 of the present book. The problems are that he omitted the large concluding chapter of *Entwicklung I* and radically revised the chapters on the mechanism of innovation and on the evolutionary essence of business cycles (see Table 5.1 on page 103). Thereby, he had largely followed Navratil's suggestion of shortening the book to half of the original size. He had also tried to overcome the criticism of the book's "super-man interpretation of economic progress" (McCrea) by focussing of the mechanism of innovation rather than on the personalities of great entrepreneurs.

Schumpeter's rewrite made clear that he did not focus on the entrepreneurial interpretation of history, which we considered in Part I of the present book. Instead, the rewrite demonstrated that he presented a model of the capitalist engine of economic evolution. According to the preface of *Development* (xi), "[t]he argument of the book forms one connected whole" (see Table 9.1 on the following page). The first chapter analyses a stationary economic system that is governed by economic routines and the mechanism producing this state. The second chapter introduces innovative entrepreneurs with the function of changing the routine system. The third chapter describes the role of credit for entrepreneurial projects; and the fourth chapter analyses the entrepreneurial profit that is used to pay back the loans. Then the fifth chapter returns to the sta-

**Table 9.1.:** ***The Theory of Economic Development*** **compared with** ***Business Cycles***

| Topic | *Development* | *Cycles* |
|---|---|---|
| Circular flow | Ch. 1: The circular flow of economic life as conditioned by given circumstances | Ch. 2: Equilibrium and the theoretical norm of economic quantities |
| Economic evolution | Ch. 2: The fundamental phenomenon of economic development | Section 3.A–B: Internal factors of change. The theory of innovation |
| Innovative profit | Ch. 4: Entrepreneurial profit | Section 3.C: The entrepreneur and his profit |
| Money and banking | Ch. 3: Credit and capital | Section 3.D: The role of money and banking in the process of evolution |
| The rate of interest | Ch. 5: Interest on capital | Section 3.E: Interest (money market; capital) |
| Waveform evolution | Ch. 6: The business cycle | Ch. 4: The contours of economic evolution |

Comment: The table demonstrates the parallelism between the accounts for the basic evolutionary theory in *Development* and the first chapters of *Cycles*. Schumpeter's restatement of his theory in *Cycles* included clarifications and extensions. This restatement applied a revised terminology (see Table 9.2 on page 247).

tionary system and tries to demonstrate that the interest on capital here is equal to zero. Finally, the sixth chapter uses the whole framework to develop a theory of business cycles as expressing the underlying process of economic evolution. It is thus not difficult to recognise that we are facing "one connected whole". Nevertheless, Schumpeter was not satisfied with the result and he did not abstain from further interference.

Schumpeter's interference with his published book also concerned its title. For its second German edition, he added a subtitle that emphasises the book's contribution to the interpretation of core economic concepts and to modelling. For the English translation, he accepted Opie's choice of words. However, neither of these interferences helped to clarify the nature of his evolutionary contribution to the literature of economics. Actually, all the four main words of the main English title of his book are problematic—and two of them are not even necessary translations of the German original. Although we cannot change the title chosen by Schumpeter, it is nevertheless helpful to consider an alternative title:

- Schumpeter's original title: *Theorie der wirtschaftlichen Entwicklung* (1912).

- Schumpeter's English title: *The Theory of Economic Development: An Inquiry into Profits, Capital, Credit, Interest, and the Business Cycle* (1934; from the German of 1926).
- Alternative title: 'A Theory of the Mechanisms of Economic Evolution: An Evolutionary Inquiry into Profits, Capital, Credit, Interest, and the Business Cycle'.

Let us focus on the main part of the title of *Development*. The first word of the English title—"*The*"—suggests that the book is presenting a whole field of theorising. However, alternative theories, like that of Marshall and Schumpeter's emerging Mark II theory, were available when the book was written and revised. Therefore, a better title would have been 'A Theory of Economic Development'. This is a possible translation of the German title since it has no definite article. The second word of the English title—"*Theory*"—is problematic in the sense that it does not signal the special nature of Schumpeter's theorising about evolutionary transformation. Although we can develop precise theories of the evolutionary mechanisms, we normally cannot make precise predictions of the outcomes of the process of transformation. Therefore, a more appropriate title might have been 'A Theory of the Mechanisms of Economic Development'. The third major word of the title—"*Economic*"—is also problematic since it might suggest that Schumpeter is covering the same ground as economic theory in general. His addition of a subtitle actually made things worse. This subtitle—"*An Inquiry into Profits, Capital, Credit, Interest, and the Business Cycle*"—suggests a general analysis of basic concepts of economic analysis while the book rather presents 'The Role in the Evolutionary Mechanisms of Profits, Capital, Credit, Interest, and the Business Cycle'. The last word of the title—"*Development*"—is, as it was emphasised in Section 1.2, the most problematic one. In a publication from the 1930s, this word should have been immediately replaced by the alternative translation of 'Entwicklung' into 'Evolution'. Therefore, the title "*The Theory of Economic Development*" should have been replaced by something like 'A Theory of the Mechanisms of Economic Evolution'. A possible improvement of the subtitle could have been 'An Evolutionary Inquiry into Profits, Capital, Credit, Interest, and the Business Cycle'.

The suggested alternative title for *Development* emphasises the need for updating the book's terminology to that of the rest of Schumpeter's evolutionary trilogy. The need for updating is caused by the fact that *Cycles* presents a modification of Schumpeter's previous terminology (see Table 9.2 on page 247). This modification is largely an improvement. His major ambition is, in *Cycles*, defined as that of developing the theory of the capitalist engine by means of the construction of operational models. His previous analytical "schemes" had been constrained by the use of an all-encompassing Statics–Dynamics dichotomy (see Section 3.4). In *Cy-*

*cles*, he used separate terms for describing the macroscopic phenomena he wanted to explain, the analytical methods he used, and the types of individual behaviour he applied in the explanatory process. Individual behaviour is divided into two classes according to whether or not it promoted a transformation of routine. While routine was previously denoted "production combination", he now used the term "production function". This shift served to relate his theory to standard terminology. However, since the underlying concept presupposed a radical reinterpretation of the standard concept that was also named "production function", the benefits of replacing the original term were hardly larger than the costs of confusion. In contrast, the clarification of the concept of "enterprise" was clearly an advantage. In *Cycles*, it had the unambiguous meaning of an innovative project that was carried out by the entrepreneur. Further clarity was obtained by defining the entrepreneur solely in functional terms rather than adding the "energetic" characteristics of elite behaviour. Finally, Schumpeter brought his terminology of business cycles up-to-date by operating with the terms of "recession" and "depression" instead of the previous fusing of these terms under the ambiguous name "depression".

The place of *Development* in Schumpeter's evolutionary trilogy becomes especially clear if we also think in terms of alternative titles for *Cycles* and *Capitalism*. The already suggested substitute of "*Business Cycles*" is 'Theory, History, and Statistics of the Waves of Capitalist Economic Evolution'. This alternative title suggests that *Development* serves to supply core parts of the theory of the waves of evolution. More specifically, *Development* supplies what was in the previous chapter called the zeroth approximation to the evolutionary theory as well as the first approximation. The zeroth approximation concerns the basic mechanism of economic evolution and the first approximation demonstrates that these mechanisms of innovation and adaptation are able to produce waveform economic evolution. Since these approximations are crucial for the analysis of evolutionary business cycles, Schumpeter gave them prominent places in his the restatement of the capitalist engine that is found in 160 pages of *Cycles*. As demonstrated by Table 9.1, this book includes rewrites of each of the chapters of *Development*. The major novelty is that Chapter 6's theory of business cycles as reflecting waveform economic evolution has been revised and radically extended. However, Schumpeter also restated and developed the arguments of the other chapters. He especially sharpened the exposition of the Mark I model of the capitalist engine—not least by extending his previous analysis of the relationship between the real and monetary aspects of the process of economic evolution. The sharpening of Schumpeter Mark I served to make clear the inherent limitations of the model. These limitations become especially clear if we, instead of the actual title "*Capitalism, Socialism and*

**Table 9.2.: Terminological shifts between *Entwicklung I* and *Cycles***

| Early term (in *Entwicklung I*) | Later term (especially in *Cycles*) |
|---|---|
| Picture, schema | Model |
| Statics as method *and* phenomenon | Statics as method |
| Static economy | Stationary economy |
| Dynamics as method *and* phenomenon | Dynamics as method |
| Dynamics as phenomenon | Economic evolution |
| Combination | Production function |
| New combination, innovation | New production function, innovation |
| Enterprise as project and firm | Enterprise as project of making a new firm |
| Entrepreneur | Entrepreneur as "New Man" |
| "Energy" as characterising the behaviour of entrepreneurs | "Energy" at the system level |
| Entwicklung, "economic development" | Economic evolution |
| Depression | Recession *or* depression |

Comment: Schumpeter's early terminology in *Entwicklung I* is reflected in Table 5.3 on page 131. Although the terminological differences between this book and *Cycles* might appear cosmetic, they have created major difficulties for the Schumpeter literature. Nevertheless, the terminological shifts demonstrate that he wanted to emphasise the construction models of the capitalist engine.

*Democracy"*, think of this book as dealing with 'Capitalist Economic Evolution and Its Socio-Political Limits'. While the limits of capitalism had by other researchers often been connected with the endogenous emergence of monopolistic structures within the capitalist economic system, *Capitalism*'s presentation of the Mark II model concludes that the limits should be found elsewhere. This conclusion is derived from Mark II's analysis of innovative oligopolistic competition. This analysis can be considered Schumpeter's last major development of the argument of his *Theorie der wirtschaftlichen Entwicklung*.

Since large parts of *Cycles* and some parts of *Capitalism* can be interpreted as the result of Schumpeter's wish to improve *Development*'s theory of the capitalist engine, the independent reading of this book is problematic. It is better to read major parts of his evolutionary trilogy as an open-ended argument. However, such a reading is difficult. The most obvious difficulty is the ambiguous and shifting terminology of the three books; but we are also facing subtle changes of a theoretical nature. The terminological and theoretical problems can be solved in several ways. It, nevertheless, seems clear that Schumpeter wished that we use *Cycles* as our main reference. This wish has only been fulfilled very partially by subsequent research; and the main reason is that *Cycles* is in several respects a problematic book. For instance, it mixes basic evolutionary theorising with a non-convincing theory of business cycles; and youthful freshness is to some extent replaced by complex arguments with many digressions. Nevertheless, we should not ignore that there is an easy way of coupling the chapters of *Development* with sections in *Cycles* and that a comparison gives some impression of new solutions and remaining difficulties (see Table 9.1 on page 244).

There is a long way from *Development*'s abstract analysis of the mechanism of the capitalist engine to the concrete historical analysis of the history of capitalist economic evolution. Although this gap can hardly be overcome systematically, it might be helpful to reconsider the already-presented image of the economic frontier (see Section 7.5). This image of a frontier between unexplored possibilities and well-established economic activities has been used to produce a rough interpretation of the early economic history of the United States. Furthermore, Kenneth Boulding (1981, 106) translated the image into ecological terms. Let us combine these contributions. The behavioural characteristics of the pioneering types of settlement and entrepreneurship were adapted to economic activities far below from their carrying capacities, while the predominant behaviour in established regions was reflecting an economic life much closer to its (preliminary) limits. These two selection environments suggest radically different solutions to basic behavioural trade-offs like quick moves versus fine tuning, self-reliance versus collaborative behaviour, or changing the environment versus adapting to the environment. Since the solutions to these trade-offs exert a strong influence on individuals and organisations, they tend to be locked to one of the two strategies. Thus, even though the pioneers in principle could make a living under more stable conditions if they adapted to the tight competition and the more regular mode of behaviour needed to function in a crowded region, they often did not succeed. Instead they moved the frontier of American economic activity westwards.

The image of a moving frontier can easily be applied to Schumpeter's waveform evolution of economic life. For instance, his mechanism of innovation may metaphorically be called the pioneering mechanism while the mechanism of adaptation can be related to more crowded conditions. However, while the movement of any geographical frontier will ultimately come to a stop, the evolution of economic life may be viewed reflecting the movement of an endless frontier. This kind of frontier shows a periodical re-emergence of possibilities of pioneering or entrepreneurial behaviour: even though a newly established frontier area will ultimately become characterised by routinised activity, new possibilities are explored by agents who prefer pioneering behaviour to the highly adaptive behaviour needed to survive in established areas. In this way, they start a new round of the Schumpeterian story. As has been emphasised in the previous chapter, this story is developed by means of successive approximations. Let us start by formulating the zeroth approximation in terms of the economic geography of the frontier:

1. We start when both the frontier area and the hinterland have reached a circular flow of economic life based on routines upheld by strong preservative forces.
2. Then an irrevocable disturbance of the circular flow is created through the movement of the frontier by S-entrepreneurs.
3. Finally a new, non-innovative state is reached in the new frontier area as well as in its hinterland; this state is reached through strong stabilising and selective forces.
4. The movement of the frontier consists of a sequence of such steps.

Although this account can be rephrased in the already considered general terms, it is presently more important that Schumpeter's favourite example of the railways can to some extent be directly related to the frontier story. He mentioned this possibility in a series of lectures called "An Economic Interpretation of Our Time". Here he emphasised that

> "there is one force [innovation] inherent in the economic process which will cause it to progress or advance in a wavelike fashion. ... Take, for instance, the railroadization of the Middle West as it was initiated by the Illinois Central. While a new thing is being built and financed, expenditure is on a supernormal level, and through a normal state of incomes we get all those symptoms which we associate with prosperity. When such a period of advance has gone on for a time ... [we see] the way in which progress is accomplished in capitalism and the old eliminated. For instance, it is easy to see that the Illinois Central not only meant very good business whilst it was built and whilst new cities were built around it and land was cultivated, but it spelled the death sentence for the [old] agriculture of the West." (*S1941b*, 349)

The case of the railroadization of the Mid-West of the USA only covers one local episode in the evolution of the economic organism, or the economic ecosystem, which is the core subject of Schumpeter's works on capitalist economic evolution. Nevertheless, it helps to clarify how he, after several successive approximations, wanted to describe the overall process of railroadization:

1. The process of railroadization started from an equilibrated system of economic routine, including the routines underlying mail-coach-based services for long-distance transportation.

2. Then the economic system became disturbed by the introduction of railway-based transport services with a large potential and with large resource needs.

3. A major reorganisation of the system of economic routine took place through several cycles of economic prosperity and recession. The end result was a relatively equilibrated state, partly based on the routines of railway transportation.

4. Then this system was disturbed by the addition of lorries for transport over long distances. Thereby, a new round of the story began.

It is the historical reconstruction of this and similar evolutionary stories that was the major goal of Schumpeter's evolutionary economics. The mechanism of innovation (the $\delta$ mechanism) covers the movement from one circular flow to the maximally disequilibrated state of the economic system, while the mechanism of adaptation (the $\sigma$ mechanism) covers the movement back to another circular flow. The sequential working of these mechanisms is related to the case of railroadization in Figure 9.1 on the next page. This choice is not least motivated by the fact that Schumpeter stated that "railroadization is our standard example by which to illustrate the working of our model" (*Cycles*, 304; *Cycles Abr.*, 202). This "standard example" covers the age of railroad construction in the nineteenth century and it fits his evolutionary theorising better that most other examples from economic history. Thereby, it serves to increase our understanding of his theory of the mechanisms of the capitalist engine. Nevertheless, this example is primarily treated elsewhere in the present book (Sections 8.2 and 12.5). One reason is that while *Cycles* includes much material on railroadization, *Development* only contains few and scattered remarks on railways (pp. 62, 64, 66, 84, 215, 230, and 250). Another, and more important, reason is that the many steps from the abstract argument of *Development* to the historical analysis of railroadization would complicate the following exposition. Therefore, we return to his highly stylised history of capitalist economic evolution as a sequence of circular flows produced by the mechanisms of innovation and adaptation.

## 9.2 The circular flow and the mechanism of adaptation

*Capitalism* had not been published when Paul Sweezy (1943) wrote his already mentioned comments on "Professor Schumpeter's Theory of Innovation". Instead, Sweezy had to rely on *Development* and *Cycles*. Since

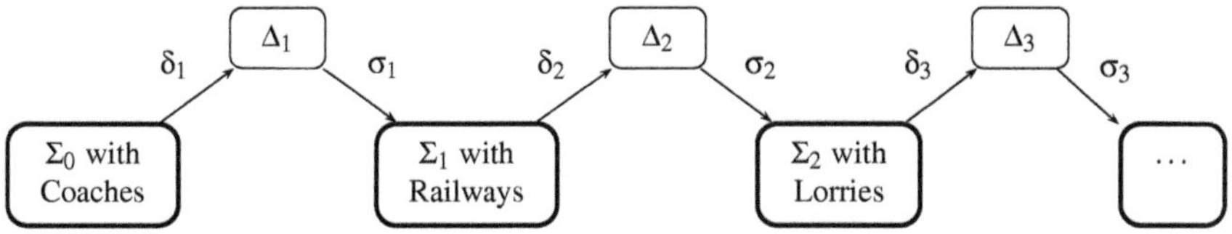


**Figure 9.1.: Combining the mechanisms of innovation and adaptation**

Comment: As we have seen in Figure 6.2 on page 149, the $\delta$ mechanism of innovation moves the economic system from a stationary $\Sigma$ state to its maximally disequilibrated $\Delta$ state; and the $\sigma$ mechanism of adaptation moves the system back to a new $\Sigma$ state. The routines of this $\Sigma$ state have absorbed previous innovation. Take, for instance, innovations for long-distance transportation. $\Sigma_0$ applies mail coaches, $\Sigma_1$ applies railways, and $\Sigma_2$ also applies lorries.

both books start by presenting the circular-flow model of a stationary economy and since the rest of these books build on this model, it was a crucial question whether it is fully satisfactory. The scientific debate had concentrated on *Development*'s first chapter on "The Circular Flow of Economic Life as Conditioned by Given Circumstances" and the closely related fifth chapter on "Interest on Capital". In the latter chapter, Schumpeter argued that his circular-flow model demonstrates that the rate of interest is zero in an economic system without innovative entrepreneurs. Such a conclusion should, of course, be derived in a consistent manner—and many economists have doubted that this is the case. Actually, it is the apparent lack of consistency of the circular-flow model that had been the major point of attack for Schumpeter's critics from Böhm-Bawerk to Lionel Robbins (1930) and Frank Knight (1935). Their criticism emerges from the assumption that such a model must include markets for loans and for the ownership of land. However, Paul Samuelson (1943) defended the logical feasibility of this model's stationary equilibrium in his contribution to the Schumpeter Festschrift. He did so by focussing on possible defences of the proposition that the rate of interest is zero. However, Samuelson's aggregative account for the circular-flow model does not provide any opening for the later inclusion of innovation and entrepreneurship. Therefore, it can hardly be interpreted as supporting the first step on the way to the Mark I model of the capitalist engine.

Sweezy (1943, 94) emphasised that "unlike many of his critics, I have no fault to find with the logic of Professor Schumpeter's argument". Thus the disappearance of profit and interest is not a result of the reasoning but of "the implicit assumptions on which it is based". These assumptions "are certainly sufficient to produce a stationary economy, and, moreover, one in which nothing is left for saving and accumulation". For instance, Sweezy emphasised that the zero rate of interest implies that the capitalised earning power of a piece of land is infinitely large and thus that there is no supply in the market for land ownership. The consequence is that "the owners of land would constitute a hereditary aristocracy to which wealth as such would not be a ticket of admission" (p. 95). However, when turning from the logic of the Schumpeterian model to its underlying assumptions, Sweezy found "more room for doubt". The background for this doubt was that he had just published an account of Marxian political economy under the title *The Theory of Capitalist Development* (Sweezy, 1942). Here we find a capitalist class of firm owners even in the stationary model of the economy. The lack of change in the Marxian model of simple reproduction is the result of abstracting from the innovative activities of the firm-owning "capitalists". When this abstraction is removed, it is easy to recognise "surpluses and accumulation ... as exercising a profound and steady pressure in the direction of economic change" (Sweezy, 1943, 95). Although a similar account could have been developed by starting from modified neoclassical models, this was definitely not what Schumpeter had done. Instead, he based his model of the circular flow on the intrinsic conservatism of incumbent firms.

Schumpeter's early formulations of the assumptions of his Mark I model have already been mentioned in Section 4.2. He based the model on three "pairs of opposites" (*Development*, 82). First, the Mark I model is based on "the opposition of two theoretical *apparatuses*: statics and dynamics". Second, the model assumes "the opposition of two real processes: the circular flow or the tendency towards equilibrium on the one hand, a change in the channels of economic routine or a spontaneous change in the economic data arising from within the economic system on the other". Third, the model implements Schumpeter's basic "opposition of two types of conduct, which, following reality, we can picture as two types of individuals: mere managers and entrepreneurs" (p. 82). These three pairs of opposites help to characterise what he, in *Entwicklung I*, had considered the overall dichotomy between "Statics" and "Dynamics" (see Table 9.3 on the facing page). He initially considered Statics as the application of the "static" method for studying the "static" routine-based circular flow and the "static" movement towards it through the action of mere managers. Similarly, he considered Dynamics as the application of the "dynamic" method for studying the "dynamic" change of the routines of the economic system as initiated by the "dynamic" activity of the Schumpeterian entrepreneur. In *Entwicklung I*, Schumpeter thus emphasised terminologically that there is something like a one-to-one relationship between the types of methods, phenomena, and behavioural conducts. In *Development* and *Cycles*, he gave up that practice. For instance, he emphasised that "statics" and "dynamics" had confusingly been used as "short expressions for 'theory of the circular flow' and 'theory of development'" (*Development*, 83n). Now he only used "statics"

**Table 9.3.: Decomposing the Statics–Dynamics dichotomy of the early Schumpeter**

| | "Statics" | "Dynamics" |
|---|---|---|
| Phenomena | Stationary equilibrium; equilibrating processes leading toward a circular flow | Disequilibrating processes that include new production functions |
| Types of agent | Routine-based agents of the process of adaptation | Creative agents as well as routine-based agents |
| Mechanisms | Mechanism of adaptation by inter-firm selection and intra-firm adaptive change | Mechanism of innovation by new firms in Mark I; and by incumbent firms in Mark II |
| Analytical methods | The static and dynamic analyses of adaptive processes | The dynamic analysis of processes that include innovation |

Comment: In *Development* and *Cycles,* Schumpeter tried to overcome *Entwicklung I*'s strong coupling of two phenomena, two methods, and two types of agent under the headings of "Statics" and "Dynamics". The decoupling especially helped him to clarify the exposition of the Mark I model. His further terminological clarification on the applied methods is covered by Table 10.1 on page 319. Source: Modified from Andersen (2008, 106).

and "dynamics" to characterise the applied analytical methods. Statics is the method that ignores time in the analysis of equilibria, while dynamics is the analytical method according to which a time subscript can be meaningfully added to each variable. The Mark I model is obviously dynamic in this sense since the explicit inclusion of time increases our understanding. For instance, it is important whether the productivity of an incumbent firm changes over time. Since Schumpeter assumed that this productivity does not change, we have found an important characteristic of his model.

Schumpeter's analysis started from "The Circular Flow of Economic Life as Conditioned by Given Circumstances"—the title of the first chapter of *Development.* The contents of this chapter had been changed very little since *Entwicklung I* and it was largely repeated in the second chapter of *Cycles.* In *Development,* he stated that the chosen title was borrowed from an expression of his Austrian teacher Philippovich (cf. 1919–22); but this hint is difficult to follow and does little to explain the nature of Schumpeter's modelling exercise. However, the title combines two elements: the circular flow of a stationary economy and the conditioning of this state of the economy by given circumstances. The development of the underlying concept is described in *Economic Doctrine and Method*, which was written shortly after *Entwicklung I*. According to Schumpeter (*Doctrine*, 40), the "discovery of the circular flow of economic life" was made by the Physiocrats; and the "first complete analysis" of this process "was directly inspired by Harvey's discovery" of blood circulation (*S1928f*, 59). Quesnay's economic model assumes that "every element of economic energy completes a definite route year in year out under the influence of definite motive forces" (*Doctrine*, 43–4). Thereby, he and other Physiocrats conceived the "constant self-renewal" of the economic system as that of "an organism with a uniform life process and uniform conditions of life". We can add that one of the major users of this circular-flow conception of economic life had been Marx (1990–92:II), who not only used it for analysing simple reproduction, but also the extended reproduction of the economic organism. In principle, Schumpeter's concept of the circular flow ranged even wider. His analogy was the continuous blood circulation (and the related physiological processes) of a biological individual with an ability to adapt its anatomical structure gradually (*S1928f*, 59). However, he did not implement all aspects of this analogy in his modelling of the circular flow.

Schumpeter changed the concept of the Physiocrats and their followers to cover an economic system in which the use of the means of production is a matter of routine and which only changes due to external circumstances. This revised concept of the circular flow has a crucial but difficult place in Schumpeter's successive approximations toward a satisfactory analysis of the realities of economic evolution. It is not only the non-evolutionary starting point for most of his subsequent evolutionary analyses, but also his bridge to standard economics. Thus his circular-flow model can be interpreted as a simplified version of the Walrasian system with added features taken from the Austrian school. The added features suggest that we should not characterise "the systems of Walras and Pareto ... with the absurdity of assuming omniscience" (*Cycles*, 53; *Cycles Abr.*, 30). Instead, the process through which the circular flow becomes conditioned by given circumstances is a very complex and troublesome one. This probably suggested to Schumpeter that he had to simplify the structure of his model as much as possible.

As we saw in Section 3.5, Walras's role list consists of the landowner, the worker, the W-capitalist who owns durable producer goods, and the W-entrepreneur. Schumpeter's list of actors for the strict circular-flow model excludes the W-capitalist and reinterprets the W-entrepreneur as a routine-following manager. Such abstract managers represent firms while landowners and workers (including managers) can be combined under the heading of households. Schumpeter's specifications of these agents might be summarised as follows: The households obtain incomes by selling the services of their land and labour. In the circular-flow model, they have given preferences and use their incomes to buy consumer goods. These goods are bought immediately, since Schumpeter ignored what he considered the inessential use of the multi-period income stream for lending and saving. The firms of his model are more interesting. They buy the services of land and labour, and they use these services to produce consumer goods according to given production functions. Although these production functions also require producer goods, no markets for such goods seem to be included in the simplest version of the model. A plausible interpretation, which unfortunately does not apply to *Cycles*, is that all firms are vertically integrated in the Austrian-style sense that each firm produces all the producer goods that are needed in the intermediate stages of the production of its output of consumer goods. In addition, the producer goods can be interpreted as intermediate goods of a non-durable character. Therefore, there is no room for W-capitalists who own durable means of production and sell their services to the firms. Investment in durable goods is thus absent as a source of demand for loans. Furthermore, firms do not need loans since they can use their current revenue to pay for the services of land and labour. The consequence is that no loan market exists in the circular-flow model. However, Schumpeter chose the formulation that the rate of interest is equal to zero.

Although Schumpeter gave no names to his theories of households and firms, they can best be understood as routine-based theories (see Nelson and Winter, 1982, Ch. 5). The economically relevant part of the reproduction of human life within a household is obviously largely a matter of routine-like acquisition and consumption of goods and services. Since this reproduction within households has to go on steadily, the production for them by Schumpeterian firms necessarily has an important element of routine. Other reasons for the routine-based activities of firms are the limited capabilities of their employees and the organisational need of integrating activities that are performed in sequence as well as in parallel (on the latter point, see Georgescu-Roegen, 1971, Ch. 9). Additional reasons become clear when we, at a lower level of abstraction, introduce markets for produced means of production. In any case, the routine-based theory of the established firm suggests a conservative behaviour that, at best, adapts to changing circumstances in very small steps. The result might be a large change if the pressure for adaptation for a long time points in the same direction: "continual adaptation through innumerable small steps" may in a large city "make a great department store out of a small retail business" (*Development*, 62). Thus Schumpeter's routine-based theory does not suggest that a population of firms cannot adapt its routines but rather that firms individually "cling as tightly as possible to habitual economic methods and only submit to the pressure of circumstances as it becomes necessary" (pp. 8–9). However, this allowance for the incremental adaptivity of the routines of a population of firms produced analytical difficulties. Therefore, he tended to fall back on more rigid formulations according to which the behaviour of firms is constrained by their fixed production functions (*Cycles*, 38–9).

According to the circular-flow model, the economic system is placed in an environment of unchanging technological and social circumstances. The economic system, furthermore, has become fully adapted to these circumstances. Therefore, it performs a stationary process that can be compared with that of the circulation of blood in a biological organism. Through this process, all goods are being reproduced in a routinised manner:

> "Such a process would turn out, year after year, the same kinds, qualities, and quantities of consumer's and producers' goods; every firm would employ the same kind and quantities of productive goods and services; finally, all these goods would be bought and sold at the same prices year after year. ... No other than ordinary routine work has to be done in this stationary society, either by workmen or by managers. Beyond this there is, in fact, no managerial function—nothing that calls for the special type of activity which we associate with the entrepreneur. Nothing is foreseen but repetition of orders and operations, and this foresight is ideally borne out by events." (*Cycles*, 40–41; *Cycles Abr.*, 17–18)

This description of the circular-flow model is constrained to the situation in which the economic system has reached what we today, in terms of game theory, might call a huge Nash equilibrium. However, Schumpeter was primarily interested in the adaptive process through which the economic system reaches such equilibrium. He also recognised that although the adaptation of the system would approach a "theoretical normal" of routinised behaviour, this normal would never be reached completely. Therefore, he often used the concept of "neighborhoods of equilibrium" as a reference point (*Cycles*, 71; *Cycles Abr.*, 45). His empirical bent meant that he wanted to include as much as possible in this concept. For instance, he emphasised the possibility of inserting at least isolated monopolies into the circular flow—provided that the monopoly gains are consumed rather than invested to enter new markets (*Cycles*, 56–7, 40; *Cycles Abr.*, 32–3, 18). He also considered the inclusion of "Growth" that derived from "changes in population" and from "savings plus accumulations". However, he rejected the possibility since "by replacing absence of change by 'balanced' or 'equilibrated progress', we arrive at a picture which really deserves to be called more unrealistic, the more it presents the misleading appearance of lifelikeness" (*Cycles*, 83, 37; cf. *Cycles Abr.*, 58).

Schumpeter's different expositions of his circular-flow model are relatively vague. This vagueness to some extent reflects the difficulties of proving the existence of a unique equilibrium of the Walrasian system—which were later overcome by Arrow and Debreu through assumptions that seem to imply omniscience (see, e.g., Debreu, 1959). Schumpeter's vagueness also reflects the serious difficulties of proving that his *tâtonnement* process leads to equilibrium. He increasingly understood that what mattered for him was not so much the analysis of the properties of a system that had already reached an equilibrium state, but rather the processes that moved the system toward such a state. To generalise these processes he needed a version of the "principle associated in physics [and chemistry] with the name of Le Châtelier"; according to this principle, the system "must tend to move, in reaction to every disturbance, in such a way as to absorb the change" (*Cycles*, 47, emphasis removed; cf. *Cycles Abr.*, 24). Schumpeter thought that this principle would largely work "provided that all actions and reactions are performed within the bounds of familiar practice that has evolved from long experience and frequent repetition" (emphasis removed). The study of these actions and reactions implies the explicit and troublesome inclusion of time in the modelling exercise. Actually, "[a]ll, or nearly all, of the difficulties we encounter will be seen to [relate] ... to the one fact that economic behaviour cannot be satisfactorily expressed in terms of the values which our variables assume at any single point of time." However, Schumpeter could not avoid this difficulty since any extended version of his circular-flow model makes essential use of a concept of time.

The primary reason for the vagueness of Schumpeter's different expositions of his circular-flow model, however, is that he designed the model to serve at least four very different purposes. First, it should bridge between the new Schumpeterian field of evolutionary economics and the core contributions to equilibrium economics by neoclassical economists. Second, the circular-flow model should provide the S-entrepreneurs with the set of equilibrated prices and quantities needed for the trustworthy calculation of the expected profits from their projects and thus for gaining financial support from Schumpeterian banks. Third, the stationary circular flow should help to define the adaptive response of the economic system to disturbances that are either related to the innovative projects S-entrepreneurs or to purely exogenous factors (like wars). This implies that we start with a disequilibrated system and follow its way back to equilibrium (see Figure 6.2 on page 149). Fourth, economic evolution can be summarised by a sequence of established circular flows. By emphasising this sequence, we arrive at a stylised economic history (see Figure 6.3 on page 151).

Schumpeter's four purposes point at different specifications of the circular-flow model. For instance, his wish to relate to equilibrium economics suggests that the model should be as close as possible to Walras's general equilibrium theory; but the Walrasian theory does not really cover adaptive responses to qualitative disturbances of the innovative type. Furthermore, the inclusion of a complex mechanism of response spoils the easiness of calculating the profitability of innovative projects. For these and other reasons, Schumpeter had to settle down with a compromise model of the circular flow that did not convince his contemporaries and gave difficulties for the extension to a full model of the capitalist engine. It was also a main source of his strange mix of economic orthodoxy and heterodoxy. It is, nevertheless, possible to overcome some of his many paradoxes by dropping the idea of a single model of the circular flow. Instead of searching for the specifications of such a model in *Development* and *Cycles*, it seems more fruitful to look for the traces of four major incarnations of the circular-flow model:

1. The model of the timeless equilibrium of a stationary economy. This model is used to find the equilibrium quantities and prices; but it is otherwise of little interest for Schumpeter (cf. *Cycles*, 42–5; *Cycles Abr.*, 20–3).

2. The model of the simple reproduction over time of a stationary economy. This model might be considered Schumpeter's core model of the circular flow. It allowed him to study the emergence of routine behaviour based on experience and selective pressures (cf. *Development*, 4–10).

3. The model of an economy that moves back to its stationary state after a small perturbation. Here the analysis is extended to cover routine-based responses to change (cf. *Cycles*, 47–56; *Cycles Abr.*, 24–32).

4. The sketchy model of the complex movement to a *new* stationary state based on the adaptation of the routine system to a radical innovative disturbance of the economic system (*Development*, 241–51; see also *Entwicklung I*, 435–47; *S1910d*, 31–39).

It is beyond the scope of the present book to discuss the Schumpeterian family of circular-flow related models systematically. However, with the possible exception of the last model, they do not assume any significant kind of foresight among established firms. The firms have accumulated knowledge about technology, consumer behaviour, and market conditions that suggest how they, for instance, adapt individually to changes in demand. This adaptation is constrained by the cumulatively strengthened network of economic relations in which they are embedded: The individual is "entangled ... in a net of social and economic connections which he cannot easily shake off. ... All these hold him in iron fetters fast in his tracks" (*Development*, 6). Both the network of firms and their hard-gained experiences promote the gradual emergence of routinised behaviour for which any rational reasons have largely become subconscious (see Table 7.2 on page 185). Furthermore, the fact that the researcher observes a behaviour that seems to reflect the rational response to given circumstances should not be used to infer that the situation has been thought through. A routine that is 'rational' from the viewpoint of the researcher is normally not the outcome of an explicit choice but rather of intra-firm selection and myopic intra-firm adaptation. Therefore, we should overcome the tradition of "bygone generations" of analysts who "have almost invariably overrated the actual range of consciously rational action" (*S1940b*, 326; cf. *Cycles*, 63n). Real action is largely guided by the limited experiences and the selection processes of the past. This type of action works at its best in the economic process that takes place in a strictly stationary economy.

Schumpeter failed to present the crucial mechanism of adaptation in a form that is easily remembered and analysed. Two explanations for this failure seem obvious. On the one hand, he probably believed that this mechanism is commonplace in any realistic version of equilibrium economics. On the other hand, he was not primarily interested in the mechanism of adaptation, but rather in the mechanism of innovation. Therefore, the task of characterising the mechanism of adaptation was largely left to later generations of researchers. Their efforts have demonstrated the difficulties involved in the characterisation of adaptation—and these difficulties may give us the ultimate reason for Schumpeter's failure. One of the most famous characterisations was actually published in the year that Schumpeter died. In 1950, Armen Alchian presented his classical exposition of the system-level mechanism of adaptation in his paper on "Uncertainty, Evolution, and Economic Theory". This paper proposes thinking by means of new assumptions on imperfect information and rule-based behaviour. While equilibrium economics had increasingly assumed that economic agents make informed and otherwise rational choices among behavioural alternatives, Alchian (1950, 213) proposed that economists should "treat the decisions and criteria dictated by the economic *system* as more important than those made by the individuals in it". In this new perspective, the object of analysis is a heterogeneous population of economic agents that is subject to a process of selection.

According to Alchian (1950, 214), we, even in economic affairs, have a Darwinian-like trial-and-error process. He illustrates some parts of his thinking by the example of the behaviour of a population of travellers: "Assume that thousands of travellers set out from Chicago, selecting their roads completely at random and without foresight. Only our 'economist' knows that on but one road are there any gasoline stations." This economic researcher "can state categorically that travellers will *continue* to travel only on that road; those on other roads will soon run out of gas. Even though each one selected his route at random, we might have called those travellers who were so fortunate as to have picked the right road wise, efficient, foresighted, etc." This example is made to demonstrate that totally random behaviour may be filtered by the selection mechanism in a way which makes the observable behaviour look as if it is informed and well-thought-out. However, the well-informed economic researcher can criticise the standard attempts to conclude on the appearances and to rationalise the behaviour of the travellers. This researcher is also able to predict changes of behaviour: "If gasoline supplies were now moved to a new road, some formerly luckless travellers again would be able to move; and a new pattern of travel would be observed, although none of the travellers had changed his particular path. The really possible paths have changed with the changing environment. All that is needed is a set of varied, risk-taking (adoptable) travellers." (pp. 214–15).

Alchian's example of a process of selection illustrates what in evolutionary biology has been called "population thinking". According to Ernst Mayr (1976), this mode of thinking confronts non-evolutionary "typological thinking". To think typologically is to focus only on the essential types of behaviour and to consider the variability of each type of behaviour as something that should be ignored. For population thinkers, this is a retreat to a Platonic world of non-evolving "ideas". This retreat is appealing from a purely logical point of view, but it makes it impossible to deal with the empirically observable evolution. The typical behaviour of a population is just an abstraction while reality is characterised by varying behaviour. More importantly, the variance of the population is the fuel of the evolutionary process. In Alchian's (1950, 213) treatment of real firms, there is an obvious "criterion by which the economic system selects survivors: those who realize positive profits are the survivors; those who suffer losses disappear." Like in the case of Darwinian natural selection, we should not presuppose that the survival of individual firms is based on full information about the rules of the game and about the other players. The question of success is decided ex post rather than ex ante: "Among all competitors, those whose particular conditions happen to be the most appropriate of those offered to the economic system for testing and adoption will be 'selected' as survivors" (pp. 213–14).

Through his criticism of standard economics, Alchian (1950, 213) arrived at "a vastly different analytical framework—one which is closely akin to the theory of biological evolution." Here the "economic counterparts of genetic heredity, mutations, and natural selection are imitation, innovation, and positive profits." Milton Friedman responded in his paper on "The Methodology of Positive Economics". Friedman (1953, 22) began by assuming that we can "[l]et the apparent immediate determinant of business behavior be anything at all—habitual reaction, random chance, or whatnot." Then selection ensures that only "behavior consistent with the rational and informed maximization of returns" survives. His conclusion was that "acceptance of the hypothesis [of profit maximization] can be based largely on the judgement that it summarises appropriately the conditions for survival." However, Friedman's argument presupposes that variance of behaviour is sufficient, that the process of selection promotes unique types of behaviour, and that selection is strong and fast enough to remove non-adapted types of behaviour (see Winter, 1971, Sect. III). Furthermore, the argument ignores the constraints on the intra-firm production of varying strategies, which might lead to path-dependent processes of change (Penrose, 1952; 1995).

If Schumpeter had known and commented on Alchian's analysis of the mechanism of adaptation, he would probably have suggested additional caveats. First, he would have warned against the too-easy use of biological analogies—as he had done previously. Second, he would have emphasised that the forces of system-level selection are not constant but variable—as exemplified by his evolutionary theory of business cycles. Third, he would have argued that the mentioned discussion on selection implied a primitive mechanism of innovation. He would hardly have failed to note that only a narrow range of variants is likely to be produced by the "random" change of routine-based behaviour. Fourth, he would have mentioned the possibility that the direction of system-level selection is to some extent influenced by his more elaborate mechanism of innovation. His innovative entrepreneurs cannot solely be seen as responding to a given selection environment. Instead, they try to change this environment. Thereby, the S-entrepreneurs, to some extent, create the environment in which routine-based firms are later selected. Although such caveats do not necessarily improve on Alchian's population thinking, they help to emphasise the characteristics of the type of mechanism of system-level adaptation that Schumpeter had proposed. The caveats also serve to explain his failure to clarify his mechanism satisfactorily. The Schumpeterian mechanism simply tended to become much too complex. His mechanism of adaptation actually covers all the phenomena that he attributed to periods of recession in the overall process of economic evolution:

> "*adaptation* to the new things created, including the *elimination* of what is incapable of adaptation, *resorption* of the results of innovation into the system, *reorganization* of economic life so as to make it conform to the data [parameters] altered by enterprise, *remodelling* of the system of values, *liquidation* of indebtedness." (*Cycles*, 137, emphasis added; *Cycles Abr.*, 113)

## 9.3 The function of the Schumpeterian entrepreneur

After having presented the circular-flow models, *Development* turns to the problem of innovation in its second chapter on "The Fundamental Phenomenon of Economic Development [Entwicklung]". The "fundamental phenomenon" of economic evolution is "the carrying out of new combinations" (*Development,* 74). However, Schumpeter used a very narrow definition of the term "new combinations of means of production"—or in his later terminology "new production functions". He thus excluded the "continual adaptation through innumerable small steps" that, for instance, could make "a great department store out of a small retail business" (p. 62). Instead, his primary example was the "building of a new railway" in the epoch when this enterprise could still be considered an innovation. Actually, "[i]t is just this occurrence of the 'revolutionary' change that is our problem, the problem of economic development [Entwicklung] in a very narrow and formal sense" (p. 63). Furthermore, his problem of economic evolution includes "only such changes in economic life as are not forced upon it from without but arise from its own initiative, from within." Schumpeter therefore had to emphasise that *if* economic change is "in practice simply founded on the fact that the data change and that the economy continuously adapts itself to them", *then* there would be no economic evolution in his narrow sense. Thereby, he secured that the studied type of economic evolution "is a distinct phenomenon, entirely different from what may be observed in the circular flow or in the tendency towards equilibrium" (*Development,* 64).

Although Schumpeter's presentation of the problem of innovation is very narrow and guarded, he nevertheless based his account on general beliefs. First, he believed that the analytical "fruitfulness" of his formulation of the fundamental problem of economic evolution could be proved by his treatment of the "problems of capital, credit, entrepreneurial profit, and crises (or business cycles)" (*Development,* 63n). Second, he believed it is a "fact that economic changes, especially, if not solely, in the capitalist epoch, have occurred thus", that is, as revolutionary changes, "and not by continuous adaptation" (p. 63). Third, he believed that his "spontaneous and discontinuous changes in the channel of the circular flow" do not appear in "the sphere of wants of the consumers of final products" (p. 65). Instead, they "appear in the sphere of industrial and commercial life". Finally, he believed that "new combinations are, as a rule, embodied, as it were, in new firms which generally do not arise out of the old ones but start producing beside them" (p. 66). For instance, "in general it is not the owner of stage coaches who builds railways."

Schumpeter's beliefs demonstrate that he in *Development*—although pretending to make a narrow account—thought that he covered the core of the general phenomenon economic evolution. He had done so by focussing on radical innovation and the subsequent change of the circular flow. Actually, his description of the circular flow as the outcome of the mechanism of adaptation served to facilitate the argument for his conclusion. Therefore, the answer to Sweezy's (1943, 93) question whether Schumpeter has "isolated and abstracted for analysis the *primum mobile* of change" to a large extent depends on how we evaluate the elimination of non-adaptive change from the circular-flow models. In *Development*, Schumpeter argued that the conservatism of established firms means that they are unable to perform innovation and that it instead has to be carried out by Schumpeterian entrepreneurs. However, Alchian's error-like type of minor innovation can hardly be said to be beyond the reach of routine-based firms. Furthermore, the oligopolistic firms of Schumpeter's Mark II model deliberately perform innovation. Since this behaviour is shared by the Marxian firms in their hunt for surplus profit, it is not difficult to realise why Sweezy (1943, 94) argued that Schumpeter had failed to prove that "without entrepreneurs we have a stationary economy". We can also say that he had failed to prove the sharp dichotomy between the mechanism of adaptation and the mechanism of innovation. It is nevertheless this dichotomy that is needed for reading *Development*. Therefore, the question of revising the Schumpeterian account shall largely be postponed to the last major section of the present chapter. The present task is rather to develop an understanding of the mechanism of innovation that was proposed by Schumpeter.

Schumpeter's theory of economic evolution has, like other evolutionary theories, a taste of circular reasoning. According to an old criticism of evolutionary theories, they argue for the survival of the fittest and define fitness by the ability to survive. Thus they seem based on the tautological proposition that survivors survive. We can also say that evolutionary arguments appear to be based on a type of logical fallacy—a petitio principii—in which the proposition to be proved is assumed in one of the premises of the argument. However, this discussion emerged from sloppy theoretical formulations and it has successfully been demonstrated that we are not facing this logical fallacy. Sweezy (1943, 93) emphasised that if Schumpeter's theory of innovation "were interpreted to mean no more than 'the cause of change is change', it would, of course, be a mere *petitio principii*". However, this theory does not argue that the cause of innovation is innovation. Instead the argument is that the cause of the innovative renewal of the routines of the circular flow is the carrying out of individual innovations. As emphasised by Sweezy, the carrying out of innovations is "the activity or function of a particular set of individuals called entrepreneurs." Since the "entrepreneur is a sociological type that can be isolated and investigated independently of the consequences which follow from the actions of the entrepreneur", Sweezy concluded that "any suspicion of circular reasoning is unfounded."

While Schumpeter had in *Entwicklung I* described the basic features of the personality of innovative entrepreneurs (see Chapter 4), he later concentrated on the entrepreneurial function of carrying out innovations. This emphasis on the innovative function in the process of evolution should not be interpreted as the result of Schumpeter's decreased interest in the individuals who perform innovative activity. However, the innovative function is what matters from the viewpoint of evolutionary economic modelling. Schumpeter's emphasis on function is also related to the functional theory of the distribution of the social product. He developed his interpretation of this theory in the paper on "The Fundamental Principle of Distribution Theory" in 1916. Here he rejected criticism of the proposition that the entrepreneur makes no profit under the conditions of perfect competition and equilibrium. His reasoning starts from the fact "that modern theory distinguishes sharply between the roles of the capitalist, the landowner and the worker on the one hand, and that of the entrepreneur on the other" (*S1916b*, 42). However, this functional division of labour is not found in reality. Here we see that even without entrepreneurial profit "our entrepreneur still always has interest on the capital belonging to himself, ground rent for the land belonging to him and an employer's salary for the work he does as well as a risk premium". Moreover, the proposition of zero profit "is related to an ideal-type understanding of equilibrium and free competition". The "nearest approximation" to equilibrium is found in "the 'calm period' which usually follows a depression and leads to a new upswing." In this situation, "the entrepreneur usually has no particular function other than routine work of the settled everyday life of the company". The possibilities of shifting from managerial work to the entrepreneurial function are a characteristic of the upswing. This upswing produces "opportunities for change and innovation in the company, and generally with opportunities for 'speculation'." Further approximations to reality are the recognition of monopoly and other "deviations from the (logical) ideal picture" (p. 42).

Schumpeter's first chance for changing the exposition of *Entwicklung I* came shortly after he had arrived at the University of Bonn in 1925. Here he initially faced two related research tasks. The first task was to write a large article on the "Entrepreneur". This article was commissioned by Schumpeter's former teacher Wieser, who, until his death in July 1926, was one of the editors of an ambitious German encyclopaedia of the social sciences (Elster et al., 1923–29). The second task was to produce a revised edition of *Entwicklung I.* Schumpeter worked with these tasks in parallel and completed them in the autumn of 1926. This completion took place under deadline pressures and very difficult circumstances (after the death of his mother and his wife), and he was not satisfied with the result. Nevertheless, they represent important steps forward in his analysis of the function of his entrepreneur.

The account for the Schumpeterian entrepreneur in *Entwicklung II* is improved in the sense that the focus is on the entrepreneurial function in the economic process. In contrast, the writing of the encyclopaedia article for an audience that was still dominated by the historical school allowed a broader treatment of the historical, sociological, and economic aspects of entrepreneurship. The focus, nevertheless, was on analytical issues. While his German-speaking audience used a loosely defined concept of the entrepreneur, Schumpeter emphasised that it should be related to the general concept of "leadership" (*S1928a*, 248). This concept should only be used "when something *new* has to be carried out, not something already established by experience and routine". Since this essential characteristic of leadership is blurred in concrete persons by their fulfilment of other tasks, the concept has to be "carved out of a more or less complicated conglomerate [of facts] by way of analysis". Then the "entrepreneurial function is nothing but the leader-function in the sphere of the economy" (p. 249). The way this function is "carved out" from the facts is not treated. However, Schumpeter's article ends with a long bibliography (pp. 256–9) that partly seems to be designed for this purpose. Thus, the list not only includes theoretical and historical works of relevance to the task, but even biographies of great entrepreneurs like John D. Rockefeller and August Thyssen as well as the families of some entrepreneurs.

Schumpeter's "Entrepreneur" article included a short account of the historical development of the economic concept of the entrepreneur (*S1928a*, 245–7; see also Hébert and Link, 2007). This account was expanded and improved in different parts of *History*, which gives a better impression of his type of concept analysis. His account can be interpreted as an attempt to demonstrate that the Schumpeterian concept of the entrepreneur served to overcome what Fritz Redlich (1966, 709) called an "unfortunate legacy". This is the legacy of hiding an important economic function in the concepts of the capitalist and of the firm. According to Redlich, it was Schumpeter who finally "freed economic theory in this area from the dead hand of Adam Smith's lack of understanding of contemporary reality and of vision of the future" (p. 716). Schumpeter's own account emphasised that "analytical progress—not only in economics—hinges in great part on making things explicit that have been implied or implicitly recognised for ages" (*History*, 555n). He considered this especially true for his core concept of the entrepreneur. Classical economics only recognised three agents of the basic model of the economic system: workers, landowners and capitalists. This did not mean that "economists had ever accomplished the impossible feat of overlooking the most colorful figure in the capitalist process" (p. 554). However, analytical progress was seriously constrained because the figure of the entrepreneur was not treated explicitly. For instance, the "entrepreneur is undoubtedly present in the Marxian drama. But he is present behind the scenes and his gain is not a Marxist problem. It can be inserted into the Marxist system only be means of an un-Marxist reinterpretation" (p. 896).

The work of French economists served to introduce the entrepreneur into explicit economic analysis. This work started with Richard Cantillon, who, when young, had made a fortune in relation to the speculative bubbles that emerged from John Law's entrepreneurial activities around 1720. Nevertheless, it was Jean-Baptiste Say who "turned a popular notion into a scientific tool" (*History*, 555). Say seems to have recognised "that a greatly improved theory of the economic process might be derived by making the entrepreneur in the analytic schema what he is in capitalist reality, the pivot on which everything turns." Nevertheless, Schumpeter missed what was crucial to him, since Say failed to realise "that the task of combining factors becomes a distinctive one only when applied not to the current administration of a going concern but to the organization of a new one." This recognition did not emerge in the Walrasian system. However, Schumpeter thought that "Walras' contribution was important though negative" since "entrepreneurs' profits can arise only in conditions that fail to fulfil the requirements of static equilibrium". This is "the proposition from which starts all clear thinking on profits" (p. 893). The proposition is only a starting point. It was, instead, the American John B. Clark who "made a great stride toward a satisfactory theory of the entrepreneur's function and the entrepreneur's gain and, in connection with this, another great stride toward the clarification of all economic problems that must result from a clear distinction between stationary and evolutionary states" (p. 868). Thereby, Clark became "the first to strike a novel note by connecting entrepreneurial profit, considered as a surplus over interest (and rent), with the successful introduction into the economic process of technological, commercial, or organizational improvements" (*History*, 894).

The concept of the S-entrepreneur had been treated in a strange manner by the early Schumpeter. He had derived the concept from "two fundamental principles": the "energy principle" and the principle of "carrying out of new combinations" (*Entwicklung I*, 170). In Section 4.2, we have already considered the "energy principle". It basically states that agents are heterogeneous with respect to their capabilities to perform socio-economic tasks. Those with limited psychic "energy" have to cling as closely as possible to routine, while the "energetic" agents tend to perform routine-changing behaviour. The first half of the original version of his core chapter on "The Fundamental Phenomenon of Economic Development" (*Entwicklung I*, 106–56) applies this principle to describe the "mass" as well as the "the man of action". Although this account gives economic examples, it provides a general treatment of the Schumpeterian "leader". This concept can be used for any sphere of social life. It is only in the second half of the chapter (pp. 156–98) that Schumpeter really turned to economics by considering the "carrying out of new combinations" by S-entrepreneurs. This delayed treatment of the core issues probably reflects his research process—but it distracted his readers' attention from the fact that he was developing an economic model. This largely explains why he radically changed his mode of presentation in *Entwicklung II* and *Development*. Here he removed most of the general account for his two types of agent and changed the sequence of the rest of the material. Becker and Knudsen (2003, 206) have suggested that the corresponding formulations in his "Entrepreneur" article demonstrate a "profound change in Schumpeter's conception of entrepreneurship" because of the "depersonalisation of his earlier entrepreneur, the strong individual of almost superhuman powers of energetic will." However, it seems more appropriate to interpret the change as reflecting a wish to make a clearer exposition that served well as a starting point for the modelling of economic evolution.

Some of Schumpeter's mature efforts to present his mechanism of innovation are reflected in the second chapter of *Entwicklung II* (and of *Development*). Here he included a new exposition of the concept of the S-entrepreneur with three steps. First, he defined the concept of carrying out of new combinations (*Development*, 65–8). Second, he described the phenomenon of credit and argued that it is strongly connected to the introduction of new combinations (pp. 68–74). Third, he turned to an extensive treatment of issues relating directly to the S-entrepreneur (pp. 74–94). To treat these issues, Schumpeter made a double definition: "The carrying out of new combinations we call 'enterprise'; the individuals whose function it is to carry out them out we call 'entrepreneurs' " (p. 74). This broadness of these definitions of entrepreneurs and their tasks was not only emphasised in *Development* (66), but also the article on the entrepreneur (*S1928a*, 250):

> "The essence of the entrepreneurial function lies in recognising and carrying out new [or, rather, untried] possibilities in the economic sphere. Such an economic leadership thus occupies itself with tasks that can be summarised in the following types:
>
> (1) the production and carrying out of new products or new qualities of products,
> (2) the introduction of new production methods,
> (3) the creation of new forms of industrial organisation (for instance trustification),
> (4) the opening up of new markets,
> (5) the opening up of new sources of supply."

While the treatment of product innovation and process innovation (see Table 9.4 on page 269) are today pretty standard, the last three types of innovation serve to emphasise special features of Schumpeter's analysis that must even be included in his circular-flow model. First, a typical outcome of an organisational innovation is a cartel that functions according to given routines. This means that such a cartel must find its stationary place in the circular flow, and this implies that its profits are consumed rather than invested (*Cycles*, 40; *Cycles Abr.*, 18). Second, the possibility of market innovations implies that the circular-flow model includes institutional hindrances against the arbitrage activities of the S-managers (*Cycles*, 9; *Cycles Abr.*, 3). Third, the input innovation points to an aspect of Schumpeter's analysis that is unclear as long as he operates with vertically integrated firms that only rely on inputs of the services of land and labour. He always considered this Austrian assumption a matter of convenience, and he often abstained from it, especially in *Cycles*. However, he lacked the analytical tools for handling the untraditional forms of innovation in a fully consistent manner. Since even product innovation brought analytical difficulties, he used process innovation as his favourite example.

All the types of innovation define tasks for Schumpeterian entrepreneurs. Their projects take the form of "enterprises" that in the end are turned into routinised firms. This is the definition of entrepreneurial enterprises in *Development* and *Cycles*. However, the definition in *Entwicklung II* (111) is broader. In 1926, Schumpeter considered "enterprise" as covering both entrepreneurial projects and their "embodiment in plants and so on". This ambiguous application of the term "enterprise" emerged in the article on the entrepreneur (*S1928a*, 237, 245), and he simply referred to this interpretation in *Entwicklung II*. A few years later, he helped Redvers Opie in producing *Development*, and by that time he must have recognised that the ambiguous use of the term enterprise might imply a return to the terminological quagmire of classical economics. The story of his solution to this problem reaches a preliminary conclusion in *Cycles* (102; *Cycles Abr.*, 77), where he restated the narrow definitions of enterprise and entrepreneur. However, these definitions were made after he had assumed that enterprise, i.e., an innovative project, implies that "New Men" are engaged in the "construction of New Plant" that is "embodied in a New Firm created for the purpose" (*Cycles*, 93–6; *Cycles Abr.*, 68–70). The last assumption is especially important. Schumpeter decided to "argue as if every innovation—as now defined—were embodied in a New Firm" and he considered the assumption "a device to bring within the reach of theory an important feature of capitalist reality". It implied a clear-cut distinction between the entrepreneurial project (the enterprise) and its result (the established firm). Thereby, he could describe the function of the S-entrepreneur as clearly as possible.

Schumpeter's striving for analytical clarity is reflected in the fact that he did not only define the function of the S-entrepreneur positively as

**Table 9.4.: Schumpeter's five types of innovation**

| Type | Description |
|---|---|
| Product innovation | A new type of product or variant of product is added to the list of products of the economic system. It requires a new production routine and a change in the consumption routines of some consumers. |
| Process innovation | A change in the the production routine for an existing product. The new technology is characterised by a change in the quantities of inputs needed for the production of one unit of output. |
| Organisational innovation | A change in the routines underlying firm behaviour that typically involves a change of market structure. For instance, it includes the reorganisation related to the creation (or destruction) of a price cartel. |
| Market innovation | A product that previously has been used routinely by one group of consumers is introduced into the consumption routines of a qualitative different group of users (for instance, in another country). |
| Input innovation | A new raw material or a new intermediate good is introduced into the economic system. It is not implemented before some firms have adapted their production routines to the new input. |

Comment: Different models are needed for analysing the effects of the five types of innovation. Schumpeter largely constrained himself to the cases of process innovation and product innovation. Source: See *S1928a* (250) and *Development* (66).

the carrying out of innovation. He also defined the function of the S-entrepreneur negatively. His wish to overcome the unfortunate legacy of classical economics meant that he wanted to emphasise that his entrepreneur is not a capitalist or the bearer of risk. Furthermore, he wanted to overcome the neoclassical mixing of the entrepreneur with the firm by emphasising the entrepreneurial enterprise. Finally, Schumpeter recognised that there are a lot of other ways of complicating the use of the concept of the entrepreneur. If entrepreneurs form a profession, their work can be considered routinised. If the entrepreneur is seen as governed by consumer tastes, then we are facing adaptive response rather than creative response. If the entrepreneur is providing the technological or organisational invention needed for the innovative enterprise, then the analysis has to leave the realm of economics and enter fields of engineering and organisation science. Schumpeter wanted to overcome such analytical difficulties through the negative definition of the S-entrepreneur. Let us consider his result:

According to Schumpeter's negative definition, the entrepreneur is

(1) not the capitalist who finances the enterprise,
(2) not the bearer of the risk related to the enterprise,
(3) not the manager of a firm,
(4) not the member of a profession,
(5) not governed by consumers, and
(6) not the inventor of the innovation that is carried out.

These six demarcations follow more or less directly from the assumption that the entrepreneurial function is innovation-based enterprise. The most crucial and difficult demarcation concern the division of labour between entrepreneurs and the capitalists in this enterprise. The S-entrepreneur is not a capitalist. Instead, Schumpeter considered the setting up of a new firm and the provision of the needed money as two very different functions. These functions were made clear by treating the S-entrepreneur as an independent agent that, in principle, is unconnected to pre-existing economic activity. Thereby, the S-entrepreneur becomes "a debtor by the nature of his economic function" (*Development*, 103). His creditor is the "capitalist", which Schumpeter identified as the "banker". Thereby he had obtained his goal: to partition the function of the "capitalist" of classical economics. This partitioning, however, only worked in relation to the creation of innovation-based firms. He avoided the possibility of the self-finance by incumbent firms by assuming that such firms are not able to perform intra-firm innovation because they "display as a rule symptoms of what is euphemistically called conservatism" (*Cycles*, 97; *Cycles Abr.*, 71).

A related part of the negative definition of the S-entrepreneur is that "risk bearing is no part of the entrepreneurial function. It is the capitalist who bears the risk" (*Cycles*, 104; *Cycles Abr.*, 79). Since this statement has created much confusion, it needs careful consideration. The early Schumpeter had spent much time studying the mathematics of insurance; and he had emphasised that insurable risk should be considered an element of the costs of economic activity (*Entwicklung I*, 49–50; *Development*, 32–33). He is obviously not talking about this kind of risk. Instead, he is referring to the bearing of the consequences of "uncertainty" in the sense of Frank Knight's "very useful emphasis on the distinction between insurable risks and non-insurable uncertainty" (*History*, 894). Like Knight later did, Schumpeter wanted to link the uncertainty to "rapid economic change" and "differences in business ability" (p. 894). This closeness of their thinking becomes less surprising when we recognise that they both started from the discussion between Böhm-Bawerk and Clark on the explanation of entrepreneurial profit and that Knight used the same literature as Schumpeter—with the exception that he could also use *Entwick-*

*lung I*. However, while Knight (1921, Part 3) thought in terms of Knightian entrepreneurs who specialised in uncertainty-bearing, this function is taken care of by the Schumpeterian bankers. The reason seems to be that the Knightian "enterprise" is an established firm that is subject to ongoing development while the Schumpeterian "enterprise" of the S-entrepreneur is an innovative project with a clear conclusion: a failure or a successfully established firm that can be run according to routine.

According to the Mark I model of *Development* and *Cycles*, the S-entrepreneur is not an S-manager who follows routines when administering an incumbent firm. The assumption that the new firm embodies an innovation also excludes the possibility that it is created by an S-manager who is hired for the purpose. However, such managers—who represent Schumpeter's version of the Walrasian entrepreneur—enter the scene when the setting up of a particular type of new firms and plants has become a matter of routine. The Mark I model also explains why nobody can remain an S-entrepreneur for the whole of his working life. A person only functions in this role during the upstart of an innovation-based firm. After he has completed the task—or failed in doing so—he might turn to the creation of another firm. Since Schumpeter considered this case as seldom, the entrepreneur nearly always turns to other functions in the economic system while enjoying the rents of his enterprise (*Development*, 78–9).

Schumpeter's idea that the mechanism of innovation is based on the creative response of the S-entrepreneur has many implications. One of them is that while the circular flow is based on the principle of consumer sovereignty, the entrepreneur cannot be governed by this principle. If this had been the case, we would be facing an adaptive response. It is, instead, the entrepreneur who influences the decisions and even the preferences of consumers. Schumpeter's example is: "Railroads have not emerged because any consumers took the initiative in displaying an effective demand for their service in preference to the services of mail coaches" (*Cycles*, 73; *Cycles Abr.*, 47). Instead, the "great majority of changes in commodities consumed has been forced by producers on consumers who, more often than not, have resisted the change and have had to be educated up by elaborate psychotechnics of advertising". However, Schumpeter is focussing on a particular type of creativity. His entrepreneur is not a thinker but a doer. This implies that the S-entrepreneur is not the provider of the innovative business opportunity that becomes embodied in the new firm. The "invention does not necessarily induce innovation, but produces of itself ... no economically relevant effect at all" (*Cycles*, 84–5; cf. *Cycles Abr.*, 59–60). The production of such an effect by the innovation is often only loosely related to any underlying invention and is dominated by the need of overcoming a large number of hindrances. The obstacles imply that even "the making of the invention and the carrying out of the corresponding innovation are, sociologically and economically, two entirely different things". The fulfilling of the function of the S-entrepreneur requires the "ability to 'walk alone' and to act on ground untried by experience" (*Cycles*, 99; *Cycles Abr.*, 73). This ability comes in grades and it might be "distributed in the population according to a Gaussian—though more plausible a skew—law". In some cases, the person able to overcome the difficulties involved in the setting up of an innovation-based firm has also provided parts of the underlying inventive activity. However, Schumpeter thought that this case is rare because very different abilities are needed. In any case, this broadly capable person would simply have filled two different functions. The entrepreneurial function is to grasp a pre-existing opportunity and to overcome the difficulties involved in its implementation.

We have now considered the six parts of Schumpeter's negative specification of the function of his entrepreneur (see the list on page 270). The general result is a concept of the entrepreneur that is radically different from the concepts applied by the analytical tradition that can be traced back to Cantillon and Say. This implies that the meaning of the concept of the S-entrepreneur has been hard to grasp for most economists. For example, Schumpeter's treatment of "risk"—in the sense of non-insurable uncertainty—has created much confusion. The reason is not least that he, after having made the statement that "risk bearing is no part of the entrepreneurial function", immediately added that "riskiness" nevertheless "enters into the pattern in which entrepreneurs work". The reason is that "every new thing is risky in a sense in which no routine action is"; and this type of riskiness "makes it more difficult to obtain the necessary capital and thus forms one of the obstacles entrepreneurs have to overcome" (*Cycles*, 104n; *Cycles Abr.*, 79n). It means that the S-entrepreneur has to persuade a banker and that "the banker should know, and be able to judge, what his credit is used for" (*Cycles*, 116; *Cycles Abr.*, 90). The entrepreneur has to persuade the banker that the remaining uncertainties concerning the outcome of the innovative project are offset by the potential profits. The banker has to consider complex issues of risk management by careful evaluation of individual projects as well as his overall portfolio of innovative finance. One of the consequences is that Schumpeter seems to consider banking in terms of what may today be called a customer-market model in which the problems of uncertainty and of asymmetric information about it are reduced. For instance, "[t]here is routine procedure for dealing with new business propositions which does not make it very easy to 'get away' with either a foolish or a fraudulent scheme" (*Cycles*, 140n). Such a procedure is necessary in face of the problem "that a majority of would-be entrepreneurs never get their projects under sail and that, of those who do, nine out of ten fail to make a success of them." Nevertheless, the imperfection in the handling of this difficulty has served "to turn the history of capitalist evolution into a history of catastrophes" (*Cycles*, 117; *Cycles Abr.*, 91). The smaller "catastrophe" for an individual project is, of course, not without consequences for the S-entrepreneur. He has failed in obtaining the expected profit, but he does not bear the financial risk: "Even though he may risk his reputation, the direct responsibility of failure never falls on him" (*Development*, 137). However, this hardly implies that behaviour towards risk is of no consequence for fulfilling the entrepreneurial function. This function is coupled with risk-seeking behaviour of an untraditional kind. The S-entrepreneur's intense effort to create an innovation-based firm is based on a "non-hedonist" psychology that includes "the will to found a private kingdom" and "the impulse to fight, to prove oneself superior to others" (p. 93). The S-entrepreneur is willing to waste substantial efforts and even his reputation for the small chance of obtaining these goals. However, his efforts can hardly be considered totally wasted in the case of failure. The reason is that the entrepreneurial psychology also includes "the joy ... of exercising one's energy and ingenuity" (p. 93).

It was not least through his negative definition of the innovative entrepreneur that Schumpeter made it relatively easy to develop his Mark I models that analyse the carrying out of new combinations by entrepreneurs as the fundamental phenomenon of economic evolution. These models vary according to the type of innovation that is carried out (see Table 9.4 on page 269). Although he often verbally described the mechanism of product innovation, the modelling of economic evolution in terms of process innovation is much easier. Here the entrepreneur obtains credit in order to create a new firm that has a better productivity than existing firms. When the new process technology has become routinised within the newly established firm, it is difficult to change it. Instead, it serves as a parameter in managerial decision-making. The basic technology of the firm only comes under question because of the threat from new firms with better productivity. Facing such a threat, the initially innovative firm might try to survive by adaptation, but often its inflexibility leads to negative profit and, sooner or later, to bankruptcy. This story could, of course, be developed in terms of an entrepreneur with multiple characteristics that include abilities to invent, manage, finance and take risks. However, the gain in realism would be followed by difficulties of model construction. Actually, this task is difficult even with Schumpeter's most narrow specification of the S-entrepreneur. Therefore, most modellers would like to make further simplifications.

## 9.4 Combining the mechanisms of innovation and adaptation

Schumpeter's insertion of the S-entrepreneur into the circular-flow models provided him with a simple version of the capitalist engine. The function of this engine is to change the routines of the economic system in a direction that basically cannot be determined in advance. Since there are always winners and losers, this economic evolution does not provide 'progress' in the sense of Pareto improvement, where no economic agent faces deterioration. Nevertheless, since all S-enterprises are designed to create a profit, they must, in a loose sense, provide results that are 'better' than those of existing routines. Under certain conditions, this means a long-term increase in average welfare of the consumers (*Capitalism*, Chs. 6–7). It is only in this sense that the capitalist engine provides 'progress'. The shorter-term functioning of the engine includes elements that work in the opposite direction. This becomes clear if we think of it as an internal combustion engine with an upswing and a downswing of a two-stroke cycle (see Figure 9.1 on page 251). The resting state is the stationary flow. The propulsive stroke is provided by "innovations which 'ignite' prosperity" (*Cycles*, 449–50). This prosperity is characterised by decreased consumption because the S-enterprises use some of the present resources of the economic system for the provision of future goods. These goods are delivered during the reactive stroke. In the case where pure process innovation carried the prosperity, average productivity is increased by the selective forces of the market system. More specifically, firms with below-average productivity are forced to exit or to adapt by increasing their productivity. This is what happens during the stressful "periods of liquidation", which Schumpeter in *Cycles* called recessions. Because of the losses incurred on many agents during recession, we should not expect any social consensus about the shorter-term functioning of the capitalist engine by means of prosperities and recessions.

As mentioned above, Paul Sweezy summarised Schumpeter's model of the functioning of the capitalist engine in terms of the circular flow and the innovative entrepreneur. On this background, Sweezy asked whether "the result of joining the two elements [is] a correct representation of the essentials of capitalist reality". By studying *Development* in the context of the two other books of the evolutionary trilogy, we recognise that even Schumpeter, to a large extent, answered in the negative. As demonstrated by *Cycles*, the idea of an equilibrated circular flow is not necessarily an adequate formulation of the background for a swarm of innovative projects. As demonstrated by *Capitalism*, the mechanism of innovation is not necessarily working by means of individual entrepreneurship. Therefore, the combination of the circular flow with innovative entrepreneurship is not the necessary way of representing the capitalist engine. The remaining question, however, is whether the critical study of the Mark I model is a fruitful starting point for the further development of Schumpeter's evolutionary economics. This question can be answered affirmatively under three conditions. First, the analytical emphasis should not be put on the circular flow as a stationary state but rather on the mechanism of adaptation that leads the evolutionary process to a halt. Second, the emphasis should not be placed on the personal characteristics of innovative entrepreneurs (see Section 4.2), but rather on the mechanism of innovation. Third, the overall Mark I model should not be interpreted as the finished representation of Schumpeter's basic vision and analysis of the capitalist engine. His model is much too constrained by the tools of equilibrium economics to be interpreted in this way. It is rather one of a whole series of models that can be used to characterise different aspects of the engine of capitalist economic evolution.

Schumpeter's image of the capitalist engine of economic evolution distinguishes him from nearly all other economists. Actually, he tried to demonstrate that equilibrium economics comes very close to thinking solely in terms of something like his circular-flow model. In this model all economic agents are equal. If we should want to define one of the agents as 'king', it would be the consumers. It is their consumption routines that determine how a given, and small, increase of income is distributed across the different parts of their budgets. However, as soon as a significant innovation is made in the previously stationary economy, then this interpretation does not hold any longer. As the one who ignites the economic engine, it is the entrepreneur who causes the change. Therefore, he is the "king" of the system who enforces change upon the other agents. It is especially the consumers who have to be pressed to change their routines. Nevertheless, like a Spartan king, the entrepreneur needs a controller, an ephor: "Like the entrepreneur is the king, so the banker is the ephor of the exchange economy" (*Entwicklung I*, 198; changed in *Development*, 74). We shall return to the role of the bankers later, but it should be remarked that Schumpeter provided his capitalist engine with a "manometer of the economy" to help their efforts (*Geldes*, 310–11). This instrument might suggest the metaphor of a steam engine rather than that of an internal combustion engine. In any case, it provides "a kind of coefficient of tension in the system, which more nearly than any other single figure expresses the degree of disequilibrium" (*Cycles*, 126, emphasis removed; *Cycles Abr.*, 101). In the basic model, this rate of interest reflects the degree to which "the financial side of capitalist evolution" (*Cycles*, 613) has mobilised available monetary resources and is approaching the more or less binding constraints on credit creation by banks.

By introducing his S-entrepreneur into the circular-flow model with no intrinsic motion (a Σ state), Schumpeter created the contour lines of his model of the capitalist engine that produced "incessant revolutions". However, Schumpeter did not consider the revolutions of the capitalist economy as "strictly incessant" but as occurring "in discrete rushes which are separated from each other by spans of comparative quiet" (*Capitalism*, 83n). In accordance with the metaphor of an engine with a two-stroke cycle, he thought that, in capitalist reality, "there always is either revolution or absorption of the results of revolution". It is this never-ending sequence of "revolutions" and "absorptions" that he claimed to cause "what are known as business cycles" (p. 83n). We have considered this explanation of business cycles in the previous chapter. Presently, it should be noted that the validity and fruitfulness of large parts of his analysis of capitalist economic evolution are completely independent of his explanation of business cycles. Therefore, we shall presently consider the sequence of circular flows that are punctuated by mutative changes as a matter of analytical technique. Schumpeter was simply unable to handle the analysis of "incessantly" occurring innovations. Instead, he analysed the activities of S-entrepreneurs and the responses to them on the background of the circular-flow model.

Schumpeter analysed the role of innovations with respect to a given circular flow by means of the assumption of the Mark I model that new men set up new plant within new firms. The simplest possible case is that one S-entrepreneur carries out a major innovation by establishing one firm. The task is then to analyse the process that transforms the initial circular flow to a new one. Let us, therefore, start with an S-entrepreneur who considers one of the many unexploited opportunities for profitable innovation that are always available in the routine-based model of the circular flow. Since such a pool of opportunities does not exist in most economic models, Schumpeter had to argue for it. However, although his basic argument concerns 'untried possibilities', he often talked about 'new opportunities'. Thus, he emphasised that "[n]ew possibilities are continuously being offered by the surrounding world, in particular new discoveries are continuously being added to the existing store of knowledge" (*Development*, 79). The crucial question is: "Why should not the individual make just as much use of the new opportunities as of the old ... [and use them] if this can be seen to be more advantageous?" (p. 79). The simplest answer is that the agent of the circular flow "never alters his routine" (p. 81n). He merely makes "passive adaptation, i.e., adaptation within the fundamental data of the system" (*Cycles*, 72). The central datum for the incumbent firm is the routine-based production function by which it combines inputs to produce output. It is the conservative upholding of given production functions that creates a pool of untried possibilities for establishing new production functions. The "creative response" (p. 9) to the available opportunities consists in the setting up of a new firm that implements one of the potential innovations. It is only through "the formation of companies for the exploitation of the new opportunities" that these opportunities are "turned into commercial and industrial reality" (p. 9). This reality confronts the agents of the circular flow when they have to choose between adaptation and destruction.

Schumpeter emphasised, as we have seen above, that the profitability of the project of the S-entrepreneur is hardly its main reason and that it seems to be promoted by an untraditional kind of risk-seeking behaviour. This point has been put even clearer by Keynes (1936, 161–2): "Enterprise only pretends to itself to be mainly actuated by the statements in its own prospectus". It is based on "spontaneous optimism rather than on a mathematical expectation, whether moral or hedonistic or economic". This optimism is "supported by animal spirits", and this means that "the thought of ultimate loss which often overtakes pioneers, as experience undoubtedly tells us and them, is put aside as a healthy man puts aside the expectation of death". However, we have also seen that the "prospectus" of the S-entrepreneur is subject to the scrutiny of his banker. Therefore, the carefulness of his study of the expected profitability of the planned firm is not without consequence. Furthermore, the conditions for calculations are much better for the S-entrepreneur than for his Keynesian counterpart. The reason is that the calculations can be performed by reference to the circular flow (or to the "neighborhood of equilibrium"). More specifically, the profitability analysis can be based on the relevant parts of the price system as well as on the consumption functions of prospective customers and the production functions of competitors. It is on this background, and for a given time horizon, that an expectation of the internal rate of interest of the project is formed. We may also say that an expectation has been made about the "surplus value" (Mehrwert) of the project (*Entwicklung II*, 207, 223).

Both the profit of the S-enterprise and the, smaller, profit of the S-entrepreneur have to be calculated. In general, such profit "is the premium put upon successful innovation in capitalist society". This premium "is temporary by nature: it will vanish in the subsequent process of competition and adaptation" (*Cycles*, 105; *Cycles Abr.*, 79–80). This competing down has by many economists been assumed to be very quick. However, the profitability of enterprise is essential for economic evolution; and "for profits to emerge it is essential that the 'suicidal stimulus of profits' should not act instantaneously". Although Schumpeter cannot rule out this possibility in the abstract, it is clear that responses in his circular-flow model are less than prompt. This sluggishness is included even when he assumed a "perfectly competitive system". Actually, "[s]ome friction may even be said to be necessary for the economic system to function at all"—just as "the physical world would be an uninhabitable chaos" without inertia (*Cycles*, 50–1; *Cycles Abr.*, 28–9). The existence of inertia is also crucial for economic evolution, but the outcome of the calculations of the S-entrepreneur is determined by the degree of inertia. Schumpeter made no general assumptions on this issue and left the determination of the degree of inertia to concrete applications of his model of economic evolution. However, the degree of inertia can only partly be described by a parameter. It also depends on the behaviour of the S-entrepreneur and his closest competitors. Therefore, Schumpeter was, from the very beginning, very interested in imperfect competition. Initially he seems to have abstained from a solid treatment of such behaviour because of the (unfair) objections by leading economists against the pioneering efforts of Cournot. In the 1930s, however, the intellectual climate was different. This inspired Schumpeter to cover imperfect competition more thoroughly (*Cycles*, 56–68, 106–8; *Cycles Abr.*, 32–42, 81–2). He even considered the efforts of the S-entrepreneur to conserve his stream of profits as well as the defensive strategies of the incumbent firms. However, his analysis did not dig deeply into these matters. The main reason seems to be that he did not want to commit himself to an analysis of one of his five very different cases of innovation (see the list in Table 9.4 on page 269). In any case, he recognised "that an enterprise in our sense almost necessarily finds itself in an 'imperfect' situation, even if the system be otherwise a perfectly competitive one" (*Cycles*, 107; *Cycles Abr.*, 81).

The next step in understanding Schumpeter's basic model of economic evolution is to recognise that the S-entrepreneur cannot function alone. Since he represents a project of an innovation-based firm, he needs financial resources that are not present in the circular flow. Therefore, the driver of the capitalist engine (the $\delta$ mechanism) has two major components: innovation and credit creation. On the one hand, innovation provides "[t]he fundamental impulse that sets and keeps the capitalist engine in motion" (*Capitalism*, 83). It consists of "the new consumers' goods, the new methods of production or transportation, the new markets, the new forms of industrial organization that capitalist enterprise creates". On the other hand, the analysis of the process of innovation and its effects on the routine activity presupposes that we include a monetary complement. Actually, "the logical relation ... between what is called 'credit creation by banks' and innovation ... is fundamental to the understanding of the capitalist engine" (*Cycles*, 111; *Cycles Abr.*, 85).

Money obviously plays a crucial role in Schumpeter's analytical movement from the stationary flow to economic evolution. The major point is that the S-entrepreneur needs money, or purchasing power, to implement his project. This money has to be borrowed "[i]n accordance to our conception of New Men setting up New Firms". Thus, "would-be entrepreneurs" cannot directly reuse the resources of old firms of the circular flow for their new purposes. If we add the characteristics of the circular flow, "we arrive at the following three propositions, which may sound strange but are tautologically true for an economic world embodying our assumptions: Entrepreneurs borrow all the 'funds' they need both for creating and operating their plants ... Nobody else borrows. Those 'funds' consists of means of payment created *ad hoc*" (*Cycles*, 110–11; *Cycles Abr.*, 84–5). The purpose of these propositions is to set the stage for the "the carrying into effect of an innovation" that involves "the shifting of existing factors from old to new uses" (*Cycles*, 111; *Cycles Abr.*, 85). The proposition of credit creation for S-entrepreneurs was, perhaps, the most radical one since "the classical theory of banking" had assumed that "what banks finance is precisely not innovation but current commodity transactions" (*Cycles*, 109n). Schumpeter rejected this assumption by proposing that "in the sphere of business, innovation is the pillar of interest" (*Cycles*, 124–5; *Cycles Abr.*, 98–9). He had two reasons for this proposition. First, "the profit it yields to the successful entrepreneur is the typical reason for a readiness to pay interest—for looking upon present dollars as a means of getting more dollars in the future". Second, "borrowing is, in the situation of the entrepreneur, the typical means of those present dollars". Since the typical income of "the capitalist class" is interest, it "lives on a return which, except for the financing of consumption, derives from innovation or processes directly induced by innovation, and would, hence, disappear if economic evolution ceased".

Although it is possible to discuss the later Schumpeter's theories of credit, interest and innovation in relation to Böhm-Bawerk's *Capital and Interest* (see Section 5.3), it seems more appropriate to relate his theories to Knut Wicksell's works from the last decade of the nineteenth century. Wicksell developed his theory of capital and interest through a critical discussion that not only covered Böhm-Bawerk's works but also those of other leading economists. Thus, he emphasised that Walras's entrepreneurs do *not* pay landowners, workers and capitalists *in advance*. According to Wicksell (1954, 95), this "fiction" is related to Walras's "fundamental mistake" of having "completely overlooked the significance of time in production". Wicksell's remedy was to assume that the W-entrepreneur pays for the services by means of a bank credit that is liquidated when a round of production has finished and the output has been sold. At the aggregate level, this process can repeat itself period after period if the bank rate of interest is equal to the "natural rate of interest", which reflects the permanent net return that emerges from the entrepreneurs' use of productive services. Any long-term discrepancy between the bank rate and the natural rate implies a disequilibrium that starts a cumulative process. If the bank rate of interest is too low, "entrepreneurs will in the first instance obtain a surplus profit" and "inevitably be induced to extend their business in order to exploit to the maximum extent the favourable turn of events" (Wicksell, 1936, 105). This means an increased demand for services and goods as well as increasing prices, and the process goes on as long as the bank rate is not corrected. According to Schumpeter (*History*, 1118n), "Böhm-Bawerk's comment on this argument was: 'Wicksell must have been dreaming when he wrote that.'" Schumpeter seems to have agreed: "For us ... there is no such thing as a real rate of interest, except in the same sense in which we speak of real wages" (*Cycles*, 127; *Cycles Abr.*, 101–2). He thought that Wicksell's concept represented a superficial viewpoint: "the money market with all that happens in it acquires for us [Schumpeter] a much deeper significance". The significance of the money market is that "[i]t becomes the heart, although it never becomes the brain, of the capitalist organism." These and other Schumpeterian insights, however, might not require a radical break with Wicksell. This is at least the experience of a number of Swedish economists who since the 1930s has tried to combine Wicksell with Schumpeter (Eliasson, 2007).

According to Schumpeter's interpretation, the monetary rate of interest is the regulator of a broadly conceived "money market" that includes the markets for money, credit, and capital. The crucial issue in this market is the transfer of real resources that is made possible by putting buying power in the hands of the S-entrepreneur. Since such buying power is not available in the circular flow, it has to be created. Therefore, there an essential relation between innovation and credit creation by banks. This way of financing innovation implies that "the new 'order of the factors' comes, as it were, on the top of the old one, which is not thereby cancelled" (*Cycles*, 112; *Cycles Abr.*, 86). Although the S-entrepreneurs bid up prices for productive services and thereby reduce the purchasing power of old firms, the real "creative destruction" begins only after the new firms have started production and the entrepreneur is repaying his debts. Thus Schumpeter covered both "credit creation" and "credit destruction". In general, he emphasised that credit "is the characteristic method of the capitalist type of society—and important enough to serve as its *differentia specifica*—for forcing the economic system into new channels, for putting its means at the service of new ends" (*Development*, 69–70). In other words, the existence is credit is a necessary condition for the existence of a capitalist economy in the Schumpeterian sense.

It is by supplying the S-entrepreneurs with newly created money that the banks alter the distribution of ownership of the productive resources. The side-effect of this supply is the short-term inflation that starts from increasing prices of productive services. The reason is that the entrepreneurial demand is added to the demand of incumbent firms and that the increased demand is confronted with a constant supply of productive services. However, the mechanism of the "capitalist engine" does not allow one-sided action of the banks. Although "our understanding of the process of the capitalist society hinges in important respects on realizing the fact that monetary capital is a distinct agent, it also hinges in no less important respects on realizing how it is related to the world of commodities" (*Cycles*, 129; *Cycles Abr.*, 104). The responsibility of relating the monetary and the real world falls on the S-banks. This is a major reason why it is "important for the functioning of the capitalist machine ... that banks should be independent agents" (*Cycles*, 118; *Cycles Abr.*, 92). The credit created by such banks not only explains short-term inflation, but also long-term deflation because of the new level of productivity obtainable by the new firms and the subsequent restructuring of the economic system.

The development of the monetary part of his model of the capitalist engine was not easy for Schumpeter. The reason is that money plays different roles in his model of the innovative transformation of the circular flow and in his pure model of a stationary circular flow. It was only with respect to latter model he could agree with the opinion of standard neoclassical economics that money is neutral and inessential. When trying to characterise the underlying theory of money, Schumpeter seems to have coined the later famous term "veil of money" (*Wesen*, 281; cf. Klausinger, 1990). However, he did not consider the removal of this "veil" allowable for his "dynamic theory of interest" that seems to have been his first approximation to a theory of economic evolution. On the contrary, he recognised that "the heart of the matter pulsates in the money market" (*Wesen*, 418). The analysis of this market continued to be a core concern for the rest of his life, and evolutionary issues lurked under the surface even when he considered cases in which "money" could be considered neutral. From this it follows, in the words of his friend Arthur Marget (1951, 112), "that it is impossible to evaluate, or even to understand, the nature of Schumpeter's contributions in the monetary field in isolation from the other parts of his magnificent over-all vision." For the present book, however, the opposite statement is more relevant: it is impossible to grasp the macroscopic aspects of Schumpeter's vision and evolutionary analysis if we abstract from his treatment of money. This treatment can be found in *Entwicklung I*, *S1917b* (translated by Marget), *S1925c*, and *Development*. Nevertheless, there is not doubt that Schumpeter hoped to present his major contribution to monetary issues in a book that, according to a statement he made shortly before he died (*BL*, 392), would have had the title "Theory of Money and Banking". This book is only available as an unfinished manuscript in German that seems to have undergone little change after the beginning of the 1930s. After Schumpeter's death, Marget only succeeded in translating the first couple of chapters (*S1930b*). Instead, most of the manuscript was later published in German with a title taken from one of the chapters. Messori (1998) has argued convincingly that this title—*Wesen des Geldes* (Essence of Money)—is misleading.

Although traditional monetary analysis had started from the familiar phenomenon of coins and gradually proceeded "to the credit transactions of reality", Schumpeter considered the opposite approach more fruitful. Instead of a "monetary theory of credit" he wanted a "credit theory of money" (*History*, 717). He emphasised that "the monetary part of our model is nothing but a device to get hold of the very facts to which the reader may feel inclined to point in refutation. ... The theoretical background of the analysis of credit ... will be developed in the writer's treatise on money" (*Cycles*, 109; cf. *Cycles Abr.*, 83). Unfortunately, this treatise (*Geldes*) was never completed. Furthermore, the unfinished manuscript does not present his theory in an easily comprehensive manner since he presents his monetary thinking by a mix of the styles of neoclassical economics and the historical school. However, it is clear that he largely thought in terms of an economic system in which all monetary transfers take place between bank accounts—like in Wicksell's (1936, 62–80) "organised credit economy" with competitive banks and a central clearing-house. Thus crucial parts of the book assume a model in which all firms and all households have bank accounts. Here a purchase of a commodity is simply debited the account of the buyer and credited the account of the seller. This "generalised concept of relations of current accounts" presupposes the existence of a reliable banking system, but it can nevertheless "be called the fundamental concept of the theory of money" (*Geldes*, 127). It also allows us to make a quick distinction between the processes of the stationary circular flow and the processes of economic evolution. The balances of the bank accounts within the circular-flow model are never negative. For instance, the balance of a firm is positive after the sales of its output, and it quickly returns to zero when the services of land and labour have been hired for a new round of production. Therefore, any non-zero rate of interest on such accounts can be ignored. In contrast, the model of economic evolution starts from the newly created bank account for the S-enterprise of an S-entrepreneur. The contract for this account provides credit in the form of overdrafts at a positive rate of interest.

Schumpeter sketched the concrete functioning of his credit economy in the chapter on "Theory of the Money Process and Functions of the Money Market" in *Geldes*. We shall consider a simplified version of this sketch. The starting point is a stationary circular flow with no stocks and with a production period of one week (the initial Σ state). The households receive their incomes at the end of a week's work and their productive services are contracted for the next week. In the beginning of the new week, they spend their incomes on the consumers' goods that are supplied by the firms. In both cases, the market for services and the market for goods clear, apparently by means of Walrasian auction in which trading is not allowed before markets clear. Although the bank accounts of households are positive during the weekend and the accounts of firms are positive during the working week, the accounting money plays no function except being an important alternative for barter trade. However, the implementation of new firms based on the introduction of new methods takes one year. Each S-enterprise obtains a bank account, and the rate of interest on its overdraft facility is negotiated continuously. According to Schumpeter, this explains the "sensitivity to disturbance of the capitalist engine" (*Geldes*, 303).

The first part of the two-stroke cycle starts when a set of S-enterprises is introduced into the stationary circular flow at the beginning of a year (*Geldes*, 296–7). The implementation of these firms is financed by using the overdraft facilities on their newly created bank accounts. To begin the implementation of their projects, the S-entrepreneurs draw on their accounts to purchase the necessary productive services at the first end-of-week market of the year. The increased demand for productive services means that their prices rise, and the increased incomes imply increased monetary demand for goods at the next week's consumer market. However, since the quantity of goods has not increased but rather decreased because some resources have been withdrawn from their traditional applications, the increased monetary demand means a rise not only of the prices of consumers' goods but also of the revenues of the old firms. During the subsequent weeks, the S-entrepreneurs increase their overdrafts and maybe the quantity of resources that they withdraw from the originally equilibrated circular flow. As a consequence, the monetary value of the national product (that, according to Schumpeter, only includes consumers' goods) increases—while its quantity shrinks. This process continues as long as the shifted means of production are being used by new firms that still produce nothing. However, the second stroke of the cycle begins in the first week of the second year. At that point of time, the supply at the consumers' goods market is increased by the output of the new firms. This increase of the national product is facing a gradually shrinking monetary demand in the weeks during which the S-entrepreneurs pay back their loans. This reduction of credit happens automatically since their revenues from the sales of output decrease the overdrafts on their bank accounts. Credit is actually reduced by more than it was originally increased, because of the payment of interest. Assuming constant efficiency of the accounting money, the immediate consequence is that prices will ultimately fall below their previous equilibrium level. The full consequence will normally be stronger, since the most adaptive of the incumbent firms repay the loans they were induced to obtain during the upswing. If the new firms and the adapted firms are successful, which Schumpeter assumed in the simple version of his model, the untransformed incumbent firms become increasingly unprofitable. They either exit or adapt. The process ends with a new stationary economy (the new $\Sigma$ state).

In the above model the phenomenon of interest basically derives from the demand for credit by S-entrepreneurs. In the hands of the S-entrepreneur, "the possession of a sum of money is the means of obtaining a bigger sum." This implies that "present sums of money—so to speak as potentially bigger sums—will have a value premium, which will lead to a value premium. And in this lies the explanation of interest" (*Devel-*

*opment,* 190, emphasis removed). Here Schumpeter reduced the many interest rates to a single one because he thought that "the rate of interest is essentially *always* a short-term one" that faces both S-entrepreneurs and some incumbent firms (that can profit from credit-demanding adaptation). As already mentioned, he considered the instrument that measures this rate of interest the "manometer of the economy" that increases rapidly as economic activity implies that credit creation by banks are approaching its limits. This approach contrasts the idea that interest provides the lender with a permanent source of income. In Schumpeter's model there is no such income since interest ultimately derives from entrepreneurial projects and since "no business venture yields eternal surpluses". The contradiction can only partly be overcome by the lender's "shifting of his money from opportunity to opportunity as each of them arises". In any case, interest "is the payment for balances with which to acquire commodities and services". On this background, it is clear that "interest actually is, not only on the surface but also essentially, a monetary phenomenon" (*Cycles,* 125–6; *Cycles Abr.,* 100).

## 9.5 Mark I and Mark II of the capitalist engine

Richard Nelson and Sidney Winter (1982, 29) have emphasised that "the intellectual coherence and power of thinking about Schumpeterian competition have been quite low, as one would expect in the absence of a well-articulated theoretical structure to guide and connect research." Nelson and Winter (1982, 39) also suggested that "evolutionary ideas provide a workable approach to the problem of elaborating and formalizing the Schumpeterian view of capitalism as an engine of progressive change". Therefore, they could state that "we are evolutionary theorists *for the sake* of being neo-Schumpeterians". However, their formalisation of the capitalist engine relates to a limited part of Schumpeter's evolutionary trilogy. Actually, the service they provided for "Schumpeterian competition" can largely be interpreted as providing the formalisation needed to study the behaviour of the Mark II model by means of computer simulation. Since Schumpeter only dealt extensively with this model in *Capitalism,* Nelson and Winter have failed to clarify the analyses of the capitalist engine found in *Cycles* and *Development.* Subsequent research has done much to handle Schumpeter Mark I. Actually, Philippe Aghion and Peter Howitt (1998) have even combined elements of Mark I and Mark II in their models of endogenous economic growth. Nevertheless, we are still missing general presentations of this model as well as systematic comparisons of the two Schumpeterian models of the capitalist engine.

Schumpeter's Mark I and Mark II models have primarily been characterised by their different mechanisms of innovation. Almarin Phillips (1971) emphasised the distinction between innovation through new firms and innovation as part of oligopolistic competition. It was, however, Christopher Freeman (1982, 211–14) who analysed the mechanisms in terms of innovation theory and labelled them by Roman numerals. He used "Mark I" to denote the linear model in which exogenous inventions are introduced by innovators and then diffused into economic life. In contrast, "Mark II" introduces "a strong feedback loop" from a successful innovation to further innovative activity. This "self-reinforcing circle" presupposes intra-firm research and development. Thereby, even inventive activity becomes an endogenous part of the process; and the result is increased market concentration. The distinction between Mark I and Mark II has led to theoretical and empirical research (Malerba, 2007); and it has also been used for interpreting Schumpeter's personal development.

Although Freeman's labelling seems clear, the use of his labels by other researchers has created some difficulties. To overcome these difficulties of characterising Schumpeter's work, we have to clarify at least three issues. First, the original version of the labelling focussed on the innovative mechanism of the evolutionary process at the expense of the mechanism of adaptation. However, there is also a need for characterising his full models of capitalist economic evolution. As it has been suggested above, the labels 'Mark I' and 'Mark II' are best used for the full models. Second, the names 'Mark I' and 'Mark II' seem to indicate that the latter model was developed to improve the former one. However, since Schumpeter had both models in mind from the very beginning of his evolutionary research programme, the numbering should rather be seen as reflecting a logical movement from a relatively simple model to a more complex one. Third, the names 'Mark I' and 'Mark II' can be seen as reflecting two historical stages of the capitalist economy. This interpretation clearly relates to Schumpeter's own reasoning. 'Mark I' relates to competitive capitalism while 'Mark II' relates to the subsequent system of big-business capitalism. This historical sequence, however, does not imply that we cannot analyse aspects of economic evolution at an earlier stage of the capitalist economic system by means of Schumpeter's Mark II model. Furthermore, it is hardly true that he ignored the continued importance of innovation through the entry of new firms in a system dominated by large corporations. The overall conclusion is that the Mark I–Mark II distinction should be handled cautiously, not only in historical analyses and when we contrast the early Schumpeter with the later Schumpeter, but also when we try to characterise his two models of capitalist evolution.

Let us start with the Mark I model. This model considers "the carrying out of new combinations" by entrepreneurs "as the fundamental phenomenon of development [Entwicklung]" (*Development*, 74). We may also say that the Schumpeterian entrepreneurs introduce new production functions into an economic system that is based on given production functions. The easiest way of characterising this model is to describe the evolutionary process as based on process innovations within an artificially isolated industry that produces a homogeneous product. Schumpeter did not apply this approach since it relates to Marshall's partial analysis rather than to Walras's general analysis. Therefore, the approach has to consider the monetary demand for the output of the industry ($D$) as given. The partial analysis, nevertheless, leads us to a quick recognition of the essential features of the Mark I model (see Figure 9.2 on the next page). Since these essential features are not related to physical capital, they can be discussed in terms of what can be called a pure-labour industry. In such an industry, output is produced solely be means of knowledge and labour. Knowledge is normally denoted by $A$ and labour by $L$; so we can speak of an $AL$ model. However, a Schumpeterian $AL$ model is not based on aggregates but on the characteristics of each firm $i$ (see Andersen, 2004b). Thus we have to measure the knowledge $A_i$ and the labour $L_i$ for each firm. We further assume that $A_i$ is the fixed productivity of the firm, at least during several steps of time. If production exploits the firm's capacity, its output $Q_i = A_i L_i$. An industry with $n$ firms has the output $Q = \sum_{i=1}^{n} Q_i$. This output defines the market-clearing price $P = D/Q$. Given this price and the wage rate ($w$), each firm can calculate its profit $\pi_i = PQ_i - wL_i$. Like Schumpeter, we shall recognise that the process of adaptation in such an industry requires a significant amount of time. Nevertheless, all surviving firms will end up having the same level of productivity—if the mechanism of innovation has been switched off for enough time.

The innovative mechanism of the Mark I model is based on "New Men setting up New Firms" (*Cycles*, 110; *Cycles Abr.*, 84). The "new men" are the S-entrepreneurs and their "new firms" are presently assumed to be based on process innovations. Let us consider the case of a single entrepreneur who, in relation to an industry in which the evolutionary process has come to a halt, considers the profitability of establishing the small and fixed-size firm # $n + 1$. The calculations of this entrepreneur are based on the new firm's expected productivity $A_{n+1}$. Since the price of the product reflects the productivities of slowly reacting incumbent firms and since the entrant is small, the expected gross profit $\pi_{n+1} = PA_{n+1}L_{n+1} - wL_{n+1}$. However, this profit has to be much larger than zero for two reasons that are both emphasised by Figure 9.2. First, the stream of profit has initially to be shared with a banker. This banker provided the S-entrepreneur with the credit needed during the setting up period of the new firm. This credit covers the cost of employing labour during the assumed long period in which no output is produced because the new routines of production are implemented and rehearsed. Second, the S-entrepreneur is driven by the expectation of a substantial increase of personal income. Actually, Schumpeter seems to have assumed that if

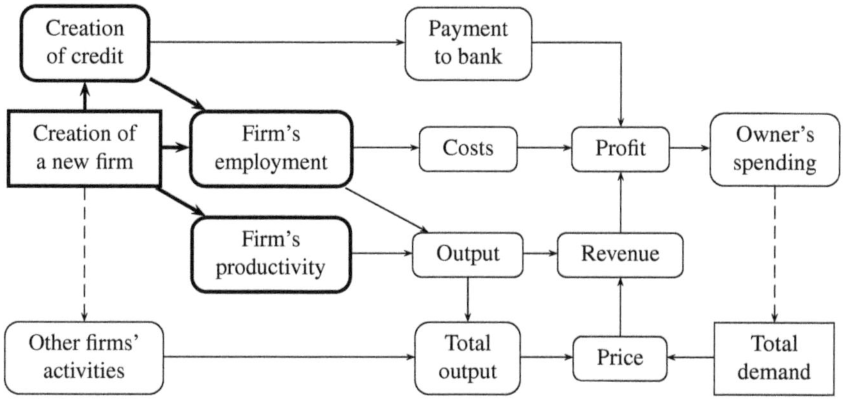


**Figure 9.2.: A pure-labour version of an industry in Schumpeter's Mark I model**

Comment: The activities of the industry is described from the viewpoint of a new firm that is based on a process innovation. The firm's productivity remains fixed after it has been established with the help of bank credit. The positive net profit is spent by the owner rather than being used for reinvestment. However, negative profit leads, sooner or later, to exit. The working of the model is indicated by arrows. An arrow from $x$ to $y$ should be read '$x$ (co)determines $y$'. For instance, Net profit = Revenue − Costs − Payment to bank.

the firm is running and its debts have been repaid, then all profit is spent and no profit is reinvested in the firm. This helps to explain his general Mark I theory of the firm as a conservative entity. Nevertheless, even conservative firms are ultimately forced to adapt or exit. Thereby, the average productivity of the industry gradually reaches $A_{n+1}$. If and when this productivity is reached, the profits of all firms are equal to zero.

The outlined $AL$ version Schumpeter's Mark I model of the capitalist engine has served to depict its mechanism of innovation. Nevertheless, mechanism of adaptation is too loosely specified to allow a full formalisation of the model and computer simulations of its dynamic behaviour. For instance, it is probably necessary to add some flexibility of established firms in order to stabilise the behaviour of the model. Therefore, it is hardly advisable to think of the mechanism of adaptation as solely taking place through entry and exit. Nevertheless, a rough understanding of the behaviour of the model might be obtained in this way. Assume that in general the founder of a firm provides the necessary knowledge and lends money for hiring a fixed set of workers from a banker. Then the new firm becomes old in the sense that it upholds its knowledge and employment and simply tries to own money for its owner's spending. The model now includes two types of entry. The non-innovative type of entry implies that the new firm is based on the best-practice knowledge. This type of entry leads to an increasing standardisation of the knowledge of the industry. The innovative entry is performed by an S-entrepreneur. His knowledge is that of a process innovation that gives the new firm a significantly better productivity than the incumbent firms. Since this process innovation is copied by a swarm of imitating S-entrepreneurs, the price of the industry's output falls so that the incumbent firms are forced to exit.

Schumpeter's Mark II model of the capitalist engine is not based on the setting up of new firms but on the innovative and adaptive responses of incumbent firms. In *Capitalism* (p. 96), he wanted to criticise the opinion that "the capitalist order becomes incompatible with progress" because "the era of big business" implies "a stop to all cost-reducing improvement". This came close to what Schumpeter had argued in *Development*. However, he had begun to emphasise that the threat of competing innovations would seriously limit such a practice. Furthermore, "the first thing a modern concern does as soon as it feels it can afford it is to establish a research department" (*Capitalism*, 96). Every employee of this department "knows that his bread and butter depends on his success in devising improvements". A success within this department is only transferred into practice if it is profitable to do so. Although existing equipment has a role to play in this calculation, the calculation is also influenced by the innovative threat from other firms. Although these remarks on "cost-reducing improvement" under oligopolistic conditions are very sketchy, they can nevertheless serve as a starting point for presenting his Mark II model. If we ignore the issue of maintaining the value of existing investment, this presentation can be formulated as an alternative version of the Schumpeterian $AL$ model.

The present version of the Mark II model describes an oligopolistic industry that produces a homogeneous output. To simplify the exposition, we shall start by removing innovative activities and instead assume that the individual firms have fixed and different productivities. Thereby, the model can be directly compared with that of a Mark I industry. Each firm still produces as much as it can; and the market-clearing price of the output is still determined by the given monetary demand of the buyers. Furthermore, relative productivities of the different firms still determine their relative profits. However, the long-term behaviour of the two models is different. In Mark II, the profit obtained by a firm in one period directly influences the size of its labour force in the next period. If the profit of a firm is positive, it is used for hiring extra labour. If its profit is negative, this problem is solved by the needed firing of labour. The resulting dynamics of the market shares of the firms is called replicator dynamics (see Appendix D). Consider, for instance, the simple replicator dynamics of the firms of the industry depicted by Figure 9.3 on the facing page. The three firms have different and unchanging productivities. Firm A

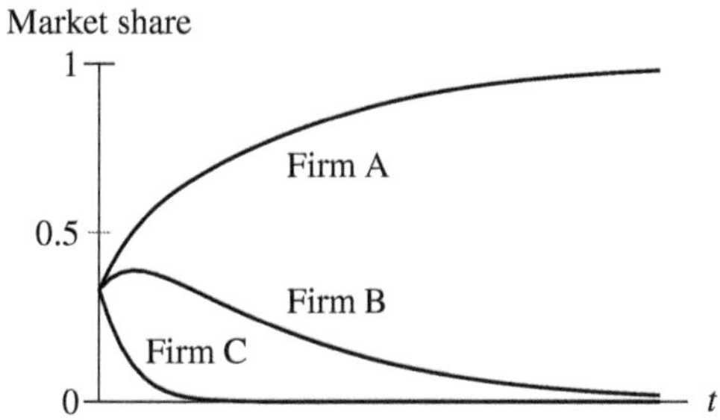


**Figure 9.3.: Simple replicator dynamics in a Mark II industry with no innovation**

Comment: The figure covers an industry in which each firm has its own fixed productivity. Firm A has the highest productivity while Firm C has the lowest one. The figure depicts the dynamics of the markets shares of the firms. Initially the three firms have the same market share and only Firm C has below-average productivity. However, Firm B soon comes below average so that Firm A ultimately takes over the whole market.

has the highest productivity while Firm C has the lowest one. The firm expands or shrinks depending on the distance of its productivity from average productivity. Initially the three firms have the same market share and only Firm C has below-average productivity. However, Firm B soon comes below average so that Firm A ultimately takes over the whole market. Here evolution—defined as the change of average productivity—has come to a halt.

The concept needed to measure the degree of evolutionary disequilibrium at any point of time during the replicator dynamics is that of statistical variance. In the present case, the relevant measure is the market-share weighted variance of productivities. In Figure 9.3, this variance moves from a positive value to zero. We may say that the engine of replicator dynamics transforms the fuel of firm-level variance into the evolution of industry-level average productivity. When variance reaches zero, the engine has used its fuel comes to a full stop. Nevertheless, we are not facing an evolutionarily stable state. The reason is that new fuel can be provided by the innovative change of incumbent firms. Since the variance of the productivities serves at the driver of the evolutionary process, we do not have to make special assumptions about the allowed times of arrival of the innovations. This means that we are not constrained to studying of innovations that occur on the background of a stopped evolutionary process. On the contrary, we may study how innovation increases any level of variance.

The mentioned version of the Mark II model can be extended to cover both the process innovations of the firms of an oligopolistic industry and

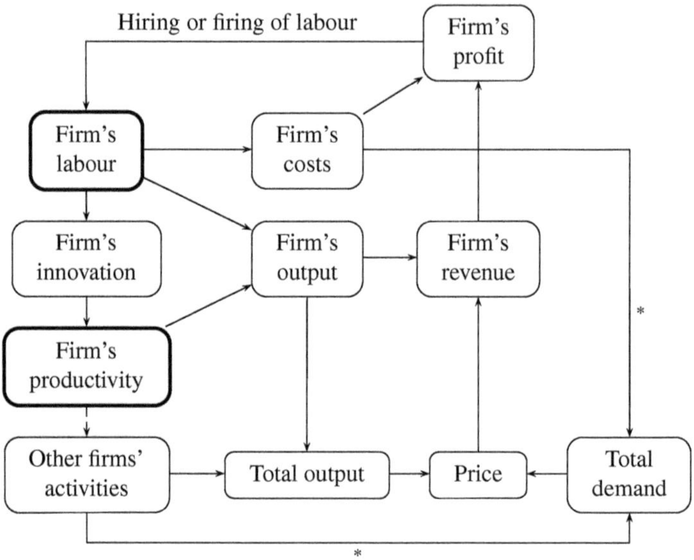


**Figure 9.4.: Innovation-based growth in Schumpeter's Mark II model**

Comment: The diagram of the *AL* growth model focusses on a particular firm, while the aggregates of the other firms are placed in the last row. The labour of Mark II firms is allocated between production and research; and the stochastic productivity of its research workers determines the improvement of the productivity of production workers. Positive profit in one period is used for hiring extra labour in the next period. In contrast, negative profit leads to firings. If the two starred arrows are removed, the evolutionary growth model is reduced to a model of a particular industry with exogenously determined demand. Source: The figure is based on Andersen (2004b, 123).

the economic growth of a highly simplified economic system. These possibilities are present in the *AL* model of Figure 9.4. Here the innovative activity is depicted directly. In contrast, the fact that we are facing a model of economic growth is demonstrated by the feedback loop from the payment of wages to aggregate demand. This feedback implies that it is only the firms of the model that generate income to the households and that these households spend all their income on the single product produced in the economy.

Let us start by considering the Mark II model as depicting the dynamics of an innovative oligopolistic industry. The assumption that innovative activities exist presuppose that some or all firms introduce a division of labour between productive workers and research workers. In the simplest case, this division of labour takes place according to fixed rules; and both types of labour are paid at the same rate. The productivity of the research workers can most easily be modelled as being of a stochastic nature. This means that firms in advance only know the expected value of the productivity of their researchers. The output of a firm's research department is process innovations. We shall assume that a successful process innovation in one period influences the productivity of all the production workers in the next period. Under this assumption, the Mark II model implies increasing returns. In other words, the same innovation generates larger profits in firms with large production departments than in firms with less productive activity. Although this feature increases the speed of market concentration compared with the case of simple replicator dynamics, the result is not necessarily a monopoly. Instead, we can change the assumptions of the model so that a few large firms can coexist under the conditions of Schumpeterian competition (see Nelson and Winter, 1982, Chs 12–14). Although the applied assumptions have largely been technological and institutional, the destiny of the oligopolies of realistic versions of the Mark II model is heavily influenced by the structure and evolution of demand (see Andersen, 2001; 2007b).

Let us finally take into account Figure 9.4's feedback loop from the payment of wages to the size of the aggregate monetary demand. This addition means that the industry model becomes transformed into a primitive Mark II model of economic growth. Actually, this $AL$ growth model is easier to handle than the truncated industry-level model. For instance, if we switch off innovation in the growth model, the dynamics of the employment shares of firms with different productivities traces the simplest possible replicator dynamics. However, Schumpeter would definitely not have liked a formal representation of his ideas that makes the emergence of an economy-wide monopoly a possible outcome of the evolutionary process. The problem is, of course, that we have reduced the oligopolistic game to concern a single characteristic of a repeated evolutionary change. The real evolutionary process that changes the routine-based structure of the economic system does not follow a trajectory of improving the technology for the production of a single product. Instead, it must formally be described as taking place along an infinite number of trajectories. Although the number of actually explored trajectories is finite at any point of time, it is one of the functions of innovative firms, and innovative S-entrepreneurs, to find hitherto unexplored trajectories. If we take this function into account, we arrive at very different versions of both the Mark II models and the Mark I models.

Although the outlined formalisations of Schumpeter's models of the capitalist engine are not satisfactory, they nevertheless serve important purposes. It is through the complexities revealed by formal modelling that researchers recognise issues that are not immediately obvious. One such issue is that an evolutionary oligopoly has an intrinsic tendency to collapse into monopoly. Schumpeter avoided this conclusion by assuming that his large firms operated in an immensely complex world of innovative opportunities. Formal modelling has to simplify this world and to confront the monopoly problem directly. The solution chosen by Nelson and Winter (1982) is that firms determine their desired level of output according to Cournot's model of oligopoly. Here each firm assumes that the other firms stick to their previous level of output. The firm can maximise its profits as if it had monopoly power at the rest of the market. This behaviour means that firms with a small market share expand their output with practically no restraints. In contrast, firms with a large market share constrain radically their output, thereby leaving room for smaller firms. This is important since the construction of the Nelson–Winter models imply that large firms obtain larger benefits from a research worker than small firms. However, there are other ways of avoiding monopoly in the economy. We can obtain the same result by assuming that entry is provided by spin-offs from dominant firms. The competitiveness of these entrants is secured if they inherit the productivity of their mother firms. This pattern of creating relatively stable oligopolies seems to be found in the history of the automobile industry of Detroit (Klepper, 2002).

Since Schumpeter seems to have thought more deeply about the problems revealed by the Mark I and Mark II models than normally recognised, it might be possible to answer the question why largely focussed on Mark I. The pragmatic reason is hardly that he wanted to avoid a Mark II-style model because it has the taste of Marshall's evolutionary economics and Marx's theory of the accumulation and centralisation of capital. More likely, he had recognised that such a model would create the same type of objections as Piero Sraffa (1926) later raised against the Marshallian system: it operates with competition but includes assumptions that create monopoly. Actually, he did not formulate Mark II until after Chamberlin and Joan Robinson had presented ways of handling monopolistic competition. A more important reason for Schumpeter's reluctance to develop the Mark II model is that his overall analysis of the problems of capitalist economic evolution seems much better reflected by Mark I. This model depicts a dramatic process of economic evolution and serves to specify a whole set of concepts on the capitalist engine. Furthermore, the Mark I model clearly emphasises the destructive side of the evolutionary process. Since this destruction provokes social and political responses, it immediately connects to evolutionary sociology and evolutionary political science. When writing *Capitalism,* he apparently recognised that the social and political reactions against the "monopolies" of Mark II could be used for the same purpose.

## 9.6 Conclusion

After having considered the later Schumpeter's modelling of the capitalist engine and its brakes, a major question is whether he succeeded in improving the basic model that he had developed before World War I. One possible answer is that the more he changed the model, the more it became clear that he was dealing with the same thing. This answer corresponds to that of Clemence and Doody (1950, 1): "His youthful vision, first reduced to a comprehensive model in 1911, has since been elaborated and refined, but it has been altered in no essential respect to the present day." Although the permanency of Schumpeter's powerful vision does not necessarily imply permanency of his theoretical models, Clemence and Doody nevertheless considered his "comprehensive model" an unchanging construct. The present part of the book has demonstrated that this assumption is not strictly correct. We have not only recognised significant differences between his early programmatic books (*Wesen* and *Entwicklung I*) and his evolutionary trilogy. We have also recognised significant differences between *Development, Cycles,* and *Capitalism.* Nevertheless, we have also recognised the basic unity of all of Schumpeter's contributions to study the engine of capitalist economic evolution. These contributions are based on the assumptions that evolution concerns the system of routines and that this evolution does not proceed continuously but by means of innovative jumps.

Schumpeter's attempts to implement his youthful vision did not proceed smoothly and they actually implied certain changes of this vision. While his original entrepreneurial interpretation of history seems carefully implemented in the Mark I model, the Mark II model seems to relate to a broader evolutionary interpretation of history. This broader interpretation is closer to the interpretation represented by classical economics; and it can easily be related to most of the contributions to modern evolutionary economics. Nevertheless, Schumpeter's main efforts concerned the elaboration of the Mark I model. Let us consider three characteristics of these efforts. First, his elaborations implied a modified terminology that connects better with modern modelling and helps to define unsolved issues. For instance, *Cycles*'s definition of innovation as the implementation of a new production function might be seen as having a double purpose. On the one hand, it suggests that we need to consider established production functions as routines before we can define innovation precisely. On the other hand, it suggests a general need for reconsidering one of the pillars of economic theorising from the viewpoint of evolutionary analysis. Second, the later Schumpeter's efforts served to clarify the assumptions underlying his basic model of the capitalist engine. *Development* and *Cycles* reached this clarification by defining narrowly the process of economic evolution as driven by the creation of innovation-based firms. This core model had been in his mind from the very beginning, but its elaboration had been disturbed by its generalisation to cover any type of social evolution. The narrow model also served to emphasise the need for an alternative model of the capitalist engine. This Mark II model covered the more and more active role of incumbent firms in innovative activity. Third, the modelling efforts in the evolutionary trilogy as a whole defined a research agenda by making explicit "many glaring lacunae" and "unfulfilled desiderata". Parts of this research programme had to be left to the "younger generation" because it was beyond the reach of Schumpeter. However, significant parts of the programme were elaborated by the evolutionary trilogy as a whole.

# Part III

# Works in Progress

# 10

# Schumpeter and the Years of High Theory

## 10.1 Schumpeterian unfinishedness

It is tempting to end the present book on Schumpeter's evolutionary economics by solving two related tasks. The first task is to provide a systematic summary of Part II's rather complex presentation of his evolutionary trilogy. The second task is to summarise and evaluate the "standard criticisms" (Lange, 1941, 192) that have been raised against the different elements of his work. The combined solution of the two tasks would serve to continue a literature that was initiated immediately after Schumpeter's death. At that time, Richard Clemence and Francis Doody published a small book called the *The Schumpeterian System*. This book, which is largely based on *Cycles*, gives a condensed and relatively systematic presentation of Schumpeter's core contributions. His two former students also had the ambition of confronting a situation where

> "a set of 'standard criticisms' has grown up that not only discourages investigation of the [Schumpeterian] model but threatens to become a part of the folklore of economics. ... The Schumpeterian System is an imposing analytical machine. It is large enough; it has sufficient power; and it is made of the best materials. But will it work? So its inventor claims, but others are less confident. Indeed, it is said that there are serious flaws in all parts of the engine, and that it can never be made to operate. Let us examine the System for ourselves, and see what conclusions we can reach." (Clemence and Doody, 1950, 5–6)

Schumpeter would never have formulated the problem in this way. He had not called for call for a closure of a "Schumpeterian System" or for defending his results against criticisms. Instead, he had argued that his evolutionary economics should be considered work in progress that needed critical scrutiny and development. When he wrote his preface to *Entwicklung I* (viii) in 1911, he stated that he only had wishes. First, that economists took seriously his "facts and arguments". Second, that "this work [*Entwicklung I*] as soon as possible shall be surpassed and forgotten." The second wish might appear paradoxical. However, it can be interpreted as suggesting that he wanted his book to become part of an evolutionary research tradition that would move far beyond its starting point. However, Schumpeter was still largely working alone on his evolutionary economics when he, in 1939, wrote the preface *Cycles*. Here he reflected on "the scaffolding which I published nearly thirty years ago":

> "It took longer than I thought to turn that scaffolding [*Theorie der wirtschaftlichen Entwicklung*] into a house, to embody the results of my later work, to present the historical and statistical complement, to expand old horizons. Nevertheless I doubt whether the result [*Business Cycles*] warrants that simile. The house is certainly not a finished and furnished one—there are too many glaring lacunae and too many unfulfilled desiderata. ... The younger generation of economists should look upon the book merely as something to shoot at and to start from—as a motivated program for further research. Nothing, at any rate, could please me more." (*Cycles*, v)

Clemence and Doody (1950, 96) interpreted Schumpeter's emphasis on the gaps and missing features of *Cycles* as relating to "the need for more and better data, studies of firms and industries, historical investigations of entrepreneurship, and innumerable researches of all sorts." However, this interpretation seems too narrow. In any case, he elsewhere suggested a deeper unfinishedness of his work and a related rejection of having would-be disciples like his two students. Thus, Schumpeter commented on the nature of his work when he, in 1932, had decided to move to Harvard University after seven years at the University of Bonn in Germany. In the farewell speech, he emphasised that "I have never strived to bring about something like a Schumpeter School. It does not exist and should never do so" (*S1932b*, 600). His reason was that "I never wish to say anything definitive. If I have a function, it is not to close doors but to open them" (p. 600, emphasis removed). His teaching along the lines of *Wesen* had served to bring formal economic theory back to Germany after it had become nearly extinct during the long reign of Schmoller's historical school; and he had also tried to open the doors for the international econometric revolution. At the same time, his research had pointed towards the theoretical conquering of the elusive topic of economic evolution. However, he had no possibility of closing the doors of evolutionary economics by creating a "Schumpeterian System"—even if he had wished to do so. The reason is that, during his lifetime, it was impossible to formulate the kind of paradigm that was necessary for the creation of a cumulative research tradition in his chosen field. Instead, he was developing the personal research programme that, in retrospect, has been called Schumpeter's evolutionary economics.

We have already recognised that Schumpeter's research programme focusses on the exploration of core problems of the capitalist economic evolution. However, this focus often becomes blurred because he actually thought that his evolutionary analysis ranged from the theory of money to the study of the evolution of Renaissance painting. Such a breadth is inexplicable unless we recognise the underlying general "theory of cultural evolution" (*Development*, xi), which Schumpeter first had sketched in a few pages of the last chapter of *Entwicklung I*. He pointed at this theory in an interview in 1944 with the students' journal *Harvard Crimson*, which had the later famous philosopher Thomas Kuhn on its graduate board. In the interview, he mentioned his research programme that "varied but always stayed on the same plane—that of evolving a comprehensive sociology with a single aim" (*S1944a*). The "single aim" seems to be that of developing an evolutionary science of society based on his entrepreneurial interpretation of history; and the preliminary results of this programme had been published in *Wesen*, *Development*, *Cycles*, and *Capitalism*. He also told the anonymous interviewer that "[m]y research program grows longer and my life shorter. My 'History of Economic Analysis' drags, and I am always hunting new hares. Time presses upon one and I am not getting younger, you know; the program lags".

The lagging of the completion of Schumpeter's research programme can hardly be described as a personal failure, however. Even an army of research workers could hardly have completed his evolutionary research in a way that was fully satisfactory for him. If we should speak of failure, it concerns his failure to recruit this army. However, he neither would nor could generate an army of Schumpeterian researchers with divisions for each sector of social life, including one division for the study of the evolution of science in general and economics in particular. He did not even try to create the core division of researchers for the study of economic evolution. Nevertheless, he repeatedly made gentle calls for help. To characterise the kind of help he needed, it is convenient to think of his implicit answer to Thorstein Veblen's (1898b) famous question: "Why is economics not an evolutionary science?" Schumpeter's answer had since *Wesen* been that economics potentially consists of two branches. One of the branches is equilibrium economics. This branch is a science but cannot be an evolutionary one to a larger degree than, for instance, biological physiology. The other branch is evolutionary economics. It is obviously evolutionary, but neither Schumpeter nor anyone else had been able to establish it as a science, that is, a tool-based area of study with a cumulative tradition.

For the evolutionary branch of economics, Veblen's question can be replaced by the very different question: 'Why was evolutionary economics not a science?' Radical answers are that 'economic life does not evolve' or 'economic evolution is based on a type of novelty that cannot be studied scientifically'. Schumpeter found such answers unconvincing and suggested instead that the main answer is that adequate tools for analysing economic evolution had not yet been found and developed. Since this was recognised implicitly or explicitly by the overwhelming majority of economic research, the attempts to develop evolutionary economics into a science became increasingly the speciality of a few stubborn researchers. While alternative approaches to evolutionary economics—like that of Marshall—largely disappeared during the 1920s and 1930, Schumpeter continued. However, he could not alone produce the foundations for the new branch of scientific economics that he had suggested in *Wesen*. This fact partly explains why he "regarded himself as in a sense a theorist *manqué*" (Samuelson, 1981a, 3), that is, an unsuccessful theorist. Actually, we have to some degree to accept the conclusion of the German economist Jürg Niehans (1981, 175) that "Schumpeter was a tragic figure in the history of economic analysis because he was unable to transform his vision of innovation into an analytical model." Nevertheless, the question of whether evolutionary economics can be established as a science is still an open one. Furthermore, there are today many arguments for an affirmative answer. Therefore, the evaluation of Schumpeter as a "tragic figure" might change.

## 10.2 The years of high theory and high econometrics

Although Schumpeter's main research strategy concerned evolutionary economics, he also supported the radical reconstruction of the much more mature equilibrium economics. Actually, he seems to have combined the prospects of equilibrium economics with those of evolutionary economics when he gave his farewell speech at the University of Bonn. Here he pointed out that

> "[t]he fascination about science is actually only the amusement one gets when one does what the best authorities have declared impossible. Only the hunt for such opportunities is worth anything. During recent years, this hunt in our field has been accompanied by some success. At hand are germs for making something else of our science. The construction up until now is conceptually becoming obsolescent." (*S1932b*, 608)

This quotation suggests that Schumpeter felt that the viewpoints of his programmatic books from 1908 and 1912 were, in the early 1930s, becoming aligned with the spirit of an emerging reconstruction of the whole of economics. The criticism of the early twentieth century had not hindered the old-style neoclassical economics of Marshall and Menger to live on as a relatively stable construct. Now things had started to change; and Schumpeter urged his ambitious students—the primary addressees of his speech—to exploit the emerging opportunities for path-breaking scientific contributions. These opportunities had relatively little to do with the elaboration of Schumpeter's evolutionary economics. According to Paul Samuelson's (1987, 52; emphasis removed) later statement, "[s]hortly after 1930 economics burst out into new life. At least four revolutions erupted: the monopolistic competition revolution, the Keynesian macro revolution, the mathematization revolution, and the econometric inference revolution." Although Schumpeter did not foresee the coming of the Keynesian macroeconomics, he was fully aware of the three other "revolutions" and strongly encouraged his many brilliant and ambitious students at Bonn to contribute to them—and he later encouraged his students at Harvard similarly. Many followed his advice. Furthermore, his American students seems to have followed the spirit of the Schumpeterian message by becoming core contributors to the "Coming of Keynesianism to America" (Colander and Landreth, 1996). In this case, Schumpeter had made an exception to his general support for scientific innovation and used his authority to declare Keynesianism an infertile detour in the development of economics. This, of course, provoked his students to prove the opposite.

George Shackle (1967) wrote a thought-provoking book called *The Years of High Theory: Invention and Tradition in Economic Thought 1926–1939*. Although Schumpeter is not present in this British account for important events in the history of economics, the period studied by Shackle is largely similar to that of the production of the evolutionary trilogy. *Development* is based on the second German edition of *Theorie der wirtschaftlichen Entwicklung* from 1926; and *Capitalism* had begun to take form in parallel with Schumpeter's work on *Cycles*, which was published in 1939. If Schumpeter had already started to write in English when he moved to the University of Bonn and if his rewrite of *Entwicklung I* had emphasised the implicit criticism of Marshall, then he might have helped to define many of the debates of the years of high theory. Actually, he seems to have been better equipped for this role than the outsider Piero Sraffa. However, it was Sraffa (1926) who got the honour of provoking the wave of analytical innovation by cutting the Gordian knot provided by Marshall's *Principles of Economics*. He did so by analysing the inconsistencies emerging from the Marshallian mix of equilibrium economics and evolutionary economics. This mix involved logical inconsistencies in the *Principles* and gave no real explanation why monopoly did not emerge in the predominant parts of real economic systems characterised by increasing returns to scale. Thereby Sraffa seems to have suggested a type of economic analysis that took its starting point in the analysis monopolistic positions. Although such an analysis might have produced an evolutionary economics, his emphasis on the logical consistency of monolithic theorising led to its increasing exclusion from economic analysis. Thereby, he and other economists apparently followed Wittgenstein's dictum: "what can be said at all can be said clearly, and what we cannot talk about we must pass over in silence".

Although Schumpeter did not reject the Wittgensteinian principle, he had serious difficulties during the years of high theory. One of the reasons is that he represented the more catholic principle of being "silent—or ... to delimit ourselves to summaries of facts—about things on which we have nothing *exact* or sufficiently *interesting* to say." (*Wesen*, 618–19). This statement (see Section 3.1) should be interpreted as reflecting the needs of an ambitious research strategy that had the aim of exactness. The strategy, however, acknowledged that this aim cannot always be reached immediately—as demonstrated by the case of evolutionary economics. Instead, the scientific handling of the facts of economic evolution started from looser historical generalisations and from an evolutionary theorising that had not reached the high principles of consistency and completeness. Although Schumpeter's strategy seems designed for conquering new land for economics, he, in *Wesen*, also supported the radical reconstruction of the relatively mature equilibrium economics. However, as emphasised by Leontief (1950, 105) in an already quoted statement, "this remarkable book remains practically unknown in the English-speaking world" although it presents "the basis of Schumpeter's whole scientific *weltanschaung*." His scientific world view suggested that the reformed equilibrium economics needed coexistence with the new evolutionary economics. Since he actively resisted the spread of *Wesen*, he missed a chance of being able to claim that he had promoted the years of high theory nearly twenty years before Sraffa did so. Furthermore, he lacked the background argument for insisting on the need of complementing formalised equilibrium economics with a less formalised evolutionary economics. Instead, not only Marshall's evolutionary economics but also Schumpeter's evolutionary economics became crowded out by the one-sided development of equilibrium economics.

Given the importance of *Wesen* for the understanding of Schumpeter's research efforts and teaching practice, it might seem a mystery why he did not insist on the English translation of this book before the publication of *Development*. His own explanation is that *Wesen* had not only become old-fashioned but also that he had changed some of his early views. Therefore, he wanted to write an improved book on "The Theoretical Apparatus of Economics". The major planned novelty seems to have been an increased emphasis on the relationship between theory and empirical evidence. *Wesen* had not really confronted this relationship although the book was designed to overcome the battle of methods between the Austrian school and the German historical school. This omission became especially problematic when he resumed his academic career in Germany in 1925. At that time, he also recognised that the relationship between theory and data is a general problem for the development of economics as a science. For instance, he remarked that "[f]rom America [e.g., from Mitchell, 1925; and Tugwell, 1924] we occasionally hear and read of symptoms of what has to be called a latent battle of methods. ... Actually, this battle is not that latent. Against the older generation, which received its characteristic tone from J. B. Clark's circle of ideas, a younger generation, more or less politely, points out a list of sins that appears well known: unreality, irrelevance, uninteresting views and results. —Does that not sound exactly like what the 'younger historical school' remarked against pre-existing economic science?" (*S1926b*, 148)

Since his early years, Schumpeter had been engaged in attempts to overcome such methodological controversies. In the 1920s, he continued these efforts; and he started by turning to the roots of the American controversy in the work of Gustav von Schmoller. According to Schumpeter, Schmoller's method had been to confront empirical evidence with a minimum of assumptions and to interact analytically with the evidence in order to derive new tools for further work. The goal of "Schmoller's programme" had been to develop "a unified sociology or social science as the mentally ('theoretically') worked out universal history" (*S1926b*, 175, 193). Although Schumpeter accepted this ultimate goal, he emphasised the necessity of scientific specialisation. Furthermore, he thought that the time was ripe to overcome the one-sided emphasis on data—and the similarly one-sided emphasis on theory. His solution was to promote and join the emerging econometrics movement.

Schumpeter's programmatic paper "The Common Sense of Econometrics" (*S1933c*, 100) states that "[w]e do not impose any credo" except "that economics is a science" and "that this science has one very important quantitative aspect". The recognition of these propositions suggests that the task is "building up the economic theory of the future" (p. 107). Thereby, the econometricians would, to some extent, implement what Marshall (1897; 1961; 1919) had been pointing at in "The Old Generation of Economists and the New" as well as in *Principles of Economics* and *Industry and Trade*. Thus Schumpeter emphasised that it is not "possible for anyone who knows how to read his [Marshall's] *Principles* in the light of his *Industry and Trade* to define what he really strove to accomplish in any but econometric terms" (*S1933c*, 104). Schumpeter wanted that the same could have been said of the relationship between *Development* and the project that led to *Cycles*. This ambition has been demonstrated by Francisco Louçã's (2007) *The Years of High Econometrics*. Schumpeter's engagement in high econometrics, however, can hardly be characterised as a success. Although he strongly supported the creation of the Econometric Society and for a period became its president, he obtained very little help for the quantitative development of his evolutionary economics. This is demonstrated by his long and intense discussions with the leading organiser of the econometrics movement, the Norwegian Ragnar Frisch (Louçã,

2007, Ch. 6). Their common aim was to make an empirically operational model of evolutionary business cycles; but neither the transformation of the routine system not the activity of the Schumpeterian entrepreneur became formalised in an adequate way by Frisch. Instead, Schumpeter became alienated in the econometrics movement and increasingly turned to economic historians in an attempt to develop his evolutionary economics. Therefore, he had little connection with the further development of the Econometric Society—that quickly came to include many of the core contributors to economic research. This is reflected in the fact that members of the society received most of the Nobel Prizes in economics for a long time—after the creation of this reward in 1969 (Simon, 1991, 319–26). Frisch and Jan Tinbergen shared the first Nobel Prize, and other laureates with a connection to both the econometric society and Schumpeter were Paul Samuelson, Wassily Leontief, Simon Kuznets, and James Tobin.

The fact that Schumpeter felt in line with the scientific *Zeitgeist* of the late 1920s and early 1930s might explain why he overlooked the fact that the events only concerned half of his early programme for the science of economics. Furthermore, he weakened the understanding of his evolutionary economics by de-emphasising his initial presentation of it as a separate branch of economics. Instead, he tended to present his underdeveloped branch as an integrated part of the ongoing reconstruction of economics and to try to contribute to nearly all parts of this reconstruction. Thereby, he confirmed the verdict of one of his early Austrian teachers, Friedrich von Wieser. According to Wieser's (1911, 54) review of *Wesen*, Schumpeter's "main error is that he wants to master too much; one gets the feeling that the author has not yet reached his equilibrium and still has to learn to delimit himself. Such youthful overexcitement is the most praiseworthy of all errors; it is a symptom of strong powers." Schumpeter stayed young in the sense of never being able to delimit himself. Instead, he wanted to continue opening doors for others as well as for himself. However, scientific progress is not only promoted by exploiting new opportunities but also by creating separate branches of economic analysis. In *Wesen*, Schumpeter had had covered both an explicit equilibrium economics and an implicit evolutionary economics; and in *Entwicklung I* he had turned to an explicit evolutionary economics. The praiseworthy error of the later Schumpeter is that he again wanted to cover and combine equilibrium economics and evolutionary economics. This ambition hampered his core task of promoting evolutionary economics as an important branch of economics. Actually, his style of research and his rejection of a Schumpeterian school hindered his evolutionary economics in becoming part of the core history of the years of high theory and high econometrics. Instead, evolutionary economics was during these years in a state of being crowded out by equilibrium economics.

Schumpeter's difficulties during and after the years of high theory and high econometrics can most easily be recognised by considering the contemporary reception of *Cycles*. We have already seen that he, in the preface, called it "a motivated program for further research". However, although quite a few of the members of the "younger generation" did "shoot at" *Cycles*, they only did so for a short while; and the book did not provide a starting point for the massive "further research" that it needed. Actually, the main conclusion of the "younger generation" was that *Cycles* did *not* provide a starting point for "further research". This conclusion was initially not unanimous, and several bright researchers tried to help Schumpeter. The collaboration started in the early 1930s, when he engaged in a long-term and intense personal communication with Ragnar Frisch. However, when *Cycles* came out, the reviews and comments of theorists, econometricians, and historians were more sceptical. The theorists emphasised that Schumpeter did not build a convincing model of business cycles. Therefore, James Marschak tried to do for Schumpeter what Hicks (1937) had done for Keynes: to bring out the underlying model. Marschak tried to formulate the theory of *Cycles* in five equations of macroeconomic aggregates, and he discussed these equations with Schumpeter before finalising his review. Schumpeter's written answer (reproduced by Stolper, 1994, 375–6) was discouraging: he disliked the use of aggregates and econometric constants as well as the treatment of innovation as an exogenous force. It was probably this criticism that induced Marschak (1940) to exclude the equations from his review of *Cycles*. Oscar Lange (1941) used a strategy more comfortable to Schumpeter by showing how innovations could be endogenised in the account for business cycles. However, neither Lange nor other theorists were enthusiastic. One of them, Paul Samuelson (1951, 98) suggested that "Schumpeter will unquestionably be labelled by future historians as a business-cycle theorist who placed primary stress on the role of the innovator." However, Samuelson later added that Schumpeter's use of simultaneous cycles—Kondratieff waves, Juglar cycles, and Kitchin cycles—"began to smack of Pythagorean moonshine" (quoted by McCraw, 2007, 253). Thus it could be compared with the unsubstantial and unreal vision of a Greek sect that tried everywhere in nature to find analogies to the waves of music.

The statisticians and econometricians were even less convinced about the solidity of *Cycles* with its large amount of largely unused statistics and its methodological attacks on aggregation. Thus, the well-known statistician Maurice Kendall (1941, 179–80) admired the richness of the book; but he also thought that the general development of statistical theory was starting to "produce the analytical instruments which will either uphold or destroy Professor Schumpeter's theories". Furthermore, Jan Tinbergen (1951, 109)—one of the leading econometricians—later remarked that although Schumpeter was one of the Econometric Society's "enthusiastic founders", "one find a mental attitude vis-à-vis econometric work which is not only rather critical, but to some extent alien to it." Furthermore, while the historically and institutionally inclined economists should have been his natural allies, they were also very sceptical. The German general historian and economic historian Hans Rosenberg (1940, 96) started his review by pointing out that Schumpeter could be seen as following Wesley Mitchell's "strong plea for the mutual penetration and permanent, harmonious co-operation of theory, statistics, and history as applied to a realistic, nonspeculative analysis of the cyclical *modus operandi* of the economic life process of capitalist society." Unfortunately, Schumpeter's attempt to perform this grand task meant that "[h]ardly a page makes sense unless he reader masters the abstract patters as well as the concrete contents of both volumes and treats them as an inseparable whole." Furthermore, this whole is defined by a "self-imposed limitation", namely that it does not "admit the significance of 'external factors' ... as generators of historic 'economic evolution'" (p. 97). Therefore, *Cycles* was of relatively little interest for those who wanted large-scale accounts for the history of capitalism (which in any case were being squeezed out by narrow and in-depth studies). Finally, we have already seen that the most devastating attack came from Simon Kuznets, who himself had had followed Mitchell's programme by combining statistical and historical analyses of business cycles.

On the background of such reviews, it is easy to understand why Clemence and Doody tried to rescue the legacy of their teacher by specifying the "Schumpeterian System" and by defending it against criticisms. They argued that the "[r]eviews of *Business Cycles* rivalled those of the *General Theory* in confusion, with the difference that there were no outstanding exceptions to the rule" (Clemence and Doody, 1950, 3). From the outset, the prevailing opinion was that "the book was hardly worth bothering about". This conclusion was maintained by "a set of 'standard criticisms'", which tended to "become a part of the folklore of economics" (p. 5). These criticisms had to "be examined soon, if they are to be examined effectively at all, and if discussion is to yield any substantial results" (p. 6). Clemence and Doody's implementation of this strategy, however, ignored the multiple purposes for which Schumpeter's analysis was designed and the unfinished character of his results. Especially, the focus on *Cycles* served to subsume Schumpeter's evolutionary economics under his very vulnerable attempts a theorist of business cycles. Although Clemence and Doody could hardly have changed this situation, a more ambitious confrontation with Schumpeter's analytical is today both possible and advisable.

## 10.3 The principle of indeterminateness

George Shackle's (1967, 133) *The Years of High Theory* summarises the predominant view by stating that "[e]conomics is about order." According to this view, "[e]conomic theory is the theory of an orderly and reasonable world, a world of a concerted interchange of knowledge, a world where actions are pre-reconciled by the great market computer and a world where what we intend is what will happen." Thus, "[i]t is not only business men, but economists themselves, who ardently subscribe to the rationality, the objectivity, the scientific well-conductedness of business reasoning." However, Schumpeter had throughout his academic life emphasised that these statements represent the world view of business managers and equilibrium economists. A radically different view is represented by Schumpeterian entrepreneurs and by Schumpeter's evolutionary economics. According to this alternative view, the orderly game of economic life is produced by a disorderly meta-game that, in the long run, changes the rules of any given orderly game. This way of formulating the problem helps us to understand why Schumpeter thought that his evolutionary economics *complements* equilibrium economics. Nevertheless, his evolutionary branch of the science of economics also serves to emphasise the narrow limits of the interpretation of economic life as being orderly and reasonable. Therefore, his efforts tended to be rejected to the extent that they were understood. Since this understanding presupposes that the standard view of economic life is to some extent rejected, his efforts met little recognition.

The idea of fundamental economic order had always been threatened by realistic treatments of monopolistic competition. John Hicks (1939, 83, 85) concluded the years of high theory by stating that "a universal adoption of the assumption of monopoly" means that "the stability conditions become indeterminate". The resulting "wreckage is that of the greater part of economic theory". This potential wreckage of their models and their world view had kept neoclassical economists from exploiting Cournot's analysis of competition among the few. It also explains the importance of Sraffa's insistence of the incompatibility of the Marshallian assumptions of perfect competition and increasing returns to the scale of production. However, while Hicks decided to concentrate on markets that do not differ much from perfectly competitive markets, other economists constrained their rescue operation to the abandonment of increasing returns and the taming of monopolistic competition. Actually, Edward Chamberlin (1951) thought that Schumpeter in *Cycles* had done his best to diminish the importance of monopolistic competition—so that he only really opened up for this issue through its informal emergence in *Capitalism*. Nevertheless, the Mark II model of innovative oligopolistic competition is not his only treatment of the relationship between innovation and monopolistic practices. What Schumpeter did reject is only the large-scale introduction of monopoly and imperfect competition in equilibrium economics (*Cycles*, 56–68; *Cycles Abr.*, 32–42).

Schumpeter's evolutionary economics is inconceivable without changing of some of the assumptions of equilibrium economics. Therefore, he had to "repeat not only that the entrepreneurial impulse impinges upon an imperfectly competitive world but also that entrepreneurs and their satellites almost always find themselves in imperfectly competitive short-time situations even in an otherwise perfectly competitive world" (*Cycles*, 160; *Cycles Abr.*, 136). Furthermore, "evolution in our sense is the most powerful influence of creating such imperfections all around." These recognitions explain why Schumpeter changed assumptions after he had presented his "Second Approximation" to waveform economic evolution as if it relates to perfect competition. He declared that "we now drop the assumption of perfect competition altogether" as well as the assumption "that there is perfect equilibrium at the start" (*Cycles*, 160; *Cycles Abr.*, 136). He immediately emphasised the grave consequences this decision. First, "propositions and proofs will be less stringent, zones of indeterminateness will emerge, sequences of events will be less prompt, and buffers will be inserted between the parts of our mechanism so that its gears will be slower to mesh". Second, this process gives "more room for individual strategy" as well as for the "moves and countermoves which may impede, although they man also facilitate, the system's struggle toward equilibrium". Third, the process will "produce many freakish patterns and the economist's engine for the production of paradoxa will be worked up to, and perhaps beyond, capacity." Finally, the evolutionary process produces "situations in which industries may, even in equilibrium, move within intervals of decreasing average costs" (*Cycles*, 160; *Cycles Abr.*, 136).

Schumpeter thus emphasised that the evolutionary process is sluggish and that its outcome is indeterminate. Since he had no solid theoretical and statistical tools for describing the sluggish working of the mechanisms of innovation and adaptation, he could not formalise this idea. What he could do was to abstain from developing the Mark II model of the capitalist engine that he had clearly pointed at in *Cycles*. Instead of exploring the complexities of this model, he used the more comfortable Mark I model. While this model was comparatively easy to handle analytically, it actually served to dramatise the problem of the indeterminate nature of the mechanism of innovation. The reason is that the Schumpeterian entrepreneur in this model is described as being totally independent of the existing structure of the established system. It is largely by means of such an entrepreneur that Schumpeter described the consequences of innovation—not only for evolutionary economics but also for his general theory of social evolution. Nevertheless, the general conclusion would have been the same for the Mark II model. This conclusion is that "[s]ocial phenomena constitute a unique process in historic time" and that "incessant and irreversible change is their most obvious characteristic" (*History*, 435). The historical uniqueness of the social phenomena observed at any particular point of time is explained by the irreversibility of their change led Schumpeter to formulate what he called a "Principle of Indeterminateness" (*S1949a*, 441). This principle rests on three characteristics of the evolutionary process. First, we have to recognise "the element of chance". To grasp this element, modern readers might think of the role of the exogenously determined mass extinction of the dinosaurs in biological evolution. However, Schumpeter's examples were the role of "major wars" and the role in the sixteenth century of the "inflow into Europe of large quantities of precious metals from the New World". Second, the need for his principle of indeterminateness also rests on the complexity of the interaction of the evolutionary processes in different sectors of social life. For instance, although "economic and political developments" are interrelated, they "enjoy a limited amount of independence". The consequence of the resulting lack of coordination is that it may be impossible to predict consequences that arise from the "temporal coincidence" of business situations and political situations. Such coincidences have to be modelled as "chance events". Third, and most importantly, Schumpeter thought that indeterminateness is the result of a fundamental characteristic of the evolutionary process. While the mechanism of adaptation can largely be described deterministically, the study of the mechanism of innovation cannot he handled in this way. His basic argument for indeterminateness of innovation is based on what he considered the fact that innovative activity is to some extent performed by "exceptional individuals" (*S1949a*, 442). His argument is that "since the emergence of exceptional individuals does not lend itself to scientific generalization, there is here an element that ... seriously limits our ability to forecast the future."

Schumpeter coined the phrase "a principle of indeterminateness" late in life. Actually, the above formulations are found in the announcement of a series of lectures that he should have started the first day after his death on 8th January 1950. However, the underlying ideas were not new. Actually, he had worked with them at least since *Entwicklung I*. Furthermore, his considerations seem to have become more explicit in the 1930s and 1940s. In 1931, Schumpeter started an intense and long-term discussion with Ragnar Frisch, who was already emerging as the leader of the econometrics movement. The discussion concerned the combination of determinism and indeterminism in the modelling of evolutionary business cycles. Frisch's (1933) result of the discussion was the production of his famous paper on "Propagation Problems and Impulse Problems in Dynamic Economics". This paper uses the model of a pendulum that will reach its equilibrium state unless it is disturbed by the impulse of random shocks. Frisch had, in 1931, interpreted his pendulum model in a letter to Schumpeter: "the more or less unpredictable innovations are those things that in my terminology world form the substance of the *impulse problem,* as distinguished from the *propagation problem*" (quoted by Louçã, 2007, 137). Schumpeter replied that he did not think in terms of "a problem *à la pendulum*" (p. 137). Instead, he wanted to include "within the economic world (= system of quantities)" an agent "which alters data and with these the economic process" (pp. 137–8). This agent is the "entrepreneurial activity" and "it makes economic things *change* instead of making them *recur.*" Furthermore, "its effects are not recurring—Ford can never be repeated—but 'historic' and definitely located in historical time. They are also irreversible." Since Frisch's oscillations are not related to economic transformation, his model did not really promote Schumpeterian analysis. However, a model that promoted evolutionary analysis was beyond the reach of Frisch and everyone else. Actually, the analysis of the behaviour of the apparently simple pendulum model was highly demanding mathematically—and Frisch's version of it might even be formally inconsistent (Zambelli, 2007). In any case, Schumpeter for the rest of his life continued communicating with Frisch. The mutual respect did not hinder Schumpeter in wondering whether "there are only two groups of economists: those who do not understand a difference equation; and those who understand nothing else" (*History,* 1168).

Schumpeter also engaged in another revealing discussion on determinism and indeterminism. This discussion took place while James Marschak, the later Research Director of the famous Cowles Commission, was preparing his review of *Cycles.* The discussion concerned a draft of Jacob Marschak's (1940) review of *Cycles.* This draft formulated Schumpeter's basic evolutionary model as a system of five differential equations that is "closed" and "causal" in the sense that in the 1930s was specified by mathematicians (Birkhoff and Lewis, 1935). While this worked fine for an aggregate version of circular-flow model, the fifth equation that defined the role of the entrepreneurs could not be specified in the standard way. In contrast to Schumpeter's argument that entrepreneurs represent endogenous drivers of basic economic change, Marschak had to model their role as exogenous shocks. These equations, however, were never published because of the negative response. Before considering the reasons for this response, it should be emphasised that Schumpeter's comments to the skilled modeller also suggested fields of collaboration. Thus Schumpeter emphasised "the highly characteristic inter-firm disequilibria" and asked Marschak: "Couldn't you do something about that—i.e., express that kind of disequilibrium as between innovating and petrified firms as an element of the total economy?" (quoted by Stolper, 1994, 376).

Schumpeter's hand-written response to Marschak (quoted at length by Stolper, 1994, 375–6) emphasised that "[e]nterprise is ... not of the nature of erratic shocks". Schumpeter's problem was that he had not found an alternative formalisation of enterprise, that is, the entry of innovation-based firms. The consequence was that his relationship to exact models "has always seemed to me to be most unsatisfactory". On the one hand, he had to admit "that my own system is not closed, if that means 'causal' in the sense of Birkhoff and Lewis". On the other hand, he emphasised that "[e]nterprise is inside the economic system"; is it "an internal, if you so please, source of energy." Actually, he thought that enterprise is "part of the system that could not live without it". Here he is referring to enterprise as providing the energy necessary for the continued working of the capitalist engine. Furthermore, he is thinking of enterprise in relation to a system "the main categories of which are keyed to its occurrence". This implies that the Schumpeterian system of concepts stands and falls with the concept of enterprise. Therefore, it was a crucial problem that the "energy" that is provided for the working of the capitalist engine by enterprise has been "refractory to quantification". While responding to Marschak's formalisation of the model in *Cycles*, Schumpeter revealed some of the pre-analytical background for his resistance against the deterministic treatment of any kind of evolutionary process. He emphasised that entreprise as an internal "source of energy" is

> "refractory to quantification so that the system subject to this impulse is ex ante indeterminate and in this sense 'open'. If you will allow me to deviate so far from your strict principles, there is something of évolution créatrice about it (as there is I think—another confessio fidei [confession of faith]—about every true evolution, a biological 'sport' for instance or de Vries' 'mutation')." (quoted by Stolper, 1994, 375)

Marschak remembered this reply to his question of Schumpeter's missing behavioural equation as: "Read Bergson" (Briefs, 1960, 3). Thus, he thought that Schumpeter did not want the emergence of enterprise (innovative projects) considered the function of well-defined variables, but instead as related to the type of philosophical speculation that is found in Henri Bergson's (1911) *Creative Evolution*. This was apparently enough to convince Marschak that further collaboration with Schumpeter was meaningless. However, we should give Schumpeter another chance. According to his interpretation, "Bergson meant that the new truth or, more generally, the new creation is not worked out by logical processes at all. This involves indeed ... an entirely new *Weltanschauung* wholly at variance, among other things, with the views then current (the Marxist one included) about cultural development" (*History*, 778). In contrast to the prevailing view that there are laws of evolution, Bergson saw the evolutionary process in all spheres of life as the unpredictable outcome of a mysterious "vital force" (élan vital). Schumpeter could in the social sphere connect this force to his innovative entrepreneurs. Thereby, he emphasised that even the process of economic evolution includes the non-deterministic emergence of novelty. However, such a process can also be modelled by a combination of random mutation and directed selection within the evolving system of competing entities. Actually, the statistician and evolutionary biologist R. A. Fisher (1934) had emphasised this point in a paper on "Indeterminism and Natural Selection", which was published in the same journal that Schumpeter seems to be referring to by recalling the contribution of Birkhoff and Lewis. Schumpeter ignored that this and similar formalisations of an evolutionary process resolved the long-term controversy between the mutationists and the selectionist followers of Darwin. His reasons might be that innovative projects are obviously goal-directed, that these projects can influence their selection environments, and that they often imply major jumps rather than gradualistic change.

Although Bergson's formulations of a mystic and immanent force of creative evolution have rightly been rejected, we should today not forget that he, in a primitive way, tried to face real problems relating to evolutionary processes. Such a process is irreversible and irrevocable. Thereby the evolutionary process, in contrast to the process in a deterministic system, defines an "arrow of time" that plays an essential role. The same role of time is found in the systems of statistical thermodynamics that inspired R. A. Fisher (1999) to formalise the theory of biological evolution. However, Bergson's ideas about time and consciousness seem of direct relevance for the understanding of the Schumpeterian entrepreneur and his insertion into the circular-flow model. These ideas might, at least for modelling purposes, be most easily understood by applying the concepts of "retrospective time" and "prospective time" (Châteauneuf-Malclès, 2002). Although Schumpeter did not use these terms, he obviously applied the underlying concepts. The concept of retrospective time is clearly related to a stationary economy. Here, boundedly rational agents can cope with the future by exploiting their experiences from the past. They can even use their retrospective behaviour to adapt to slightly changed situations. In contrast, the S-entrepreneur must operate with "prospective time". He *has* to look into a future that is not defined by the experiences of the past. Schumpeter also thought that the innovative entrepreneur is, in a certain sense, the *creator* of the future. This idea, however, does not have to be accepted to appreciate the concept of the S-entrepreneur. The purposeful innovation might often, from the viewpoint of the overall evolutionary process, be considered a random variable. Furthermore, less conspicuous examples of the breaking of routine may play an important role. Finally, an extended account for the relatively independent selection processes in relation to something like the circular-flow model could also be included. Nevertheless, the purposes of the present book suggest that we should follow Schumpeter's own analysis of the function of the S-entrepreneur.

Although Schumpeter tried to create some degree of collaboration with several skilled theorists, the major attempt concerned a planned book on "The Theoretical Apparatus of Economics". This project was planned to be performed by Schumpeter and Nicolas Georgescu-Roegen. Although the book never materialised, the idea seems to have promoted further reflections by the co-author on the kind of analytical difficulties that had also emerged in the discussions with Frisch and Marschak. Georgescu-Roegen (1971, xiv) later emphasised that Schumpeter's vision "combined in a harmonious manner quantitative and qualitative evolutionary analysis." Major parts of his *The Entropy Law and the Economic Process* may be seen as a response to some of Schumpeter's problems in making his vision operational. According to Georgescu-Roegen, the analytical problem is that Schumpeter was as an economist of discontinuity rather of continuous growth. This is emphasised by his dictum: "Add successively as many mail coaches as you please, you will never get a railroad thereby" (*Development*, 64). Georgescu-Roegen (1988, 294–5) saw this idea of discontinuous change within the economic system "from mail coaches to railway engines, to automobiles, and to rockets" as a plea for the recognition of "speciation" within economics. On this background, he remarked: "While insisting that biology is the Mecca of the economist, Marshall in his vision used only a couple of superficial biological parables. Schumpeter, by contrast, never mentioned any sisterhood between economics and biology; instead he worked out a perspective applicable to both because both are life sciences" (p. 296). Although this statement is exaggerated, Georgescu-Roegen's point is that both sciences have to handle qualitative leaps.

In biology, Darwin had been unable to handle this indeterministic problem of speciation analytically; and it was not before the 1940s that solutions started to emerge. Georgescu-Roegen emphasised the highly controversial proposal by Richard Goldschmidt that speciation takes place by discrete jumps: although most mutations produce hopeless "monsters", a few of them lead to successful "monsters" that serve as the starting points for new species. Georgescu-Roegen (1988, 295) considered Schumpeter's theory to have a similar structure: "Evolution, which is what economic development really is, needs saltations, needs the emergence of successful 'monsters' ". These leaps (saltations) cannot be deducted from any analysis of growth within pre-existing structures. Georgescu-Roegen seems to have thought that as soon as these jumps were treated convincingly, the rest of the mathematical formalisation of the evolutionary process could easily be provided. However, his works did not provide a systematic analysis of these jumps.

Since several students of economic and social evolution are still struggling with the problem of "speciation" as well as with the deeper problem of the necessity of combining chance and necessity, it might be relevant to emphasise that Georgescu-Roegen's reflections are expressed in very old-fashioned terms. They actually seem to relate to the controversy between mutationists and selectionists in the beginning of the twentieth century rather than the state of the art after the neo-Darwinian Synthesis had become predominant in the late 1940s (Provine, 1971). Georgescu-Roegen ignored that modern biology had moved from "hopeful monsters" to a more complex conception of speciation (see e.g. Futuyma, 1998, 24–5). Some might give the counterexample of Stephen Jay Gould (2002, 451–66). Although Gould recognised the failure of Goldschmidt's theory with respect to individual heredity, he tried to rescue the overall conception of relatively quick jumps in the process of macroevolution. However, even these jumps take time and they are fully explainable in standard neo-Darwinian terms. Schumpeter gave the reason: "a new apparatus poses and solves problems for which the older authors could hardly have found answers even if they had been aware of them" (*History*, 39). In any case, the principle of indeterminateness is still essential—and Georgescu-Roegen's reflections help us to remember that it is especially important with respect to the emergence of new technologies and new industries. However, while the issue of the emergence of new species of the biological system is precise because it involves separate populations, it is more difficult to specify the more or less analogous issue of the emergence within the "industrial organism" of new and relatively distinct specialities.

As we have seen, Schumpeter's interpretation of the mechanism of innovation was a major hindrance for his recruitment of support for the development of his evolutionary economics. His way of characterising this mechanism also hindered his own formalisation of the functioning of the capitalist engine. He was so fascinated by the characteristics of propulsive stroke that he de-emphasised the more operational task of characterising the reactive stroke precisely. In retrospect, it seems clear that he would have performed better by defining a first approximation to the functioning of the capitalist engine. This approximation could have combined clusters of innovations that emerge randomly with the deterministic response to these clusters. In this setting, the task of his collaborators would have been to model the deterministic process of evolution produced by an already created disequilibrium. We have seen that he actually asked Marschak to contribute to this task by expressing "that kind of disequilibrium as between innovating and petrified firms as an element of the total economy". However, this task was not operationalised. Instead, it became hidden in the cloud of semi-philosophical issues that emerged from Schumpeter's reference to Bergson's strange book on *Creative Evolution*.

## 10.4 The theoretical apparatus of economics

The present book has tried to demonstrate that *Wesen* does not only contain Schumpeter's interpretation of equilibrium economics, but also essential perspectives for his evolutionary economics. Nevertheless, he actively resisted the spread of this book. In the above mentioned interview with the *Harvard Crimson,* he told interviewer about his dislike for *Wesen*: "I have no copy and have been trying to atone for this effort of my youth since it was issued" (*S1944a*). Although this remark is not explained in the short interview, Schumpeter had for quite some time planned to replace his very first book. Thus, he, in 1934, in the preface to the new German reprint of *Entwicklung II,* had referred to "a book on the 'Theoretical Apparatus of Economics' that shall serve in place of a second edition of my 'Wesen und Hauptinhalt' " (*S1935f,* p. xx). This book project, which Schumpeter probably considered his most important one, also working titles like "The Mathematical Apparatus of Economics" and—later—the "Preliminary Volume". The task of the project was initially simply to rewrite *Wesen* in a way that emphasised the concrete problems of quantitative economic analysis rather than abstract theoretical considerations. To prepare the writing of this book, Schumpeter in the beginning of the 1930s had wanted to replace his lost copy of *Wesen.* One of his students found a second-hand copy; but since the book was in high demand from collectors, Schumpeter had to pay an "outrageous" sum for it (Allen, 1991:I, 279). The planned rewrite included not only the new developments within equilibrium economics—especially those of non-evolutionary dynamics; its treatment of evolutionary economics should probably, in accordance with the programme of econometrics, include "a quantitative theory of evolution" (*BL,* 184).

Since Schumpeter quickly became aware that he lacked the technical skills to solve his task, he looked for researchers that could help him. They were ready at hand—especially among the international researchers that came to work with Schumpeter. Although Schumpeter, in contrast to Keynes, did not wish to impose his ambitious and difficult ideas upon the young researchers, he found a solution. Georgescu-Roegen was a brilliant Rumanian statistician and mathematician who—like the young Schumpeter—had studied in London with Karl Pearson, the famous statistician and evolutionary biologist. Under Schumpeter's guidance, Georgescu-Roegen was at Harvard becoming an economist whose preliminary exercises served to bring him international fame; but he also had ambitions to formalise some of the Schumpeterian ideas. Georgescu-Roegen's stay at Harvard was extended to cover the period 1934–36, and Schumpeter applied for money to allow for another stay at Harvard so that he could become the co-author of the book on the tools of economic analysis (*BL,* 283–4). Unfortunately, he felt obliged to work in Rumania, so their collaboration did not develop and gave no direct results. We have already considered the indirect result that he became interested in economic "speciation". The general result seems to have been that Georgescu-Roegen (1992, 130) became a heterodox economist who continued to consider himself Schumpeter's only real disciple: "Every one of his distinctive remarks were seeds that inspired my later works. In this way Schumpeter turned me into an economist—the only true Schumpeterian, I believe. My degree in economics is from Universitas Schumpeteriana." Even Schumpeter got food for thought and an understanding of the difficulties of his project, which he shelved for quite some time.

During the World War II, Schumpeter decided to resume the project on "The Theoretical Apparatus of Economics" in modified form. The purpose of the project had originally been to develop a book that defined the foundations of modern economics in general as well as Schumpeter's extension of it to handle economic evolution. It is not surprising that this project had showed up to be unmanageable. Therefore he, in the early 1940s, decided to concentrate on a "Preliminary Volume", in which he seems to have wanted to focus on the formalisation of his own theory. Thus, he told Arthur Smithies (1950, 630) about "what I described to interested publishers as a book that should do from my standpoint what Keynes' *General Theory* did from his". He jokingly admitted his exaggeration by asking "which Washington agency now prosecutes dishonest advertising?" These plans were of much interest to Schumpeter's theoretically minded friends and colleagues, but the plans failed to substantiate. Thus, Smithies, after Schumpeter's death, searched in vain for the results, and he remarked that "[w]hat there is of this book probably exists only in his undecipherable Austrian shorthand; no manuscript has yet been found." Later researchers have continued the search, and they have concluded that no manuscript exists, because "even to produce a sketch, the nearly complete theory must be in hand" (Allen, 1991:II, 111). Since Schumpeter did not succeed in producing a formal theory of economic evolution, he stayed with an intensive work on the mathematical preliminaries both during and after the War. This work was frustrating because he did not find an adequate mathematical formalism for handling his verbal analysis of economic evolution. Actually, he tried to recruit help from mathematically skilled economists like Richard Goodwin; but he insisted on characteristics of his evolutionary scheme that they were not even able to recognise. Goodwin (1988, 6) remembered that this problem emerged, after the War, during a presentation of mathematically formulated theories: Schumpeter "patiently listened to a series of lectures I gave on Keynesian-type cycle theories, but he would have none of it." Godwin admitted that he had "quite failed to take adequate account of the continuing, innovational re-structuring of the productive economy." Since neither the help Schumpeter got nor his own work produced convincing formalisms, it is understandable why he felt frustrated. This state of dissatisfaction arising from the unresolved problem of finding adequate formalisms was not a new one. Even in 1933, he had told Gottfried Haberler that "I sometimes feel like Moses must have felt when he beheld the Promised Land and knew that he himself would not be allowed to enter it" (*BL*, 240). During the War, he reflected on his difficulties when answering a letter from a colleague:

> "I experienced a moment of real pleasure when I read your brief reference to your own family history. This is, indeed, the one thing of my theoretical (so far that it is not purely technical) writing on which I pride myself; it is all seen, and in this sense there is nothing in my structures that has not a living piece of reality behind it. This is not an advantage in every respect. It makes, for instance, my theories so refractory to mathematical formulations. They can never be so cut and dried as Keynes' schema is; but there are compensating advantages, and one of them is that so many people have told me, as you have done: 'Yes, that is so. I know it from my own experience and observation'. Your family history seems to be a particularly typical case." (*BL*, 343)

Schumpeter's failure in producing a formalisation of his evolutionary analysis meant that the "Preliminary Volume" ended up as scattered, and underexplored, notes in the Harvard University Archives. Instead, we largely have to look in *Cycles* and *History* for detecting some of the probable characteristics of his unsuccessful book project. Even *History* is influenced by the fact that Schumpeter wrote it in parallel with his continued efforts to develop his systematic treatment of economic dynamics in general and evolutionary dynamics in particular. These efforts seem not least to be revealed by *History*'s clarification of the relationship between the static methods of equilibrium analysis and the different dynamic methods. The study of this clarification can start with Samuelson's (1943, 58; 1947, 311) contribution to the celebration of Schumpeter's 60th birthday that stated: "Often in the writings of economists the words 'dynamic' and 'static' are used as nothing more than synonyms for good and bad, realistic and unrealistic, simple and complex. We damn another man's theory by terming it static, and advertise our own by calling it dynamic. Examples of this are too plentiful to require citation." The works of the early Schumpeter had contributed to the misuse of the concepts by developing an all-encompassing dichotomy between Statics and Dynamics (see Sections 3.4 and 4.2). Although this dichotomy was crucial for the arguments in *Wesen* and *Entwicklung I*, he largely applied the terms in quotation marks and he pointed out that the terminology is "very unfortunate" (*Wesen*, 182). Nevertheless, he had not least used Statics to denote the circular flow and Dynamics to denote the process of its innovative disturbance. In 1934, he stated that "I first used the terms 'statics' and 'dynamics' for these two [theoretical] structures, but have now (in deference to Professor Frisch) definitively ceased to use them in this sense" (*Development*, xi).

In 1929, Ragnar Frisch had in Norwegian published his attempt of clarifying the terminology, the next year he spread his view as mimeographed versions of his lectures in the US; and he, probably, informed Schumpeter in relation to their collaboration on the founding of the Econometric Society. By applying the tradition of mechanics to economics, Frisch (1992, 392) pointed out that the difference between static and dynamic propositions (laws) is simply a question of whether the concepts "rate of growth" or "response rate" (with respect to time) are used or not. This means that "the distinction between statics and dynamics refers to the analytical method, not to the nature of the phenomena. We may speak of static or dynamic analysis, but not of a static or dynamic phenomenon" (p. 400). However, "phenomena as such may be *stationary* or *evolutionary*." Furthermore, there is "the distinction between what may be called *analytical* dynamics and *historical* dynamics in economics." Frisch captured many of Schumpeter's ambitions by stating that "[h]istorical dynamics can be said to be an attempt to analyse those phenomena which have not yet been incorporated in, or which it is not possible to incorporate in, rigorously formulated theoretical laws" (p. 400; emphasis removed). While Frisch's broad use of the term "evolutionary" to cover all non-stationary phenomena gradually disappeared, his use of the terms "statics", "analytical dynamics", and "historical dynamics" convinced both Schumpeter and the rest of the economics profession (partly through Samuelson, 1947, 311–17). This usage decouples method and phenomenon so that, for instance, "evolutionary" phenomena can be studied by both static and dynamic methods; and dynamic analysis can use both abstract time and historical dating of events. However, neither Frisch nor Samuelson was interested in the type of evolutionary processes that engaged Schumpeter throughout his academic life. Actually, he remarked that they and most other economists concentrated on the study of the simple dynamics of the economic variables within an unchanging economic framework (*S1946c*, 424–5n). This might be called "evolution" in the "wider sense" and it includes "growth", but he was obviously interested in the "narrower sense" of evolution with abrupt changes of "institutions, tastes, or technological horizons" (*History*, 964). It was not before the last couple of years of his life that Schumpeter made the necessary clarification on this point and obtained nine cases (see Table 10.1 on the facing page)—ranging from the static analysis of stationary phenomena (Case 1) to the historical dynamical analysis of mutative evolutionary phenomena (Case 9). This result was obtained in the unfinished section in *History* (963–7) on "Statics, Dynamics; the Stationary State, Evolution". The terminological discussion in

**Table 10.1.: Using the terms statics and dynamics to denote methods**

| Processes | | Method of Analysis | | |
|---|---|---|---|---|
| | | Static | Dynamic | |
| | | | Analytical | Historical |
| Stationary | | (1) | (2) | (3) |
| Non-Stationary | Growth Processes | (4) | (5) | (6) |
| | Evolutionary Processes | (7) | (8) | (9) |

Comment: Schumpeter recognised that evolutionary processes can not only be studied by the methods of analytical and historical dynamics (Cases 8 and 9), but also by the static method of comparative statics (Case 7). To emphasise this point, he gave up his early strong coupling of phenomena and analytical methods (see Table 9.3 on page 253). The matrix of possibilities demonstrates that the three methods can also be used for analysing stationary phenomena and growth processes. Source: The table is based on *History* (963–7).

this section indirectly served to emphasise that his work focussed on mutative evolutionary phenomena and that it was performed by means of analyses that applied the methods of statics, analytical dynamics, and historical dynamics. Furthermore, these characteristics distanced him from practically all other practitioners of dynamic analysis. Although he, in 1946, remarked "that during the last twenty years or so an economic dynamics has emerged", he had to emphasise that "this dynamics has nothing to do with the factors that are incessantly at work to change the structure of economic life" (*S1946c*, 424–5n). However, the static and dynamic analysis of stationary and growing economies provided him with starting points for his own work.

A stationary economic system can, from the viewpoint of evolutionary analysis, be characterised as a system in which no evolution takes place. Here we see neither creation nor destruction. Such a system is obviously an ideal starting point for discussing what evolution is. However, this discussion presupposes a subtle change of the assumptions of standard equilibrium analysis. Stationarity is *not* the result of decisions made by fully rational agents that are fully informed about the conditions in an economic system that is placed in an unchanging environment. Stationarity is instead the result of an evolutionary process that has come to a halt. This process is based on routinised behaviour. The stationary state implies a system of routines in which each routine is the best response to the other routines. Thus, we are facing a hugely complex Nash Equilibrium of routinised behaviour. Those accustomed to the analysis of such equilibria will recognise that Schumpeter's stationary state implies very unrealistic assumptions. The systematic analysis of the Schumpeterian

Game requires the more modest structure of partial equilibrium analysis. Even for a well-specified version of the Schumpeterian Game, we are likely to discover multiple Nash equilibria. Therefore, we look for refinements that might allow us to predict the actually obtainable equilibria. One such refinement is an evolutionarily stable equilibrium. This concept was developed by the evolutionary biologist John Maynard Smith (1982), and it has even in economics been the starting point for the development of evolutionary game theory. However, the idea of evolutionary stability does not fully capture Schumpeter's concept of the circular flow. This state of the economic system can only be considered an evolutionary stable equilibrium for the routine-based agents. In contrast, the Schumpeterian entrepreneur takes it as the starting point for his equilibrium-breaking activity.

If the use the $3 \times 3$ matrix to describe major developments in the years of high theory and high econometrics, we can say that equilibrium economics moved from the static analysis of stationary processes to the dynamic analysis of such processes (i.e., from Case 1 to Case 2 of Table 10.1). Furthermore, equilibrium economics moved into Case 5, if we consider it as covering the dynamic analysis of both growth processes and oscillating processes. Schumpeter had, from the very beginning, argued against this way of transcending the limits of equilibrium economics. For instance, he had argued that dynamical extensions of the Walrasian system would not create insight, but only confusion. This explains his rejection of the model of dynamic general equilibrium by the Swedish economist Gustav Cassel (1923, 34–42, 148–52). Cassel had tried to demonstrate that the elements of the Walrasian system could grow in a uniform manner; and thereby he became a pioneer of the modern theory of economic growth. Schumpeter, however, considered the study of growth without structural change an example of the misleading artefacts that emerge when the proper boundaries of equilibrium economics were transgressed: "by replacing the absence of change by 'balanced' or 'equilibrated progress', we arrive at a picture which really deserves to be called the more unrealistic, the more it presents the misleading appearance of lifelikeness" (*Cycles*, 37).

Schumpeter instead tried to deepen his analysis of evolutionary processes by both static and dynamic methods. His basic procedure can be described as a static analysis of the evolutionary process combined with some analytical dynamics. The static analysis of an evolutionary process is made in terms of comparative statics (Case 7). We start from an evolutionarily stationary state and compare it with the state that emerges after an innovation has been absorbed. Since the initial state and the innovation are not sufficient to predict the subsequent equilibrium state of an otherwise undisturbed system, Schumpeter put much emphasis on the methods of historical dynamics (Case 9). To the extent that his early assumption that the actual evolutionary process periodically comes to a halt is right, then the analysis can proceed by comparing two subsequent equilibria of the real economic system. For instance, he compared the relatively equilibrated state that included mail coaches with the subsequent state that had adapted to the railways. This comparison is not designed as the exercise of forgetting the latter state and instead predicting it by means of information on the initial state and the railway innovation. The task was, instead, to analyse the evolutionary mechanisms that transform the economic system between the two evolutionary equilibria. Although these mechanisms can be studied in terms of creation and adaptation, they show up to be very complex. Some of this complexity is due to the fact that Schumpeter emphasised that the railway innovation reshaped its own environment. This reshaping includes basic changes of consumer preferences, financial procedures, and the institutional setting for the economic system. He obviously preferred this broad conception of the evolutionary mechanisms, but the result is that he reduced his possibility of handling the mechanisms theoretically.

Ragnar Frisch considered analysis of "historical dynamics" as a potential starting point for developing "rigorously formulated theoretical laws". Schumpeter's experience seems to demonstrate that it is much harder to formulate "laws" about the creative side of evolution than about its adaptive side. Furthermore, it has shown up to me much easier to formulate "laws" about the creative aspects of Schumpeter Mark II that it is to clarify fully the functioning of the entrepreneurs of Mark I. Therefore, it is surprising that he did not spend more time on Mark II, which he seems to have had in mind at an early point of his career as a theorist. It is even more surprising that he did not study the mechanism of adaptation more thoroughly than he actually did. The basic reason seems to be found in his vision of the process of creative destruction. This vision considered the creative force of the innovators as largely shaping the subsequent process of destruction (including adaptation). However, his theoretical analyses of evolution as well as his historical studies give another impression. The Schumpeterian entrepreneurs are modifying a system of routines that they largely take as given and they are facing selective forces that are largely out of their control. Although routines and selective forces are themselves the outcome of the long-term evolutionary process, their shorter-term character of parameters ease the study of the evolutionary mechanisms that in a relatively incremental manner transforms the economic system. This humble approach to economic evolution does not imply that the concept of creative destruction is unimportant; but Schumpeter apparently wanted to start with the greatest issues of economic evolution. His failure to move gradually from the simplest parts of evolutionary analysis to the most difficult tasks largely explains his failure to provide a richer toolbox for this type of analysis.

## 10.5 Schumpeter's "final thesis"

After finishing his studies at the University of Vienna in the beginning of 1906, Schumpeter had published an article on "On the Mathematical Method of Theoretical Economics" (*S1906a*). Near the end of his career at Harvard University, he presented a paper on "The Historical Approach to the Analysis of Business Cycles" (*S1949g*). Fritz Machlup (1951, 145) dramatised these facts by starting his memorial on Schumpeter's economic methodology in the following way:

> "The very first paper which Schumpeter published—in 1906—was a plea for the *mathematical* method in economic theory. Two months before his death—in November 1949—he delivered a paper pleading for the use of the *historical* method in business cycle analysis. Can the fact that the 23-year old Schumpeter stood up for mathematical economics and the 67-year-old Schumpeter for economic history be taken as indicative of a trend in his development?"

Although "[s]uch a trend would be typical of the great minds in our field", Machlup's rejected this explanation. For him, it was evident that Schumpeter did not represent "an evolution from youthful keenness of a mathematical turn of mind to the mature perspective of a historical one." Instead, he "never lost the one and never lacked the other." Machlup (1951, 145) described this "methodological tolerance" as the outcome of Schumpeter's early attempts to overcome the battle of methods. These attempts did not only lead to the idea of a peaceful coexistence and collaboration between formal theorising and historical inquiry. They also lead to Schumpeter's movement from Statics to Dynamics, that is, from equilibrium economics to evolutionary economics. It is in his definition of these complementary branches of economics that we find the "really revolutionary change which Schumpeter proposed for economic theory" (p. 150). This change was "to banish from Statics some of the most important economic concepts, such as profit, entrepreneurship, and interest". However, this revolution did not succeed. Instead, most economists "have continued to find some useful work which these concepts can do for Statics, from which they were to be locked out."

Although Schumpeter's revolutionary proposal of a division of labour between equilibrium economics and evolutionary economics might be seen as a call for battle, he thought differently. Actually, he not only called for peaceful coexistence and collaboration between theory and history but also between equilibrium economics and evolutionary economics. A closer look at his paper on "The Historical Approach to the Analysis of Business Cycles" demonstrates that it contains a plea for both types of coexistence and collaboration. However, his message would have been clearer if he had chosen a title like 'The Historical Approach to the Analysis of Waveform Economic Evolution'. The immediate background for the paper was that economic history was becoming increasingly marginalised within economics departments. According to Samuelson (1987, 52), this marginalisation was especially clear from the 1940s, where "something had to give in the economics curriculum. What gave, and gave out, was history of thought—followed quickly by attrition of foreign language requirements and of minima of economic history." This was indeed a troublesome side effect of the development of modern economics since the history of economics, the access to foreign literature, and the study of economic history had been major sources of evolutionary thought. Therefore, Schumpeter's plea for the historical approach was not least a plea for upholding the preconditions for the future development of evolutionary economics.

Such considerations were probably at the centre of Schumpeter's attention when he, at the end of November 1949, for a short while left what has been called his "unusual ivory tower" (Leontief, 1950, 110) to contribute to the important Conference on Business Cycles. This business-cycle conference was organised by a committee from the leading American universities and the National Bureau of Economic Research. At this conference, Schumpeter addressed many of the most talented theorists, statisticians, and model builders. Several of these young researchers would later receive Nobel Prizes. Among them, Paul Samuelson, Wassily Leontief, Jan Tinbergen and Simon Kuznets were closest to Schumpeter; but the audience also included the later Nobel Laureates Milton Friedman, Lawrence Klein and Tjalling Koopmans—as well as several other famous economists. Schumpeter was known to all of them as one of the founding members and past President of the prestigious Econometric Society with its "alliance between statistics and theoretical economics" (*History*, 1141). The audience expected him to repeat his ardent support for mathematical modelling and related statistical studies. They should have known better, since they were developing the kind of macroeconomic analysis that Schumpeter disliked: an analysis that abstracted from innovative competition and waveform evolution. Nevertheless, he had normally kept his worries to himself. Therefore, the audience was greatly surprised when he presented the already mentioned paper on "The Historical Approach to the Analysis of Business Cycles".

Schumpeter's paper questioned the role of econometrics and mathematical economics as the all-dominant forms of the progress of modern economics. These research areas were actually, according to Schumpeter, moving in the wrong direction because of their lack of attention to the evolutionary forces that shaped the economic aggregates under investigation. For him, "the most serious shortcoming of modern business-cycle studies is that nobody seems to understand or even to care precisely how industries and individual firms raise and fall and how their raise and fall affects the aggregates" (*S1949g*, 329). This underlying evolution was, for Schumpeter, reflecting the essence of the capitalist engine; but for the competent young researchers, the study of the evolutionary process represented uninteresting complexities that they, furthermore, could not handle systematically in their models and econometric investigations. In this context, he chose to "let the murder out"—to spoil his own performance and create an unpleasant and intellectually troublesome state of affairs—by presenting his view in a provocative form. The conclusion of the paper included the following statement:

> "To let the murder out and to start my final thesis, what is really required is a large collection of industrial and locational monographs all drawn up according to the same plan and giving proper attention on the one hand to the incessant change in production and consumption functions and on the other hand to the quality and behavior of the leading personnel." (*S1949g*, 328)

The purpose of this apparently idiosyncratic proposal of "detailed historical case studies" (*S1949g*, 325) was obviously to elucidate the evolutionary mechanisms underlying much of the cyclical behaviour of economic aggregates. Schumpeter thought that even the cyclical behaviour of investment is in itself a surface phenomenon and that we have to investigate "the actual industrial process that produces it and in doing so revolutionize existing economic structures" (*S1949g*, 326; emphasis removed). The mechanisms of this evolutionary process had not been studied by Schumpeter's ambitious audience of modellers and econometricians; and he must have known that they would not even dream of contributing to the production of "a large collection of industrial and locational monographs". Even his formulation of the "final thesis" was provoking. The idea of a large collection of monographs served to remind of the work of the German historical school, which from the 1870s and onwards had seen historical monographs as the starting point for the development of economic theory. This viewpoint had been shared by American institutionalism. Practically all theoretically minded economists agreed that the strategy of the historical and institutionalist schools was hopeless; and Schumpeter had apparently been the strongest proponent of this view. Now he seemed to state that he had changed his mind.

The audience did not miss Schumpeter's point since "[a]s usual, he spoke without notes, with great vigor and temperament" (Haberler in Universities–NBER, 1951, viii). Therefore, he succeeded in provoking—and even shocking—those present at the conference on business cycles. At a conference that dealt with the complex issues of macroeconomic business cycles, Schumpeter suggested a postponement of this task in order to create a new foundation through case studies of the evolutionary process. This suggestion not only was impractical; it also pointed to a return to a past when American institutionalists had dominated the National Bureau of Economic Research. Samuelson (1951, 100) remarked on this aspect of Schumpeter's "uncharacteristic performance" that it "represents one of those rare occasions when Schumpeter gave what was from his viewpoint 'comfort to the enemy'." Samuelson emphasised that to Schumpeter the real enemy "was first and foremost those who opposed—and I now employ the words uttered a thousand times by him in the classroom—'the use of exact methods in economic analysis'." This research programme was that of ambitious young economists, especially those connected to the Econometric Society; and it has later proved crucial for the development of modern equilibrium economics into a tool for the analysis of major problems of economic life. However, although Schumpeter supported the work of the young economists, he did do as he preached. His ardent support for exact analysis did not exclude his parallel support for evolutionary studies of a more historical nature.

Schumpeter's paper on the historical approach was published as "unrevised draft which ... he intended to revise later" (Universities–NBER, 1951, viii). The reason for this unfinishedness is that he was busy writing *History* and died before he found time to make the necessary revisions. The unfinished nature of the paper opens up two apparently alternative interpretations of his underlying strategy. According to the first interpretation, his strategy was solely to develop his emerging alliance with economic historians. This strategy would have implied that his revisions helped to specify of the common plan for the proposed case studies—or at least provided hints about the set of concepts he wanted to be used by carefully planned case studies. The problem with this interpretation is that Schumpeter's audience did not consist of economic historians, but of economic theorists and econometricians. Nevertheless, he might have tried to get funding for a research programme that had been suggested by his collaboration with economic historians interested in entrepreneurship and industrial change. According to the second interpretation, Schumpeter's strategy was to renew his well-established alliance with theorists and econometricians to include the study of economic evolution. He must have recognised that the suggested collection of case studies of the evolutionary change of selected industries and regions would hardly give any direct input to the study of business cycles. However, the suggested monographs would provide information on the evolutionary mechanisms of innovation and adaptation; and this information could in turn provide the stylised facts for theoretical modelling and econometric studies. Although Schumpeter did not give these intermediate arguments in the unrevised paper, he did state the conclusion: "Theoretical and statistical work is as necessary as is the historical work. In fact they are inseparable because there is an incessant give and take between them" (*S1949g*, 329). Therefore, it is obvious that two interpretations Schumpeter's strategy of emphasising the historical approach do not exclude each other. He wanted to "indicate how theory and statistics fit in with the historical approach" (*S1949g*, 322). This would be a first step in creating an alliance that combined all the three fundamental fields of research in order to understand capitalist economic evolution.

# 11

# Evolutionary Analysis and the History of Economics

The coherence of the present book would have increased if Schumpeter's *History of Economic Analysis* could, in some way, have been presented as an extension of *Development, Cycles,* and *Capitalism.* However, he did not extend the evolutionary trilogy into a quatrology. Actually, the 1260 pages of *History* do not even contain the elements of a smaller book on the 'History of Evolutionary Economic Analysis'. In order to exploit the potentials of *History,* we have to step back from Schumpeter's evolutionary trilogy and consider his double programme for the science of economics. This programme had in *Wesen* been described as the refinement of equilibrium economics and the creation of evolutionary economics. Since he could not describe the history of something that does not yet exist, the historical addendum to his research programme had to focus on equilibrium economics. The first version of Schumpeter's complement emerged before World War I in *Doctrine.* His early task, commissioned by Max Weber, had been to overcome the confusing battle of methods. The self-assigned task of *History* seems to have been to overcome the confusion of the period 1926–39, which, in Section 10.2, was called the years of high theory and high econometrics. After this period, "[t]he field looks like a big building plot all covered with ruins and half-completed new structures" (*S1948b,* 95). Schumpeter could not describe this "building plot" directly, since "[n]o well-rounded picture could be drawn to be presented to the layman or beginner with a comfortable sense of authority. Also, much of the most worthwhile work is being done for reasons of method quite outside the range of a majority of the profession." However, it was possible for him to make sense of the situation by presenting a retrospective study of the history of economic analysis. This study was motivated by the fact that when "the methods of the economic research worker is undergoing revolutionary change", "older authors and older views acquire (rightly or wrongly) an importance which they would not have under more fortunate circumstances" (p. 95).

Although *History* does present "older authors and older views" in detail, it also tries to make sense of the "big building plot" of economics by means of an evolutionary interpretation of the overall history of this science. Schumpeter's explanation of scientific evolution is largely analogous to his explanation of economic evolution. As we shall see, his explanation of the evolution of science can, metaphorically, be described as being produced by a scientific engine with a mechanism of innovation and a mechanism of adaptation. However, *History* is not designed to become part of a literature on the working of the scientific engine within the science of economics. It is, instead, designed for ambitious economic theorists who want to orient themselves before making great contributions to their science. This design has, to some extent, worked. Thus *History* has obtained a safe place among the handbooks of a significant number of economic theorists. However, the book has an even safer place in the handbook collection of most historians of economics. This function of *History* can easily be understood. If we ignore its many apparent idiosyncrasies and turn to the index of the book, we find 1240 names and a similarly impressive number of subjects. Furthermore, most of the individual entries not only reveal Schumpeterian opinions but also provide useful information. Nevertheless, the use of *History* as a handbook is seriously impaired by the lacking understanding of the underlying evolutionary theory of science that determines both its structure and many of its apparently idiosyncratic opinions on the contributions of individual economists.

The reason why Schumpeter's theory of the evolution of economics has been overlooked can be found by a quick comparison of Schumpeter's *History of Economic Analysis* with a book that also deals with the working of the engine of scientific evolution. This book is Thomas Kuhn's (1970) *The Structure of Scientific Revolutions*. Kuhn's book became, for quite a while, very influential among philosophers and sociologists of science because it, with great pedagogical skill and the use of case stories, develops a simple theory of a, more or less, evolutionary history of science. According to this theory, scientists are mostly engaged in the routinised work of normal science. While this ordinary activity produces an expansion of knowledge within a given scientific paradigm, it cannot transcend the limits of this paradigm. Therefore, the long-term evolution of science depends on the infrequent revolutions in science that produces new paradigms for ordinary research. Schumpeter's *History* is based on a roughly similar theory. Since this theory preceded and, in some respects, complements Kuhn's theory, the fact that that it has largely been ignored is unfortunate (Jensen, 1987, 145). The reason, however, is obvious. Schumpeter concentrated on analysing the facts of the history of economics and largely presented the applied evolutionary theory implicitly. His detailed historical work did not leave time and space for describing the underlying theory, which would have had to be presented in terms of successive approximations. Furthermore, he seems to have grown tired of the overselling of theoretical contributions in the style that was later represented by Kuhn. Nevertheless, the reception of *History* would probably have been very different if its contents had been presented under the heading 'The Structure of Scientific Revolutions in Economics'. Even a boring, and more fitting, title like 'The Theory and History of the Evolution of Economic Analysis' might have saved *History* from becoming a mere handbook.

According to Schumpeter, the scientific engine does not produce economic thought or economic policies. Instead, each of its cycles is based on the given set of techniques for economic analysis; and the outcome of a cycle is new or modified techniques. As soon as this is understood, the implication of *History* for Schumpeter's evolutionary economics becomes clear. His theory of the history of economics only concerns the evolution of the tools of scientific analysis; and it places the visions and ideas of great economists as part of the environment of this evolutionary process. Furthermore, Schumpeter emphasised that no semi-cumulative history of visions and ideas can be written. Since the study of economic evolution had not moved from visions and ideas to professional analysis within a community of economic researchers, he could not write a history of evolutionary economics. Instead, the evolutionary visions and ideas of great economists like Adam Smith, Marx, and Marshall had to be presented as part of the environment that influenced their development of tools for economic analysis. Nevertheless, Schumpeter did emphasise the analytical difficulties that had hindered great economists in implementing their evolutionary visions and ideas. For instance, he considered the concepts of division of labour and increasing returns as reflecting the unsatisfactory development of the tools for evolutionary analysis. Thereby, Schumpeter's book in several respects points beyond more recent contributions like Geoff Hodgson's (1993) history of evolutionary ideas within economics and Jack Vromen's (1995) methodologically oriented account for evolutionary economics. However, it should be noted that Schumpeter's implicit explanation of the failure of evolutionary economics to emerge as a semi-cumulative research tradition within the science of economics was not his main reason for writing *History*. Instead, his evolutionary interpretation of the overall history of economics primarily serves to explain the crucial role of equilibrium economics in the evolution of the science of economics before 1950. Nevertheless, most of the general arguments of *History* cover not only equilibrium economics but also the conditions of existence of an evolutionary economics.

## 11.1 The gradual development of *History*

Hardly anyone reads *History* from cover to cover. Its 1260 densely printed pages are a sufficient reason for most researchers, and those who nevertheless give the book a try are quickly stopped by its highly complex contents. The present account is not intended to change this situation. We, nevertheless, have to recognise the main outlines of Schumpeter's presentation in order to appreciate his evolutionary interpretation of the history of economics and its implications for his evolutionary economics. One way of recognising the outlines of *History* is to study how this book emerged. This study brings us back to three of Schumpeter's early works. The first is "How does one study social science?". The second work is "The Past and Future of the Social Sciences". The third work is "Economic Doctrine and Method". These works prepare a study of the lectures on the history of economics that Schumpeter gave at Harvard University during the 1940s. Finally, we shall arrive at the outlines of *History*.

Schumpeter's works demonstrate that he was an intellectual craftsman with a complex analytical strategy. Since this strategy applied "modes of research which had been presented as mutually exclusive", it often was "misjudged as weak eclecticism" (Machlup, 1951, 146). However, Schumpeter's "methodological tolerance" was crucial not only for overcoming the battle of methods, but also for the development of his double research programme. His somewhat old-fashioned type of craftsmanship is not well covered by *Wesen*, and this might be one of the reasons why he came to dislike that book (see Section 3.1). A complementary description is found in a small booklet from 1910, which he produced while working on *Entwicklung I*. At that time, he addressed his students at the University of Czernovitz with the question: "How does one study social science?". For this purpose, he emphasised the need of understanding that "[s]ocial science is the study of social processes: the science of what holds state and society together, of what determines the conduct and fate of individuals and social classes, in short, the science of man's social existence and development" (*S1910b*, 58; *AÖT*, 556). The study of these social processes require specialisation, but here we have to recognise that "individual social sciences ... did not arise through the logical division of some originally unified realm of knowledge; they arose by chance ... from some particular problem or method". In general, however, "social sciences do the same as the natural sciences. They collect factual material and then attempt to discover regularities, that is, to order and analyze the material data." Therefore, the learning of these sciences starts with reflections on the "data" provided by experience as well as by "historians, ethnologists, and statisticians". However, since "unanalyzed facts are dumb", we need "to learn to think theoretically, to develop a sense for scientific abstractions". Actually, it is only after "we have become practised in theory" that our "contemplation of social reality teach us something" (*S1910b*, 60–1; *AÖT*, 557–8).

The main problem is thus to learn the theoretical aspects of "*craft* of science". This learning starts from recognising that "theory can obviously never present a precise picture of reality". We also have to take into account that the immature state of the social sciences means that there are competing theories. Therefore, it is crucial to "ascertain the preconditions which lie behind a particular line of analysis". This can be done by focussing on the basic structure of each theoretical "building". An additional advantage of this focus is that we recognise that "any particular theory [or proposition] is never valid in itself, but is always part of a theoretical structure and can only be understood as such" (*S1910b*, 62; *AÖT*, 560). This does not imply that we should abstain from "analyses of concepts", provided that they are not purely terminological but deal with "the essence of real processes". Such studies lead us to questions like "what is an entrepreneur, a banker, and so on?, or, what is capital?" (*AÖT*, 563; not in *S1910b*). The next step is to study the discussion of individual theoretical theorems, and here Schumpeter suggested to use the many "brilliant examples of theoretical work" in Böhm-Bawerk's (1921:I) account for and criticism of theories of interest. Through such specialised studies we gain a deeper understanding of more complex theoretical structures. Independent research work of the theoretical type, however, is very risky. While historians, ethnologists and statisticians will always have collected some materials, "unfruitful theoretical work" leaves no result. Schumpeter had only one advice for those of his students who wanted to do theoretical research: "Have talent, open the eyes, and leave the rest to your good star" (*AÖT*, 564).

Schumpeter's students at the University of Czernovitz hardly wanted to do theoretical research at his own level of ambition. However, it is his ambition to make a major theoretical breakthrough for economics that largely explains his life-long study of the history of economics.This way of opening the eyes was already clear in his farewell lecture at Czernovitz. Here Schumpeter had talked about the "The Past and Future of the Social Sciences"; and he later developed this lecture into a small book called *Vergangenheit und Zukunft der Sozialwissenschaften*. *Vergangenheit* makes clear that the strong commitment to his roles as a promoter of the science of economics and an explorer of economic evolution was closely related to his overall vision of the development of the social sciences. He describes their past in terms of centuries. The eighteenth century was the heroic period during which previous insights on "natural law" brought forth the social sciences (*Vergangenheit*, 9–59). These sciences studied the essence of societies, states and economies; and they emphasised the general laws of evolution. In principle, the social sciences of the nineteenth century could have been built directly on the results produced by the rationalism of the previous century. However, its political results (enlightenment, liberalism) were unpopular and its scientific results had run into difficulties. This largely explains why the nineteenth century became a period of reaction (pp. 59–81). The scientific difficulties undermined the universal social science based on natural law. Instead of universality and integration, the Romanticist reaction created a diversity of social sciences with strong national links. This meant, on the one hand, neglect of theoretical research and, on the other hand, a radical strengthening of the collection of historical and statistical data. The battle of methods and other controversies near the end of the nineteenth century demonstrated that the research programme of the eighteenth century was not dead (pp. 81–108). However, the controversies also demonstrated that the lessons of the past had not been learned. The obvious task in the beginning of the twentieth century was to return to the research programme of the eighteenth century with the arms provided by the empirical research of the nineteenth century (*Vergangenheit*, 109–36).

Schumpeter's next step is found in the 200 pages of *Economic Doctrine and Method*. *Doctrine* was a contribution to Max Weber's ambitious handbook of economics in the broad sense (*Grundriss der Sozialökonomik*). *Doctrine* pointed out that when economics took its shape near the end of the eighteenth century, it not only built on the tradition of the philosophers of natural law. It also included the empirically oriented studies and suggestions that were developed in relation to the practical problems of public finance; and these contributions should not be dismissed as expressions of old-fashioned mercantilism (*Doctrine*, Ch. 1). The scientific breakthrough, however, came with the theoretical discovery by the French physiocrats of the circular flow of economic life—considered as a model of a self-organised economic system (Ch. 2). The development of post-Smithian classical economics in the first half of the nineteenth century established economics as a science; but the lack of empirical relevance and analytical rigour led, in the end, to decay (Ch. 3). This decay paved the way for two types of reactions (Ch. 4). On the one hand, the historical school reacted against the lack of empirical sensitivity, and here it could build on previous mercantilist and cameralist thinking; this school also helped to establish "[t]he point of view of evolution" (p. 178; emphasis removed). On the other hand, the emerging neoclassical economics reacted against the lack of analytical rigour, which it overcame through marginalist analysis. In spite of "the bitterness of the struggle" between these two types of economics, "our science does not lack an organic development [Entwicklung]" (*Doctrine*, 201). Furthermore, "the misdirection of energy will abate [end] as time goes on" and thereby promote the intrinsic evolution of economics.

Although Schumpeter studied the works and lives of many individual economists between World War I and World War II (see Appendix C.4), his next major step was to produce his final account for the history of economics. He spent most of his time for the last eight years of his life writing the 1200 pages of *History of Economic Analysis*; and he left for his widow, Elizabeth Schumpeter, the huge task of editing this book from rather disorganised manuscripts. The immediate background for his work is obvious. From 1939 to 1948, Schumpeter's teaching of the graduate students at Harvard included his course on the history of economic thought—with an impressive and demanding reading list. According to a student's thorough notes from the 1947 version of this course (see Staley, 1983), Schumpeter started with a summary and Reader's Guide to Adam Smith's *Wealth of Nations*. He depicted this book as a synthesis of the previous efforts to understand economic phenomena; and he went back in time to detect the underlying ideas. In doing so, he repeated much of the material that he had already presented in *Doctrine*; but there were several novelties. For instance, he provocatively argued that the economic thinking of the scholastics of the Middle Ages and the Renaissance was similar to that of Smith—and "more correct" (Staley, 1983, 28). He also tried to rescue the intellectual reputation of the thinkers of mercantilism, which were a major target of criticism in the *Wealth of Nations*. After such studies of its background, Schumpeter again moved his students' attention to Adam Smith—and depicted him as one of the exponents of the synthesis of economic thinking that emerged after a long period of "rapid and uncoordinated progress" (p. 29). In other words, Schumpeter tried to convince his students that Smith's work had no originality and that his work easily could have spelled intellectual stagnation.

After this thought-provoking exposition, which covered nearly half of his course, Schumpeter turned to subsequent developments. When doing so, he did not deny that Adam Smith's work was of a high quality and served as the starting point for a new period of progress in economic thinking. The advance of classical economics, for which the work of Ricardo was a mixed blessing, lasted until works as John Stuart Mill's *Principles of Political Economy* successfully expressed a new "classical period". This period was characterised by a feeling that the great work on economic theory largely was completed. Therefore, progress came by opposing Mill's pedagogical synthesis from two viewpoints. On the one hand, an emphasis on empirical detail and moderate economic reform emerged in both England and Germany. On the other hand, theoretically minded economists in several countries developed marginal utility theory into a new foundation for economic analysis. Since other courses covered the most recent history of economics, Schumpeter only followed these developments until the eve of World War I. At that time, a new "classical situation" had emerged—and it was possible to "discern signs of decay, of new breaks in the offing, of revolutions that had not yet issued into another classical situation" (*History*, 754).

Schumpeter could easily have used his lectures on the history of economic thought to make a rewrite of his *Doctrine*; and this was how he initially envisaged his new book project. It gradually developed into something entirely different, however (see Elizabeth Schumpeter's appendix to *History*, 1185–1204). During World War I, he seems to have concentrated on the period from the rudimentary economics of classical

**Table 11.1.: *History of Economic Analysis* [1954]**



Comment: *History of Economic Analysis* consists of five parts. Part I, which exists in an alternative version (*S1948b*), contains relatively explicit formulations of Schumpeter's theory of scientific evolution. The importance of this theory for the present book is emphasised by the inclusion of details of the contents of Part I. Schumpeter's theory of the scientific engine is not only applied, but also developed implicitly in Parts II–V. His book contains a very detailed table of contents as well as extensive indexes (*History*, xv–xxv, 1209–60).

Greek philosophy through Adam Smith to Stuart Mill. By studying this period, Schumpeter found refuge from current events, and the description of his materials in Parts II and III of *History* swelled to more than 700 densely printed pages with lots of scholarly footnotes (see Table 11.1). Although he always had modern economics in mind, it was mainly after the war that he focussed on the development from the works of neoclassical economics and the historical school to the later emergence of modern economics. He, for Part IV of *History*, produced some 350 pages on the period 1870–1914, with some outlooks into later events; and he made sketches for the concluding Part V on the most recent events during the years of high theory and high econometrics. However, it was not only the conclusions but also much of the rest of the work that ended up in unfinished form. He undoubtedly had planned to revise his book so that it formed a coherent whole; but he died before he had produced a full version of the manuscript.

Schumpeter wanted to inform the readers of *History* thoroughly about what they could expect. Since this preparation included an account for his theory of the history of economics, his introduction is of major importance for the present book on his evolutionary theorising. Unfortunately, he never finished the introduction to his book. The manuscript that was included as Part I of *History* covers the "Scope and Method" of the history of economics in 44 pages. Its four chapters are listed in Table 11.1. In addition, the table includes the details of Chapter 1, which Schumpeter had designed as three chapters. Elizabeth Schumpeter's editorial comments (*History*, 1190) emphasise that we are facing an unfinished expansion and revision of a former manuscript. This first version of the introduction, from 1948, was labelled "Chapter I: Some Questions of Principle". It has later been published (*S1948b*) and shows up to be much more readable than the rewrite that Schumpeter made shortly before his unexpected death. The editor decided to use this rewrite because it included matured formulations and much new material. Thereby, she secured that most of Schumpeter's arguments got published immediately, but the straightforward approach of the first version was for a long time lost—and so were its alternative formulations. In any case, even the unfinished state of Part I of *History* calls for the further development of his theory of the history of economics. This development is supported by the alternative formulations of the first version as well as by his many remarks in relation to the application of the theory in the main part of the book.

Schumpeter obviously used his comparative advantages and his enormous energies to produce a book that is unique in the literature on the history of economics. This has given *History* the status as a much-used reference book for both economic theorists and historians of economics. Nevertheless, we are facing the description of a journey through the history of economic *analysis*, "half of which is given over to narrative history, political theory, and philosophical climates of opinion" (Blaug, 1996, 5). Even the other half of the text includes many apparently idiosyncratic opinions. These are probably the reasons why Lionel Robbins (1955, 22) ended his long review by comparing *History* with a picnic on the River Thames that Schumpeter happened to join in the middle of the 1930s. At that picnic of the graduate students of the London School of Economics, there had been "discoursing with urbanity and wit on theorems and personalities". The book includes the same elements: it appears as "a splendid excursion down the river of time, with good talk and magnificent vistas". Jacob Viner (1954, 894) was more blunt by emphasising that there is "much in this book which is redundant, irrelevant, cryptic, strongly biased, paradoxical, or otherwise unhelpful or even harmful to understanding." Nevertheless, Viner added that "there still remains enough to constitute, by a wide margin, the most constructive, the most original, the most brilliant contribution to the history of the analytical phases of our discipline which has ever been made." A way of overcoming this dualistic interpretation of *History* is to study Schumpeter's theory of the evolution of the science of economics.

## 11.2 Long waves in the evolution of economic analysis

There are many ways of studying the history of economics (see, e.g., Samuels et al., 2003) and Schumpeter did not constrain himself to a single approach. Nevertheless, he primarily seems to have approached the study of the history of economics by means of his general evolutionary theory. Let us start by explaining the overall structure of *History* by means of this theory of the scientific engine. The structure of the book has some similarity with the structure of the historical parts of *Cycles*. While *Cycles* had described two complete Kondratieff waves followed by an incomplete one, *History* describes two complete waves of analytical work in economics and ends with an unfinished wave. Part II does not describe a wave but covers the prehistory of economics that ends around 1790 when Adam Smith's *Wealth of Nations* had helped to establish a separate group engaged in the analysis of economic problems. Part III studies the wave of subsequent analytical work and the routinisation that became obvious around 1870—not least promoted by John Stuart Mill's synthesis in *Principles of Political Economy*. Part IV follows the wave of neoclassical economics from its innovative emergence to the mature phase that Marshall's *Principles of Economics* helped to establish in the first decade of the twentieth century. Finally, Part V records the wave of innovative contributions of the 1920s and 1930s that did not result in routinisation until after Schumpeter's death.

The overall structures of *Cycles* and *History* apply Schumpeter's idea of dividing macroevolutionary processes into periods that are separated by situations in which the evolution has largely stopped (see Figure 9.1 on page 251). As we have already seen, these "neighbourhoods of equilibrium" are difficult to handle analytically. Actually, it is only evolution in Schumpeter's radical sense that has come to a full stop. In contrast, adaptive and incremental innovation can for a while continue. This specification of the resting points of Schumpeterian evolution (the $\Sigma$ states) is crucial for the understanding the already mentioned similarity between his theory of scientific evolution and the theory of Thomas Kuhn (1970). Schumpeter's $\Sigma$ states are largely identical to the Kuhnian states of "normal science". In such states of science, problem solving continues in the form of adaptive and incremental innovation within a given paradigm. However, radical scientific innovators overcome the constraints of any given paradigm. The consequence of such disturbances is not the immediate abandonment of the old paradigm; but sooner or later a series of radical innovations becomes integrated in a scientific synthesis that can serve as a new paradigm.

This macroevolutionary interpretation of the working of the scientific engine serves to explain the periodisation applied in *History*. Consider the process described by Figure 11.1 on the facing page. The first classical

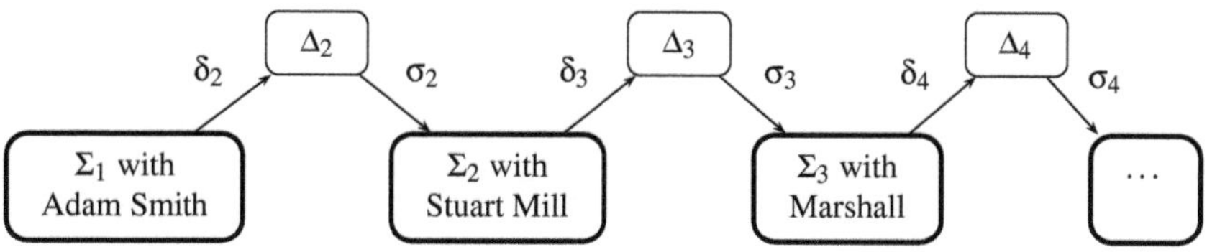


**Figure 11.1.: Evolution of economic analysis by innovation and adaptation**

Comment: Schumpeter's most abstract model of the capitalist engine can also be used to depict his basic model of the evolution of the science of economics. Here Σ states are called "classical situations". The first classical situation can be represented by the adaptive synthesis of previous work provided by Adam Smith. This synthesis was disturbed by the analytical innovations of Ricardo and others (the $\delta_2$ mechanism). The next classical situation was produced by the adaptive $\sigma_2$ synthesis that is symbolised by the work of Stuart Mill. Similarly, the innovations of Jevons, Walras and Menger were integrated into Marshall's synthesis.

situation (the $\Sigma_1$ state) is characterised by the paradigm that emerged in relation to Adam Smith's *Wealth of Nations*. Schumpeter emphasised that such a paradigm is essentially conservative. The possibilities of analytical progress were found in two ways. First, Adam Smith had made a very incomplete synthesis of the previous work of the pre-paradigmatic period. Second, his work included inconsistencies that provoked Ricardo and others to engage in radical innovation. However, the innovative work created confusion. Thereby, it set the stage for the synthetic work of Stuart Mill and the emergence of the second classical situation (the $\Sigma_2$ state). In turn, this new paradigm provoked the discovery of non-included contributions as well as the radical innovations of Jevons, Walras and Menger. These innovations were included in the synthesis provided by Marshall's *Principles of Economics*. A new cycle of the scientific engine started during Schumpeter's youth and had not yet produced a classical situation at the time of his death.

While the capitalist engine works on a relatively well-defined economic system, the demarcation of the range of the scientific engine of economics is problematic. The easiest demarcation is that of the borderlines of a community of researchers. Although Schumpeter did not develop the metaphor explicitly, he to some extent seems to have been writing a history of "the republic of the economists" (*History*, 765). Since this metaphor provides a quick way of grasping much of the contents of *History*, we shall ignore that he would hardly have accepted any extended use of this metaphor as well as the fact that the notion of the republic of science has later been used ideologically by Michael Polanyi (1962). We simply start by noting that Schumpeter's criterion of citizenship in the republic of the economists is the mastery of a set of analytical tools that serves "to *understand* economic phenomena" (p. 3). This criterion implies that the republic could not be established before a toolbox was created that clearly distinguished citizens from non-citizens. According to widespread folklore, Adam Smith was the founding father of economics because he provided such a toolbox. However, this folklore is hardly better than the Greek legend of Lycurgus as creating the laws of the state of Sparta. In both cases, we recognise a gradual evolution of the relevant rules of the game. In the case of economics, these rules were the results of the efforts of philosophers and people engaged in discussions about economic policy. However, in contrast to Lycurgus, Adam Smith was a real person with a real function: to provide the " 'codification' " (*History*, 307) of pre-existing analytical results. Although others had also contributed to this codification, the sociology of economics demonstrates that he had a core role.

The further history of the republic of the economists can also be written in terms of the evolution of the analytical toolbox. For ambitious citizens of the republic, this evolution defined three possible functions. The first function is that of upholding tradition. This function was largely taken up by " 'elder statesmen' " who "laid the foundations of their reputations in the preceding period and exerted considerable influence" (*History*, 844). Such "statesmen" were the obvious target of attack of members of younger generations who wanted to obtain reputations and positions of their own. Although they could make such attacks in different ways, the lasting effects and their real role were based on their rejection of parts of the analytical toolbox as well as on their improvement of existing tools and addition of new tools. Thereby, these radical innovators performed the second function, that of increasing the power of the analytical toolbox available to all citizens in their attempts "to *understand* economic phenomena". Nevertheless, the performance of this function also created confusion and controversy. Therefore, a third function was also crucial. This is the synthesising and codifying type of innovation, which had already been present in the efforts of Adam Smith. Similar synthesising innovations were provided by the major works of Mill (1848) and Marshall (1890). Although for instance Marshall's *Principles* was "the result of a creative effort and not a synthetic one" (*History*, 837), its objective role in the history of the republic of the economists was that it provided a synthesis of previous contributions to the analytical toolbox. Thereby, it helped to conclude "an important phase in the history of economics" (p. 837). Nevertheless, this contribution did not immediately overcome the conflicts among leading economists. This means that the important date in the history of economics is not the year in which Marshall's treatise was published, but the period during which it had a coordinating role in the republic of the economists.

Schumpeter entered the republic of the economists in the period that was characterised by the reign of the synthesising efforts of Marshall and other leading economists. Although the system of economic analysis "was established by Jevons, Menger, and Walras in the 1870's and 1880's", it "found its classic form in Marshall's *Principles*" (*History*, 953). Actually, the "breaks with tradition around 1870" was followed by "two decades of struggle and more or less heated discussions" (pp. 753–4) before the analytical system was described in a sufficiently comprehensive form. Even in this form, the system "came to most theorists as something new and unfamiliar" (p. 953). However, resistance was gradually overcome and we recognise "the Classical Situation that emerged roughly around 1900". This situation was characterised by a codification of economic analysis that reigned until World War I (and to some extent beyond 1918). Schumpeter considered the reign of this synthesis "a typical classical situation in our sense, the leading works of which exhibited a large expanse of common ground and suggest a feeling of repose, both of which created, in the superficial observer, an impression of finality—the finality of a Greek temple that spreads its perfect lines against a cloudless sky" (p. 754). Schumpeter was no superficial observer, however. He had already in *Wesen* recognised that the beautiful architecture of the analytical constructs of leading neoclassical economists had been obtained at serious costs. Now he emphasised that "in the last decade or so before the outbreak of the First World War, even the superficial observer should have been able to discern signs of decay, of new breaks in the offing, of revolutions that had not yet issued into another classical situation" (*History*, 754). Nevertheless, the republic of the economists was hardly influenced by radical analytical innovations before the 1920s.

The easiest way of understanding Schumpeter's theory of the history of economics is to interpret it as the generalisation of these events. This generalisation has a certain similarity with theory of the waveform history of the capitalist economy. We have already seen how he considered a Kondratieff wave as being characterised by an innovation-based upswing that is followed by a troublesome period of adaptation before a relatively equilibrated state sets the stage for another Kondratieff. With respect to the science of economics, he described the relatively equilibrated states by "the concept of Classical Situations" (*History*, 51). A period is characterised as a classical situation if "the leading works" demonstrate "a large expanse of common ground and suggest a feeling of repose" (p. 754). We may also "discern signs of decay" as well as signs "of new breaks in the offing, of revolutions that had not yet issued into another classical situation". Such a new situation only emerges after a few "decades of struggle and more or less heated discussions."

Since Schumpeter thought that the history of social life is really an indivisible process, it is tempting to try to combine his theory of the history of economics with his theory of the history of the capitalist economy. Thus, Yuichi Shionoya (1997, 266–7) compared Schumpeter's dating of the neighbourhoods of equilibrium that define the end of Kondratieff waves with his dating of classical situations. This comparison is suggested by the fact that Schumpeter found the same number of relatively equilibrated situations in his two studies. However, Schumpeter did not propose that there exists a close coupling of economic and scientific evolution. Furthermore, he emphasised that "[p]eriodizing, as we know, is a necessary evil" (*History*, 379). It is an "evil" for four reasons. First, "historical developments are always continuous and they can never be cut into pieces without arbitrariness and loss." Second, his "particular way of periodizing" was "dictated" by his concentration "on the history of economic analysis" and did not reflect the broader issues of the history of economic thought. Third, "there are well-founded objections to a method that puts A. Smith near the end of the preceding period instead of at the beginning of the one that might be said to have been dominated by his influence." Fourth, his treatment of a particular period did not cover "all the authors who belong in it chronologically". With respect to the post-Smithian period that ended with Mill, the most important omission is Cournot. The reason is that Cournot's contribution to mathematical economics, published in 1838, did not really influence economic analysis until after 1870. While Schumpeter did not ignore these four objections, he considered the definition of periods a "necessary evil" for *all* historical accounts. Furthermore, he held "that our periodization brings out essential truth. This will be for the reader to judge" (*History*, 379).

The periodisation chosen by Schumpeter decomposes the history of economics into four parts by means of three classical situations. As already mentioned, this periodisation is reflected in the basic architecture of *History* (see Table 11.1 on page 334):

- "Part II. From the Beginnings to the First Classical Situation (About 1790)": The 328 pages of this part concern the emergence of economics. Here Adam Smith's *Wealth of Nations* from 1776 is central only because it represented the "codification" (*History*, 307) of preexisting analytical results by policy-oriented writers and abstract theorists. The classical situation emerged when this codification had been accepted by leading economists; and this is the reason why it is dated to 1790 and not to 1776. Schumpeter's task was not only to describe the codification, but also to trace the elements of the toolbox backwards in time. One line of analytical results related to attempts to present policy recommendations clearly. This line of analysis was coupled with the emergence of modern states in a capitalist environment. The other major line was of a more abstract nature. Its roots could be traced, via the early philosophers of Natural Law and medieval scholasticism, backwards to Aristotle's rudimentary economic analysis.

- "Part III. From 1790 to 1870": The 374 pages of this part are centred around on second classical situation that can be represented by John Stuart Mill's *Principles of Political Economy*. Although this book was published in 1848, its widespread acceptance was present during the 1860s. The core question of Part III is how the analytical toolbox of this second classical situation emerged through a series of innovative contributions. The first of contributions were provided by Thomas Malthus and David Ricardo, both of whom took the stagnating state of the first classical situation during the 1790s as the starting point for their contributions. Mill's synthesis of Post-Smithian classical economics was much more encompassing. Its acceptance implied "a state in which 'those who knew' were substantially in agreement" (*History*, 380). They thought that, since "'the great things having been done'", what remained was "only elaboration and application".

- "Part IV. From 1870 to 1914 (and Later)": The 386 pages of this part point towards the third classical situation that emerged around the beginning of the twentieth century. Schumpeter mainly characterised this classical situation by means of Alfred Marshall's *Principles of Economics* from 1890, but he added Knut Wicksell's works for aspects of the situation that are not well covered by Marshall. The period started with the "stagnation" and "decay" of post-Smithian classical economics (*History*, 380). On this background, the "breaks with tradition around 1870 were meant to be breaks by the men whose names are connected with them" (p. 753). However, "[u]pon these 'revolutions' followed two decades of struggle and more or less heated discussions". Then Marshall provided his synthesis, but another decade followed before the third classical situation could start its reign that lasted to about 1914. This reign included not only neoclassical economics but also some of the results of the historical school.

- "Part V. Conclusion: A Sketch of Modern Developments": This unfinished part covers, in 46 pages, the emergence, after World War I, of economic dynamics, macroeconomics, and other elements of modern economics. Since Schumpeter could not discern a fourth classical situation, he was writing during a period of intellectual disequilibrium and could not present his materials in the same way as he had done in the three previous parts of his book. If we want to make a rewrite of Part V, we might use Paul Samuelson's major works as a rough approximation for the fourth classical situation. However, the emergence of this situation is not identified by the publication dates of Samuelson's (1947; 1948) *Foundations of Economic Analysis* and his famous textbook called *Economics*. The new classical situation rather emerged in the late 1950s and the early 1960s. The two books, however, can be used as a starting point for defining how the previous years of high theory and high econometrics influenced the new synthesis.

Although the above account of Schumpeter's periodisation of the history of economics covers much of the contents of *History*, it should be noted that he included much background information on social and political issues as well as on developments in other social sciences and in philosophy. This material was probably meant to emphasise that life in "the republic of the economists" had been influenced by the fact that its citizens were also part of other social structures. Marshall is an obvious example. Although Marshall emphasised that "economic theory is not eternal truth but 'machinery of universal application in the discovery of a certain class of truths'" (*History*, 954n), his search for these truths was motivated by his broad interests in social reform and social philosophy. Since the subject matter of *History* is the evolution of the analytical toolbox, these motivations were of no direct relevance for Schumpeter. However, since "this book is intended to be an introduction to economic analysis from the genetic standpoint" (*S1948b*, 94), information of relevance for the motivation of contributors to the analytical toolbox was hardly irrelevant.

## 11.3 Why do we study the history of economics?

The proposed interpretation of the overall structure of *History* as reflecting Schumpeter's application of his general evolutionary theory can be used as a starting point for a search for the presence of other applications of that theory. Thereby, *History* might be related to the literature that has emerged in relation to the contributions of Thomas Kuhn (1970) and Imre Lakatos (1970). However, Schumpeter did not write his book for philosophers and historians of science. Instead, he addressed economic researchers in their role as scientists. Such scientists do not normally consider the history of science an integrated part of their science. On the contrary, they ignore this history and thereby leave it to philosophers and historians of science—like Kuhn. Nevertheless, some evolutionary scientists have demonstrated an unusual interest in the history of their field. R. A. Fisher, the great statistician and evolutionary biologist, provided an important example in relation to the 1959 centenary of Darwin's *Origin of Species*. At that occasion, Fisher (1999, 318) made the following provocative statement:

> "More attention to the History of Science is needed, as much as by scientists as by historians, ... and this should mean a deliberate attempt to understand the thoughts of the great masters of the past,

to see in what circumstances or intellectual *milieu* their ideas were formed, where they took the wrong turning or stopped short of the right track."

Fisher's statement was probably based on experiences related to his development of much of the foundation for modern evolutionary analysis. Actually, the development of his path-breaking mathematics and statistics of evolution in the 1920s was partly based on a solid study of Darwin's theory of natural selection, a study that took place in some collaboration with one of Darwin's sons. This contribution combined selectionism with mutationism, and thereby it had a large role in overcoming the crisis of Darwinism by transforming it into neo-Darwinism. Other researchers gave this evolutionary synthesis concrete content, but the formalisms of Fisher and a few other theoreticians were crucial (Depew and Weber, 1995, Part II).

Schumpeter's studies of the history of economics could not be as focussed as those R. A. Fisher. Schumpeter's most obvious problem was that he considered himself the founder of a serious treatment of economic evolution. However, his main problem was that he felt responsible for the double program of strengthening equilibrium economics and establishing evolutionary economics. Furthermore, the emergence and evolution of equilibrium economics provides the bulk of the history of economics. Therefore, he could not argue narrowly for studying the history of evolutionary economics. Instead, the first version of Schumpeter's introduction to *History* started with the general question "Why do we study the history of economics?" (*S1948b*, 93). The question was urgent because a crowding out of the history of economics from the curriculum of graduate students at Harvard University was clearly visible in the late 1940s. It is still important since the crowding out has continued at Harvard and elsewhere to the present day.

Schumpeter changed the question on the relevance of the history of economics to the even more general question: "Well, why do we study the history of *any* science?" (*S1948b*, 93). The standard viewpoint among scientists is that there is no reason for such a study since "[c]urrent work, so one should think, will preserve whatever is still useful of the work of preceding generations". A later discussion of the attitude is found in Mark Blaug's (2001, 145) paper with the title: "No History of Ideas, Please, We're Economists"; and George Stigler (1969) has developed a strong argument for ignoring the history of economics. According to Schumpeter, such an argument implies that "the old stuff" can "safely be left to the care of a few specialists who love it for its own sake" (*S1948b*, 93). He admitted, characteristically, that "there is more to be said for discarding old wood resolutely than there is for sticking to it." He, nevertheless, thought that "a strong case can be made on behalf of research as well as of teaching for the study of the work of preceding generations in every field of scientific endeavor" (p. 93). This case includes the recognition that the "kind of problems we set ourselves are never the only possible ones" and that the "way in which we go about solving them is never the best possible one in any absolute sense" (p. 94). Furthermore, "our own work is so largely conditioned by the work of those who went before us that we cannot fully understand what we are about without understanding this background as well." This background is important irrespective of "[w]hether we work with the tools that are handed to us or discard them in favor of others which we forge ourselves, whether we stress continuity or innovation" (*S1948b*, 93).

Schumpeter's considerations provide four major arguments for the study of the history of economics:

1. The pedagogical argument: "Every student and every teacher who cares to do so is bound to discover what a powerful ally the history of a problem is" (*S1948b*, 94). This statement holds "even in the purest of pure mathematics where, more than everywhere else, the newest treatise out seems to be all that we need." The difficulty of using systematic accounts is that "propositions are not as a rule found in the way in which they are expounded." Thereby, a degree of refinement is obtained "which obliterates all the lessons of the preceding struggle and of the scientific motivation." This approach is not warranted in a book that "is intended to be an introduction to economic analysis from the genetic standpoint."

2. The discovery argument: "The history of science is also a powerful source of inspiration" (*S1948b*, 94). Although different researchers "experience its stimulating and refreshing influence to a very unequal degree", it is "probably few ... which it will not benefit at all." For instance, the "fundamental ideas that eventually issued in the theory of relativity [by Einstein] first emerged in a book on the history of mechanics [Mach (1919)]."

3. The argument of the historical difficulties of economics: These difficulties are of two types. First, "the subject matter of economics is itself a unique historical process" and this fact lends "increased interest to doctrinal history" (*History*, 6). Second, although "scientific economics does not lack historical continuity", the "filiation of ideas has met with more inhibitions in our field than it has in almost all others." For example, "results have been lost on the way or remained in abeyance [temporary inactivity] for centuries. We shall meet with instances that are little short of appalling."

4. The argument of learning "logic in the concrete": The general "logic in the concrete, logic in action, logic wedded to vision and to purpose" can be studied in any "field of human action" (*S1948b*, 95).

> However, it is only in science that "people take so much trouble to report their mental processes". Actually, "scientific performance itself is in its nature more or less revelatory." Therefore, "a sort of pragmatic or descriptive logic may be abstracted from observation and formulation of scientific procedures—which of course *involve*, or merge into, the study of the history of sciences" (*History*, 5). This "general methodology of scientific procedure" (p. 776) is the subject matter of "the general science of science" (p. 5).

Schumpeter concluded that "task of the historian of economics then seems to be to describe the process of scientific advance in the field of economics, to write is it were, another story of intellectual conquest" (*S1948b*, 95). This story is relevant for four reasons. First, the story serves as a pedagogical introduction to the science of economics. Second, any story about a science might induce scientific discoveries. Third, the story of economics is especially likely to induce discoveries. Fourth, this story helps to achieve a realistic understanding of the genesis of scientific contributions. This sounds like a general argument for the importance of any kind of study of the history of economics.

Historians of economics should be warned that Schumpeter's argument for their area of specialisation is not as general as it might appear. He designed the argument immediately after the years of high theory and high econometrics. At that time he, as we have already seen, suggested that the field of economics looked "like a big building plot all covered with ruins and half-completed new structures" (*S1948b*, 95). When facing the ruins and unfinished structures, there was an obvious need of surveying the history of economics. A similar need was felt when he, before World War I, had written *Doctrine*. However, the recognition of such needs tend to disappear during "classical situations". During such periods, the interest in the history of economics is mainly represented by those with excluded visions and ideas. Although it might be possible to ignore many of these visions and ideas as representing the fringe of economics, Schumpeter had tried to demonstrate that his evolutionary economics could not be presented as being at the fringe of equilibrium economics. He had, in *Wesen*, argued that his evolutionary economics is a crucial and relatively independent complement to equilibrium economics. On this background, we recognise that *History* does not only provide the story of how the difficulties of equilibrium economics and the related econometrics might be overcome. It also provides a much shorter story on the difficulties of establishing the fundamental branch of evolutionary economics.

## 11.4 Economics as a tool-based science and its evolution

If we accept the importance of the history of economics, the next question is how to analyse this history. Schumpeter implicitly answered the how question by pointing at his general evolutionary theory and its implementation for the analysis of scientific evolution. However, this answer presupposes that economics is a science and that citizenship of "the republic of the economists" is determined by the command of an evolving set of analytical techniques. With these assumptions, it is not difficult to understand the crucial importance of the improvement of the analytical apparatus available for the work of each research worker. Schumpeter used "the analytical engine" as a metaphor for these analytical techniques (see Section 6.3). Therefore, we may say that the community of scientific economists is not least engaged developing their analytical engine. Nevertheless, the long-term "development" of this engine is not the result of an overall plan for the science of economics. It is, instead, the unplanned outcome of the competitive struggle among the scientific economists. Since this struggle involves both innovative and adaptive behaviour, it produces an evolutionary process that metaphorically can be described as the outcome of the scientific engine. In other words, the scientific engine produces improvements of the analytical engine available for individual economic researchers. The Schumpeterian history of economics is basically an account for this evolutionary process.

The logical structure of Schumpeter's evolutionary argument implies that he had to start with the questions whether economics is a science and whether this science is based on analytical techniques. He answered both these questions in the introductory section of *History* (6–11) that he called "But Is Economics a Science?" (see also *S1948b*, 96–9). This question is difficult because the economic profession covers many non-scientific activities. Since he did not want to cover these activities, he could make little use of most of the available definitions of economics. Furthermore, these definitions were normally static while he wanted to cover scientific economics as an evolving entity. Actually, Schumpeter found all fixed definitions of economics of very limited interest for his purposes. Although "they are indicative of an author's range of interests and ideas", he rejected to "bother with them much or try to frame one for ourselves" (*S1948b*, 111). The reason seems to be that economics is an evolving entity with potentially defining characteristics that change over time. For instance, "[i]t is impossible to tell not only where economics ends and sociology begins but also what the frontiers are of that intermediate field of economic sociology that has come into existence *via facti*." The evolution of the borderlines between the different sciences and scientific fields emerge *not* by means of design and planning but by means of practical acts. Actually, we see a "splitting older units into a ever-increasing number of specialties in a haphazard way", and this specialisation has "produced the splendid disorder of tropical growth". The consequence is that definitions are "rough approximations made for practical convenience and without particular logical dignity." In other words, "definitions of individual sciences are much the same sort of thing as the definitions that are to define individual industries for the purposes of census" (*S1948b*, 111).

Schumpeter wanted to make a demarcation between scientific economics and non-scientific economics. For this purpose, it was important to distinguish between, on the one hand, "economic thought" and, on the other hand, "economic analysis" or "scientific economics" or simply "economics". He used "the term Economic Thought in order to designate all the views on, and all the knowledge about, economic life" (*S1948b*, 108). Here he included "prevailing judgments on what is desirable or undesirable" as well as "systems of economic policies" in his definition of economic thought. This kind of thought "is revealed by historical and contemporaneous documents or observation." It can be divided into "common-sense knowledge" and "more general ideas about the economic world" (p. 109). Since these kinds economic thought "is essentially extra-scientific", we recognise that "it belongs in the field of economic history and not in the field of history of economics." In contrast, Schumpeter by "economics" designated "that part of economic thought which is the product of scientific method within our meaning of the term." Since the "scientific method" is evolving, its meaning has to be treated cautiously.

When making his demarcation of scientific economics, Schumpeter focussed on "economic theory" considered, with Joan Robinson's expression, as "a box of tools" (*History*, 15). It is solely the evolution of this toolbox of analytical techniques that can be characterised by the dangerous term of "progress". We could, of course, try to work without the help of the available toolbox in our attempt to understand individual economic phenomena. However, the tools represent "a tremendous economy of mental effort" (p. 16; emphasis removed). Furthermore, experience teaches us that "a new apparatus poses and solves problems for which the older authors could hardly have found answers even if they had been aware of them" (p. 39). The increased efficiency in posing and solving problems "defines in ... a perfectly unambiguous manner, in what sense there has been 'scientific progress' between Mill and Samuelson." This judgement is based on "a widely accepted standard, confined, of course, to a group of professionals" (p. 40). However, "while it is thus possible to speak of analytic progress and impossible to deny the facts that this word is to denote, there is nothing corresponding to this in the field of economic thought or even in any of the historical array of systems of political economy." Actually, preference for one or another system of economic policy "comes within the same category of subjective evaluation as does, to plagiarize Sombart, a man's preference for blondes over brunettes" (*History,* 40).

Schumpeter used the idea of the analytical toolbox to formulate three variants of his definition of "science" (*S1948b,* 96; *History,* 7):

(1) "a science is any kind of knowledge that has been the object of conscious efforts to improve it";
(2) "a science is any field of knowledge that has developed specialized techniques of fact finding and of interpretation or inference (analysis)";
(3) "a science is any field of knowledge in which there are people, so-called research workers or scientists or scholars, who engage in the task of improving upon the existing stock of facts and methods and who, in the process of doing so, acquire a command of both that differentiates them from the 'layman' and eventually also from the mere 'practitioner'."

These three definitions of science are closely interconnected. Let us start with the second definition of science as being knowledge provided by "specialized techniques". These techniques are the "habits of mind" (*History,* 7) that emerge because of "conscious efforts to improve" upon knowledge, that is, they are the result of what is covered by the first definition. Furthermore, it is the possession of special techniques that brings forth the sociological differentiation between a community of research workers and other people; and this differentiation is emphasised by the third definition. Therefore, Schumpeter's basic definition is that "science is tooled knowledge" (p. 7), and his core concern was that of the "increasing efficiency of the scientific engine" (*S1948b,* 99). This "increasing efficiency" is the result of "the emergence of pure sciences". These sciences produce "arsenals of concepts and propositions that do not apply directly to the problems with which practice struggles in their corresponding fields." Their goal is, instead, a never-ending improvement of analytical toolboxes. Schumpeter's corresponding definition of science might be seen as reflecting the methodology of instrumentalism that he had already supported in *Wesen* (see Shionoya, 1997, Ch. 5). This instrumentalism can to some extent be related to a widespread but implicit modern methodology and this methodology was made explicit by Paul Samuelson and Milton Friedman. However, if Schumpeter was an instrumentalist, he subscribed to evolutionary instrumentalism.

Schumpeter did not develop an explicit model of the evolution of the analytical toolbox that is directly comparable with his Mark I and Mark II models of the capitalist engine. It is, nevertheless, not difficult to discern the main features of his model of scientific evolution as soon as we realise that they can be related to his unusual concept of "filiation" (*History,* 6, 32, 56, 245, 778, 1080, 1109, 1125, 1135, 1150). In most cases, Schumpeter seems to have used the term 'filiation' as a synonym of 'affiliation' in the sense of "fixing of the paternity of a child" or "the assignment of anything to its origin" (Oxford University Press, 1989). However, he was obviously not thinking of the unmodified descent of scientific ideas, but rather of what Darwin called descent with modification. Schumpeter did not mention the affiliation of his concept of filiation; but he might have found it in Auguste Comte's *Positive Philosophy* or in Herbert Spencer's (1908) note on "The Filiation of Ideas". Among Schumpeter's examples of filiation, his comments on Darwin are of special interest. Without mentioning the term, he presented the "Historical Sketch" added by Darwin to the later editions of the *Origin of Species* as "a case study about one of the objects of our interest—the ways of the human mind and the mechanisms of scientific advance" (*History*, 444n). He even considered this document "one of the most important pieces of scientific history ever written" as well as an example of the correct "Acknowledgement of Priorities". Normally, scientists make much less complete accounts for the affiliation of their concepts and propositions. Instead, the historian of science has to try to detect major patterns of affiliation in the evolutionary sense.

Schumpeter thought that the process of modified scientific descent as well as the process of scientific selection improves the power of the analytical toolbox. Thereby, this evolutionary process produces some sort of scientific progression. He denoted the broader concept of filiation by "the process of the Filiation of Scientific Ideas" (*History*, 6). Schumpeter emphasised that in *History* it was his "main purpose to describe what may be called the process of the Filiation of Scientific Ideas—the process by which men's efforts to understand economic phenomena produce, improve, and pull down analytic structures in an unending sequence." It was partly in these terms that he studied the evolution of the analytical tools that are at the disposition for each successive generation of scientists. The individual tools are not only improved cumulatively. The evolutionary process also serves to add new types of tools and to "pull down" other types of tools. Although this evolutionary process of innovation and selection is not normally supported by those whose tools are driven to extinction, it nevertheless tends to improve the toolbox available for the next generation of scientific economists.

The demarcation of science in general and economics in particular by means of an improving toolbox of analytical techniques is not an easy one. Theologians also represent a kind of "tooled knowledge". For instance, for "a theologian of the twelfth century revelations and miracles were facts—facts of experience—and appeal to scriptural authority was a perfectly valid method of establishing a proposition" (*S1948b*, 97). However, "one of the most important factors in the advance of scientific knowledge was and is an increasingly critical attitude as regards what should be accepted as a fact and as a proof." As we have seen in Section 7.5,

Schumpeter largely ascribed this critical attitude to the socio-cultural evolution within capitalist societies. When writing his history of economics, he re-emphasised "the particular kind of rationality the capitalist process eventually impressed upon the human mind although science was ... at first slow to announce allegiance to theology." It was the "spirit and technique of mundane empiricism" that came to define the realm of science (p. 97). Nevertheless, the historian of science has to recognise that "our standards" have emerged gradually and that we cannot apply them when studying the whole evolution from magic to science (*History*, 7–8). The sciences "emerge by slow accretion, gradually differentiating themselves from their common-sense backgrounds under the influence of favorable and inhibiting environmental and personal conditions" (*S1948b*, 97). During this process of emergence, it is possible to find time ranges "within which it is about equally justifiable to aver or to deny the existence of a body of scientific knowledge." This means that a science like economics has no founding date. Actually, "bias or ignorance are the only reasons that can be adduced for such statements that A. Smith or F. Quesnay or Sir W. Petty or anyone else 'founded' that science or that the historian should begin his report with any of them" (p. 97). Even for established sciences, it is clear that all definitions of science are merely "rough approximations" (p. 111).

## 11.5 The brakes of the scientific engine

Schumpeter's account of the history of economics after the first classical situation around 1790 can be interpreted as being closely related to the widespread view among modern economists. According to this view, we have gradually obtained a relatively independent process of scientific improvement beyond the quagmire of non-scientific economic thought. We economists are, with some right, proud of the similarities between present-day economic science and the natural sciences with respect to level of formality and degree of cumulativeness. To emphasise this relationship, we might even compare ourselves with Newton's metaphorical dwarf who, by standing on the shoulders of a giant, may see farther than the giant himself (Merton, 1965). This use of "the institutional norm of humility" (Merton, 1957, 646) implies that we can at the same time acknowledge the greatness of economists of the past and ignore their work. The reason is that we base our research on the most recent contributions to an impressive cumulative tradition that has already included the contributions of previous generations of economists. Nevertheless, Schumpeter produced nearly 1200 dense pages on the history of economics and emphasised that "the process of the Filiation of Scientific Ideas" had not worked smoothly within economics. On the contrary, "this filiation has met with more inhibitions in our field [economics] than it has in almost all others" (*History*, 6).

Schumpeter's remark suggests that he was not engaged in writing a story of unconditional scientific advance. Although he was interested in understanding the unhindered functioning of the scientific engine, he was just as interested in understanding the brakes of this engine. We have already encountered a couple disturbing facts. First, the demarcation of economic analysis is not a fixed and safe one. Second, the criteria by which individual tools are included in or excluded from the analytical toolbox are themselves changing. These facts are among those on which Schumpeter based his warning against a simplistic view of scientific progress within economics. He strengthened this warning by emphasising that economics is "a field which deals with a unique process in historic time" (*S1948b*, 100). The historical nature of the subject matter of economics suggests that we should use with caution "the misleading term Laws" since "our relations are less stable and that they work out in very different ways in different historical environments". Furthermore, we should learn "to specify the institutional framework to which our analytical apparatus is intended to apply and to stress the corresponding limitations of our results." Finally, we should recognise that "economic history ... is the great storehouse of both the economist's facts and the economist's problems" (p. 101).

The necessary cautions and historical specifications do not emerge automatically. On the contrary, even scientific economists often fail to recognise the limits of their "tooled knowledge". Instead, they tend to exaggerate their knowledge to make it relevant for economic policy, economic ideology, the other social sciences, and philosophy. They also tend to form scientific schools; and the monopolistic competition between these schools tends to increase the number of transgressions of the limits of the science of economics. Since such transgressions can hardly be avoided, Schumpeter suggested that the task is to delimit the degree to which they inhibit the evolution of the toolbox of economic analysis. Although these inhibitions are much stronger in economics than in the natural sciences, lessons can be learned from the history of natural science. Take, for instance, the problem of ideological influences. Although some degree of ideological influence can hardly be avoided, it does not necessarily hinder the evolution of the analytical toolbox: "Statements that proceed from an ideological bias may be suspect, but they may still be perfectly valid" (*S1948b*, 103). Although both "Galileo and his opponents may have been swayed by ideologies", the ideology of Galileo "does not prevent us from saying that he was 'right'." The difficult tasks are "to establish both the validity of a proposition and its scientific filiation no matter what its ideological associations are".

It is not only ideology that often constrains the functioning of the scientific engine. Similar constraints, or brakes, on the evolution of scientific economics emerge from the influences of economic policy, philosophy, and scientific schools. Since we shall consider most of these brakes in the rest of the present section, it is convenient to present them in list form.

According to Schumpeter, the major brakes of the scientific engine are:

(1) the influence of economic policies
(2) the influence of ideologies
(3) the influence of philosophies and of other sciences
(4) the influence of sociological phenomena like scientific schools

Let us start with the issue of economic policy. By emphasising the relevance of their theories for economic policy, scientific economists tend to engage in value judgements. Although we often protect our policy advices against the suspicion that they presuppose particular value judgements by stating them in conditional form, our if–then statements on economic policy nevertheless tend to influence the theoretical development of economics. Some economists consider this influence positively. However, Schumpeter considered the policy orientation a major brake on the scientific engine. His reason is not least that concepts, propositions, and other analytical tools tend to become tagged by the derived policies. This means that many economists become sceptical about the possibility of using the tools of other economists. Although Schumpeter did not study this obstacle systematically, he might have wished to include it in *History*'s introductory chapter on "The Sociology of Economics". He actually had planned in this chapter to include a section with the title "The Motive Forces of Scientific Endeavor and the Mechanisms of Scientific Development" (*History*, 44). This section was never written; but it is not difficult to find out what he considered "essential for a proper understanding of the evolution of our science" (p. 369). For instance, if we study Dudley North's contribution from the late seventeenth century, it becomes obvious that his work was motivated by policy considerations. His scientific contribution, nevertheless, cannot be described in such terms. Actually, "we can discard his free-trade convictions as irrelevant in an appraisal *of his analytical apparatus*" (p. 370). Thereby, we recognise "that it is more correct to say that his analytical work grew out of that of the 'mercantilist' than to say that the relation was one of head-on clash." However, this aspect of his contribution was not recognised by North's contemporaries. Actually, many of them abstained from using the analytical tools of the opposing school of economic policy. Since similar inhibitions have been present after the emergence of scientific economics, the emphasis on policy relevance represents an important brake on the scientific engine.

Schumpeter's views on the negative role of the emphasis on economic policy for the evolution of the science of economics had emerged at an early state of his life. They are thus clearly present in *Wesen*. Nevertheless, they became especially conspicuous in the aftermath of famous, and notorious, Vienna meeting of the Verein für Sozialpolitik, that is, the association for German-speaking economists—that was dominated by the historical school and its leader, Gustav von Schmoller. Although it is unclear whether Schumpeter attended the 1909 general assembly of the German economic association, it was a major event that influenced his thought. The meeting was not only opened by Schmoller but also by the Austrian Prime Minister, the Minister of Trade, and the Mayor of Vienna. Nevertheless, it ended in scandal. Schumpeter's teacher Philippovich had convinced Schmoller that the meeting should discuss the relevance of economic theory. While Philippovich tried to clarify the concept of the productivity of the economic system, Wieser covered the value of money. Thereby they tried to bridge between economic theory and economic policy. The younger members of the historical school argued strongly against this mixture—just as Schumpeter had done in *Wesen*. Sombart argued that the use of theory for economic policy was based on value judgements that were of the same type as found in discussions "whether the blondes or the brunettes are more pretty" (Verein für Sozialpolitik, 1910, 572). According to him, the discussion served to reveal that economics was in a state of conspicuous immaturity. Max Weber agreed, and he added that the scientific "fall of mankind [Sündenfall]" began to occur when empirical statements and theoretical deductions mixed with practically oriented value judgements (Verein für Sozialpolitik, 1910, 583). Such statements started a controversy—the "battle of value judgements" (Nau, 1996)—that had some similarity with the battle of methods. In this battle, Schumpeter sided with Sombart and Weber. The final round of the battle of value judgements took the form of an internal debate in written form within the German economic association. Many German-speaking economists emphasised that they had been heavily engaged in policy-making and had based their scientific work on this fact. However, in their internally distributed papers, Weber (in Nau, 1996, 145–86) and Schumpeter (*S1913b*; see also *Wesen*, 574–80) argued strongly for the strict separation of science and politics.

Schumpeter's argument against the mixing of science and politics primarily served to protect the science of economics. This point is emphasised in his programmatic paper for the Econometric Society: "No science thrives ... in the atmosphere of direct political aim, and even practical results are but the by-products of disinterested work at the problem for the problem's sake" (*S1933c*, 101). However, he also wanted to protect politics against advice directly deduced from abstract theorising. Economic models are much too crude to provide solid foundations for economic policy, and, in addition, any policy needs to be based on controversial value judgements. Therefore, he thought that economic science and economic policy must be kept strictly separate. Those who specialise in economic policy-making should only use economic theory as one of many relevant inputs—while economic theorists should ignore any immediate policy implications of their work. Those theorists who, nevertheless, propose policies based on their crude models are suffering from something that comes close to medical quackery—the "Ricardian Vice" (*History*, 473). Schumpeter named this vice after David Ricardo's problematic way of using his radically incomplete theory for making policy advice in the turbulent aftermath of the Napoleonic Wars. Ricardo shared the vice that got his name with many other economists. For instance, the French economist Jean-Baptiste Say "was an addict to the Ricardian Vice" (p. 618). Say was, "[l]ike many other economists at all times, ... much more anxious to exploit it [his theory] for practical purposes than to formulate it with care." However, Keynes served as Schumpeter's primary contemporary example of addiction to the Ricardian Vice (*History*, 472–3, 1171).

Schumpeter's criticism of the mixing of science and politics has been emphasised above because he seems to have considered this mixing the primary brake on the evolution of scientific economics. However, he also suggested and criticised other brakes. The brake most closely related to economic policy is ideology. His criticism of the ideological influence on science does not suggest that this influence can be totally removed. Furthermore, the emergence of "pure sciences" does not exclude ideological motivations of the work of scientists. This becomes clear from the field that might be called "the psychology of scientific research" (*S1948b*, 98). If this field is combined with the sociology of science, we become able to "analyse the scientific process itself in order to see where ideological elements may enter it and what are our means of recognizing and perhaps eliminating them" (*History*, 41). Schumpeter performed this analysis by means of a thought experiment. Although "we all start our own research from the work of our predecessors", the experiment is to "suppose we did start from scratch". Then we would "first have to visualize a distinct set of coherent phenomena as a worthwhile object of our analytic effort". It is this "preanalytic cognitive act that supplies the raw material for the analytic effort". Schumpeter called this act the provision of a "Vision", and he emphasised that vision is a crucial part of the scientific process. He noted that it "must precede historically the emergence of analytic effort in any field" and that it "also may reenter the history of every established science each time somebody teaches us to *see* things in a light" that cannot be derived from "the facts, methods, and results of the pre-existing state of the science" (*History*, 41).

We already know the importance of a pre-analytic vision for Schumpeter's own work; but *History* primarily uses the example of Keynes. It is argued that the "*General Theory* presented an analytical apparatus" that can be traced back to "a conception or vision in our sense" (*History*, 42). We read about this "vision, as yet analytically unarmed, in a few brilliant pages of Keynes's *The Economic Consequences of the Peace* (1919)." On this background, "the whole period between 1919 and 1936 was then spent in attempts, first unsuccessful, then increasingly successful, at implementing the particular vision of the economic process of our time that was fixed in Keynes's mind by 1919 at latest." The general point is that innovative scientific work moves from vision to the implementation of the vision by means of analytical tools that are combined into scientific models. However, Schumpeter became increasingly aware that "vision is ideological almost by definition" (p. 42). Such a vision is not only pre-analytic but also ideological in the sense of reflecting "preconceptions about the economic process" that are "beyond our control in a sense in which value judgements and special pleadings are not" (*S1949e*, 274). This use of the concept of ideology implies an "amended version of the Marxist definition" (*History*, 36). In this version, it is recognised that personal ideologies are not "exclusively shaped by well-defined class or group *interest*" and that "ideologies are not lies." In any case, the personal ideology may give a vision of important economic phenomena that is not served by existing economic theories and tools. Although the results of the modelling process are scrutinised by "increasingly more rigorous standards of consistency and adequacy", "ideology enters into this process … on the very ground floor" (p. 42). Such a personal vision is not unproblematic because it may serve as an entry point for social or personal "ideologies" and "delusions" (*S1949e*, 278), the vision also gives meaning and motivation to the creation of scientific novelty: "No new departure in any science is possible without it. Through it we acquire new material for our scientific endeavours and something to formulate, to defend, and to attack." In other words, the work based on pre-scientific vision is subjected to scientific debate and competition whereby "the stock of facts and tools grows and rejuvenates itself". Hence, "though we proceed slowly because of our ideologies, we might not proceed at all without them" (p. 286).

Schumpeter's criticism of the influence of ideology did not aim at the impossible task of removing this influence on individual scientific contributions. Instead, he tried to change the focus of historians of economics from visions to analytical tools. By focussing on analytical tools, we recognise that the common techniques of economic analysis can be improved through the competition between economists with different visions and ideological biases. For instance, Pareto's elitist world view was radically different from Walras's more egalitarian viewpoint. Nevertheless, Pareto improved the analytical tools provided by Walras. The same approach can be used to handle the influence of conflicting philosophies. Schumpeter asked "how far economic analysis has experienced influences from philosophy"; and he answered provocatively that "economic analysis has not been shaped at any time by the philosophical opinions that economists happened to have" (*History*, 28, 31). His reason is "that the garb of philosophy is removable" without influencing the analytical contents of the contributions of individual economists. By ignoring these influences, we overcome "pseudo-explanations of the evolution of economic analysis" that "obliterate the [real] filiation of scientific ideas" (p. 32). Such "pseudo-explanations have a strong appeal for many historians of economics who are primarily interested in philosophical aspects and therefore attach an undue weight to the references to such aspects which in fact abound in the literature". Schumpeter's favourite example was the removal of the "phraseological influence of Hegel upon Marx" (p. 438). If we ignore this influence, it becomes clear that Marx represented "a separate type of evolutionism". Schumpeter emphasised that his arguments for removing philosophy were of an empirical nature. He did *not* argue that "because the philosophical and theological garb is removable from propositions belonging to the physical sciences, it must therefore also be removable from propositions belonging to the social sciences" (p. 31). Neither did he argue that "human action itself and the psychic processes associated with it ... are uninfluenced by, or uncorrelated with, philosophical or religious or ethical convictions." He, finally, did not try to establish "the logical autonomy of the economic proposition or theorem from philosophy". He simply argued that "even those economists who held very definite philosophical views, such as Locke, Hume, Quesnay, and above all Marx, were *as a matter of fact* not influenced by them when doing their work of analysis" (p. 32). This proposition simplified the tracing of scientific filiation within the history of economics; but it sometimes required a reconstruction of the works of the contributors to that history. Further reconstruction was needed because "Filiation of Scientific Ideas is an objective process which may, but need not, involve any subjective relation" (*History*, 1125n).

The discussion of the influences of policy considerations, ideologies, and philosophies on the works of economists has made clear that Schumpeter's tool-based definition of economics as science "implies nothing about the motives that compel man to exert himself in order to improve his knowledge in any field" (*S1948b*, 98). He, nevertheless, emphasised that the predominance of practical and ideological motives had inhibited scientific advance in economics. For him, "the comparative weakness of the purely scientific impulse explains a great deal of what is universally felt to be the unsatisfactory achievement of economics" (p. 99). His sociological explanation of the "weakness of the purely scientific impulse" serves as an additional antidote against the purely cumulative image of scientific evolution. This antidote emerges from the "sociology of science" and the "sociology of economics" (*History*, 45, 33). The sociological mechanisms within economics do not only promote the competitive struggle that steadily improves the analytical toolbox. We also have to recognise that the "[t]eaching in any established science stereotypes the mind of the tyro and may stunt such originality as he may have" (p. 46). Furthermore, "the resistance that an existing scientific structure offers" causes a characteristic pattern of change. The introduction of "major changes in outlook and methods" are "at first retarded, then come about by way of revolution rather than of transformation". This means that "elements of the old structure that might be permanently valuable or at least have not yet had time to yield their full harvest of results are likely to be lost in the process." Consequently, there is "plenty of justification ... for a certain type of mind to emphasize continuity and to defend old insights against new ones." Finally, the fact that "the professionals that devote themselves to scientific work in a particular field ... tend to become a sociological group" has both positive and negative consequences for scientific evolution (*History*, 47). The negative consequences are not least "associated with the phenomenon of scientific schools."

Schumpeter wanted to emphasise "scientific schools" since they "inevitably play a considerable role in our story" (*History*, 47). However, his unfinished manuscript of the chapter on "The Sociology of Economics" breaks off with this statement. Therefore, we find no systematic treatment of the sociological phenomenon of scientific schools. However, we have already met Schumpeter's opinion of such schools in relation to the battle of methods of his youth (see Section 3.3). He later upheld his negative evaluation of schools within the science of economics. Furthermore, he thought that the evolution of economic analysis would, in the long run, decrease the importance of such schools. However, he considered the building of schools an unavoidable phenomenon. This is especially clear in a paper he wrote for a Japanese journal. This paper has the title "The Present State of Economics, or On Systems, Schools and Methods" (*S1931e*; see also Shionoya, 1996). Here he stated that "I am sometimes credited with the saying that there are no schools in economics. By this I mean that there are now no differences as to fundamental standpoints among serious economists." However, this does not imply that schools no exist. "There is *one* meaning to the word 'school', which is incident to the very life of science: groups of disciples gather round some teacher or some institution." Such "disciples" acquire "a sort of mental family likeness" in three ways: "[b]y being interested in similar problems, by being taught similar ways of handling them, by exchanging and assimilating their views and results". Such schools are an important subject for the study of the science of economics. This study "explains to a considerable degree why we have just the sort of science which we do actually have and why it is that no other lines of thought, just as promising in themselves, have been followed". Therefore, we have to study in detail "[h]ow such schools arise and decay, how and why they fight each other and how their success or defeat determines the directions in which scientific endeavour moves". This study is "intimately linked up with the fundamental sociological phenomenon of Leadership" (*S1931e*). Nevertheless, Schumpeter considered such schools a necessary evil that not least served as a brake on the scientific engine. This is especially clear in his description of the Keynesian school as a sociological entity, that is, "a group that professes allegiance to One Master and One Doctrine" (*S1946d*, 288). Such a school "has its inner circle, its propagandists, its watchwords, its esoteric and its popular doctrine."

## 11.6 The fundamental fields of scientific economics

The above description of the scientific engine and its brakes has made clear that Schumpeter did not think that science works within fixed scientific disciplines; and he rejected the ideas that its outcomes are the successive discovery of a given reality or the unfolding of a logical process. First, "[s]cientific analysis is not simply a logically consistent process that starts with some primitive notions and then adds to the stock in a straight-line fashion" (*History*, 4). Second, the scientific process cannot be described as the "progressive discovery of an objective reality—as is, for example, discovery in the basin of the [River] Congo." Third, the history of science teaches its students that "the frontiers of the individual sciences ... are incessantly shifting and that there is no point in trying to define them either by subject or by method" (p. 10, emphasis removed). Instead, individual scientists, and groups of them, compete within the environment created by their predecessors and contemporaries. The pedagogical task of the history of economics is to give "*direction and meaning*" by presenting the present state of the art in terms of "methods, problems and results that are historically conditioned and are meaningful only with reference to the historical background from which they spring" (p. 4). The shifting focus of economic research is one the reasons why "results been lost on the way or remained in abeyance [temporary inactivity] for centuries" (*History*, 6). Here, Schumpeter could have used economic evolution as his primary example. However, he did not do so; and the main explanation is probably that evolutionary economics had not yet been established as a branch of the science of economics. Instead, he exemplified his statement on lost or inactive results with numerous examples from the history of equilibrium economics and from the prehistory of econometrics.

Schumpeter's definition of "science as tooled knowledge" and his association of it "with particular groups of men is almost the same thing as emphasizing the obvious importance of specialization" (*History*, 9). However, his evolutionary interpretation of science meant that "this process of specialization has never gone on according to any rational plan—whether explicitly preconceived or only objectively present" (p. 10).

Therefore, "science as a whole has never attained a logically consistent architecture; it is a tropical forest, not a building erected according to blueprint." This statement also covers the division of labour that has emerged within the science of economics. Schumpeter discussed the structure of this division of labour in *History*'s introductory chapter on "The Techniques of Economic Analysis" (see Table 11.1 on page 334). This chapter includes a short description of the bewildering set of "applied fields" of economics (see also *S1948b*, 111–3). However, its main concern is the "fundamental fields" of economic analysis (*History*, 12–21). He emphasised that "[w]hat distinguishes the 'scientific' economist from all the other people who think, talk, and write about economic topics is a command of techniques that we class under three headings: history, statistics, and 'theory' " (p. 12). Later in the chapter, he added "a fourth fundamental field to complement the three others, ... economic sociology" (p. 21). Since hardly any individual economist but himself mastered all these four fields, it appears surprising that he described them as "fundamental". This term implies that each of the four fields provides part of the basis for economics as a science and that this part is an essential component of the erected structure. It also implies that he, as a historian of economics, did not accept the opinions of most of the economists that he was studying. Instead, he relied on his evolutionary theory of the history of economics. Furthermore, he might have wanted to create room for his own branch of evolutionary economics by his definition of the fundamental fields.

According to Schumpeter, the fundamental fields of scientific economics are:

(1) economic history
(2) economic statistics
(3) economic theory
(4) economic sociology

Schumpeter's wide definition of the science of economics probably had two primary purposes. On the one hand, he wanted to develop the abovementioned thought experiment of an economic researcher that started from scratch. On the other hand, he wanted to promote reconciliation among the basic specialities of economics. Let us start with Schumpeter's thought experiment. We already know that it is not purely hypothetical. He was convinced that major breakthroughs emerged from new ways of seeing economic phenomena. The ultimate implementation of such visions requires the use and renewal of economic theory; but it also suggests the use and renewal of economic history, statistics, and economic sociology. We have already considered his personal strategy of combining these four fundamental fields (see Figure 8.1 on page 198). He had, from the very beginning, been convinced that an ambitious theorist needs

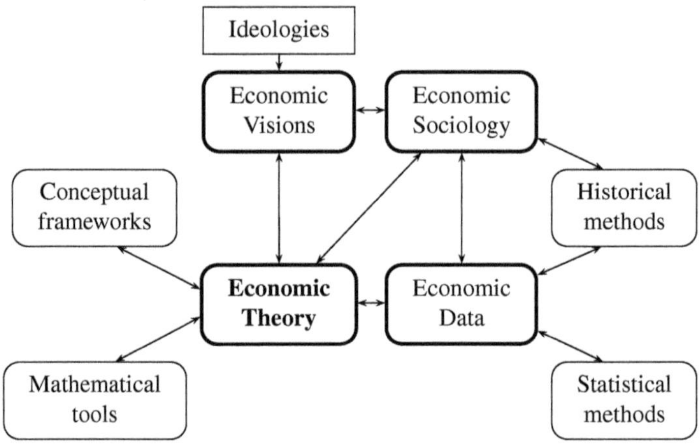


**Figure 11.2.: Schumpeter's scheme of the scientific process**

Comment: The figure covers the competitive scientific endeavours within a community of researchers. The competition means that individual contributions based on fixed ideologies provide analytical tools that are potentially useful for all members of the community. The production of individual contributions are depicted by Figure 5.1 on page 100. Source: The figure summarises propositions found in *History* (Part I), *S1948b*, and *S1949e*.

to build his work on a vision that focus on novel aspects of economic phenomena. These aspects are novel in the sense that they are not adequately covered by existing economic theory. This is the starting point for a personal research programme that begins from the existing toolbox and often leads to innovative efforts with respect to that toolbox. Since we are presently focussing of the economic theorist, the innovative activity primarily concerns concepts and mathematical tools (see Figure 11.2). However, scientific theories have, in some way or another, to be related to data, and these data are not necessarily available. In the attempt to provide them, or to persuade other researchers to do the job, the theorist may become aware of the need of innovating existing methods for providing historical or statistical data. Furthermore, he might need to develop the implications of the vision and the theory for economic sociology.

The discussion presupposes a definition of the concept of theory. We have already seen that Schumpeter suggested that "economic theory is a box of tools" (*History*, 15). It should be noted that he thought in terms of scientific toolboxes that are to some extent highly structured: "Economic theory ... cannot indeed, any more than can theoretical physics, do without simplifying schemata or models that are intended to portray certain aspects of reality and take some things for granted in order to establish others according to certain rules of procedure". To be more specific, "the things (propositions) that we take for granted may be called indiscriminately either hypotheses or axioms or postulates or assumptions or even principles, and the things (propositions) that we think we have established by admissible procedure are called theorems." Although the hypotheses we take for granted are normally suggested by facts, they are, in strict logic, "mere instruments or tools framed for the purpose of *establishing* interesting results" (p. 15). It is in this sense that he thought of economic theory and not in the sense of creating an unstructured collection of concrete explanatory hypotheses that are "essential ingredients of historiography and statistics also" (p. 14).

As long as we emphasise the viewpoint of an individual economic theorist, the ambition to innovate the existing statistical and historical methods is an auxiliary one. However, the priorities are different if we consider the genesis of scientific contributions from the viewpoints of statistical or historical researchers. For instance, both Gustav von Schmoller and Wesley Mitchell wanted to recreate the science of economics by promoting the production of huge amounts of novel data; and these efforts were presumably guided by personal visions. As demonstrated by the battle of methods between the history-oriented Schmoller and the theory-oriented Menger, they both wanted their own field to be the one that defined the science of economics and that determined the conditions for the development of the other field. Thereby, they "not only created a lot of bad feeling but also set running a stream of literature, both of which took decades to subside" (*History*, 814). This "is substantially a history of wasted energies, which could have been put to better use." The alternative was to develop an understanding of the actual existence of mutual interdependence, and Schumpeter's favourite example was the situation within physics with its division of labour between "the experimental and the theoretical physicist" who "work toward a common end" (*S1931c*, 289). Nevertheless, the complexity of the task is "calling for very different mentalities, aptitudes, and likings" (*S1930e*, 76). For instance, the experimentalist dislikes the theorist, but he needs his services. Thus Schumpeter quoted an experimental physicist as saying: " 'If I call in a theorist to hear what he has to say about the results of my experiments, and if the man shows so much experimental aptitude as implied in being able to switch off the light, I begin to doubt his competence as a theorist'." Nevertheless, the experimentalist increases his chance of succeeding in contributing to the evolution of physics by consulting the despised theorist.

Although economics is very different from physics, Schumpeter was convinced that it is a science. The science that serves to promote the understanding of economic phenomena has, unfortunately, few possibilities of developing specialities for experimentalists; but their substitutes are economic historians and statisticians. The services of their fields have, in principle, been needed during the whole evolution of economics from

Aristotle to Keynes. In practice, the slow and relatively independent emergence of the empirically oriented fields inhibited the evolution of theoretical economics as well as economics in the broad sense. Schumpeter, nevertheless, wanted to study the primitive forms of co-evolution between theory, history, and statistics as well as the unexploited possibilities that had often existed. He also wanted to point at the more ambitious forms of collaboration that he, with respect to statistics, had already suggested in his extensive comments on Wesley Mitchell's *Business Cycles*. On the one hand, theoretical "economics has to learn from the statistician's approach and, incidentally, to struggle with its static fetters" (*S1930e*, 86). On the other hand, "statistical methods will have to be remodelled at the suggestion of economic theory and get as much as possible over its own static fetters, which are no less stringent than ours."

On this background, it is not difficult to understand Schumpeter's ideas of three fundamental fields of economics and of the need of promoting reconciliation among these fields. He summarised his ideas by comparing scientific economics with "a big omnibus which contains many passengers of incommensurable interests and abilities" (*History*, 827). These "passengers" make contributions that are "incommensurable", that is, were lacking a common quality on which to make a comparison. Nevertheless, he wanted to appreciate not only Léon Walras's abstract model of general equilibrium as the greatest contribution to pure theory (p. 827). He also characterised the "rather pedestrian" work of Schmoller as representing "a tremendous advance in accuracy of knowledge about the social process" (p. 810). More generally, he thought that the contributions of both neoclassical economics (economic theory) and the historical school (history, statistics, and economic sociology) had been crucial for the evolution of economics. The fundamental fields cannot individually produce the long-term advance of scientific economics. Nevertheless, the practitioner of each fundamental field tends to consider the contributions of all the other fields as auxiliary. This approach would be without problems if all fundamental fields were accepted as auxiliary to each other and if means of coordination were provided. However, this methodological pluralism had not normally been applied. Therefore, it was important for Schumpeter to introduce his *History* with a discussion of the three, or four, *fundamental* fields of economic analysis.

Schumpeter's emphasis on the statistical and historical tools of economic analysis had a polemical twist. His studies of the long-term evolution of economics seem to demonstrate that economic theory is the main driver of this evolution; but he also emphasised that this dominance can be detrimental for scientific advance. This is clearly the reason for the sequential ordering of his "three headings: history, statistics, and 'theory'" (*History*, 12). He put theory after history and statistics in his truncated list of "fundamental fields" because he wanted to emphasise the first item. Although he suggested that a graduate student needed "a tolerably good undergraduate training in history *or* mathematics" to become "an all-round economist" (p. 14n), history was his main concern. Thus he emphatically ended the first version of his introduction to *History* by stating that "I wish to testify to my belief that if for the training of an economist I had to choose one of the three fundamental fields to the exclusion of the two others, my choice would not be theory but economic history" (*S1948b*, 115). His main reason was that this field can reveal "the great contours of past developments" as well as "an adequate amount of details" that are "pregnant with important truth about economic life which, though refractory to cut and dried formulation, may convey us an intimate sense of organic necessities which we can never derive from mere theorems." However, this way of using history is not easy. Schumpeter pointed out that his "American students ... have plenty of historical material pushed down their throats" (*History*, 472n). Nevertheless, "[t]hey lack the historical *sense* that no amount of factual study can give. This is why it is so much easier to make theorists of them than economists."

Schumpeter included "economic sociology" as "a fourth fundamental field" of economic analysis (*History*, 21). Its task is to mediate between theory and history by providing "a sort of generalized or typified or stylized economic history" (*History*, 20). The division between economic sociology and economic analysis, in the narrow sense, is not difficult to define. On the one hand, "economic analysis deals with the questions how people behave at any time and what economic effects they produce by so behaving" (p. 21). On the other hand, "economic sociology deals with the question how they came to behave as they do." The latter question had normally been treated by the theorists themselves—the last major example being the efforts in this field by Alfred Marshall. However, Schumpeter had, in *Wesen*, argued for the exclusion of this type of work from works within theoretical economics. He had instead glimpsed an alternative way of developing the field of economic sociology by studying the efforts of members of the historical school like Schmoller, Sombart and especially Max Weber. Since these efforts had not developed into a tool-based field of economics (or sociology), his possibilities of describing the field in *History* was largely limited to the by-products of the works of economic theorists and to what he considered the unconvincing works of American institutionalists. His thus lacked good examples that motivated the inclusion of economic sociology as a fundamental field. Instead, he remarked that "[t]he proof of any pudding is in the eating and hence I refrain from saying anything in its defence just now" (*History*, 21).

Schumpeter's inclusion of "economic sociology" as one of the fundamental fields of the science of economics was based on afterthought. Ever since he wrote *Wesen*, he had emphasised that economic sociology cannot be considered part of economics in the narrow sense. Instead, it is either an independent science or part of an independent sociology. In both cases, the necessary coordination depends on the long-term co-evolution of relatively independent entities. Throughout the years that he had spent on *History*, he had thought of only three fundamental fields of economic analysis: economic theory, economic history, and economic statistics. Even when he drafted the first version of the introductory part, he omitted economic sociology and apparently had no intention of including it. It was only shortly before he died that he, according to Elizabeth Schumpeter, considered "Economic Sociology as a *possible* fourth" fundamental field (*History*, 1190). This field is, to some extent, a misfit. According to Schumpeter, scientific analysis requires that we take something as given. In relation to his economic analysis, we normally consider basic behaviour and basic institutions given by assumption. However, we would not like to leave the discussion of these assumptions to sociologists. Since Schumpeter was also a sociologist, the problem only surfaced when he had to define the borderlines of economic analysis in general. However, he did not really solve the methodological problems connected with the inclusion of economic sociology. On this background, the safe strategy is to pass over the problem in silence. However, economic analysis rests on shaky ground if it simply makes arbitrary assumptions about basic behaviour and basic institutions.

In Schumpeter's latest account for the science of economics, he emphasised "that our three fundamental fields, economic history, statistics and statistical method, and economic theory, while essentially complementing each other, do not do so perfectly" (*History*, 20). What is missing are especially systematic analyses of "the facts of economic behavior" and "the institutions that characterize the economic organisation of the societies to be studied" (p. 544). Although economic theorists often did not bother to analyse these frameworks, Schumpeter had come to the conclusion that this task is essential. The institutional frameworks of the models introduce "social facts that are not simply economic history but are a sort of generalized or typified or stylized economic history" (p. 20). The same is the case for "the general forms of human behavior which we assume either in general or for certain social situations but not for others." These institutional and behavioural backgrounds for theoretical modelling mean that economics textbooks often include an "institutional introduction that belongs to sociology rather than to economic history as such" (pp. 20–1). Schumpeter had, in *Wesen*, suggested that such introductory topics were better left to the sociologists. However, he, in *History* (783), concluded that "[a]ll social sciences run up against certain fundamental problems of society, and none of them can afford to surrender its claims to some competence in matters of motors and mechanisms of social life". This explains "the necessity under which we have found ourselves of recognizing an Economic Sociology." He was thus thinking of '*economic* sociology' and not of 'economic *sociology*'. To emphasise this point, we might interpret 'economic sociology' as covering a combination of what is presently called 'behavioural economics' and 'institutional economics'. This interpretation, however, ignores the fact that much of the relevant work is today called 'economic sociology', 'organisation sociology', 'organisation science', 'behavioural science', and so on. Therefore, it seems appropriate to use Schumpeter's term "economic sociology" in an open-ended sense.

Since the inclusion of economic sociology among the fundamental field of scientific economics has created some confusion, it is relevant to summarise four of the reasons that seem to have led to his decision. The first reason is that economists have always added pieces of economic sociology into their works. Although *Wesen* had criticised the mixing of economic theory and economic sociology that is exemplified by Marshall's *Principles*, Schumpeter later emphasised that "[t]here is nothing surprising in the habit of economists to invade the sociological field" (*S1949h*, 134). Actually, he recognised that a "large part of their work—practically the whole of what they have to say on institutions and on the forces that shape economic behaviour—inevitably overlaps the sociologist's preserves. In consequence, a no-man's land or everyman's land has developed that might conveniently be called economic sociology." The second reason seems to be that we wanted to bring ideological influences on economic analysis under control. Since ideologies and visions are often closely related to the behavioural and institutional assumptions of theoretical work, a tool-based study of these assumptions would serve as an additional check on the constructs of theoretical economics. The third reason is that economic sociology might help to provide the historical sense that is not automatically promoted by theoretical work. This is clear from Schumpeter's idea that economic sociology provides "stylized economic history" (*History*, 20). He was probably thinking of the works of Schmoller and Max Weber, which he characterised as belonging to "economic sociology" and "Wirtschaftssoziologie" (*S1926b*, 181; *History*, 819); and he was writing before most sociologists apparently lost just as much of their historical sense as economists had already done. In any case, the function of economic sociology in his system of fundamental fields of the science of economics seems to require that it relates closely to economic history. The fourth reason for including economic sociology among the fundamental fields of economics is that it would allow many of Schumpeter's broader research interests to be pursued within his definition of economics. His need for room was not only related to what may be called his evolutionary economic sociology, which tended to be excluded by the increasingly stricter demarcation of the borderlines of economics. Even his work on basic evolutionary economic theory was threatened by the same destiny. If his inclusion of economic sociology was accepted, this problem would be solved since his economic theories of evolution could for a while be classified under the heading of evolutionary economic sociology. However, it is hardly advisable to use the term in this sloppy way.

## 11.7 Conclusion

Schumpeter's *History of Economic Analysis* has normally been considered a very valuable but also highly idiosyncratic handbook for the study of the history of economics. This chapter has tried to overcome this classification of his last, and largest, book. The chapter has tried to demonstrate that Schumpeter was right by emphasising "the evolution of technical economics that this book is intended to present" (*History*, 384). He focussed on "the kind of analytical apparatus ... the evolution of which is the subject matter of this book" (p. 365n). He fulfilled these ambitions by developing a general evolutionary theory of the history of science and by applying this theory in his concrete study of the history of economics.

Schumpeter's theory of scientific evolution helps to explain at least four characteristics of his book. First, it explains *History*'s presentation of the contours of the history of economics. These contours take the form of long waves that start with innovative periods of confusion and end with classical situations of relative equilibrium between the different components of economic analysis. The architecture of *History* is designed to emphasise these contours: the prehistory until the first classical synthesis around 1790; the period of post-Smithian classical economics until the second classical synthesis around 1870; the period of neoclassical economics and historical economics that culminated in the third classical situation before 1914; and the years of high theory and high econometrics that had not yet produced a classical situation when Schumpeter died in 1950. Second, the theory of scientific evolution explains why Schumpeter excluded economic policy and philosophy and instead concentrated on the "filiation" of the attempts of improving the analytical toolbox. It is neither the economic thought on policy issues nor economic philosophy that has an evolutionary history. What evolve are instead the techniques of economic analysis. Third, the theory of scientific evolution to some extent explains why *History* includes large amounts of background material. Individual contributors to the science of economics are often motivated by an ambition to implement personal visions of an extra-scientific nature. The provided background material seems to help reconstructing these visions as well as other personal motivations. It also helps to analyse the major brakes on the scientific engine. Fourth, the theory of scientific evolution explains Schumpeter's somewhat paradoxical definition of the "fundamental fields" of economic analysis as including not only theory and statistics but also history and economic sociology. The reason is that these fields are parts of the actual evolutionary process and that an isolated history of economic theory would be radically incomplete. The actual evolution of economics is better described as the co-evolution between the four fields. However, since economic sociology has never been established as an independent field, it has mainly manifested itself in the visions and verbal accounts of economic theorists.

*History* does not explicitly confront the evolutionary economics that Schumpeter had tried to develop. Nevertheless, the book's theory of scientific evolution seems appropriate for analysing his failure of establishing this branch of the science of economics. His theory of the role of leadership in science can be combined with the fact that he did not want to provide this leadership himself. As a result, the critical mass needed for any take-off of evolutionary economics was not obtained. However, it is not obvious that leadership was a sufficient condition for take-off. The main explanation for the lack of success at the level of scientific communities is rather that Schumpeter did not succeed in finding or producing sufficiently operational techniques for analysing economic evolution. This failure is not only a theoretical one. Since economic evolution is a unique process in historical time, the theoretical concepts of evolutionary economics have to be closely related to historical and statistical analysis. Furthermore, the long-term evolution of the system of economic routines is heavily influenced by the parallel evolutionary processes within other sectors of social life. The consequence is that evolutionary economics has to include evolutionary economic sociology in order to define its borderline with, and its relation to, the evolution of other sectors of social life. The overall conclusion is that the development of all Schumpeter's four fundamental fields is crucial for the establishing evolutionary economics as a branch of economics. In contrast, the necessary condition for the further evolution of equilibrium economics might be constrained to the development of theoretical and statistical tools. If we combine the two last statements, we recognise that Schumpeter had his evolutionary economics at the front of his mind when developing one of the conspicuous arguments of *History*. It is not difficult to recognise the same background for many of the other arguments of his largest book.

# 12

# Beyond Schumpeter's Evolutionary Economics

In the beginning of the present book, Paul Samuelson (1981a, 1) was quoted suggesting that Schumpeter at the end of his life "would have traded his Popeship for a Keynesian revolution." The preceding chapters have reconstructed the type of scientific revolution Schumpeter tried to promote. His main ambition was to complement the reformed branch of equilibrium economics with a new evolutionary branch of the science of economics. Since Schumpeter's evolutionary economics was largely a personal affair, he only became the "Pope" of the reformation of equilibrium economics. He became a 'heterodox Pope', however, since his dual research programme not only led to praise of Samuelson's (1947) *Foundations of Economic Analysis*, but also to the recognition of its inadequacy with respect to his wished-for establishment of evolutionary economics as a viable branch of the science of economics. Actually, Schumpeter's ambition seems to have been to write a book that could have had the title 'Foundations of Evolutionary Economic Analysis'. Such a book would probably have emphasised not only evolutionary economic theory, but also the related statistics and history. As a co-founder of the Econometric Society, he seems to have hoped that the econometric alliance between theoretical and statistical studies would help to develop evolutionary economics. The underdevelopment of his evolutionary analysis, however, served to drive ambitious theoreticians and econometricians away from him. The major thing he could do was to point out that they thereby ignored crucial facts about capitalist economic evolution. His work caught more serious attention among economic historians; and collaboration with some of them was emerging in the late 1940s. Nevertheless, even this alliance seems to have been unstable, because the historians recognised the insufficient operationalisation of his basic concepts of economic evolution as well as of his related work in economic history and economic sociology.

The scale and scope of Schumpeter's work as well as the unfinished nature of his evolutionary economics serve to emphasise that it is impossible to establish what might have been called 'Schumpeterian evolutionary economics'. Instead, we might try to think in terms of 'neo-Schumpeterian' or 'post-Schumpeterian' evolutionary economics. However, even such labels seem to be too narrow. For instance, even though we have seen that Georgescu-Roegen considered himself a student of the "Universitas Schumpeteriana", he developed the original inspiration from Schumpeter far beyond what can reasonably be labelled 'neo-Schumpeterian' or 'post-Schumpeterian'. Such labels can be of more use in characterising, for instance, the analysis of "Schumpeterian competition"—like that of Nelson and Winter. Nevertheless, the theorem that history-specific labelling of scientific activities is always inadequate in the long run can be derived from Schumpeter's theory of scientific evolution. Therefore, it seems more appropriate to say that evolutionary economics has moved beyond Schumpeter's evolutionary economics. We should add two facts. First, evolutionary economics has also moved beyond Marshall and Veblen and many other pioneers. Second, evolutionary economics has more formally moved beyond classical game theory and the mathematical theory of linear systems. These facts help explain the complexity of the present-day situation for evolutionary economics. Nevertheless, Schumpeter's theory of scientific evolution might help us clarify important aspects of the situation. One of the services provided by this theory is the thinking in terms of the fundamental fields of the science of economics. In the present chapter, his idea of fundamental fields is used to characterise the situation of modern evolutionary economics. To make this characterisation manageable, it is constrained in three ways. First, the characterisation starts from Schumpeter's work. Second, it is delimited by the knowledge of the present author. Third, it has to be relatively short and coherent.

## 12.1 The fields of evolutionary economics

Since Schumpeter's evolutionary economics must be interpreted as work in progress rather than as an established paradigm, the best way of ending the present book seems to consider some of the ways in which we have moved and can move beyond his analysis of capitalist economic evolution. He had in the preface of *Cycles* (p. v) suggested that the "younger generation of economists should look upon the book merely as something to shoot at and to start from—as a motivated program for further research." He could have said the same of the rest of his contributions to the analysis of economic evolution. If we consider his major books as a whole, we recognise the core features of his motivated research programme. This is not a programme for doing "normal research" in the sense of Thomas Kuhn (1970). Therefore, the otherwise useful attempts to reconstruct Schumpeter's evolutionary economics in terms of Kuhn's concept of "scientific paradigms" and Lakatos's concept of "scientific research programmes" are of little direct relevance for present-day research. Schumpeter's "motivated research programme" can rather be deducted from *Cycles*'s subtitle: *"A Theoretical, Historical, and Statistical Analysis of the Capitalist Process*". Since he defined the "capitalist process" as essentially an evolutionary one, his research programme consisted in an attempt to coordinate three types of evolutionary analysis—evolutionary economic theory, evolutionary economic history, and evolutionary economic statistics. He did not present the resulting research programme with three fundamental fields as being strictly coordinated. On the contrary, he fully accepted the relative independence of each of the fields of evolutionary economics. This independence became blurred by the fact that he covered all the mentioned fundamental fields—as well as what can be called evolutionary economic sociology. However, as soon as we move beyond Schumpeter's evolutionary economics, we have to recognise that the necessary coordination is largely provided by the co-evolution of three (or four) relatively independent fundamental fields of evolutionary economics.

The present chapter is limited to covering only the three fundamental fields of evolutionary economics. The discussion concerns theory, statistics and history—and their interconnections (see Figure 12.1 on the next page). Since the presentation begins with basic evolutionary economic theory, the reader might get the impression that this field of evolutionary economics is more fundamental than the other fundamental fields. However, this misuse of the term 'fundamental' is not suggested: all three fields are fundamental in the sense of being necessary and relatively independent elements of evolutionary economics. Ideally, "they are inseparable because there is an incessant give and take between them" (*S1949g*, 329). The alliance between theory and statistics not only helps to produce evolutionarily relevant statistics; it also promotes the operationalisation of theoretical concepts and enforces the development of new theoretical tools. The alliance between theory and history can confront economic evolution as a complex historical process and study the degree to which it can be decomposed adequately. We should also add the alliances between evolutionary theory and the many fields of applied evolutionary economics. Take, for instance, evolutionary organisation studies. This field has not only applied given theoretical schemes but also developed analytical tools that are important for basic evolutionary theorising. Nevertheless, the development of basic evolutionary theory is crucial because it helps to coordinate the research efforts conceptually. Another type of coordination might be provided by the different evolutionary economic visions—of which the vision underlying Schumpeter's Mark I model is a conspicuous one. Since extra-scientific visions seem to provide guidance not only for personal work, but also for the work of groups of researchers, these visions have to be made explicit. The study of the visions may also help to reveal and, perhaps, correct the resulting biases. Let us, however, return to the three fundamental fields of evolutionary economics.

Since the fundamental fields of evolutionary economics have not

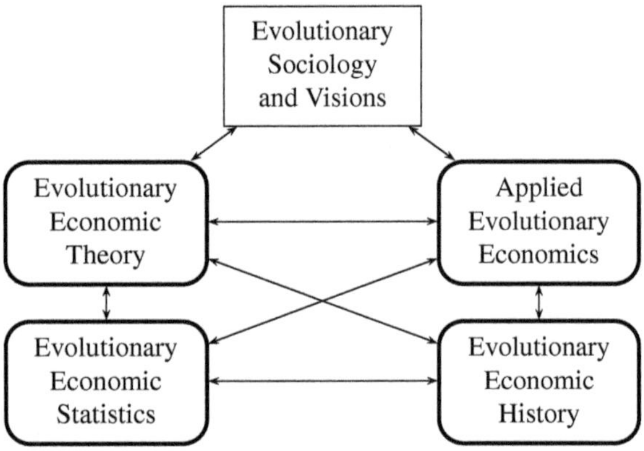


**Figure 12.1.: The fields of evolutionary economics**

Comment: The three fundamental fields are theory, statistics, history. The many actual and potential fields of applied evolutionary economics have been joined under one heading. Explicit considerations of the relationships with evolutionary sociology and evolutionary visions seem important. Furthermore, evolutionary economic sociology might be considered a field of evolutionary economics.

presently established themselves and since some of their analytical tools of are still insufficiently developed, even the naming of the fields might seem premature. The idea of fundamental fields is, nevertheless, tried out in the present chapter. If we add the many fields of applied evolutionary economics, we are basing the discussion on the following definition:

The development of evolutionary economics needs a combination of

(1) evolutionary economic theory
(2) evolutionary economic statistics
(3) evolutionary economic history
(4) applied evolutionary economics

Before considering these fields of evolutionary economics in the next four sections, the general difficulties of analysing the problems of the division and coordination of scientific labour have to be emphasised once more. Schumpeter thought much about these problems in relation to the general specialisation of the social sciences. His description of the impossibility of solving the problems of specialisation and coordination in a fully satisfactory manner started in the following way: "When specialization ... had produced the splendid disorder of tropical growth, the idea suggested itself of making coordination or a general survey into another specialty" (*S1948b*, 111). The pioneer was Auguste Comte, and many philosophers followed his lead. However, their "attempts to create logical order where none exists can never be entirely successful." In this respect, there is some similarity between the evolution, branching and co-evolution of scientific specialities and the evolution and branching of industries. According to Schumpeter, "definitions of individual sciences are much the same sort of things as the definitions that are to define individual industries for the purposes of census—rough approximations made for practical convenience and without particular logical dignity." This conclusion is based on Schumpeter's theory of scientific evolution. It is especially important for analysis of the fields of evolutionary economics. Actually, there is one field that is so difficult to handle that it shall not be treated explicitly below. This field is evolutionary economic sociology. What could have been described under this heading is simply treated as part of evolutionary economic theory and evolutionary economic history.

## 12.2 Evolutionary economic theory: general problems

The contours of the reformation of equilibrium economics were becoming clear to Schumpeter when members of the younger generation, in the late 1940s, tried to reap the fruits of the preceding years of high theory and high econometrics. We can, by applying the Schumpeterian theory of scientific evolution, say that they were moving economics toward a new classical situation. Their new synthesis applied the assumption that economics has to be a mathematical science based on solid econometric results. By producing such a synthesis, leading members of the younger generation created a type of equilibrium economics that is extensible into non-evolutionary dynamics and that helped to confront central problems of economic life. However, the classical situation that established itself in the late 1950s and in the 1960s had no room for evolutionary economics. Since Schumpeter foresaw the coming classical situation, he regretted that his 'evolutionary revolution' had not been developed sufficiently to be included in the synthesis. The immediate problem was that his type of evolutionary analysis was formulated in a way that violated the increasingly higher standards of the science of economics. These standards, which Schumpeter accepted, suggested a negative evaluation of his loosely formulated evolutionary economics. The more basic reason for ignoring his evolutionary analysis, however, is that it was not considered a relatively independent complement to equilibrium economics. The younger generation was striving at relatively monolithic foundations of economic analysis. In this perspective, Schumpeter's evolutionary economics had to be ignored because it is built on assumptions that create analytical trouble for equilibrium economics and for its dynamic extensions.

The fact that it was equilibrium economics that served to define the borderlines of the science of economics is not based on general scientific necessity. The major goal of any science is to discover and explain real phenomena—while systematic consistency is only a heuristic and mainly for limited scientific domains. Let us imagine that the borderlines of the science of biology were defined by its anatomical and physiological branches. Since these branches focus on static structures and stationary processes, they would not allow any room for the study of biological evolution (cf. *Cycles*, 36–7; *Cycles Abr.*, 14). Actually, evolutionary biology is still struggling for a safe place in the curriculum for the training of medical doctors. Nevertheless, the branch of evolutionary biology is a crucial part of the science of biology. It has even been argued that "Nothing in Biology Makes Sense Except in the Light of Evolution". This is the title of a small paper by the prominent evolutionary biologist Theodosius Dobzhansky (1973). However, although Dobzhansky's proposition has become the motto of evolutionary biology (Barton et al., 2007, 1), it has to be interpreted cautiously. Anatomy and physiology can, of course, be studied without any understanding of biological evolution. However, the "ultimate explanation" of biological structures and processes has to be found in the process of biological evolution. Furthermore, the evolutionary perspective helps to focus attention on the imperfections and continued change of biological phenomena. Schumpeter considered the relationship between equilibrium economics and evolutionary economics in a similar way. Although the branch of equilibrium economics produces the bread and butter for professional economists, it seriously needs an evolutionary complement. This need is primarily scientific; but the fulfilment of the need will also help economists in their professional practice.

Although the leading role of equilibrium economics in defining what economics is about has not been based on general scientific necessity, the underdevelopment of evolutionary economics is a historical fact that cannot primarily be explained by exogenous factors. According to Schumpeter's theory of scientific evolution, the powerfulness of equilibrium economics is the outcome of the cumulative improvement of the tools for the analysis of economic equilibria and the related dynamic processes. The underdevelopment of evolutionary economics is the result of the lack of a progressive development of its analytical tools. Modern evolutionary economists are, thus, not in a situation similar to that of Newton's metaphorical dwarf that can stand on the shoulders of the giant produced by a long and cumulative tradition. Let us crudely, and in sharp contrast to Schumpeter, consider Adam Smith (1986–99) as the founding giant of economics who presented two major contributions in his *Wealth of Nations*: (1) the well-developed theory of market co-ordination within a given economic system and (2) the sketchy theory of the evolution of the economic system through a gradually increasing division of labour. The first shoulder has been solid enough to form the foundations of economic knowledge that is, more or less, cumulative while the second shoulder has been so slippery that it is difficult to discern any cumulativeness until very recently. The problem was left unsolved by Marshall's evolutionary economics and by Veblen's evolutionary economics. Even though Schumpeter worked consciously and hard to produce a more solid tradition that broke with Smith and Marshall and largely ignored Veblen, he did not succeed in producing a cumulative research tradition that could move far beyond his preliminary results. This proposition about Schumpeter's legacy should not be misunderstood. As emphasised by Wolfgang Stolper's (1994, 377) already quoted statement, "many of his 'visions' have become so commonplace that the present generation can hardly imagine how radical they once were". However, the widespread use of the Schumpeterian visions and concepts of innovation and entrepreneurship can hardly be said to have become part of a cumulative research programme for the study of economic evolution before the work of Nelson and Winter (1982). Furthermore, alternative approaches to the modern study of economic evolution have been developed independently of Schumpeter's work. For instance, evolutionary game theory became established by the evolutionary biologist John Maynard Smith (1982) almost simultaneously with the Nelson–Winter type of evolutionary economics. Nevertheless, the application of evolutionary game theory in economics has not yet become connected with the analysis of Schumpeterian competition and other of Schumpeter's core topics (Andersen, 2007a).

Edmund Phelps (2008) has suggested that the modern analysis of the great problems of capitalism should not put too much emphasis on Schumpeter's work. Phelps pointed at the "modeling capitalism along the modern (or modernist) lines proposed at various times by Knight, Keynes, Hayek and M. Polanyi—and inevitably Schumpeter, though many of his concepts remained unnaturally classical." What, according to Phelps, makes Schumpeter less modern than the other contributors is largely related to the Mark I model in which any invention has to await the effort of the right innovative entrepreneur to become an integrated part of the economic system. Although Schumpeter used this model to produce concepts that are still crucial, he built it on a distinction between invention and innovation that was dropped by the "early moderns". He also closed his model in a way that is "strikingly premodern" by assuming that "bankers can discern the worth of the projects submitted". Thereby, he ignored the crucial problems of asymmetric information. According to Phelps (2007, xvii), the result is that the "Schumpeterian model makes determinate (at least probabilistically) the rate of innovative activity and the time path of productivity—as if the creativity is all science and no commerce." Here he might be thinking of neo-Schumpeterian models of endogenous growth, since we are hardly considering a model that emphasises the creative response to business situations and the Schumpeterian principle of indeterminateness (see Section 10.3). In any case, we are presently not considering Schumpeter's relevance for general macroeconomic modelling but rather for the analysis of economic evolution. This analysis can be greatly enhanced by the insights of the "early moderns" and by analysis of asymmetric information pioneered by Phelps, Akerlof, Spence and Stiglitz. Nevertheless, Schumpeter's Mark I and Mark II models can hardly be substituted by the contributions of Knight, Keynes, and Hayek when we want to confront the problems of capitalist economic evolution.

The understanding of the theoretical toolbox can be promoted by considering how evolutionary theorists explain economic phenomena. In Section 6.3, it was suggested that evolutionary explanations differ from the standard type of causal explanation. While the standard of scientific explanation is phrased by reference to scientific laws, the complexity of evolutionary processes implies that evolutionary explanations are normally made by reference to evolutionary mechanisms. As emphasised by Jon Elster (1989, 168), "[w]hat we have is a toolbox of mechanisms, not a set of laws". The set of mechanisms available for evolutionary explanations has grown enormously since Schumpeter died. Nevertheless, it is still possible to consider his mechanisms of innovation and adaptation as relevant for the construction of satisfactory evolutionary explanations. For instance, we may still follow Schumpeter in explaining the predominance of railway transportation at the end of the nineteenth century by means of the basic mechanisms of the capitalist engine. However, modern evolutionary theorists need to translate and interpret these mechanisms. The major problem is apparently related to Schumpeter's Mark I model. This model was originally developed to implement his dualistic vision of economic evolution as the outcome of two opposing forces: the force of the minority of innovators and the force of the majority of people with adaptive behaviour. According to Ulrich Witt (2008b), the Schumpeterian idea that the minority transforms the behaviour of the majority can best be described by the terms novelty, emergence, and dissemination. However, Witt's way of conceptualising economic evolution is not constrained to the implementation of the dualistic interpretation of economic evolution. On the contrary, it seems possible to abandon Schumpeter's strange dualism without losing his major results. His Mark II model points in this direction by locating both conservatism and innovation within incumbent firms. Furthermore, the Mark I model can be reinterpreted in a similar vein. In any case, evolutionary economic theorists tend to assume that the difficulties are resolvable. Moreover, many of them do not apply the terms novelty, emergence and dissemination. They instead operationalise the evolutionary mechanisms under headings like the innovation, preservation and selection of routines.

The concept of the mechanism of innovation was troublesome for Schumpeter. He recognised that this mechanism has indeterministic characteristics; but did not want to model it solely be means of a random variable (see Section 10.3). Through his "principle of indeterminateness" he tried to capture basic characteristics of evolutionary explanation. However, most innovations are not as enigmatic as they might appear. They do not take place in the Walrasian field of competitive forces in which the structural components are described so abstractly that they do not portray the situation faced by a Schumpeterian entrepreneur. Instead, an innovation can often be modelled as a *localised* change of the circular flow of a reproducing "economic organism" with a complex ecological structure. In this setting, classes of potential innovations can be modelled precisely. This is especially the case for process innovations, organisational innovations and minor product innovations. Therefore, it is possible to analyse the introduction and major effects of such innovations by means of flexible, and computer-implementable, models of Schumpeterian competition (Winter, 1984). Such models, however, should only be considered simplistic approximations to the analysis of the complex processes of economic evolution. They serve to define the important strategy of moving from the simple to the complex. The alternative strategy for evolutionary economic theorising is not to start from an unstructured economic system, since innovation would become incomprehensible in this context. It is rather to start from the structured "economic organism". When we add the financial component of this system, we recognise that innovations in a particular component can influence the evolutionary processes that take place in all the other components of the system.

Schumpeter normally emphasised the mechanism of innovation and de-emphasised adaptation. Moreover, he mixed what can be called firm-level adaptive inertia with population-level adaptation by the selection between firms with different routines. Thereby, he left a huge task to modern evolutionary economics. As already mentioned, two alternative strategies have been proposed. One strategy is to emphasise firm-level inertia in terms of the mechanism of reinforcement. The other strategy is emphasise population-level response in terms of the mechanism of selection. The obvious question is whether these two mechanisms can be analysed within a single theoretical framework. A possible solution is to apply the framework of multi-level selection (see Appendix D.2). This framework is most easy to describe if we think in terms of a population of firms in which each firm consists of multiple plants. If the evolving characteristic of this population is reflected by productivity, then the evolution of the average level of productivity can be considered the outcome of three effects. First, average productivity increases because of inter-firm selection: firms with above average productivity expand and firms with below average productivity shrink. Second, since the productivity of each firm is really the average of the productivities of its plants, the firm-level productivity can increase because of its inter-plant selection. Third, the productivity of each plant can increase because of its internal innovative and adaptive activities. This example demonstrates how the concept of selection can be used both at the level of the population and at the level of the firm.

Since most of the adaptive activities at the level of the firm cannot be described as selection between plants, the question remains whether it is adequate to think of firm-level adaptation in terms of selection. This question seems to have been answered in the negative by Ulrich Witt (Witt and Cordes, 2007). His sceptical view serves to emphasise an important difference between biological evolution and economic evolution. While the offspring of biological individuals only differ from their parents because of the recombination and mutation of genes, a firm can change much over time and a spin-off can differ much from the mother firm. Evolutionary economic statistics (see Section 12.4) seems to demonstrate that intra-firm change is a major factor of economic evolution. Furthermore, theoretical considerations indicate that part of this change is due to strategic attempts to avoid the negative consequences of inter-firm selection and to exploit the forces of selection positively. These considerations have to some extent been developed in relation to Schumpeter's Mark II model. They help to explain why economic evolution is much faster than biological evolution.

The inclusion of myopic foresight severely complicates the analysis of economic evolution by means of the evolutionary mechanisms. However, even if we ignore this complication, there are other challenges for the use of the mechanisms. A major one was provided by Schumpeter's suggestion that the evolutionary mechanisms are not independent of each other. Instead, he repeatedly emphasised that the mechanism of innovation directly influences the mechanism of selection (see Figure 12.2 on page 380). His idea is that a set of innovation-based firms is sometimes able to influence its selection environment. One case is that the innovation-based firms influence the preferences of their potential customers. Another case is that they promote a change of the selection rules that banks use in their evaluation of the projects that need financial support. Yet another case is that the preferences of the potential employees are influenced. Thereby, the way in which the markets for goods, finance, and labour select firms will change. The analytical consequence is that the study of economic evolution becomes more complex and difficult to handle. As usual, the method of successive approximations provides a way of handling these difficulties. We start by analysing a model in which there is no feedback from innovation to selection. Then we add the simplest possible feedback and study the consequences. If this model is implemented for computer simulation, it is relatively easy to continue to add complexity. The end result will be that it becomes nearly impossible to analyse the behaviour of the model. Nevertheless, the exercise will have provided us with knowledge not only of the importance of time lags but also of ways in which the process of economic evolution may drift. Thereby, it will have improved our understanding of the consequences of Schumpeter's "principle of indeterminateness".

## 12.3 Evolutionary economic theory: specific mechanisms

The evolutionary theorist has to proceed by means of simplifying assumptions and successive approximations to the complex reality of economic evolution. In this perspective, Schumpeter's work seems to be too complex to provide a starting point for evolutionary research. As already suggested, the starting point can instead be the simplest possible schemes of evolutionary explanations. Two major schemes are available: explanation by reinforcement and explanation by natural selection (Elster, 1983; Parijs, 1981). Explanation by reinforcement locates adaptation at the level of individuals. For instance, when a population of firms has found a common standard in an evolutionary standardisation game, the choice of this standard by each individual is reinforced since shifts to other standards are punished. Explanation by natural selection locates adaptation at the level of the population. While the individual firms are unable to change their chosen standard or slow in doing so, the average choice of standard will nevertheless change until a common standard has been found because of the system-level selection between firms. Schumpeter seems to have applied a mix of these two types of explanation and so does modern evolutionary economics. Nevertheless, a modified form of the natural-selection mechanism seems to have become predominant while the reinforcement mechanism has become auxiliary. This tendency was overemphasised by Andersen (1994, 14, emphasis removed), who only admitted explanation by the generalised mechanisms of natural selection:

> "An evolutionary-economic explanation is an explanation of a fact of economic life by reference to previous facts as well as to a causal link which (immediately or in reconstructed form) may be shown to include
> (1) a mechanism of preservation and transmission,
> (2) a mechanism of variety-creation,
> (3) a mechanism of selection,
> and which includes or may be enhanced by introducing
> (4) a mechanism of segregation between different 'populations'."

Since this definition does not cover evolutionary economic explanation in general, its first words ought to have been: 'a particular type evolutionary-economic explanation'. This change also serves to emphasise the open-ended nature of evolutionary economic theory. Nevertheless, the improvement of theoretical work sometimes requires that particular approaches are developed one-sidedly while their later unification

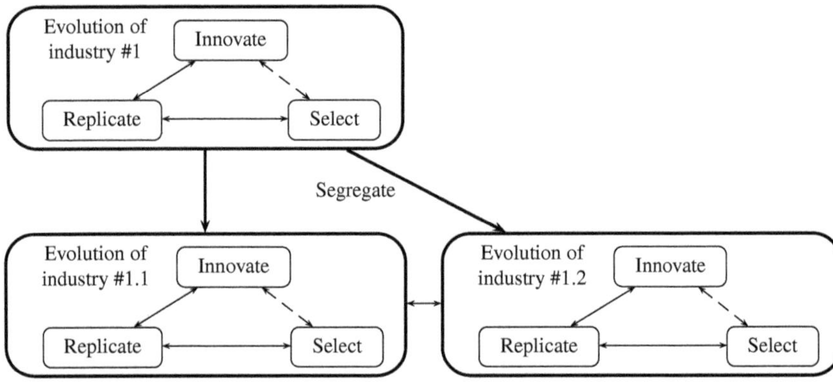


**Figure 12.2.: A version of the basic mechanisms of evolutionary processes**

Comment: An evolutionary process takes place within a population characterised by a certain balance between three mechanisms: (1) the mechanism of innovation that creates new routines; (2) the mechanism of inertia through the conservation and replication of routines; and (3) the mechanism of selection of routines. Different populations are to some extent segregated from each other. New populations are created by (4) the mechanism of segregation through which an integrated population branches into two populations. The figure depicts how the industry #1 splits into two industries through the creation of industry #1.2.

seem to require a formal study of evolutionary dynamics that cannot be discussed in the present book (Nowak, 2006; Hofbauer and Sigmund, 1998). Furthermore, the above definition points at the practice of important research traditions that have emerged in relation to the modern modelling of Schumpeterian competition and Schumpeterian economic growth (including the tradition derived from Nelson and Winter, 1982). Finally, a discussion of the definition can serve to exemplify how a set of evolutionary mechanisms provides a heuristic for producing the first approximation to an evolutionary explanation. Actually, since we have no generally acknowledged and relatively simple mechanisms of the type found in the study of biological evolution, any kind evolutionary economic explanation needs to specify the mechanisms that are referred to. The set of abstract mechanisms provided by a generalisation of those of neo-Darwinian biology is a well-known and often applied specification in evolutionary economics. Therefore, we shall consider these mechanisms and their interaction—as depicted by Figure 12.2.

When compared with Schumpeter's general account for economic evolution, the major difference is that Figure 12.2 suggests that the study of economic evolution should *begin* at the level of individual populations ("industries"). This partial analysis has to assume that the behaviour of the rest of the economic system is determined exogenously. Furthermore, it has to assume that the borderlines of the industry have been determined by the evolution of the past. Under these assumptions, we can study the three basic evolutionary mechanisms and their interaction. The mechanism of preservation and transmission (inertia due to replication) secures that the routines of the individual firms do not change over time. Furthermore, the routines can also be upheld by establishing new firms that are clones of old ones. The mechanism of industry-level selection leads to the differential growth of sets of firms characterised by the same routine behaviour. Some groups of firms grow while other groups contract. Thereby, the average behaviour of the industry changes over time. Finally, the mechanism of innovation returns us to the level of individual firms. Innovation can take place either through the establishment of new firms or through the change of routine of already established firms. However, both the mechanisms of innovation and selection are heavily influenced by exogenous factors. The mechanism of innovation is influenced by the markets for finance and labour. The mechanism of selection is influenced by competition from other industries and by the general state of the economic system. For instance, the selection pressure will be stronger during recession than during prosperity.

Schumpeter would probably have argued that this picture is even worse than the one provided by Marshall's partial equilibrium analysis. The reason is that the evolving characteristics of any particular industry is likely to influence its place in "the actual structure of the industrial organism" (*Cycles*, 143; *Cycles Abr.*, 119), that is, in the industrial ecosystem. However, this issue can in principle be covered by adding a mechanism of creating new industries (see Figure 12.2). This mechanism of segregation of populations of firms might be represented by the gradual differentiation of incumbent firms because repeated influence of a small pressure to differentiate. Such a small pressure can have powerful effects (Schelling, 1978, 147–55); but it is easier to think in terms of a 'speciation' event. This event is the establishment of a new industry by one or a few innovation-based firms. This industry can be defined by a relatively independent mechanism of selection. However, the industry can also be defined by its mechanisms of replication and innovation. These definitions allow us to make partial analyses of the evolutionary processes in each of the industries. They also allow us to analyse the ecological interaction between the two industries. Since this analysis is rather complicated, even the summary account has been transferred to Appendix D.1. In any case, by repeated speciation events we may arrive at a complex industrial ecosystem. If we switch off innovation in all industries and assume that the mechanism of selection has done its work in each industry, we can study the non-evolutionary dynamics produced by the linkages between the industries. We may also switch off the linkages and study the transformation of the system of industries as a pure selection process.

Moreover, since each of the industries of the industrial ecosystem has its own evolutionary processes, we can study the co-evolutionary dynamics of the system. However, this study can easily become too complex to be handled analytically. Therefore, the analysis of bilateral processes of co-evolution is the appropriate starting point.

Before proceeding along the path suggested by Figure 12.2, it is important to note that the evolutionary mechanisms are not as easily and securely specified for economic evolution as they are for the biological evolution based on "blind" genes (Dawkins, 1991). Moreover, although the suggested mechanisms have to some extent been adapted to the reality of economic evolution, they might still be considered as tainted by their emergence from biological analogy. Schumpeter was a staunch opponent of uncritical interchange between different sciences. However, he ultimately seems to have accepted the cautious use of "the Darwinian concepts of Struggle for Existence and Survival of the Fittest to the facts of industrial and professional life in capitalist society" (*History*, 789). Thus he did not deny that "it may be ... that certain aspects of the individual-enterprise system are correctly described as a struggle for existence, and that a concept of survival of the fittest in this struggle can be defined in a non-tautological manner." However, he emphasised that "these aspects would have to be analyzed with reference to economic facts alone and no appeal to biology would be of the slightest use". Even Marshall (1898, 43, 39) seems to have thought similarly since he not only suggested that "[t]he Mecca of the economist is economic biology" but also "that analogies may help one into the saddle, but are encumbrances on a long journey. It is well to know when to introduce them, it is even better to know when to stop them off." Therefore, it should be emphasised that the presently discussed concepts of the mechanisms of economic evolution are no longer based on analogy. They cannot be interpreted or supported or criticised by reference to biological evolution. Concepts are not evaluated by their names but by their contents; and these contents have to be evaluated in the context of evolutionary economics.

One way of coming to grips with the mechanisms of Figure 12.2 on page 380 is to relate them to Schumpeter's two models of economic evolution. As mentioned repeatedly, his Mark I model is based on the carrying out of innovations by individual entrepreneurs who create new firms with borrowed money. In contrast, established firms are conservative and show much inertia. Schumpeter's Mark II model is very different. Here, large corporations perform repeated innovations in their oligopolistic competition; and these innovations are prepared in their departments of research and marketing. These models have, in Part II, been presented in terms of mechanisms of innovation and adaptation. The question is how to interpret them in terms of the augmented set of mechanisms, which includes inertia and segregation. An attempt to answer this question is presented in Table 12.1 on the next page. The Mark I model explains inertia in terms of the conservatism of established firms and their ability to replicate fixed routines; innovation takes only place through the founding of new firms by Schumpeterian entrepreneurs who need external finance. In this setting, the creation of innovation-based firms normally leads to the destruction of old ones. The mechanism of segregation is also related to innovation: radical innovation can lead to the creation of new industries. The Mark II model is different. Here the mechanism of inertia is not clearly specified, since established firms have both adaptive and innovative capabilities. Nevertheless, Schumpeter's upholding of his concept of innovation suggests that he thought that even the behaviour of large firms is normally characterised by a significant degree of inertia. It is on this background that their R&D departments engage in the preparation of innovations. Nevertheless, the selection mechanism becomes more difficult to handle analytically than in the Mark I model. The problem is that incumbent firms often adapt before they are selected out of business. The mechanism of segregation is also more complex. The problem is that each firm can operate in several markets. Some of the firm's strategic moves concern the location in market niches where competition is limited; and some niche positions can lead to the creation of new industries.

An important literature on evolutionary economic theory concerns the process of economic evolution within industries. This literature can be roughly divided in three parts be means of the Schumpeterian models. The first part of the literature relates to Mark I; the second part relates to Mark II; and the third, and smallest, part combines the two models. Let us quickly consider the two pure cases. Nelson and Winter (1982, 49) clearly related what they called the "Organization-Theoretic Foundations of Economic Evolutionary Theory" to their Mark II models of economic evolution. The preliminary version of these foundations was largely provided by Herbert Simon (1982) and by the related work by Richard Cyert and James March (1963). However, the behavioural theory of firms as characterised by temporarily fixed procedures and routines was not designed to provide a description of the evolutionary mechanisms of replication and innovation. This meant that Nelson and Winter had to reinterpret the theories of routinised behaviour and its change. Thereby, they provided a complex research agenda that included theoretical and empirical issues. This agenda became complex because of the complexities of Mark II models. In such models, firms are supposed to innovate and adapt. Therefore, it is not obvious how we should specify intra-organisational change in terms of the innovation and replication of routines. Actually this distinction can be described in several ways. Nelson and Winter (1982, 98) started "by considering the analogue of Schumpeter's 'circular flow' at the level of the individual organization." Since they considered this sys-

**Table 12.1.: The evolutionary mechanisms and the Schumpeterian models**

| Mechanism | Mark I model | Mark II model |
|---|---|---|
| Inertia | Incumbent firms are upholders of routine | Incumbent firms can adapt significantly by metaroutines |
| Innovation | New routines are introduced by new firms | New routines are introduced by incumbent firms |
| Selection | Selection between new and old firms | Selection between incumbent firms, which can to some extent preempt selection |
| Segregation | Innovative entrepreneurs can create new industries | Incumbent firms can establish new industries in their attempts to avoid competition by entering market niches |

Comment: Schumpeter's two models of economic evolution have been discussed at some length in the present book. The reconstruction of Mark I is summarised by Figure 9.2 on page 287; Mark II is depicted by Figure 9.4 on page 290. Since the previous discussion was performed in terms of the mechanisms of innovation and adaptation, we need a translation to the mechanisms of modern evolutionary analysis. The table presents the solution suggested by the present book.

tem of routine unrealistic, they turned to a series of approximations to routine-based, but relatively flexible, behaviour of real firms. They also provided a theory of the routinisation of the innovative and imitative behaviour of firms. Thereby, they provided important research questions for organisation science and for the literature on business strategy (Dosi, Nelson, and Winter, 2000). Nevertheless, it is an open question whether the organisation-theoretic foundations of evolutionary economic theorising have really been established (Knudsen, 2002). Especially, evolutionary theorising seems to need more behavioural inertia than provided by the flexible reinterpretation of procedural rationality. Actually, the more general theorising on, and investigation of, organisational routines might help to localise the main sources of inertia (Becker, 2008). However, there seems to be a need of retrying Schumpeter's Mark I model as well as its explanation in terms of the evolutionarily necessary architecture of complexity—as it had been suggested by Simon (1996, Ch. 8; 2002).

While Nelson and Winter clearly related their work to the Schumpeterian Mark II model, Schumpeter's name is seldom mentioned in the literature that otherwise can be seen as representing either a mix of Mark I and Mark II or as implementing a pure Mark I model. The reason seems to be that this literature presents itself as organisation studies (covered by Aldrich and Ruef, 2006)—and that Schumpeter can hardly be said to have contributed directly to this field. It is, nevertheless, possible to consider organisational theoretic literature as providing contributions to evolutionary economic theory in general and to the analysis of Mark I in particular. Let us consider the subset of this literature that has imported its major analytical tools more or less directly from evolutionary ecology and other parts of evolutionary biology. Core books are Michael Hannan and John Freeman's (1989) *Organizational Ecology* and Glenn Carroll and Freeman's (1999) *Demography of Corporations and Industries*. Even the titles of these books indicate the application of tools of evolutionary biology. Their basic question is: why are there so many different types of organisations? The answer is: because of the complexity of industrial environment (ecology). Their next question is: how do organisational forms emerge? The answer is: through something close to Darwinian natural selection based on heterogeneity and inertia.

The analytical procedure of organisational ecology can be described as involving four steps. First, the adequacy of the tools for analysing industrial evolution rests on the degree to which the behaviour of individual firms is characterised by inertia. Therefore, the theoretical and empirical analysis of behavioural inertia is a core element of the research programme; and the result seems to be that organisational inertia is enormous with respect to some evolutionarily relevant characteristics. Second, the programme has to answer why novelty nevertheless emerges. The obvious answer is that novelty is connected to new firms that are not simple copies of incumbent firms. This answer is supported by theoretical and empirical arguments. Third, the question is how incumbent firms with given characteristics are selected. Even the name of "organizational ecology" suggests that the literature gives a very rich answer to this question. It not only studies competitive selection, but also other types of "ecological" interdependence. Moreover, it emphasises that selection pressures in new industries are very different from those in mature industries characterised by overcrowding. Thereby, organisational ecology applies concepts similar to those presented in Appendix D.1. Fourth, the literature can turn to a large number of concrete problems like those of classifying organisational forms in a way that is consistent with evolutionary analysis. The simplicity of the procedures also allows their formulation in axiomatic format (Hannan et al., 2007).

Although the mentioned literature is called organisational ecology, it should have become clear that it is largely presenting a variant of evolutionary economics. Moreover, even though the simplicity of this approach has been criticised strongly, simplicity is actually its main strength. First, the approach is so simple that it is easy to detect how it relates to Schumpeter's Mark I model of the capitalist engine. It obviously includes no systematic application of the Schumpeterian concepts of entrepreneurial profit, credit, capital, and interest; but it instead helps us to overcome some of the difficulties involved in the analysis of the complex industrial ecology. Second, the simplicity of the approach supports the coordination of the different fields of evolutionary economics. Although the approach of organisational ecology can primarily be characterised as evolutionary economic sociology, it has also produced contributions that have to be classified as covering evolutionary economic theory, evolutionary economic statistics and evolutionary economic history. Third, the approach allows a quick and efficient teaching of a first approximation to evolutionary economics. Nevertheless, the one-sided teaching of organisational ecology might also promote the crowding out from evolutionary economics of the study of the "magnificent dynamics" produced by the capitalist engine—as well as its difficult interaction with the evolutionary processes that take place in other sectors of social life. Since the approach of organisational ecology has also been used for the analysis of the evolution of organisational forms within political life and cultural life, this might not be a necessary conclusion. However, because of the complexity of co-evolutionary processes, it clearly has to be complemented by other approaches to the analysis of evolutionary processes. Otherwise, we might not even be able to understand the theory of socio-cultural evolution that the early Schumpeter sketched in the later-omitted chapter of *Entwicklung I*.

Since the Mark II model focusses on organisational routines rather than on individual behaviour, it has served to de-emphasise the relationship between evolutionary economics and evolutionary biology. Peter Hammerstein and Edward Hagen (2005, 608) have suggested that the "modern Schumpeterian branch of evolutionary economic theory is relevant to economics but remote from biology". In contrast, there is a "second strand" that is directly relevant for biologists because it "incorporates behavioural ecology into economics, seeking to root human preferences and beliefs in human evolutionary history". One consequence of this approach is that bounded rationality and adaptive behaviour become firmly rooted in social and biological evolution (Gigerenzer and Selten, 2001). Researchers in the Nelson–Winter tradition have been sceptical about the relevance of such a biological rooting of their theory. Nevertheless, the development of the Mark I model allows relatively easy access to a literature that, for instance, is treated by Samuel Bowles (2004) and Herbert Gintis (forthcoming). This literature—which combines evolutionary game theory with experimental economics—has hitherto emphasised the study of the evolution of behaviour and institutions through the simplest possible mechanisms and shown little direct interest in the problems raised by the evolution of the highly complex industrial system. However, the underlying approach seems extendible to the study of the evolution of this system through the multi-level mechanisms of innovation, replication, selection and segregation. Whether time is mature for such an extension is a controversial question.

The above description of some mechanisms and models of evolutionary economic theory has largely been constrained to what Kurt Dopfer and Jason Potts (2008) have called "evolutionary microeconomics" and "evolutionary mesoeconomics". Although these types of evolutionary analysis were crucial for Schumpeter, his works give us the impression that he was primarily interested in "evolutionary macroeconomics". Actually, it is only in the historical parts of *Cycles* that we find his explicit treatments of the microevolution in individual markets and the mesoevolution in clusters of markets. Therefore, the present emphasis on the micro and meso levels can be considered a reflection of the state of the art of modern evolutionary economics, where we only find sporadic attempts to develop evolutionary macroeconomics (like Foster, 1987). Nevertheless, the emphasis on microevolution and mesoevolution can also be seen as an attempt to rescue important parts of Schumpeter's evolutionary economics from his premature attempts to overcome the difficulties of evolutionary macroeconomics. Although economic life is based on money and finance, and although macroeconomic cycles and policies heavily influence the process of industrial evolution, the inclusion of these factors presupposes a solid understanding of microevolution and mesoevolution. This understanding might be emerging; but we still have to confront explicitly many Schumpeterian challenges. Since economic evolution is a unique process in historical time, even evolutionary macroeconomics has to be sensitive to this process—that even involves the evolutionary change of the basic mechanisms of the capitalist engine. Furthermore, evolutionary macroeconomics will have to take into account that the functioning of the capitalist engine presupposes a certain balance between the different mechanisms. If the mechanism of the conservation of routine predominates, the result is a stationary state of economic affairs that cannot be described as being adapted. If the mechanism of selection predominates, an adapted stationary state emerges. If the mechanism of innovation predominates, the result is chaos. It is only if the three mechanisms have reached a certain balance that we observe an evolutionary process. Therefore, it is important to describe the relative strengths of the evolutionary mechanisms.

## 12.4 Evolutionary economic statistics

Schumpeter emphasised that we "need statistics not only for explaining things but also in order to know precisely what there is to explain" (*History*, 14). He emphasised the importance of statistics by adding that we need to understand "how they have been compiled" and how we "extract information from them". Thus we need to understand how statistical offices produce statistics and the methods that researchers use to analyse statistics (including "the epistemological backgrounds of these methods"). These requirements provide "a necessary (but not a sufficient) condition for preventing the modern economist from producing nonsense". These statements reflect statistical interests that had engaged Schumpeter since his youth. Actually, his first three papers concerned statistical methods. He had produced these papers at the University of Vienna in relation to the statistical seminars conducted by Inama-Sternegg and Juraschek in the period 1903–05 (*S1901–05*, 70–2). The papers, which were published in the journal of the Austrian Statistical Office, cover population statistics (*S1905c*), international prices (*S1905a*), and index numbers (*S1905b*). The latter paper includes the characteristic warning: "Beware of Statistical Aggregates" (quoted by Haberler, 1950, 338). In general, the papers demonstrate that Inama-Sternegg's seminar concentrated on the advanced exploitation of the data that the statistical office produced under his leadership. His office had just completed the most sophisticated census of the time; and the students had a chance of visiting the office and inspecting its electrical calculation machines. The seminars also seem to have discussed novel issues like business cycle research. Such introductions to a new world of data, statistical methods, and interesting research topics made a lasting impression on Schumpeter. Like the historical school, to which Inama-Sternegg belonged, he came to see statistics as a fundamental part of the science of economics.

Schumpeter formulated an important theme of *History* by stating that the fundamental fields of "history, statistics, and theory ... have, until comparatively recently, shown a lamentable tendency to travel in separate compartments" (*S1948b*, 112). According to his interpretation of the econometrics programme, its task is to overcome the split between theoretical analysis and statistical analysis. Actually, he found the wished-for integration at the very beginnings of the emergence of modern statistics. Thus he praised William Petty's pioneering efforts in the seventieth century as "illustrating to perfection what Econometrics is and what Econometricians are trying to do" (*History*, 209). Petty "was one of those theorists for whom science is indeed measurement: who forge analytic tools that will work with numerical facts and heartily despise any others; whose generalisations are the joint product of figures and reasoning that are never allowed to part company." Nevertheless, this inspiring research programme "wilted in the wooden hands of the Scottish professor [Adam Smith] and was practically lost to most economists for 250 years". However, the period between Petty's death in 1687 and the founding of the Econometric Society in 1930 had not been without efforts. Schumpeter emphasised that "Marshall's was one of the strongest influences in the emergence of modern econometrics" (*S1941a*, 107). He had not only presented "the first pronouncement by a leading theorist in favor of an econometric program" (*History*, 962). More importantly, "Marshall *theorized* with a view to making his concepts numerically operative, and his occasional appeal to statistical figures has more than illustrative importance."

The "highly regrettable" disappearance of the original alliance between economic theory and statistics meant that the major impulse for the collection and analysis of statistical data came from the political needs of the state. However, statistics also engaged in scientific alliances—not least with the theory of human populations and their growth. This alliance became more ambitious when Quetelet introduced the concepts of "Average Man" and of the abnormal deviations from this norm as being comparable to probabilistic errors of observation (*History*, 525–6). Although Nietzsche's opposition against this kind of thinking (see Section 4.2) did not confront the real problems, Francis Galton and his follower Karl Pearson were able to forge an alliance between evolutionary population theory and statistical analysis (*History*, 790–1; see also Stigler, 1986). This alliance, which considered deviations from the statistical mean as the fuel of evolution, served to produce much of modern statistical analysis as well as much of modern evolutionary analysis. Schumpeter's interest in elite theory meant that he became interested in this alliance. Therefore, it is hardly by chance that he, during his postdoctoral studies in London, attended the lectures of Pearson, who had emerged as a leading statistician and evolutionary biologist as well as the editor of *Biometrika*. In 1909, Schumpeter designed a university course on the "problems of modern statistics" (see Table 2.2 on page 34). This course would not cover the "easily available performances by Quetelet or Galton" (*S1909a*, 78). Instead, it would present "that route of statistics that has developed in relation to biometrics and anthropology and that presently is connected with the name of K. Pearson" (pp. 77–8).

Schumpeter's exploitation of this type of statistics for the purposes of his evolutionary economics was constrained by Pearson's sole focus on selection based on continuously varying characteristics. Actually, Pearson seems to have been unable to understand how discretely varying characteristics could be subject to a process of selection; and the "mutationists" rejected selection as an important explanation of evolution (Provine, 1971). The battle between the selectionists and the mutationists had started immediately before Schumpeter enrolled at the University of Vienna. Actually, the Viennese Philosophical Society, which he joined and which was led by his teacher Höfler, arranged *seven* sessions "On the Crisis of Darwinism" in 1901–02 (Blackmore, 1995, 298, 289). The eclipse of Darwinism was not least overcome by R. A. Fisher, who, in the period before the 1930s, not only created much of modern statistical analysis, but also core foundations for modern evolutionary analysis. However, this contribution was not acknowledged by Pearson, who happened to consider Fisher his personal enemy—although their statistical analyses were closely related and although they both tried to promote eugenics. This fact might, together with the intrinsic difficulty of the task, help to explain why Schumpeter failed to recognise the possibilities of generalising Fisher's statistical approach to make his own evolutionary economics operational.

When the econometrics movement finally emerged in the late 1920s, it needed a name. According to Schumpeter, "[t]he word Econometrics is, I think, Professor Frisch's, and it had been coined by analogy with Biometrics, statistical biology" (*History*, 209). Similarly the name Frisch chose for the journal *Econometrica* was suggested by Pearson's journal *Biometrika*. However, while the journal of biological statistics focussed on evolution, the journal of the econometrics movement ignored economic evolution. Schumpeter tried in vain to change this situation by producing his *Business Cycles*. The failure of this book was actually especially conspicuous with respect to its application of statistics for the analysis of waveform economic evolution. The problem is that the book's statistical analysis is solely performed by means of longitudinal data of aggregate variables. Schumpeter stated that his statistical analysis "reduces to analysis of time series which reflect economic growth and the cyclical process of evolution as distorted by the influence of external factors" (*Cycles*, 193–4; *Cycles Abr.*, 163). However, it is practically impossible to capture economic evolution by means of the statistical analysis of aggregate data since evolution concerns the transformation of these aggregates. Therefore, the statistical analysis of economic evolution has to rely primarily on longitudinal microdata. Schumpeter's unconvincing statistical performance in *Cycles* reflects the fact that such data were practically absent in the 1930s. Many of his statistical arguments can be interpreted as reflecting an attempt to specify the dangers involved in building theories of business cycles in terms of aggregates: "Such reasoning is at the bottom of much faulty analysis of business cycles. It keeps analysis on the surface of things and prevents it from penetrating into the industrial processes below, which is what really matters" (*Cycles*, 43–4; *Cycles Abr.*, 21). However, he spoiled this message by relying solely on aggregate data in his attempt to date historically the evolutionary equilibria that split the long-term process of capitalist economic evolution into separate periods.

Schumpeter's difficulties serve to explain the contrasting evaluations of his efforts by Jan Tinbergen and Ragnar Frisch. According to Tinbergen (1951, 111, 109), Schumpeter was "to some extent alien to" econometric work; he "lived another life" although he expressed "warm sympathy" for econometrics. Frisch, who in economics shared the first Nobel Price with Tinbergen, formulated a very different evaluation of Schumpeter. According to Frisch (1951, 89–90), Schumpeter used econometric tools wherever possible and had an immense "intuition" of what to model, an intuition that is "the vital part of our science and *the true criterion of an econometrician.*" If we accept Frisch's view, Schumpeter's problem was that he needed new statistical data and new statistical methods. He had already in *Wesen* (182) emphasised that equilibrium economics and evolutionary economics "are completely different" since "they concern not only different problems but also different methods and different materials". Consequently, they also need different types of econometrics: "statistical methods ... must grow out of the theory of the patterns to which they are to apply" (*Cycles*, 199; *Cycles Abr.*, 170). Since the term 'econometrics' has become practically synonymous with the alliance between equilibrium economics and statistics, it seems inappropriate to name the complementary alliance 'evolutionary econometrics'. Alternative names are "evolumetrics" (Cantner and Krüger, 2007) and "evometrics" (Andersen, 2004c).

If we think in terms of evolutionary economic statistics, it is difficult to read *Cycles* without adding the statistical component that Schumpeter was missing: evolutionarily relevant population statistics. His evolutionary model defines a cycle as a basic evolutionary step that starts from a neighbourhood of equilibrium characterised by routine behaviour. Then a swarm of product innovations and process innovations disturbs the equilibrium and creates the extra credit and demand that leads to an upswing. In the following downswing (due to the stoppage of innovation-induced investment), there is a competition between new and old routines; and the selected routines form the basis for a new equilibrium. Population statistics can help to explain roughly what goes on during such a period. In the initial equilibrium, the population of firms has a minimum level of variance with respect to the applied routines. Then the innovations of the upswing increase variance and the expansion of demand slacken the selection pressure. In contrast, the harsh selection during the downswing reduces variance to a new minimum. Given this interpretation of Schumpeter's theory, it is possible to perform a partial—and admittedly naive—testing of it. We collect statistics on different industrial populations; we measure their variance with respect to important selection criteria in the two more or less equilibrated states as well as at the top of boom; and we check whether they behave as expected by the Schumpeterian model of the cyclical process of economic evolution.

We do not have to accept Schumpeter's idea that evolutionary economics is closely related to business cycle analysis. In any case, the clarification of many basic issues has to start at the level of individual industries; and here we shall presently use productivity as a proxy for an evolving variable. Moreover, the statistical analysis of these industries does not have to choose between Schumpeter's Mark I and Mark II models. Any comparison of two censuses of a population of firms will have to take into account that the second census does not contain the same firms as the first census. Between the two points of time, there have been mergers, spin-offs, entries and exits. To simplify the initial analysis, mergers in the second census can be handled by also merging the firms of the first census statistically. Similarly, spin-offs can be statistically merged with their mother firms. However, entries and exits are best treated separately. The reason is not only that we thereby connect to Schumpeter's Mark I model. The reason is also that entry from the outside cannot be handled by the analytical framework that we shall below use to analyse the mark Mark II model. In any case, it is not difficult, first, to study the part of total productivity change that is explained by entry and exit before we turn to the study of the residual evolution in terms of inter-firm selection and intra-firm change. Let us define total evolution as the change of the mean value of an evolutionarily relevant characteristic of a population of firms. Thereby, we arrive at the following statistical description of the evolutionary process that takes place between the two censuses:

$$\begin{matrix}\text{Total} \\ \text{evolution}\end{matrix} = \begin{matrix}\text{Entry} \\ \text{effect}\end{matrix} + \begin{matrix}\text{Selection} \\ \text{effect}\end{matrix} + \begin{matrix}\text{Intra-unit} \\ \text{change effect}\end{matrix} + \begin{matrix}\text{Exit} \\ \text{effect}\end{matrix}$$

"Total evolution" is here the total change of the narrow evolving variable that we have chosen to study. If we used the mean productivity of an industry as the evolving variable, we might imagine that it is possible to perform an analysis of the aggregate total evolution of the whole economy. However, this aggregate concept seems to imply a transgression of the boundaries within which evolutionary analysis can be performed safely. That evolutionary macroeconomics might be able to say something important about the change of the aggregate productivity of a country is another matter—but this issue cannot be discussed in the present book.

If we assume that the incumbent firms of the industry are extremely conservative, the intra-unit change effect disappears. Then we arrive at a simplified version of Schumpeter's Mark I model. This model can easily be made statistically operational. If we reduce the coverage of this model to an arbitrarily defined industry, we recognise that it is simply a model of evolution by entry and exit. Old firms stick to their given characteristics while new firms may either be clones of old ones or based on innovation. Therefore, the basic statistical task is simply to collect and analyse the vital statistics of the industry over a period of time. For each firm, we register its time of birth and death as well as its evolutionarily relevant characteristic, which is assumed to be constant within an established firm. Then we define evolution of the industry as the change of the average value of the chosen characteristic and study the movement of this average over time. Furthermore, we study the degree to which births and deaths are correlated with the value of the characteristic. Finally, we check whether we observe the typical pattern of the industry life

**Table 12.2.: Theoretical concepts of evolution and statistical procedures**

| Theoretical concept | Statistical specification |
|---|---|
| Evolution of a characteristic of the industry | The change between two censuses of the weighted average of the firm-level information on that characteristic. The weights are the resource shares, or the market shares, of individual firms. |
| Effect of inter-unit selection | The covariance between relative fitnesses and characteristics. The absolute fitness of a firm is defined by the ratio of its sizes in the two censuses. The relative fitness of a firm is found by dividing the absolute fitness by the average absolute fitness of the population. |
| Effect of intra-unit adaptation and incremental innovation | The average of intra-firm change weighted by fitness: we multiply the size of the change of the characteristic of each firm by its fitness; and then we calculate the average of that measure. |
| Effect of exit | This effect can be included in the effect of intra-unit adaptation by setting the fitness of defunct firms to zero. The effect can also be calculated separately. |
| Effect of entry by simple imitation | This effect can be included in the effect of inter-unit selection by merging the new firm with the firm it copies. The effect can also be calculated separately. |
| Effect of entry based on incremental innovation | The effect normally has to be calculated separately. However, if the entrant is a spin-off from an incumbent firm, the two firms can be merged statistically. The consequence is that we increase the effect of intra-unit adaptation. |
| Creation of a new characteristic of the industry | Extend the list of characteristics for which information is collected in the next census. |
| Creation of a new industry by the first entrant | Extend the standard industrial classification of the next census. |

Comment: The coupling between evolutionary theory and the statistics of evolutionary processes is difficult. Although the coupling primarily serves to clarify the process of selection, it also helps to make explicit the processes of intra-unit adaptation and innovation. See also Table D.1 on page 438.

positive and does not change much over time, it does not have to be large to influence productivity significantly in the longer run. The only thing that is needed is that the firms actually vary with respect to productivity. If the variance is zero, no selection can take place. This brings us to the issue of intra-firm change. Since the data collected by the two censuses did not include information on the underlying routines of the firms, we have to stay at the uncomfortable level of the productivities. However, we are able to determine whether the variance of the productivities have decreased or increased. The variance must have decreased between the two censuses if no intra-firm change has taken place and if the regression coefficient is different from zero. The variance will ultimately also decrease if firms obtain 'new' productivities by copying the routines of other firms. Therefore, the upholding or increase of variance presupposes that some firms, by positive or negative 'innovation', obtain productivities that are new in the sense that they did not exist in the first census.

We have above only considered the simplest issues of evolutionary economic statistics. A few additional issues serve to remove this impression of simplicity. First, if the firms of the industry consist of multiple plants and if the censuses register to which firm each plant belongs, then the intra-firm change effect can be split into an inter-plant selection effect and an inter-plant change effect. This decomposition can be made mechanically, and the results will probably demonstrate whether the plants of a firm are sufficiently independent to make the exercise analytically useful. Second, the choice of the productivity race as the example of an evolutionary process might give a wrong impression about the selection mechanism (and the mechanisms of intra-firm change). This impression can be removed by returning to thinking in terms of homogeneous firms and by extending the set of characteristics that is registered by the two censuses of the industry. Then it becomes clear that productivity provides us with a special example of a "fitness function" (Conner and Hartl, 2004, Ch. 6). This fitness function displays positive directional selection while characteristics of firms are subject to negative directional selection. Moreover, the emergence of a standard represents stabilising selection while the segregation of an industry into two industries is sometimes the outcome of disruptive selection within the originally unified industry. Third, the analysis of multiple characteristics will also reveal that some of them are correlated. Thereby, trade-off problems emerge; and different industries seem to have evolved into different solutions to such problems. The three mentioned problems might suffice to demonstrate that the toolbox of evolutionary economics needs to be extended by new statistical methods and new theoretical concepts; but we cannot stop here. We also have to consider how evolutionary economics can move beyond the limits of a given population of firms.

Let us conclude the discussion of evolutionary economic statistics by connecting to the conceptual image of the evolutionary process that was given by Figure 12.2 on page 380. We have considered a statistical analysis of the evolutionary process within a single industry. The selection effect has been treated with some care; and we have also obtained an impression of the adaptive and innovative processes that take place within individual firms. However, we have ignored two crucial mechanisms: the mechanism of segregation and the mechanism of replication. It is the mechanism of segregation that creates the industrial and economic "ecosystem" with its complex web of interconnections. It is of crucial importance to make statistical analyses of the establishment of new industries, their destiny in their ecological context, and the macroscopic behaviour of the "economic organism". However, even the discussion of a few of the statistical and theoretical tasks will overburden the present exposition. Therefore, the comments on the ecological approach to economic evolution are found in Appendix D.1. In contrast, the mechanism of replication will engage us below. Therefore, it is important to note that we have not considered whether the inertia of the routines of firms is sufficient to allow for a process that can properly be called evolutionary. The chosen example of productivity might give a false impression of flexibility. Therefore, it seems better to think in terms of basic standards or basic technologies or basic solutions to complex trade-off problems. In any case, the statistical task is to study whether the studied characteristic shows sufficient inertia over a long period to allow an evolutionary process that is directly controlled by the inter-firm selection effect. If this is not the case, the next and more difficult task is to study the degree to which the process of intra-firm change mimics the selection that could have taken place through inter-firm selection. A large degree of firm response in accordance with the industry-level forces of selection will produce change that can be called evolution. However, it is also possible to think of cases in with the industry is not characterised by evolution but rather by a process that comes close to random drift.

## 12.5 Evolutionary economic history

In Schumpeter's original definition of the fundamental fields of the science of economics, he included economic theory, economic statistics, and economic history. He added that "[o]f these fundamental fields, economic history—which issues into and includes present-day facts—is by far the most important. I wish to state right now that if, starting my work in economics afresh, I were told that I could study only one of the three but could have my choice, it would be economic history that I should choose" (*History*, 12). He added that "[h]istory must of course be understood to include fields that have acquired different names as a consequence of specialization, such as prehistoric reports and ethnology" (p. 13). Actually, the economic history of the development of the capitalist economy in relation to the tax state *was* Schumpeter's original choice of research area (see Section 2.2). He had later expanded this interest in several directions. First, he had during his postgraduate studies in London emphasised the statistical anthropology of Karl Pearson and studied under the anthropologically oriented sociologist Edward Westermarck and the ethnographer Alfred Cort Haddon (*S1909a*, 75). Second, he had, to some extent, developed his evolutionary economics in relation to the research tradition of the German historical school—represented by Spiethoff, Sombart, and Max Weber. Third, he had continued his historical studies in relation to his sociologically oriented papers (*S1918b; S1919b; S1927g; S1928a*) and, especially, in relation to the production of *Cycles*. Therefore, he had a wide and relatively solid understanding of the function of economic history in the science of economics.

Since *History* concerns the fundamental fields of economics in general, Schumpeter could not present his arguments for the primacy of economic history relation to his evolutionary economics. However, it is in relation to the analysis of economic evolution that his three arguments obtain their real strength. His first argument is that "the subject matter of economics is essentially a unique process in historic time. Nobody can hope to understand the economic phenomena of any, including the present, epoch who has not an adequate command of historical *facts* and an adequate amount of historical sense or of what may be described as *historical experience*" (pp. 12–13). This argument "does not render 'theory' ... either impossible or useless" (p. 13n). Actually, "economic history itself needs its help." Nevertheless, economic history is "pregnant with important truth about economic life which, though refractory to cut and dried formulation, may convey us an intimate sense of organic necessities which we can never derive from mere theorems" (*S1948b*, 115). Schumpeter's second argument for the primacy of economic history is that it helps to understand the broader social framework in which the economic process takes place: "the historical report cannot be purely economic but must inevitably reflect also 'institutional' facts that are not purely economic" (*History*, 13). Actually, economic history provides "the best method for understanding how economic and non-economic facts *are* related to one another and how the various social sciences *should* be related to one another." He added that "I personally believe the study of history to be not only the best but the only method for this purpose" because of "the unreliability of 'theories' on this subject" (p. 13n). Schumpeter's third argument for economic history is that "the fundamental errors currently committed in economic analysis are due to lack of historical experience more often than to any other shortcoming of the economist's equipment" (p. 13). This is also the basic point of the paper on "The Historical Approach to the Analysis of Business Cycles" (see Section 10.5). In this paper, it becomes clear that Schumpeter did not think of "the historical approach" in general, but rather evolutionary economic history.

Paul Samuelson (1951, 100) was right in suggesting that Schumpeter delivered an "uncharacteristic performance" at the 1949 conference on business cycles; but Samuelson did not describe correctly what was uncharacteristic about the performance. It is not correct that Schumpeter gave "comfort to the enemy" since he had always supported theory-related historical studies. Samuelson probably missed this point because he did not understand the needs of rethinking the mechanisms of economic evolution by means of novel historical facts. However, Schumpeter's performance was indeed uncharacteristic because it lacked a clear definition of his message and because it failed to present his provocative message in a form that would not allow his audience to pass it over in silence. Schumpeter could, nevertheless, have tried to avoid this fate by rewriting his paper. Since he died before he had time to make the rewrite, we can only speculate about his possibilities. According to the evolutionary interpretation of Schumpeter, he could have changed the title of the paper to something like 'The Importance of the Historical Approach to Economic Evolution'. In the text of the paper he could then have emphasised three points. First, he could have emphasised that the production of a large collection of industrial and locational monographs is a long-term research programme that cannot deliver immediate applicable results for theoretical and statistical work. To motivate such a programme, he could have discussed the limited degree to which the already existing historical literature covered crucial details of economic evolution. Second, the change of the scientific climate after publication of *Cycles* had made impossible the unmodified revival of its research programme on evolutionary business cycles. Therefore, he could—as he actually did in *S1947d*—have tried to sell the historical approach as the necessary complement to the emerging studies of economic growth and economic development. However, he could have made an even better performance if he presented the historical approach as the necessary complement to the evolutionary economic theory that he had started to develop in *Wesen* and *Development*. Third, he could have developed his argument that any type of aggregative modelling, including the modelling of business cycles, will ultimately have to confront the problem of the evolution of the underlying industrial structure. Since he had already developed the details of this argument, it would not have been difficult to provide persuasive examples.

Although Schumpeter could have used the three mentioned points for rewriting his paper on the historical approach, he hardly would have done so. The problem is that a presentation of his evolutionary economics as the solution to the study of business cycles could never have performed well. It is more likely that he would have wanted to specify his plan for the production of detailed case studies. Thereby, he would have exchanged his audience from that of leading economic theorists and econometricians to that of practical economic historians. Maybe he could have found another conference for this purpose. What would his arguments have been at such a conference? As already emphasised, the suggested industrial and locational monographs do not support isolated historical studies of entrepreneurship. When specifying the common plan for such monographs, he would have had to define the evolutionary processes in which the historical function of the S-entrepreneur can be defined clearly. For this purpose, his general definition of the evolution of the overall economic system of routine is not adequate. However, if the historical study of evolution instead focusses on particular industries or locations, then the innovative function can be specified in a way that might be operational for historians. Actually, Schumpeter had already pointed at this solution in his unpublished "Comments on a Plan for the Study of Entrepreneurship" (*S1946c*, 420). Here, he had not suggested studies of individual entrepreneurship, but rather the need of coordinating the production of "the usual monograph of an individual industry or locality" according to a "comprehensive plan". Therefore, the proposal did not reflect the type collaboration Schumpeter had already developed with American economic historians.

In the 1940s, Schumpeter found historians who apparently were willing to apply his evolutionary concepts. They were led by the economic historian Arthur Cole. Cole (1942; 1946) was engaged in organising historical studies of entrepreneurship. The result was the establishment of the Harvard Research Center in Entrepreneurial History, which functioned under Cole's leadership from 1948 to 1958. Schumpeter provided crucial support for Cole's project and became a member of the centre who especially emphasised the theoretical underpinnings of the research (*S1946c; S1947c; S1947d; S1949c*). However, the major help Schumpeter could have got from the research of the many economic historians that worked with his concept on entrepreneurship at the Harvard research centre and elsewhere is rather limited. The help might be summarised as a pressure for removal of elitist connotations and for developing a more operational concept. This pressure was later formulated by Peter Kilby (1971) by his comparison of the problem of finding real entrepreneurs with the problem of "hunting the Heffalump". Readers of the children's books on Winnie-the-Pooh will know that the Heffalump is a strange animal that is never seen and so ill-defined that it could hardly have been recognised. Similar pressures for specification emerged from studies of the history of individual industries and the waveform evolution of whole countries. However, the difficulties were not really overcome even after Schumpeter's death. Actually, it was not least the vagueness of Schumpeter-inspired studies of economic history of, for instance, railroadization that helped to set the stage for the "new economic history" of researchers like Robert Fogel. By applying some of the tools of equilibrium economics, new economic history provided more limited, but also more precise, analyses of phenomena like the effects of the railway innovation than those researchers who had been working along Schumpeterian lines (O'Brien, 1977). Although Fogel's analysis gives very little help for the development of a precise analysis of the process of economic evolution, it serves a useful warning against the too-easy acceptance of, for instance, the railway innovation as the major cause of the Second Kondratieff.

From the viewpoint of evolutionary economics, a "research center in entrepreneurial history" is problematic because entrepreneurship can only be defined adequately in the context of the mechanisms of economic evolution. The entrepreneur becomes an undefined Heffalump when studied in isolation from this context. Schumpeter's contributions to the Harvard Research Center in Entrepreneurial History seem to have made this point clear. However, the point can also be derived from his "final thesis" that "what is really required is a large collection of industrial and locational monographs all drawn up according to the same plan" (*S1949g*, 328). Schumpeter died before he could develop a plan that gave "proper attention on the one hand to the incessant change in production and consumption functions and on the other hand to the quality and behavior of the leading personnel". However, even his sketchy remarks demonstrate that he was not trying to gain support for isolated studies on entrepreneurial history. Instead, he seems to have promoted a research effort that could have been called a research centre for the historical study of the mechanisms of economic evolution. This centre could have been divided in two sections. The first section would have studied the evolution of individual industries; and these studies would have had to relate to the types of problems that we above focussed on in relation to the theoretical and statistical approaches to economic evolution. The second section of the centre would have studied the more complex evolutionary process that takes place within geographically defined locations; and they would have had to confront the problems of evolutionary industrial ecology. It can, of course, be argued that these two lines of research have been promoted by the rapidly increasing literature on entrepreneurship and innovation. However, a large number of important topics have been left underexplored because of the lack of an analytical framework that systematically combines the mechanisms of innovation, replication, selection, and segregation.

Schumpeter's proposal of industrial and locational case studies seems related to *Cycles*. In this book, he had performed a wide-ranging historical study in order to come to grips with his proposition that business cycles reflect long-term economic evolution within the capitalist system. He had been especially happy with his results on the Second Kondratieff Wave (see Section 8.2). Actually, he had concluded that his theoretical approach served to emphasise

> "how railroad construction produces both prosperities and recessions—with the latter, situations which easily slide off into depressions—and, in particular, simultaneous cycles of different span. ... [T]he reader should not fail to work this out again step by step. For railroadization is our standard example by which to illustrate the working of our model. ... [Many factors] combine to make the essential features of our evolutionary process more obvious in this than they are in any other case. More easily than in any other can the usual objections to our analysis be silenced by a simple reference to obvious facts." (*Cycles*, 304; *Cycles Abr.*, 201–2)

This statement seems to suggest macroscopic historical studies of the process of railroadization. Such studies are very different from the detailed monographic works suggested by Schumpeter's "final thesis". When presenting this thesis, he must have recognised that the readers of *Cycles* had failed to work out his "standard example" of railroadization. Furthermore, he must have recognised that a comprehensive study of how railroadization transformed the economic system would not supply the necessary details on the working of the mechanisms of economic evolution. This seems to be the major background for his idea of coordinated industrial and locational monographs. The proposed large collection of monographs could have started with a subset of books that explores important details of the age of railway construction—both in its pioneering and its more mature stages. The pioneering period was the time when the horse-driven mail coaches were out-competed, schemes for financing railway projects blossomed and failed, industries supplying and using the railways were set up, railway towns mushroomed, and so on. The period of maturation was not least characterised by the routinisation of what earlier had been novelties. The maturation period also included the emergence of early forms of the modern corporation, partly as the forced outcome of financial crises and conspicuous examples of creative destruction. This grand vision of the age of railroadization suggests topics for monographs. The proposed locational monographs could have been exemplified by a study relating to his comments on the process of "the railroadization of the Middle West" of the United States (see the quotation on page 249 of the present book). Similarly, his industrial monographs could have been exemplified by a study of the evolution of the transport industry during the age of railroadization. One aspect of the evolution of the railway industry is covered by Alfred Chandler's (1977, Part II) history of organisational innovation and adaptation in the railway business. Actually, Chandler's study emerged from efforts that initially had relationships with members of the Harvard Research Center in Entrepreneurial History. Nevertheless, his study can hardly be considered a full-blown example of the application of Schumpeter's historical approach to economic evolution.

One of the reasons for Schumpeter's proposal of carefully designed monographs was probably that they could promote the understanding of his evolutionary theory. This pedagogical aim is well supported by nearly any evolutionary study of important aspects of the railway age. Actually, no other well-known period in the history of capitalist economic evolution leads so directly into the core of his mode of thinking and his analytical scheme. The writing of the historical parts of *Cycles*, however, must have convinced Schumpeter about the limitations of the pedagogical argument for the historical approach. Instead, he probably envisaged a give and take between the production of historical monographs and the theoretical specification of the mechanisms of capitalist economic evolution. Let us quickly consider some of the potential results of studies of aspects of the process of railroadization. First, the studies might promote a clear-cut distinction between S-entrepreneurs (railway promoters) and managers (of established railway companies). Second, the studies might explain why Schumpeter considered not only the first, but also the immediately following railway promoters to be innovators (and not simply imitators): they all had to overcome serious hindrances. Third, the studies might make clear the distinction between the early technological inventions (in the mines) and the primary economic innovations (like the Liverpool–Manchester railway). It would also clarify his enigmatic example of the jerkiness of innovation: "Add as many mail coaches as you please, you will never get a railway thereby" (*Development*, 64n). Actually, each of the passenger wagons of the Liverpool–Manchester line from 1830 consisted of three added mail-coach cabins; but the innovation also involved new engines, rails, and wagons for the transportation of goods. This new combination provided the evolutionarily relevant innovation. Fourth, the studies serve to demonstrate the degree to which it, for an epoch of economic evolution, is "easy to locate the ignition of the process and to associate it with certain industries and ... with certain firms" (*Cycles*, 102; *Cycles Abr.*, 76). Fifth, the studies demonstrate the conditional nature of theoretical propositions. For instance, the so-called Schumpeterian hypothesis of the positive relationship between firm size and innovation (Kamien and Schwarz, 1982) is obviously related to industries that can be described by means of the Mark II model and does not hold for the large firms of the age of railroadization. Sixth, and finally, the studies might support the more general understanding of the working of Schumpeter's Mark I model—both in its abstract version and in the versions provided by *Cycles*'s three approximations to the realities of economic evolution. In contrast, his more loosely specified Mark II model is often needed for the study of more recent processes of evolution. Although the best exemplification of Mark II seems to be provided by the chemical and electrical industries, we can also use this model for studying aspects of the process of "automobilization" or "motorization" (*Cycles*, 774, 1022; *Cycles Abr.*, 394). In any case, Schumpeter remarked that the "[t]he automobile industry" may qualify "for the role of standard example for the processes embodied in our model" (*Cycles*, 772; *Cycles Abr.*, 310).

By emphasising case studies of industries and locations rather than individual entrepreneurship, Schumpeter suggested that he wanted to combine the creative side and the destructive side of his "process of creative destruction". He defined this concept sloppily and it seems to have worrying links to Nietzsche's *Zarathustra* and Werner Sombart's writing during World War I (Reinert and Reinert, 2006) as well as to Bergson's *Creative Evolution*. Therefore, it is understandable why Ernst Helmstädter and Mark Perlman (1996, 1) considered it a "careless slogan" that should be dismissed: "Schumpeter wrote in 1942 of creative destruction as being the heart of progress, but by 1947 he had reconsidered his earlier careless slogan and substituted and stressed 'the creative response' in place of destruction." This way of avoiding the concept, however, ignores the fact that Schumpeter never abandoned his idea that "the process of creative destruction" describes the essence of capitalism and, probably, other types of economic and social change. In the mentioned paper from 1947 (*S1947c*), he emphasised the "creative response" because he was involved in the establishment of the Harvard Research Center in Entrepreneurial History. Nevertheless, he still considered this type of response to be an element of an overall process of creative destruction. Even at the above mentioned Conference on Business Cycles, he stated that "we must investigate historically the actual industrial processes that ... revolutionize existing economic structures" (*S1949g*, 326, emphasis removed).

Schumpeter's concept of the process of creative destruction points at an important task for evolutionary economic history; and the process of railroadization provides a large number of examples that need careful investigations. These investigations can hardly avoid combining the economic process of creative destruction with socio-political change. Actually, each of the early railways in the United Kingdom needed to be accepted by the Parliament. There was also, here and elsewhere, political lobbying against concrete railways, movements for protection against increased imports due to the railways, and attempts to organise labour to delimit the social consequences. By confronting this type of issue, the historical studies serve to solve problems that are often ignored by specialised theorists. The economic historian Douglas North (1998, 23) wanted, in his own terms, to confront questions "akin to those that Schumpeter raised in *Capitalism, Socialism and Democracy*. He did so by comparing the long-term growth of "the costs of transacting" with the "productivity gains from improvements arising from the increments to the stock of knowledge". Since transaction costs to some extent reflect the strength of the reactions against the social consequences of the capitalist engine, the outcome is uncertain. Christopher Freeman (2007, 139) seems to have been closer to the perspectives of evolutionary economic history when he recently discussed the idea of a "Schumpeterian renaissance". This "renaissance" includes "the resurgence of ideas about innovation, including industrial revolutions" and the main results "have enriched evolutionary theory in economics." Freeman especially based his conclusion on contributions to the analysis of the long-term economic fluctuations that are still often called Kondratieff waves. This analysis focusses on the interrelations between technological change, institutional change, transition periods and crises. The examples he provided included the works of Perez (2002) and Freeman and Louçã (2001); and we may add the millennial perspective on historical statistics (Maddison, 2003) and the thinking of economic transitions in terms of general purpose technologies (Lipsey et al., 2005). It is a major challenge for those engaged in developing the analytical toolbox of evolutionary economics to support this type of work on the grand questions of Schumpeter's "magnificent dynamics".

Historical studies of the mechanisms of economic evolution have to confront serious methodological problems. The above account has hinted as several of them. Let us consider one of the problems more carefully—the problem of counterfactual statements. We have already met counterfactual speculation in relation to the question 'what would Schumpeter have done with his paper on the historical approach if he had not died in January 1950?' This speculation could have continued by trying to answer the question 'what would have happened to the historical analysis of the mechanisms of evolution if Schumpeter had improved his paper?' The answers to such questions are based on the production of counterfactual history. Since these virtual histories are not based on solid theory, we quickly move into a quagmire of speculation. Nevertheless, the historical analysis of economic evolution actually involves the production of counterfactual histories. The more general need for counterfactuals was recognised by Max Weber. Weber (1906, 275) emphasised that "judgements of possibility" is an essential aspect of the writing of history. He argued that this type of evaluation is used to advance knowledge "[i]n every line of every historical work, and indeed in every selection of archival and source materials for publication". Let us, however, concentrate on the role of counterfactual history in the study of the process of railroadization and its consequences. Robert Fogel (1964) famously tried to answer the question: what would have happened to the GDP of the United States if the railway innovation had failed to take off around 1830? To answer this question, Fogel reconstructed the situation in 1890 so that it was without railways; in other words, he produced a virtual history that can be compared with the actual history. The comparison of the two histories led to the conclusion that economic historians had overrated the role of railways as an engine of economic growth. Jon Elster (1978, Ch. 6) has studied the logic of Fogel's procedure. Elster (p. 204) argued that, by taking the analytical starting point in 1830, "one could legitimately assume a branching point without railroads". However, Fogel used the naive procedure of removing from the actual 1890 economy all features directly linked to the railways. Although this procedure makes comparison relatively easy, it fails to cover the influences of the railways during the process of railroadization. This issue has been treated generally by Robin Cowan and Dominique Foray (2002) in their paper on the role of counterfactuals in evolutionary economics.

Cowan and Foray (2002, 548) accepted Elster's interpretation of counterfactuals as relating to a branching view of history. According to this view, history is comparable with the depiction of decisions by means of a decision tree. Each decision is represented by a branching of the tree; and the higher we move up in the tree, the more branches we find. Actual history can be compared with the movement in the decision tree from the root and upwards. At each branching point, actual history has followed one of the branches; and we end in a particular branch at the top of the tree. History could instead have followed other trajectories in its upward movement in the tree. Even remotely realistic versions of the tree of potential histories have to include a huge number of alternatives. Since previous historical events cannot be undone, history is irreversible and irrevocable. It represents a huge number of "decisions" at different branching points. In contrast, Fogel's study of the role of the railways was based on the assumptions that American economic history branched in 1830 and that the consequence of this branching could be studied by removing railways from the economic system of 1890. Cowan and Foray (2002, 550) emphasised that "to ask what would be the effect on GDP if railroads did not exist in 1890 is not to ask about the world in the instance that railroads were vaporised on the first of January 1890". The problem is that "some of the remaining physical and institutional structures would be historically incongruous". Therefore, we cannot avoid the detailed reconstruction of virtual histories without railways. To reduce the number of relevant virtual histories to a manageable level, we need to assume that the movement upwards in the tree is constrained and that we have theoretical knowledge about the constraints. However, a "central tenet of Evolutionary Economics ... is that there are many sources of indeterminacy in any economy" (p. 552). These sources of indeterminacy include details about the interacting agents, the process of learning, and the timing and characteristics of innovations. The result is that the theoretical model of the movement in the tree of alternative histories is underdetermined and has to be complemented by actual historical information. Thereby, the theoretical link between one state of the economy and a later state is weakened. Nevertheless, evolutionary economic theory does help the reconstruction of evolutionary economic history since it does specify some of the relatively loose constraints on the historical process. This description of the analytical situation explains why Cowan and Foray were sceptical about the applicability of comparative statics as an analytical tool. The problem is that the evolutionary process between the initial and the final state does not produce a unique outcome. This means that evolutionary "theory does not tell us what will happen, it only restricts us to a set of possibilities" (p. 553).

Cowan and Foray's paper seems to formulate a very general framework that is close to Schumpeter's way of thinking about economic evolution. This framework suggests many areas of collaboration between evolutionary economic theorists and evolutionary economic historians. The results produced by this alliance cannot be expected to be directly comparable to the type of "new economic history" that emerged from the work of Robert Fogel and other historians inspired by equilibrium economics and by relatively simple quantitative methods. However, as demonstrated by evolutionary economic statistics, the evolutionary complement does not have to be dominated by qualitative reasoning. Moreover, the collaboration cannot be expected to be asymmetrical in the sense of a stream of ideas from theorists to historians. On the contrary, evolutionary theorists seriously need to be confronted with new historical facts that can force them to improve the analytical toolbox of evolutionary economics. This work might return cautiously to Schumpeter's "magnificent dynamics" of capitalist economic evolution in terms of "Kondratieff waves". Presently, it nevertheless seems to be Schumpeter's type of detailed historical case studies of the major mechanisms of economic evolution that are most crucial for the further development of evolutionary economics.

## 12.6 Evolutionary economics as a whole

In relation to equilibrium economics, it has been suggested that Schumpeter must be "placed in the category 'footnote economist', that is, an economist whose works are mentioned in footnotes but are seldom reported and applied more directly in a theoretical interpretation or further development" (Jensen, 1988, 97). He has obtained a very different role within parts of the evolutionary economics that started to take shape in the 1970s and 1980s. Especially within the part of evolutionary economics that studies the processes of innovation and technical change, Schumpeter's work could be considered the main text, while subsequent research served to annotate this text. The economic historian Nathan Rosenberg (1982, 106) remarked that "the study of technological innovation", in the broad sense, "still consists of a series of footnotes upon Schumpeter. Although the footnotes may be getting longer, more critical and, happily, richer in the recognition of empirical complexities, we still occupy the conceptual edifice which Schumpeter built for the subject." Since this remark was written, we have been facing a Schumpeter revival in which larger parts of his work has been confronted and developed. Christopher Freeman (2007, 139) concluded "that the 'Schumpeterian renaissance' has been a real phenomenon" and that its main results "have enriched evolutionary theory in economics." Nevertheless, the continued strong presence of Schumpeter in evolutionary economic research is a sign of the immaturity of this important branch of economics. Even within evolutionary economics, Schumpeter will ultimately have to be moved to the footnotes. This is actually the place he had hoped for, at least since 1911. At that time, his preface to *Theorie der wirtschaftlichen Entwicklung* stated that he wished "that this work as soon as possible shall be surpassed and forgotten" (*Entwicklung I*, viii). Evolutionary economics seems to be approaching a state in which this wish has been fulfilled.

It is especially difficult to surpass Schumpeter's grand vision of the working of the capitalist engine and his idea of evolutionary economics as providing a comprehensive toolbox for analysing this engine. With respect to vision, the task is to make explicit the alternative visions that can actually be found underneath much of present-day research. With respect to the toolbox of evolutionary economics, the first task is to recognise its comprehensive nature. In 1908, Schumpeter defined evolutionary economics ("Dynamics") as an emerging complement to equilibrium economics ("Statics"). Although he initially pointed at the evolutionary economic *theory* that he later developed in his evolutionary trilogy, he became increasingly aware that a viable evolutionary economics also needs to include evolutionary economic statistics and evolutionary economic history. The co-evolution of these three fields cannot be taken for granted. Schumpeter provided means of coordination by his vision of capitalist economic evolution and by his core concepts. While his basic vision has become commonplace and we need to try out alternative visions, his models might still have a coordinating role—especially if we are able to develop the mechanisms and structural features of Mark I, Mark II, and combinations of them. The strategy suggested in the present chapter might be called coordination by simplification. This strategy suggests that the Mark I model still has an important role in defining related tasks of theory, statistics and history. Moreover, Schumpeter's Mark I model should not be abandoned too easily, since Mark II is not a single model, but a huge family of models that does not necessarily provide the needed coordination across research fields. Furthermore, it is on the background of the careful study of Mark I that we can formulate difficult questions about the evolutionary mechanisms of the family of Mark II models. Finally, the strategy of successive approximations will allow us to judge the degree to which we have moved beyond the Schumpeterian models.

The present chapter has abstained from commenting on the large number of actual and potential fields of applied evolutionary economics. The applied fields include evolutionary organisation science, evolutionary finance, evolutionary industrial dynamics, evolutionary economic geography, evolutionary development studies, evolutionary environmental economics, and so on. The emergence of such applied fields clearly demonstrates that we have moved beyond Schumpeter's evolutionary economics. Although these fields cannot be called "fundamental" in Schumpeter's sense, they are nevertheless crucial for the viability of evolutionary economics. Schumpeter never contributed to applied economics, but he remarked that "it is impossible to divorce any of the applied fields from the fundamental ones" (*History*, 24). From the perspective of the fundamental fields, his major reason for avoiding divorce is that "the applied fields not only apply a stock of facts and techniques that lies ready for their use in general economics but also add to it." Many of these additions are of little general relevance. However, the applied fields of economics "have repeatedly developed accumulations of facts and conceptual schemata that should be recorded as contributions to general economic analysis". Since such contributions are most likely in immature areas of analysis, the alliance between the fundamental fields and the applied fields is of special importance for evolutionary economics.

# Appendices

# A
# Chronology

**Personal information**

| | |
|---|---|
| 1883 | Born on 8th February in Triesch (now: Třešť) in the Moravian part of Austria-Hungary (now in the Czech Republic). Father: the textile manufacturer Josef A. K. Schumpeter; mother: Johanna M. Schumpeter. |
| 1887 | Death of the father. Next year the mother and her son move to Graz in modern Austria. |
| 1893 | The mother marries the retired lieutenant general Sigmund von Kéler, and the family moves to Vienna. |
| 1907–20 | Marriage with the English Gladys Ricarde Seaver/Schumpeter. They, in practice, separated in 1914. |
| 1925 | Obtains German citizenship. |
| 1925–26 | Marriage with the Austrian Anna Josefina Reisinger/Schumpeter; but after less than a year, she dies, together with the baby, under the birth. |
| 1933 | Applies for American citizenship, which is obtained in 1939. |
| 1937–50 | Marriage with the American economic historian Elizabeth Boody Firuski/Schumpeter. |
| 1950 | Dies from cerebral haemorrhage on 8th January in Taconic, Connecticut, USA. He thus became nearly 67 years. |

**Education and degrees**

| | |
|---|---|
| 1888–93 | Elementary school in Graz. |
| 1893–1901 | Grammar school education at the prestigious Theresianum Academy of Knights in Vienna. |
| 1901–05 | Studies law and economics at the University of Vienna (see Table 2.1 on page 31). In the beginning of 1906, the title Doctor of Laws is obtained. |
| 1906–07 | Research education at the University of Berlin and in England (based at the London School of Economics). |
| 1909 | The licence to teach at universities (*Habilitation*) is obtained based on the book on the essence and contents of economics (*Wesen*). |

**Employment**

| | |
|---|---|
| 1907–08 | Associate in an Italian law firm at the International Tribunal in Cairo (Egypt). |
| 1909 | Unpaid lecturer (*Privatdozent*) at the University of Vienna. |
| 1909–11 | Extraordinary Professor at the University of Czernovitz (now: Tjernovtsy, then in Austria-Hungary, now in Ukraine). |
| 1911–21 | Professor at the University of Graz (Austria). |
| 1913–14 | Guest Professor at Columbia University in New York. |
| 1919 | Minister of Finance of the Republic of Austria, from March to October. |
| 1921–25 | President of Biedermann Bank in Vienna. |
| 1925–32 | Professor at the University of Bonn (in Germany). |
| 1927–28 | Guest professor at Harvard University. |
| 1930 | Guest professor at Harvard University. |
| 1932–50 | Professor at the Department of Economics of Harvard University. |

**Core academic functions**

| | |
|---|---|
| 1916–17 | Dean of the Faculty of Law at the University of Graz. |
| 1919 | Member of the Socialisation Commission of Germany. |
| 1940–41 | President of the Econometric Society. |
| 1948 | President of the American Economic Association. |
| 1949 | Elected to become the first President of the International Economic Association in 1950. |

**Academic memberships and co-editorships**

Philosophical Society of the University of Vienna (organised by Alois Höfler).
International Statistical Institute of Paris.
Royal Statistical Society, London.
Verein für Sozialpolitik (i.e., the German Economic Association).
Academic Association for the Social Sciences, Czernovitz.
Sociological Society in Graz.
*Archiv für Sozialwissenschaft und Sozialpolitik* (co-editor with Max Weber, Sombart, Lederer, et al.).
American Economic Association.
Econometric Society (founding member).
*Econometrica* (co-editor with Ragnar Frisch et al.).
*Review of Economic Statistics* (co-editor with Crum, Haberler, Hansen, Harris, Leontief, Mason, et al.).

Harvard Research Center in Entrepreneurial History (with Arthur Cole et al.).
*Kyklos* (co-editor with Edgar Salin et al.).
*Quarterly Journal of Economics* (co-editor with Edward Chamberlin et al.).

**Publications**

| | |
|---|---|
| 1905–50 | Nearly 200 papers, especially in German and English. See the list of Schumpeter's "Papers and Reviews" that starts on page 449. |
| 1908 | *Das Wesen und der Hauptinhalt der theoretischen Nationalökonomie* (Wesen; see Table 3.1 on page 46). |
| 1912 | *Theorie der wirtschaftlichen Entwicklung* (Entwicklung *I*; see Table 5.1 on page 103). |
| 1914 | "Epochen der Dogmen- und Methodengeschichte"; translated as *Economic Doctrine and Method: A Historical Sketch* (Doctrine). |
| 1915 | *Vergangenheit und Zukunft der Sozialwissenschaften* (Vergangenheit). |
| 1926 | Radically revised second edition of *Theorie der wirtschaftlichen Entwicklung* (Entwicklung *II*). |
| 1934 | *The Theory of Economic Development* (Development; see Table 5.1 on page 103 and Table 9.1 on page 244); it is the slightly revised translation of Entwicklung *II*. |
| 1939 | *Business Cycles: A Theoretical, Historical, and Statistical Analysis of the Capitalist Process* (Cycles; see Table 8.1 on page 194). |
| 1942 | *Capitalism, Socialism and Democracy* (Capitalism; see Table 7.1 on page 158). |
| [1951] | *Ten Great Economists: From Marx to Keynes* (*TGE*), ed. Elizabeth B. Schumpeter. |
| [1951] | *Essays on Entrepreneurs, Innovations, Business Cycles, and the Evolution of Capitalism* (*EE*), ed. Richard V. Clemence. |
| [1954] | *History of Economic Analysis* (History; see Table 11.1 on page 334), ed. Elizabeth B. Schumpeter. |
| [1970] | *Das Wesen des Geldes* (Geldes), ed. Fritz K. Mann. |
| [1991] | *The Economics and Sociology of Capitalism* (*ESC*), ed. Richard Swedberg. |

# B

# Literature on Schumpeter

The present book includes very little biographical information on Schumpeter. The primary exceptions are found in Chapters 2 and 10, where some information is given on the contexts in which he worked before and after World War I. The present appendix gives some pointers to the Schumpeter literature.

The literature on Schumpeter's life and work consists of specialised papers and monographic treatments. The contributions published before 1990 have been catalogued by Massimo Augello (1990), and Cunningham Wood (1991) and Horst Hanusch (1999) have collected large selections of the papers. The monographs are of two types. On the one hand, we have the personal biographies by Loring Allen (1991), Richard Swedberg (1991), Wolfgang Stolper (1994), Thomas McCraw (2007) as well as accounts that include much biographical material like the books by Erich Schneider (1975) and Eduard März (1991). On the other hand, the accounts for Schumpeter's works include books by Clemence and Doody (1950), Khan (1957), Perroux (1965), Oakley (1990), Bottomore (1992), Vecchi (1995), Shionoya (1997), Reisman (2004), and Heertje (2006). The extensive account for neo-Schumpeterian economics that is edited by Horst Hanusch and Andreas Pyka (2007) also includes much of interest for the interpretation of Schumpeter's work.

The first wave of biographical and interpretative work started immediately after Schumpeter's death in 1950. This was a period during which the main issue for ambitious economists was the development and application of the tools of equilibrium economics and the related econometrics. Apart from trying to collect facts about his life and work, the early literature emphasised Schumpeter's promotion of these kinds of activity through his ardent support for the efforts of his students—as well as the paradoxical nature of his evolutionary contributions. Such a picture was developed by Schumpeter's friends, colleagues, and students who made joint efforts to commemorate him as a great economist and scholar, and one of them, Seymour Harris (1951b), collected many of their papers for the book *Schumpeter, Social Scientist*. Both the title of the book and Harris's introductory paper reflect the difficulty Schumpeter's memorialists had in finding and defining his core scientific contribution. Most of them were economists—including Edward Chamberlin, Ragnar Frisch, Gottfried Haberler, Alvin Hansen, Paul Samuelson, Erich Schneider, Wolfgang Stolper, and Jan Tinbergen—but the title of the book suggests that Schumpeter's greatness largely was due to his efforts as a social scientist of broad interests.

The second wave of biographical and interpretative work started in connection with 1983 centenary of Schumpeter's birth; and here a new generation of researchers joined forces with his contemporaries. Since the memory of Schumpeter had waned, this wave of accounts used the simultaneous centenaries of Keynes's birth and Marx's death to produce comparative accounts (Bös and Stolper, 1984; Helburn and Bramhall, 1986; Wagener and Drukker, 1986; and Bharadwaj and Kaviraj, 1989). These comparisons were very general; and they normally failed to make clear specifications of the scientific tasks in "The Age of Schumpeter", to use the formulation by Giersch (1984). The larger works that were published some time after the centenary tried to overcome this problem. However, each of these works tended to concentrate on one of the many incarnations of Schumpeter—the modernising teacher, the neoclassical economist, the evolutionary economist, the economic historian, the sociologist, and the historian of economics. Thereby, the image of Schumpeter became rather diffuse.

One of the reasons for producing specialised accounts is that Schumpeter was a very versatile researcher who worked in very different environments. When contributing to the analysis of his efforts, each author has had to concentrate on a main perspective even if it meant a concentration of certain parts of Schumpeter's works and academic efforts. His many close colleagues and research students have been major contributors; but as already mentioned, most of them got surprisingly little information from him on his core research activities. Instead, they found it natural to emphasise his effort for *Opening Doors*. This is the title of the most extensive Schumpeter biography; and it is written by one of his research students, Loring Allen (1991). Allen's book provides detailed biographical information; but we find little on the development of Schumpeter's core scientific concerns. Instead, the book promotes the image of Schumpeter as an economist who found his own way during his creative youth and who helped others to do the same by "opening doors" for them. In addition, the book paints an image of Schumpeter as a paradoxical and ultimately frustrated economist with serious psychological problems. On the other hand, the policy-oriented economist Wolfgang Stolper was one of those students how shared Schumpeter's core concerns and who tried to correct Allen's (1991:II, 262) picture of "a multifaceted man of paradox". Stolper (1994, 373, 377) also suggested that "Schumpeter's is a truer vision than most, even all, others"; and that "many of his 'visions' have become so commonplace that the present generation can hardly imagine how radical they once were". However, Stolper's book concentrated on documenting Schumpeter's policy-oriented efforts.

Stolper had been one of Schumpeter's students both at Harvard University and at the University of Bonn. Other German and Austrian research students and colleagues have also contributed to the Schumpeter literature. For instance, the general economist Erich Schneider joined Schumpeter at Bonn. In his biography, Schneider (1975) emphasised Schumpeter's general efforts to promote economics as a theoretical science. In contrast, Eduard März (1991) was an Austrian who studied under him at Harvard; and März emphasised that Schumpeter was not least an economic historian like himself. This interpretation was further developed by Thomas McCraw (2007). As a business historian who emphasises entrepreneurship, McCraw has been able to expose important aspects of Schumpeter's work. However, the very old contribution by the versatile French economist François Perroux (1965) still seems one of the best in critically catching the basic research questions that underlie much of Schumpeter's research. Perroux concentrated on Schumpeter's core concerns because he originally designed his intellectual biography as an excessively long introduction to the French translation of *Entwicklung II* (*Théorie de l'évolution économique, S1935e*). Perroux's introduction included a short biographical account to explain the role of this book in Schumpeter's overall research programme. According to Perroux, this programme served to combine the Austrian school with the contributions of Walras and Pareto as well as with the German historical school.

Sociologists with an interest in economic institutions and their change have focussed on some of the aspects of Schumpeter's work that were emphasised by Perroux. Even though Schumpeter was a worshipper of hard-core economics, it is clear that he considered sociological investigations as a necessary complement to the economic analysis of economic evolution. This view is not only present in *Capitalism,* but throughout his writings—with especially important contributions in his minor works (like *S1918b, S1919b,* and *S1927g*). An interest in how Schumpeter functioned as a theorist who pointed toward a comprehensive treatment of capitalism and its institutions has inspired the short intellectual biography by Tom Bottomore (1992). It has also inspired the book by Richard Swedberg (1991), which is presently the standard account for Schumpeter's life and work. Both of Bottomore and Swedberg have interpreted Schumpeter as a central social theorist who, like Marx and Max Weber, sought to grasp and analyse the core of capitalist society in an interdisciplinary way. Swedberg especially emphasised that Schumpeter's efforts can largely be described as an attempt to implement Weber's large-scale programme for a broadly defined "social economics" ("Sozialökonomik").

# C

# Accessing and Grouping Schumpeter's Works

While practically all of Keynes's (1971–89) works are found in the easily available *Collected Writings*, we have nothing like the "Collected Writings of Joseph A. Schumpeter". Furthermore, some of his core contributions are still only available in German language—a few even printed with old-fashioned German types, which might look strange in the eyes of some readers. A short bibliography was produced by Elizabeth B. Schumpeter (1950)—and she also promoted the collecting and translation of Schumpeter's writings. Massimo Augello (1990) provided what is presently the standard bibliography of works by and on Schumpeter; and Ulrich Hedtke (2007a) has made available an updated and easily available list of Schumpeter's works.

Since the world of economics is increasingly an Anglophone world, the present listing of "Schumpeter's Works" (starting on page 447) emphasises their English availability. This implies that reference is made to English translations whenever they are available. Furthermore, the present author has translated a significant number of quotations from the German works. Finally, the present book's references to Schumpeter's papers emphasise easy availability by pointing at collections of papers.

## C.1 The Schumpeter Archives

There are still many unpublished letters and notes by Schumpeter. These documents are mainly found in the Harvard University Archives as the "Papers of Joseph Alois Schumpeter (1883–1950)", which was described by Loring Allen (1991:II, 277–9). Some documents not found elsewhere have been included in Ulrich Hedtke's (2007b) list of Schumpeter's letters.

A special problem of using the Schumpeter archives is that many notes and parts of manuscripts are written in Gabelsberger shorthand. This system of stenography was probably taught at some Austrian grammar schools and it was used extensively by many Austrian researchers (see Fasol-Boltzmann, 1990, 9–11; Dawson, 1995). However, the Gabelsberger system was hardly used in Anglo-Saxon countries; it was radically reformed on the European Continent around 1925; and stenography in general is a craft that today is nearly extinct.

Few of the many paradoxes that attach to Schumpeter's name are likely to be solved by studying the surviving archives. The main problem is that Schumpeter did not bring his personal archive with him when he left Germany in 1932. Instead, this pre-1932 archive was left with a German family; and the archive was destroyed during World War II. Therefore, we largely have to solve the paradoxes by studying the available publications.

## C.2 Collections of Schumpeter's papers and letters

Collections with English papers

- *ATP* = *Aufsätze zur Tagespolitik*. The volume has a somewhat misleading title. It not only contains many papers on current affairs from Schumpeter's period as Bank President but also papers from his career as an academic researcher. Some of the collected papers are in English language. They include *S1906d, S1928c, S1930c, S1932f*, and *S1949b*.

- *BL* = *Briefe/Letters*. This volume collects 245 letters, half of them in English. These letters have not been included in the present book's list of papers (starting on page 449). Therefore, the reader will, for instance, have to check that *BL* (pp. 321–2) refers to the letter written on 18 March 1940 to Simon Kuznets.

- *EE* = *Essays on Entrepreneurs, Innovations, Business Cycles, and the Evolution of Capitalism*. This volume contains 25 papers written in English. The included papers are: *S1909b, S1927i, S1928f, S1930e, S1931g, S1933c, S1934a, S1934c, S1934b, S1935d, S1936b, S1936c, S1937, S1940a, S1943, S1946b, S1946f, S1947c, S1947d, S1948c, S1949f, S1949c, S1949d, S1949g*, and *S1949e*. The volume also includes Elizabeth B. Schumpeter's (1950) bibliography of Schumpeter's works.

- *ESC* = *The Economics and Sociology of Capitalism*. This volume contains 12 sociologically oriented papers in English, some of which are newly translated or published from English manuscripts. The major papers are: *S1918b, S1919b, S1927g*, and *S1946c* (large version of *S1947c*). The other papers are: *S1920a, S1931c, S1936a, S1940b, S1941b, S1946g, S1948d, S1949a*.

- *TGE* = *Ten Great Economists*. This volume contains 10 papers on famous economists: Marx (*S1942a*), Walras (*S1910c*), Menger (*S1921*), Marshall (*S1941a*), Pareto (*S1949h*), Böhm-Bawerk (*S1914c*), Taussig

(*S1941c*), Fisher (*S1948a*), Mitchell (*S1950c*), and Keynes (*S1946d*). It also contains three small papers: *S1926a*, *S1927c*, and *S1932d*.

Collections with untranslated German papers

- *AÖT* = *Aufsätze zur ökonomischen Theorie.* This volume contains 16 papers on economic theory. The untranslated papers are *S1906a*, *S1907a*, *S1908a*, *S1913a*, *S1913c*, *S1914a*, *S1920b*, *S1924*, *S1925c*, *S1927a*, *S1928e*, *S1929c*, and *S1932b*.

- *AS* = *Aufsätze zur Soziologie.* This volume contains 4 papers on sociology. The only untranslated paper is *S1929a*.

- *ATP* = *Aufsätze zur Tagespolitik.* This mixed volume contains not only papers on current affairs from his periods as Bank President but also academic papers. It includes *S1911*, *S1913b*, *S1918a*, *S1923a*, *S1923b*, *S1926c*, *S1929d*, *S1931a*, and *S1933a*.

- *AWP* = *Aufsätze zur Wirtschaftspolitik.* This volume documents Schumpeter's thinking about public policy. It includes *S1916–17*, *S1919a*, *S1927j*, *S1928d*, *S1929b*, and *S1931b*.

- *BL* = *Briefe/Letters.* This volume contains 245 letters in two languages, half of them in German.

- *BSÖ* = *Beiträge zur Sozialökonomik.* This is a collection of 18 of Schumpeter's papers that are either in or translated into German. Only one paper has not been included elsewhere: *S1914b*.

- *DBA* = *Dogmenhistorische und Bibliographische Aufsätze.* This volume contains 14 papers, which are largely untranslated, on the history and biography of economics. The papers include treatments of Oppenheimer (*S1917a*), Edgeworth (*S1925e*), Schmoller (*S1926b*), economic theory in Germany (*S1927b*), Wicksell (*S1927d*), Sombart (*S1927f*), and Cassel (*S1927h*).

- *PR* = *Politische Reden.* This volume documents Schumpeter's period as Austrian Minister of Finance in 1919.

- *RBZ* = "Reden in der Bonner Zeit". This is a collection of the full texts or summaries of 11 speeches held in German by Schumpeter in the period 1926–32. It includes *S1928b*.

## C.3 Translating Schumpeter's German texts

Until 1933, Schumpeter largely wrote in German and many of his works have not yet been translated into English. The present book tries to give an impression of this part of his work by including many translated quotations—but it is not easy.

The first problem is that in some German journals (but not in the Austrian ones) his writing was typeset in old German fonts. The typeface used in *Schmollers Jahrbuch* was a version of the Schwabacher font (see Mittelbach and Goossens, 2004, 394–6; Haralambous, 1991). Thus the beginning of Schumpeter's paper on Schmoller in Schmollers Jahrbuch, more or less, has the following appearance (*S1926b*, 148):

> Gelegentlich hören und lesen wir aus Amerika Symptome von etwas, das man nur als latenten Methodenstreit bezeichnen kann. [...] Und dieser Kampf ist nicht einmal so latent. In mehr oder weniger urbaner Form hält eine jüngere Generation der älteren, die ihre charakteristische Note vom Ideenkreise J. B. Clarks empfing, ein Sündenregister vor, das uns bekannt anmutet: Unrealität, Irrelevanz, Interesselosigkeit der Auffassungsweisen und der Resultate — klingt das nicht ganz wie das, was die „jüngere historische Schule" gegen die ökonomische Wissenschaft anzuführen hatte, die sie vorfand?

This quotation has to be converted into modern typeface. In slightly anglicised typesetting it reads:

> "Gelegentlich hören und lesen wir aus Amerika Symptome von etwas, das man nur als latenten Methodenstreit bezeichnen kann. [...] Und dieser Kampf ist nicht einmal so latent. In mehr oder weniger urbaner Form hält eine jüngere Generation der älteren, die ihre charakteristische Note vom Ideenkreise *J. B. Clarks* empfing, ein Sündenregister vor, das uns bekannt anmutet: Unrealität, Irrelevanz, Interesselosigkeit der Auffassungsweisen und der Resultate—klingt das nicht ganz wie das, was die 'jüngere historische Schule' gegen die ökonomische Wissenschaft anzuführen hatte, die sie vorfand?" (*S1926b*, 148)

Then we turn to the real problem, the translation:

> "From America [e.g., from Mitchell, 1925; and Tugwell, 1924] we occasionally hear and read of symptoms of what has to be called a latent battle of methods. ... Actually, this battle is not that latent. Against the older generation, which received its characteristic tone from J. B. Clark's circle of ideas, a younger generation, more or less politely, points out a list of sins that appears well known: unreality, irrelevance, uninteresting views and results. —Does that not sound exactly like what the 'younger historical school' remarked against pre-existing economic science?" (*S1926b*, 148)

It is obvious for those reading German that we are not facing a direct translation. First, sentences have been rearranged and words have been added to increase readability. Furthermore, several translational issues have emerged. For instance, "Ideenkreis" could be translated by 'ideology' but it has cautiously been translated by 'circle of ideas'. Third, the German-style setting of authors with italics (like *J. B. Clark*) has been changed without notice. Finally, omitted text is simply marked by three dots while other insertions are placed in brackets.

These principles have in the present book been used for translating (without notice) quotes by Schumpeter and by other authors.

## C.4 Subjects of Schumpeter's works

Augello (1990, 37–111) classified the works on Schumpeter in the categories "biography", "methodology", "development", "money", "cycle", "sociology", "politics", and "history". The integrated nature of much of Schumpeter's work excludes a similar treatment of his individual books and papers. However, a rough and very incomplete classification shall nevertheless be attempted.

- The state of economics: *Wesen, S1910a, Doctrine* (Ch. 4), *Vergangenheit* (81–136), *S1926b, S1927b, S1931c, S1931d, S1931e, S1932b,* and *History* (Pt V).
- The study of economics: *S1909b, S1910b, S1928e,* and *S1929c.*
- Mathematics and statistics in the science of economics: *S1905a, S1905b, S1905c, S1906a, Wesen, S1927d, S1930e, S1933c, S1946a, S1947a,* and many places in *Cycles* and *History.*
- The basic analysis of economic evolution: *Wesen* (especially 176–86, 414–30, 614–22), *S1910d, Entwicklung I, S1912a, S1912b, S1913a, Entwicklung II, S1928a, S1928f, S1931a, S1932c, Development, S1935d, S1937, Cycles* (especially 72–129), *Capitalism* (Pt II), *S1946c, S1947c, S1947d, S1949c,* and *S1949a.*
- Distribution, profit, entrepreneur, monopoly, and economic evolution: *S1906b, S1907a, Wesen* (313–440), *S1908a, S1910a, S1911, Entwicklung I* (Chs 2 and 4), *S1914d, S1916b, S1917a, Development* (Chs 2 and 4), *S1927j, S1928a, S1928d, S1934b, Cycles* (56–68, 103–9), and *S1948d.*
- Money, credit, interest, and economic evolution: *Wesen* (276–97), *Entwicklung I* (Chs 3 and 5), *S1913a, S1913c, S1917b, S1923b, S1925c, S1925d, S1927a, Geldes, S1930b, S1930f, Development* (Chs 3 and 5), *Cycles* (109–29), and *S1948c.*

- Business cycles and waveform economic evolution: *S1906c, S1909c, S1910d, Entwicklung I* (Ch. 6), *S1914b, Entwicklung II* (Ch. 6), *S1926c, S1927i, S1928f, S1930e, S1931b, S1931g, S1931h, S1932f, S1933a, Development* (Ch. 6), *S1934a, S1935d, S1935b, S1935f, Cycles, S1946f, S1949g, S1950a,* and *S1950c.*

- Socio-economic and political evolution: *S1907b, S1911, Entwicklung I* (492–548), *S1912a, S1912b* (109–42), *S1918b, S1919b, S1920b, S1924, S1926b, S1927g, S1927f, S1927j, S1928b, S1928f, S1929a, S1929b, S1929d, S1932e, S1936a, S1941b,* Capitalism, *S1942a, S1944b, S1946b, S1946g, S1946e, S1948d, S1949a, S1949b, S1949f, S1949d,* and *S1950b.*

- Public finance, economic policy, and general policy: *S1916a, S1916–17, S1916–18, S1918b, S1919a, S1925c, S1925d, S1927a, S1928c, S1928–29, S1931f, S1932f, S1934c, S1940a, S1945, S1946g, S1948c,* and *S1948d.*

- History and methodology of economics as well as the Schumpeterian theory of scientific evolution: *Wesen* (7–21), *Entwicklung I* (87–101), *S1913b, Doctrine, Vergangenheit, S1914a, S1927b, S1931c, S1931d, S1931e, S1940b, S1948b, S1949e,* and *History.*

- Individual economists: *History* covers all major economists and numerous minor ones. Further treatments include: Auspitz (*S1906d, S1930d*), Böhm-Bawerk (*S1914c, S1925a, S1930a*), Bortkiewicz (*S1932d*), Cassel (*S1927h*), Clark (*S1906b, S1908b, S1910a*), Edgeworth (*S1925e*), Irving Fisher (*S1948a*), Hayek (*S1946e*), Keynes (*S1933b, S1936c, S1946a, S1946d*), Knapp (*S1926a*), Marshall (*S1941a*), Marx (*S1918a, S1942a, S1949f*), Menger (*S1921*), Mitchell (*S1930e, S1950c*), Oppenheimer (*S1917a*), Pareto (*S1910e, S1949h*), Rist (*S1932a*), Joan Robinson (*S1934b*), Schmoller (*S1926b*), Sombart (*S1927f*), Stigler (*S1942b*), Taussig (*S1936b, S1941c*), Walras (*S1910c, S1935c*), Max Weber (*S1920a*), Wicksell (*S1927d*), Wieser (*S1927c*), Allyn Young (*S1935a*), and Zeuthen (*S1930c*).

- Information on Schumpeter: *S1901–05, S1909a, S1916a, S1925b, S1927e,* and *S1944a.*

# D
# Some Tools for Evolutionary Analysis

The present appendix deals with some of the analytical problems that Schumpeter faced in his treatment of the evolutionary processes that take place between and within 'industries'. The suggested solutions are treated under the headings the ecological approach (Section D.1) and the statistical approach (Section D.2). Although these solutions do not include economic variables like profit and credit, they can nevertheless be classified as relating to, respectively, evolutionary mesoeconomics and evolutionary microeconomics (Dopfer and Potts, 2008). The theoretical, historical, and statistical analysis of the capitalist engine requires a lot of other analytical tools. For instance, there is an obvious need of developing the tools for evolutionary macroeconomics. Furthermore, the algorithmic reconstruction of evolutionary processes and their implementation in computer simulation models have proved crucial (Nelson and Winter, 1982; Andersen, 1994). Nevertheless, a short presentation of the ecological approach and the statistical approach serves to demonstrate that formalisation can clarify existing concepts of evolutionary analysis and suggest new concepts.

## D.1 The ecological approach to evolutionary analysis

Schumpeter's treatment of the process of creative destruction in the historical parts of *Business Cycles* (see Section 8.2) included much attention to the structure of the "industrial organism". When we take the structure of this "organism" into account, it becomes clear that he was considering a much more complex process of creative destruction than the one covered by the reconstructions of the Mark I and Mark II models of the capitalist engine (see Section 9.5). It also becomes clear that he seriously needed analytical tools for handling this complexity. The nature of some of these tools becomes clear if we translate his old-fashioned term "industrial organism" by the more appropriate term "industrial ecology". It is not difficult to find out that tools for analysing this industrial ecology have to some extent been provided by biologists. The study of the relevance of the ecological approach for the study of creative destruction is helped by the fact that Schumpeter made some comments on its applicability and to some extent seems to have used it to guide his historical narrative in *Cycles*. The approach has obtained more precise contours due to its explicit application by the economic sociologists and organisation theorists Michael Hannan and John Freeman (1989) and due to the toolbox for this approach presented by Glenn Carroll and Freeman (1999). The ecological approach may also be formulated in terms of qualitative change (Saviotti, 2007) and self-organising complex systems (Arthur, 2007; Allen, 2007; Pyka and Fagiolo, 2007). Moreover, the formal analysis of evolutionary dynamics in ecological context by Josef Hofbauer and Karl Sigmund (1998, xviii–xix) has demonstrated that "ecology is the godfather of evolutionary game theory". However, the present section shall largely present the ecological approach in relation to Schumpeterian models and related historical accounts along the lines suggested by Andersen (2002; 1994, Ch. 3).

The ecological approach can be represented by the simple formalisms of biological ecology, which emerged as basic analytical tools in the 1920s (Kingsland, 1985). These formalisms—the logistic equation and the Lotka–Volterra equations—were initially used for the study of the growth and interaction between homogeneous populations. These studies were made in some opposition to the biologists who studied the evolution of heterogeneous populations in statistical terms, but it ultimately became clear that a marriage between ecology and evolution is possible and fruitful (Pianka, 1999). Schumpeter especially met the economic applications of the ecological approach in the work of Simon Kuznets (1930). Actually, Kuznets extensively used a version of the logistic equation for his study of the "secular trends" of American industries as well as a reference point for his study of business cycles. His work, for instance, included full data on the development of the railways of the US as well as an estimation of the logistic function that seemed to express these data rather well. Furthermore, the materials are presented in a form which is much easier to follow than Schumpeter's scattered presentation of railway construction. Since he knew and criticised Kuznets's book, it might seem strange that he did not exploit this possibility. An obvious explanation is that he assumed that his readers knew Kuznets's material, but he was also sceptical about the possibilities of using the logistic equation for his purposes. Nevertheless, the equation seems helpful for the clear expression of aspects of his thinking.

Schumpeter's comments on the ecological approach are mainly found in his section on "The Analysis of the Trend in Total Industrial Output" (*Cycles*, 491–500). This section confronts the micro-foundation of the "secular-stagnation thesis" as it had been provided by Simon Kuznets and Arthur Burns. Although Schumpeter never developed his criticism of Kuznets's work in a systematic manner, it seems clear that he was less concerned with the use of the logistic function for individual industries than for its use to cover the "economic organism" as a whole. It was at this aggregate level that Alvin Hansen (1938) and others could use his work to develop a general "stagnation thesis". Such a use implied that the opportunities of creating new industries had disappeared so that performance depended on established industries. Schumpeter rejected this assumption. Furthermore, the existence of innovative opportunities within existing industries was also a theme in *Cycles*. Schumpeter's scattered remarks on this matter do not refer directly to Kuznets. Schumpeter was fully aware of the possibilities of arriving at a "descriptive trend" (*Cycles*, 201) through such an attempt to fit the empirical curves of growth. However, he was even more aware of the dangers, especially for aggregate data. He pointed out that at this level any revealed exponential growth rate has little meaning. Instead Schumpeter denoted the logistic function as "Verhulst's formula", which had entered economic analyses like those of Kuznets via biological treatments of typical processes of population growth:

> "Still more treacherous and pregnant with danger of speculative temerity may be the application of Verhulst's formula, $y = \frac{a}{be^{-t}+1}$, which was intended (1838) to present certain features of organic or of similar types of growth. Even a perfect fit in the least square sense would not prove anything. We are, however, on somewhat safer ground when applying such expressions to the behavior in time of quantities of individual commodities." (*Cycles*, 492–4)

Schumpeter also mentioned the application of the logistic equation in ecological analysis: "Verhulst's formula is used (slightly generalized) by A. J. Lotka. The form $y = \frac{a}{10^{-bt}+c}$ is known as the Pearl–Reed curve" (*Cycles*, 492n). In any case, Schumpeter had already pointed out that "[o]utput of a new commodity may easily trace out a Verhulst curve" (p. 205). This fact should not be interpreted "as a trend special to that commodity and distinct from any cycles that may run their course in the same period". Schumpeter's conclusion was that the logistic curve expresses the obvious fact that

> "no industry can go on expanding output at the rate of its innovation stage. Each reaches maturity in the sense that [1] it finds its place in the economic organism and the amount of output beyond which it cannot profitably go, unless that amount be increased [2] by some further innovation within it or [3] in some 'complementary' industry and [4] by the general effects of ... Growth." (*Cycles*, 497)

Through these remarks, Schumpeter is actually applying an economic combination of the ideas of ecology and evolution. His term of an "economic organism" is apparently best expressed as an 'economic ecosystem'. First, this system defines the maximum size of an innovation-based industry. Second, this size can be increased by further innovations that emerge in the industry, often of the incremental type that emerge when it is close to its previous maximum. Third, the maximum size of the industry can be increased or decreased by innovations in other industries. Fourth, the maximum size can be increased or decreased by growth (and the phase of the cycle). Although these possibilities does not confront the basic question whether the capitalist economy is stagnating or progressing, their formalisation may help to preserve Schumpeter's extended analysis of innovation as well as to provide an alternative to the increasingly shaky concept of equilibrium that he developed in *Cycles*. When making such a formalisation of routine-based expansion and innovative change in the ecology of the "economic organism", we have to move beyond even what Schumpeter considered the core of his evolutionary models. The ecological approach to economic evolution is, nevertheless, sketched out in the present section. Here several types of adaptive-response 'innovations' are defined; but we should remember that Schumpeter argued that long-term evolution is only sustained by the addition of more radical forms of innovation.

The logistic equation can be discussed in relation to expansion of the railways during the nineteenth century. The length of the railways of Great Britain for the period 1825–1913 is depicted by Figure D.1 on the next page. The curve can be described as being loosely S-shaped. The deviations from the S-curve in the period until 1860 are essential for the Schumpeterian analysis of waveform economic evolution. In contrast, the smoother expansion after 1860 reflects the fact that railways had become integrated parts of the economic organism and that the establishment of new railway lines had become a matter of routine. Therefore, the further expansion is of a qualitatively different type than the early expansion. According to Schumpeter, it reflects adaptation to the growth of the economic organism rather than radical innovative activity. The curve after 1860 would have been more flat if we, instead of the length of the railways, had measured this length relative to some measure of their potential length at any point of time. The increasing carrying capacity of the economic system with respect to railways not only reflect ordinary economic growth but also an adaptive type of railway innovation: "Kondratieff downgrades ... display a wide variety of induced or complementing innovations" (*Cycles*, 339; *Cycles Abr.*, 229). In the following, we shall think in terms of curves that are transformed in this way. Furthermore, we shall ignore the problem of smaller waves of evolution within the overall wave. This means that we, to some extent, can think in terms of the type of S-shaped curves that has been called logistic curves.

Logistic curves can be described by the Replicator Equation, which can be traced back to R. A. Fisher's (1999) statistical analysis of evolution. Assume that two technologies, $z_1$ and $z_2$, are competing. To take Schumpeter's favourite examples of transport technologies, the mail-coach tech-

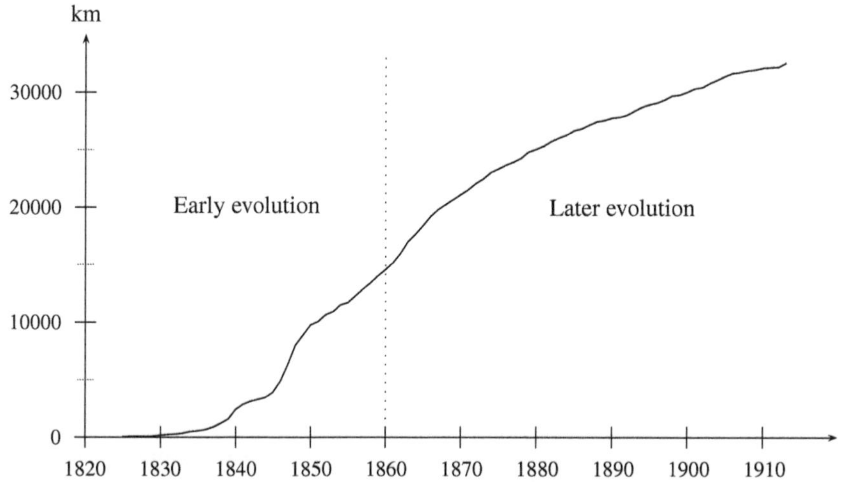


**Figure D.1.: Total kilometres of railway line open in Great Britain 1825–1913**

Comment: The overall appearance of the curve describing the expansion of the length of British railways is loosely S-shaped. A closer look suggests that we are facing the combination of two or three smaller S-shaped curves. The same British data were logarithmically transformed in Figure 8.3 on page 207 to allow comparison with the developments in the USA and Austria-Hungary. Source: Brian Mitchell (1988, 588).

nology $z_1$ competes with the railway technology $z_2$. Assume furthermore that $z_1 = 1$ and $z_2 = 2$. Assume finally that the resource shares of the two technologies are $s_1$ and $s_2$. Then the weighted average technology is $\overline{z} = \sum s_i z_i$. The difference between the average indicator and the indicator of the railway technology can be taken as a measure of its propensity to expand. However, we need to add the speed of selection $\alpha$ to describe the dynamics. Thereby, we obtain a version of the Replicator Equation for the dynamics of the change of the resource share of the railway technology:

$$\Delta s_2 = \alpha s_2 \left( \frac{z_2 - \overline{z}}{\overline{z}} \right). \tag{D.1}$$

The study of the dynamics of the Replicator Equation presupposes that we treat explicitly each sector of the population. However, it sometimes suffices to study the population as a single unit that is homogeneous with respect to its characteristic. In this case, it is the 'density' of the population with respect to its given environment that determines its absolute fitness. This density is measured relative to the maximum size of the population in its environment. It is convenient to use the standard notation of formal ecology and evolutionary ecology to present the Logistic

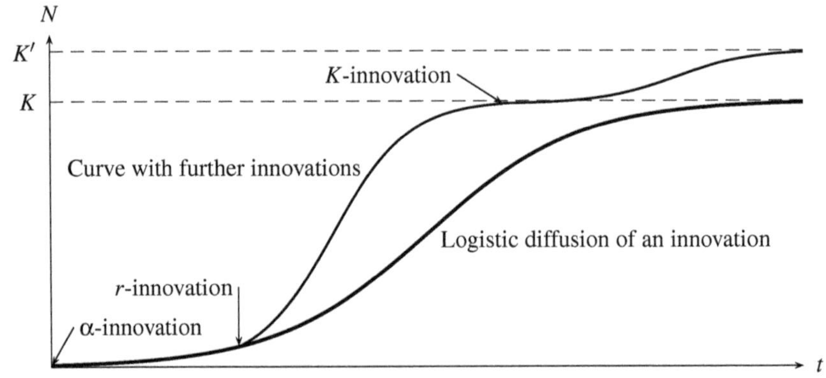


**Figure D.2.: Logistic diffusion with added $r$-innovation and $K$-innovation**

Comment: The logistic curve is produced by the logistic differential equation (D.2). This equation for the process of replication has two parameters: $r$ is the potency of replication and $K$ is the the maximum number of copies. The description of the simple dynamics starts when a Schumpeterian innovation (the $\alpha$-innovation) has produced the first applications of a new economic routine. The expansion of the number of applications ($N$) is described by the logistic curve. However, the economic routine can be improved by two additional types of innovation during its density-dependent process of diffusion. The early $r$-innovations increases the potency of spread while the later $K$-innovations moves the carrying capacity upwards. In contrast to Figure D.1, the present figure assumes a given environment and thus a given level of $K$ unless further innovation takes place.

Equation. The use of this equation for the replication of an innovation describes the movement of the number of its applications, $N$ (see Figure D.2). The increase of $N$—for instance, the number of applications of railway technology—can be approximated by the logistic differential equation, which includes two parameters, $r$ and $K$. In the form that is normally used by ecologists, the equation of density-dependent replication is

$$\frac{\mathrm{d}N}{\mathrm{d}t} = rN\left(\frac{K-N}{K}\right). \tag{D.2}$$

If we divide this equation (D.2) by $N$, we see that relative growth rate depends on two factors. The first, $r$, is the maximum growth rate. It can be considered to be the 'potency of spread' of a particular innovation (like railways). The second factor represents the influence of the total number of railway line units on the growth rate. As $N$ approaches $K$, the growth rate becomes smaller and smaller. For $N = K$, the growth rate is zero. Thus, $K$ can be considered the 'carrying capacity' of the economic system. However, because the expansion occurs in steps, we might see an overshooting of this capacity so that $N > K$, and this leads to subsequent oscillation around $K$. For extreme values of $r$, the pattern of the discrete oscillation will express what has been called deterministic chaos. We shall not consider this phenomenon, but oscillations around $K$ might be considered the main cause of Kitchin cycles. If this is the case, Kitchin cycles should be most visible when the economic system reaches its highest degree of coordination while innovation is a cause of economic stability.

The logistic differential equation has to take its starting point in an innovation that is already introduced in the economic system. Thus we may define a Schumpeterian innovation as an innovation that produces a dynamics that can, in principle, be described by means of the equation (D.2). Furthermore, as demonstrated by Figure D.2 on the preceding page, the logistic differential equation can also be used to define two other types of 'innovation', which are inspired by evolutionary ecology (Pianka, 1999, Ch. 9). By using the names of the parameters, these innovations are called $r$-innovations and $K$-innovations. The $r$-innovations may be seen as relating to competition for being the first to enter underexploited economic space. In this respect, they relate closely to the behaviour of the S-entrepreneur. However, the increase of $r$ is not only the result of the innovative activities of these entrepreneurs; it may also reflect the increasing readiness of the financial system to support innovation. In contrast to these types of change, the $K$-innovations may, to a smaller or larger degree, be related to the activities that takes place when the new industry is close to its carrying capacity. Here the managers of the extended Schumpeterian model are assumed to be able to carry out minor innovations that are motivated by the crowded situation. The spread of their $K$-innovations leads the industry to its new carrying capacity $K'$. Another cause for the change of $K$ is changes in the macroeconomic environment. The outcomes of economic growth and cycles can both be an increase and a decrease of the carrying capacity. These changes may also be caused by "external factors".

While the logistic equation only covers the growth of a single industry, Schumpeter's extended model of evolutionary business cycles also dealt with the interaction between different industries and different innovations. The analysis of these issues can be supported by an extension of the equation into the Lotka–Volterra differential equations. These equations had been proposed for the study of the dynamics of the competitive interaction between biological species by Lotka and Volterra in the 1920s. Later, Richard Goodwin (1982), one of Schumpeter's friends, transferred this formalism to the realm of macroeconomic analysis. However, the primary Schumpeterian use of these equations seems to be the analysis of the effects of innovations on related parts of "economic organism"—considered as an ecosystem. Assume that we have two clearly separate technologies, Technology 1 and Technology 2. Then $N_1$ and $N_2$ are measures of the numbers of their economic applications; $r_1$ and $r_2$ are their intrinsic growth rates (abilities to multiply); and $K_1$ and $K_2$ are their carrying capacities if they are left alone in the economic environment. We now add two interaction coefficients that describe the degree of competitive interaction. $a_{1,2}$ represents the influence of a unit of Technology 2 on the number of applications of Technology 1, while $a_{2,1}$ indicates the opposite direction of influence. Then the system of Lotka–Volterra Equations that describes the dynamics of competing technologies is

$$\begin{aligned} \frac{\mathrm{d}N_1}{\mathrm{d}t} &= r_1 N_1 \left( \frac{K_1 - N_1 - a_{1,2}N_2}{K_1} \right) \\ \frac{\mathrm{d}N_2}{\mathrm{d}t} &= r_2 N_2 \left( \frac{K_2 - N_2 - a_{2,1}N_1}{K_2} \right). \end{aligned} \tag{D.3}$$

The dynamics of the Lotka–Volterra Equation (D.3) will normally display non-innovative cycles of the type that Schumpeter disliked. Actually, this can only be avoided if one technology becomes extinct or if the number of applications of the two technologies is kept at their equilibrium levels. Let us ignore this problem. Instead, we note that Lotka–Volterra equations can easily be reduced to the Logistic Equation (D.2). For instance, the second equation becomes a logistic differential equation if $a_{2,1} = 0$ while $a_{1,2} > 0$. The meaning becomes clear if Technology 1 is a population of mail-coach firms while Technology 2 is a population of railway firms. The reason for the asymmetry is that the competitiveness of railways is so overwhelming that the size of the population of mail coaches has hardly any influence on railways. In contrast, the opposite influence, and thus $a_{1,2}$, is large. The dynamics of the system simply means that the railways spread according to their own logic while the mail coaches are forced out of business. This seems to have been Schumpeter's favourite case. However, there will often be possibilities for the old Technology 1 to react innovatively to the challenge, while the new Technology 2 may also respond. This means that we will see both $a_{2,1}$-innovations and $a_{1,2}$-innovations. Both these types of innovations can emerge in firms related to both technologies.

The Lotka–Volterra equations of competition represents a special case of ecological interaction. A qualitative description of all types of interactions can be represented by pairs of pluses and minuses. For instance, $(-,0)$ represents the case where Technology 2 influences Technology 1 negatively, while there is no opposite effect. This case was crucial for Schumpeter's "Pure Model". However, in his analysis of the effects of the "carrying innovations" of the Kondratieffs, the cases of $(+,0)$ and $(+,+)$ also played a considerable role. This notation may even be used to define a whole set of types of innovations, but if we take the viewpoint of a particular technology, then $+$-innovations are complementary while —-innovations are competing. Such types of innovation help us to approach difficult problems relating to the co-evolution of distinct populations defined in terms of technologies or firms.

Although the brief discussion of the evolutionary-ecological formalisms has been technical as well as incomplete, these formalisms seem to lend some support to Schumpeter's most elaborate modelling of economic evolution. Especially, they serve to specify the types of incremental innovation that Schumpeter used to characterise Kondratieff downswings. Furthermore, they may serve to specify the "process of Creative Destruction" that Schumpeter considered "the essential fact about capitalism" (*Capitalism*, 83). Each Schumpeterian innovation can be considered as adding a new equation to the Lotka–Volterra system of equations. Although such an innovation might interact positively with some of the established activities, it practically always has negative effects on other activities. Actually, he seems to have considered his industrial revolutions in a way that comes close to the Great Extinctions of biological evolution. However, while the extinction of the dinosaurs was the result of an exogenous meteor, economic routines are brought to extinction by radical innovations that are endogenous to the economic system. Therefore, they provoke the type of socio-political reactions that might bring capitalist economic evolution to a halt. However, we are presently concerned with the perspectives for capitalism that can be derived from an isolated analysis of its process of economic evolution.

The efforts to develop the toolbox of evolutionary analysis not only concern the addition of new tools but also an analysis of the degree to which the tools correspond to each other. An example of how to confront this challenge is found in a short but very condensed paper by Karen Page and Martin Nowak (2002). The title of the paper is "Unifying Evolutionary Dynamics", and it proves that the formalisms of the ecological approach can be translated into those of the statistical approach (and vice versa). The proof of the basic unity of the major mathematical formalisms that are used in evolutionary analysis does not reduce the analytical toolbox to a single tool. On the contrary, each tool serves to focus analysis on a particular aspect of what Schumpeter called creative destruction. The fact that the paper was published in *Journal of Theoretical Biology* demonstrates that unification with respect to tools tends to transcend the boundaries between the different sciences. Again, this does not imply any total unification. Social scientists engaged in evolutionary analysis will with envy read the excellent account for biological evolution in advanced textbooks with titles like *Evolution* (Barton et al., 2007) and *Evolutionary Biology* (Futuyma, 1998); but they will also recognise most of the presented tools and facts have no immediate relevance for their own areas of study. The unity of the basic formalisms for treating evolutionary processes nevertheless suggests that there are many possibilities of interchange between very different disciplines.

## D.2 The statistical approach to evolutionary analysis

Statistical methods can help evolutionary analysis in different ways. Uwe Cantner and Jens Krüger (2007) have suggested that empirical analysis of evolution needs an extensive development of the field of "evolumetrics". More narrowly, Koen Frenken (2007) emphasised the help that can be obtained from measures of entropy and information. These empirically oriented tools can also be of help when formulating evolutionary theories. However, as emphasised by Stanley Metcalfe (2007), evolutionary theorising is seriously in need of an explicit statistical system of accounting for evolutionary change. This accounting system is what matters in the present section since it serves to define the concepts of evolution, selection, and intra-unit change. Although the concept of intra-unit change includes radical innovations in the Schumpeterian Mark II model, it also allows for change caused by organisational learning and organisational decay. We are, nevertheless, facing a very general approach that was originally developed by the statistician and evolutionary biologist R. A. Fisher (1999). This approach has proved crucial—especially since the 1970s. At that time, George Price (1972b; 1972a; 1995) clarified Fisher's fundamental theorem of natural selection, developed a general and very fruitful partitioning of *any* evolutionary change, and helped to overcome the fruitless controversy on group-based selection versus individual selection. The following presentation builds on Steven Frank's (1998) *Foundations of Social Evolution* and previous work by Andersen (2004c).

The clarification of the statistical approach to evolutionary analysis is the topic for a whole book rather than for a short section. It, nevertheless, is important to give an impression of the ways in which statistical tools can specify the conceptual analysis. Especially, it should be noted that the statistical tools are not only needed for the testing of evolutionary models, but also for the construction of these models and broader evolutionary theories. The starting point is that evolution is a unique process in historical time, and this is the main reason why the analysis of evolutionary change has proved difficult. This analysis presupposes a number of definitions and notational decisions that can be more or less scientifically fruitful. According to the solution of George Price, we start by selecting points of time in the unique evolutionary process. Our partitioning of time into steps is sometimes quite natural, like in the case of agricultural crops, but often we have to enforce discrete time upon our data to allow for a simple treatment. In any case, we have a sequence of points of time, $t, t', t'', \ldots$ Evolution may then be described in terms of the states of the evolving system at subsequent points of time as well as by the function that transforms the state of the system between two points of time. In the simplest case, we have a transformation mechanism $T$ that works on the state of our focal population $P$ (called the pre-selection population) and the given state of the environment $R$ to bring forth a new population $P'$ (the post-selection population). Thus we have

$$(P; R) \xrightarrow{T} (P'; R). \tag{D.4}$$

It is obvious that the core concept of the transformation (D.4) is that of a population of units. This concept is not only difficult because we have to specify the borderlines of the population but also because the units can be defined in many different ways. It might be convenient to think that the units are the firms of an industry, but we are facing a general concept in which units can just as well be the plants of a firm or the industrial districts of a country. They can even be arbitrarily defined groupings of firms or plants or industrial districts. However, although arbitrary groupings are allowed by the approach under consideration, they hardly serve to reveal much about the actual evolutionary process. Another difficulty is that by assuming an unchanged environment for the population, the transformation equation (D.4) obviously defines a simplified step in an evolutionary process. This is, of course, a simplification, which tends to be undermined by the ecological approach to evolution (see Section D.1). Presently, we shall concentrate on the evolutionary change in the focal population (which may consist of sub-populations) as it is brought forth by the transformation mechanism under the condition of an unchanging environment (which to a large extent consists of other populations). In this context, we may consider two different questions. The first question presupposes that we know $P$ and $T$. Then the question is which population $P'$ will emerge. Our knowledge of $T$ normally has the form of a theory. Therefore, the use of this theory to determine $P'$ has the form of a theoretical prediction. This prediction may be falsified by means of experiments that often have the form of 'natural experiments', i.e. simple comparative cases of evolutionary change from real life. The second question can be put if we know $P$ and $P'$. Then the question is what transformation $T$ has brought about this change. In the present context we shall concentrate on this question about the details of the evolutionary transformation that brings about an observed change of the population.

In order to describe the change from the pre-selection population $P$ to the post-selection population $P'$, we, in principle, need full unit-level information. Since each unit of a population is characterised by a large number of evolutionary relevant characteristics, this is a very demanding requirement. In practise we may, however, concentrate on the evolution of a single or a few characteristics. Another requirement for our analysis is that we in an evolutionarily relevant way connect each unit of the post-selection population to a unit of the pre-selection population. In some

**Table D.1.: Notation for defining and describing evolutionary processes statistically**

| Variable | Description | Definition |
|---|---|---|
| $X, X'$ | Format of variables relating to the pre-selection population and the post-selection population | |
| $x_i$ | Size of unit $i$ | |
| $x$ | Size of population | $\sum x_i$ |
| $s_i$ | Population share of unit $i$ | $x_i/x$ |
| $z_i$ | Value of characteristic of unit $i$ | |
| $\Delta z_i$ | Change in value of characteristic of unit $i$ | $z_i' - z_i$ |
| $\overline{z}$ | Mean value of characteristic | $\sum s_i z_i$ |
| $\Delta\overline{z}$ | Change of mean value of characteristic | $\overline{z}' - \overline{z}$ |
| $\text{Var}(z)$ | Variance of characteristics | $\sum s_i (z_i - \overline{z})^2$ |
| $w_i$ | Absolute fitness (or reproduction coefficient) of $i$ | $x_i'/x_i$ |
| $\overline{w}$ | Mean absolute fitness | $\sum s_i w_i$ |
| $f_i$ | Fitness (or relative fitness) of unit $i$ | $w_i/\overline{w}$ |
| $\overline{f}$ | Mean fitness | $\sum s_i f_i = 1$ |
| $\text{Cov}(f, z)$ | Covariance of fitnesses on characteristics | $\sum s_i (f_i - \overline{f})(z_i - \overline{z})$ |
| $\beta(f, z)$ | Regression of fitnesses on characteristics | $\text{Cov}(f, z)/\text{Var}(z)$ |
| $\text{E}(f\Delta z)$ | Expected value of change of characteristics within units weighted by their fitnesses | $\sum f_i \Delta z_i$ |

Comment: The coupling between evolutionary theory and the statistics of evolutionary processes is difficult. R. A. Fisher and George Price have demonstrated that an adequate notation is crucial for this coupling. The suggested notation is close to the one that has emerged within mathematical biology (see Frank, 1998). Although the notation primarily serves to clarify the process of selection, it also helps to make explicit the processes of intra-unit adaptation and innovation. Source: Modified from Andersen (2004c, 134).

cases, this is an even more demanding requirement, but in practice the connection can normally be done. Table D.1 shows the information and the calculations needed for analysing evolutionary change of a population with respect to a single characteristic.

For concreteness, we may think of $P$ as consisting of firms. For exiting firms and for firms that are present in both $P$ and $P'$, the coupling between the two populations is unproblematic. In contrast, we cannot make this coupling for radically new firms. However, it is sometimes possible to connect new firms to old ones (like in the case of spin-offs). Given that we have solved this problem, we turn to the description of the population of firms and the structural change of this population. First, firm $i$ is described in terms of its resources $x_i$ and their population share $s_i = x_i/x$, where $x$ is the aggregate resources of the population of firms. Second, the firm is described by the value of an evolutionarily relevant characteristic $z_i$, like productivity, and the change in this productivity, $\Delta z_i$. This example is hardly the best one, but it has the advantage of being subject to change within the incumbent firms of Schumpeter Mark II. In contrast, change that is largely due to the entry, as in Schumpeter Mark I, is more difficult to handle by the present statistical approach. This is the reason why we do not take Schumpeter's favourite example of mail coaches ($z_i = 0$) and railways ($z_i = 1$). Third, the firm is described by its fitness $f_i$. To specify the meaning of this controversial term, we start by defining the concept of the absolute fitness of a unit as what may be called its reproduction coefficient, $w_i = x_i'/x_i = 1 + \Delta x_i/x_i$. The absolute fitness is thus one plus the growth rate of the unit. Then we turn to the more fundamental concept of relative fitness, or simply fitness. The use of this concept requires that we know the weighted mean of the absolute fitnesses, $\overline{w} = \sum s_i w_i$. Then we define the fitness of a unit as $f_i = w_i/\overline{w}$. It is this relative concept of fitness that we normally use in evolutionary analysis. It implies that we ignore the overall growth of the population of firms and only study its structural change.

Given this information, it is fairly easy to describe and analyse how $P'$ is brought forth from $P$. One strategy is to make an explicit study of the change of the population shares of the entities by means of the Replicator Equation (Metcalfe, 2007). The simplest form of this equation assumes that it is the difference between the characteristic of the firm and the mean characteristic of the population that determines the change in its resource share. We have met this replicator approach in Section D.1. Presently we use the complementary approach of George Price to the analysis of the evolutionary process. More specifically, we shall consider Price's Equation that decomposes evolution into the sum of two effects. This equation has the following structure:

$$\begin{matrix}\text{Total} \\ \text{evolution}\end{matrix} = \begin{matrix}\text{Effect of selection} \\ \text{between units}\end{matrix} + \begin{matrix}\text{Effect of change} \\ \text{within units.}\end{matrix} \qquad \text{(D.5)}$$

The first task in relation to this verbal form of Price's Equation (D.5) is to define evolution as the change of the population with respect to a characteristic like productivity change. If there is no aggregate change of productivity, then there is no evolution. The evolutionary process within the population of firms concerns the weighted mean of the values of the characteristic, $\overline{z} = \sum s_i z_i$. Actually, evolution is *defined* as the change in this mean, that is, $\Delta\bar{z} = \bar{z}' - \bar{z}$. This definition of evolution means that if $\bar{z} = 0$, then no evolution takes place in the population with respect to the studied characteristic. We have thus defined the left side of the above equation (D.5):

$$\text{Total evolution} = \text{Change of the weighted mean of a characteristic} = \Delta\bar{z}. \tag{D.6}$$

Given that we observe evolutionary change, we turn to the analysis of the elements of the mechanism of evolutionary transformation. This mechanism has two major components: transformation by selection between units and transformation by more or less innovative change within the units. Let us first consider transformation by selection, which in a certain sense is the most crucial part of our analysis. Here we shall treat the values of the characteristics of firms of the pre-selection population as parameters for the selection process. Thereby, it becomes possible to specify the nature of a selectively relevant characteristic. Such a characteristic must have a degree of inertia that allows the composition of the population to change quicker than the characteristics of the units of the population. The task is then to couple the characteristics of firms with their respective fitnesses. Normally the fitness of a unit is influenced by a large number of its characteristics. However, we shall assume that it is only the values of a single characteristic that, apart from an error term, influences fitness. This assumption is solely made for the sake of simplicity. It allows us to describe quickly the fuel for the selection process as the variance in the pre-selection population with respect to the characteristic, $\mathrm{Var}(z)$. The description is obvious since selection presupposes variance: if $\mathrm{Var}(z) = 0$, no selection can take place. Selection is the process that promotes and demotes firms based on their characteristic values. This promotion and demotion is described by the fitnesses of the firms. Therefore, selection is defined statistically as the covariance between the fitnesses and the values of the characteristic, $\mathrm{Cov}(f, z)$. If $\mathrm{Cov}(f, z) = 0$, no selection takes place. The reason can either be that no variance is present or that this variance is not exploited for selection. These two aspects of covariance are clear because it can be rewritten as the product of population variance and a regression coefficient. This is the regression coefficient of fitnesses on characteristics, $\beta(f, z)$. This coefficient is not defined if $\mathrm{Var}(z) = 0$. It describes the degree to which the variance present in the population is used for selection between the units. We may also say that this regression coefficient defines the selection efficiency, that is, the degree to which variance is exploited to produce evolution of the characteristic. However, Price's basic definition is that the selection effect is a covariance. Thereby, he specified the selection effect of the previous equation (D.5):

$$\begin{matrix}\text{Effect of selection} \\ \text{between units}\end{matrix} = \begin{matrix}\text{Covariance between} \\ \text{fitnesses and characteristics}\end{matrix} \tag{D.7}$$
$$= \mathrm{Cov}(f, z) = \beta(f, z)\mathrm{Var}(z).$$

When considering this definition, it is important to note the context in which it is made. We are simply studying the single step in the evolutionary process from the pre-selection population $P$ to the post-selection population $P'$. The covariance is defined by the data for these two populations and no assumptions are made about the next step from $P'$ to $P''$. Especially, we do not assume that the covariance stays constant. On the contrary, selection serves to reduce variance. Furthermore, selection serves to change the mean value of the characteristic of the population and this might imply a change of the regression of fitnesses on characteristics. The fitness function might be linear in the case of productivity, but in other cases it is inversely U-shaped or even directly U-shaped. A fitness function that is linear represents directional selection, while the inverse U-shape represents a selection that stabilises the population and the direct U-shape represent disruptive selection in the sense that it tends to split the population into two groups with different characteristics. The fact that Price's statistical approach leaves these issues open implies that it is compatible with the ecological approach.

Although the definition of the selection effect as a covariance is Price's basic contribution, he made a further contribution to the analysis of mechanism of evolutionary transformation. One might think that evolution is covariance plus a residual that in the above equation (D.5) was called the 'effect of change within units'. However, Price made a useful statistical definition of this effect and, at the same time, he also made his equation a mathematical identity. He obtained these results by defining the effect of intra-unit changes as the mean of the product of each unit's fitness and its change of the value of its characteristic, $\mathrm{E}(f\Delta z) = \sum f_i \Delta z_i$. According to this definition, intra-unit change is the component of the total evolution that is determined by the weighted influence of the degree to which the units of the post-selection population have changed their characteristics when compared to the pre-selection population. Let us summarise this in relation to the verbal version of Price's Equation (D.5):

$$\begin{matrix}\text{Effect of change} \\ \text{within units}\end{matrix} = \begin{matrix}\text{Mean of characteristic changes} \\ \text{weighted by fitnesses}\end{matrix} = \mathrm{E}(f\Delta z). \tag{D.8}$$

In this definition we are obviously not assuming that the change in productivity or any other characteristic has to be described like Schumpeter would have liked it or like it is described in the neo-Schumpeterian tradition of innovation studies (Fagerberg et al., 2005). While innovation in these studies is seen as the introduction of a positively valued novelty with respect to the overall population, we presently apply a neutral concept that covers any kind of local-level change. It simply means that a change has been made of characteristics at the unit level of the evolving population. Thus there is no assumption that the change is good for its carriers, so the value of the characteristic for individual units may have increased or decreased. In the case of the productivity of firms there are, of course, many potential reasons for both negative and positive values, but let us concentrate of the knowledge issue. In this respect productivity change may be positive because of innovation, imitation or learning processes. It might be negative because the firm does not have an effective system of reproduction of its knowledge. The expected aggregate effects of both learning and forgetting are, of course, influenced by the resource shares of the firms in the post-selection population.

By combining equations (D.6), (D.7), and (D.8), we obtain a formal version of George Price's equation. This version is:

$$\Delta\bar{z} = \mathrm{Cov}(f, z) + \mathrm{E}(f\Delta z). \tag{D.9}$$

George Price proved that his equation is an identity (Andersen, 2004c, 148), and this also holds for the present version of it. Price's Equation (D.9) tells that any evolutionary change can be partitioned into an inter-unit selection effect and an intra-unit change effect, provided that we are able to perform the descriptive work assumed by Table D.1 on page 438. This partitioning of evolution might seem an obvious and rather trivial result, but this is hardly the case. Generations of evolutionary biologists and evolutionary economists have had the possibility of deriving this surprisingly fruitful partitioning, but they have failed to do so. The major reasons are that they have not had a sufficiently general concept of selection and that they have not been willing to make the necessary coupling of the pre-selection population with the post-selection population. George Price overcame these difficulties, but his equation is even more general than the version of it that we have developed above. The version that he actually used is formulated in terms of absolute fitnesses. They are obtained by multiplying both sides of the above version of Price's Equation (D.9) by the mean absolute fitness. Thereby we reach what is normally called Price's Equation:

$$\overline{w}\Delta\bar{z} = \mathrm{Cov}(w, z) + \mathrm{E}(w\Delta z). \tag{D.10}$$

This version of Price's Equation (D.10) can be used for multi-level selection processes. For instance, as long as we consider the units of the population as firms, we will probably obtain a large effect of intra-unit changes that is, furthermore, difficult to interpret. The reason is that many firms have multiple plants that individually improve their productivity. One way of handling this problem is to make two sets of studies, the first in terms of firms and the second in terms of plants. However, Price wanted his decomposition to cover populations that consisted of groups that consisted of individual units, and this possibility had been used to resolve the long-lasting controversy in so-called group selection. To allow for multi-level analysis, Price's Equation (D.10) has a recursive structure. The recursive nature of this equation derives from the fact that the left hand side ($\overline{w}\Delta\overline{z}$) is structurally identical to the contents of the expectation term on its right hand side ($w\Delta z$). This means that Price's Equation can be used to expand itself—provided that the units of the population are groups with members (Frank, 1998, 15). This is the case when firms consist of plants and when the productivities of firms are means of the productivities of their plants. This interpretation gives notational problems, but we mark members of groups by the additional subscript $j$ and the operators that sum over members of individual groups by the subscript $G$. Furthermore, we use capital letters for variables relating to group members. For instance, $w_i = \overline{W_i} = \sum W_{i,j}$. The consequence is that Price's Equation (D.10) can be expanded into

$$\begin{aligned}\overline{w}\Delta\overline{z} &= \mathrm{Cov}(w,z) + \mathrm{E}(\mathrm{Cov}_G(W,Z) + \mathrm{E}_G(W\Delta Z)) \\ &= \underbrace{\mathrm{Cov}(w,z)}_{\text{Inter-group selection}} + \underbrace{\mathrm{E}(\mathrm{Cov}_G(W,Z))}_{\text{Intra-group selection}} + \underbrace{\mathrm{E}(\mathrm{E}_G(W\Delta Z))}_{\text{Intra-member change}} .\end{aligned} \tag{D.11}$$

If we compare this multi-level equation (D.11) with the above equation (D.10), we see that what was at the level of firms considered a change effect is now partitioned into the expectation of the selection effects within firms and the expectation of the more narrowly defined change effects within the plants the firms. We can easily add additional levels if it gives any meaning. For instance, productivity at the level of the manufacturing sector suggests a selection between industries, a selection of the firms within each industry, and a selection of the plants within each firm. Whether studies in terms of three and even more levels of selection give new insight into the evolutionary process might be doubted, but the power of Price's Equation is obvious. This power has not really been exploited within Modern Evolutionary Economics. However, Stanley Metcalfe (2002, 90) has remarked that "[f]or some years now evolutionary economists have been using the Price Equation without realising it"; and similar statements have been made in a more developed form by Thorbjørn Knudsen (2004). Such statements hold for Metcalfe's (1998; 2001) contributions to a statistically oriented evolutionary economics and for the discussion of group selection within evolutionary game theory, but they also have some truth for Nelson and Winter's (1982) pioneering contribution to the field of "neo-Schumpeterian economics". Even in applied economics with no evolutionary pretensions, we seem to find a groping towards a statistical approach that seems similar to that of George Price.

If we move back from "neo-Schumpeterian economics" to Schumpeter's own work, the present statistical approach appears to represent a very truncated view of creative destruction. The basic problem is that it only treats incumbent units of a population. It only studies the selection between the units and the change that takes place within these units. It should, however, be noted that the units can be freely defined. Especially, they can be aggregates of simpler units. Let us take the example of the transformation of a stylised version of the sector for transportation of passengers in the nineteenth century. Then the pre-selection population can be seen as consisting of only two units: Unit 1 is the mail coach unit and Unit 2 is the railway unit. The size of Unit 1 might be studied in terms of the number of passenger kilometres performed by mail coaches, $x_1$, and its market share is $s_1$; and we obtain similar data for Unit 2. We also need to define the characteristics numerically. A simple solution is to set $z_1 = 1$ and $z_2 = 2$. The mean behaviour of the population is $\overline{z} = s_1 + 2s_2$. The mean is thus one if only mail coaches are used and two if only railways are used. In these extreme cases, $\text{Var}(z) = 0$.

While the description of the pre-selection population is relatively simple, this simplicity does not extend to the description of the post-selection population. Take, for instance, Unit 1. The old firms within this unit might have stayed or gone bankrupt, but they might also have switched to railway technology. Furthermore, new firms might have entered the unit; and some of these might have come from the railway unit of the pre-selection population. Before we can define $x_1'$, we need to know how to handle these cases. The stayers obviously contribute to the size of the unit; and bankrupt firms make no contribution. The mail coach firms that have switched to railway technology are treated as intra-unit change. If we take the highly unlikely case that all firms of unit 1 switched during a single evolutionary step, then their size would still be accounted for in Unit 1. However, in this special case $\Delta z_1 = 2 - 1 = 1$. Similarly, railway firms that switch to mail-coach technology are accounted for in Unit 2. Finally, we come to the entrants to unit 1 that come from the outside. We simply add their size to that of the other contributions to the size of the unit in the post-selection population and obtain $x_1'$. After having performed similar calculations for Unit 2, we also obtain $s_1'$.

Although the above procedure for studying an apparently simple phenomenon like the evolutionary transformation from mail coaches to railways may appear obscure, it is dictated by the logic of Price's Equation. If we accept this logic, we obtain a new way of seeing the process of creative destruction by splitting the process at a number of points of time, $t, t', t'', \ldots$. Then we use the formula for each of the periods. We could, in principle, start from an initial evolutionary equilibrium based on mail coaches. Here we have that $\overline{z} = 1$ and $\text{Var}(z) = 0$. However, Schumpeter emphasised that "in general it is not the owner of stage-coaches who builds railways" (*Development*, 66). When we recognise that the railway innovation did not emerge from the mail-coach unit, we have to start our statistical analysis *after* the basic creative act and assume that at least one railway firm has emerged in the railway unit. In this initial situation $\overline{z} > 1$ and $\text{Var}(z) > 0$. During the process of transformation, the mean value of the characteristic increases because of the expansion of old railway firms and because of the entry of firms using railway technology. The variance increases initially, but it moves back towards zero during later parts of the process. The final equilibrium is obtained when $\text{Var}(z) = 0$ and $\overline{z} = 2$. The evolutionary prosperity phase might be defined by $\text{dVar}(z)/\text{d}t > 0$, while the evolutionary recession phase has $\text{dVar}(z)/\text{d}t < 0$. This definition is only loosely related to Schumpeter's idea and it does not solve the task of creating a relevant index of the type of highly multidimensional macroscopic evolution that is at the focus in his treatment of business cycles. However, the study of the change of variance had the advantage of being operational for well-defined processes of evolutionary change.

The discussion in terms of variance does not provide an adequate analysis of the creative destruction of the mail-coach unit. For this purpose, we have to turn to the former version of Price's Equation (D.9). If we assume that no switching takes place between our two units, then $\text{E}(f\Delta z) = 0$. In this case, we are dealing with a dramatic form of destruction in which it is not only the routines of Unit 1 but also its firms and jobs that are destructed. Since the equation simplifies to $\Delta\overline{z} = \text{Cov}(f, z)$, the task is to understand how to interpret the covariance between fitnesses and values of the characteristic. This can be done in at least two ways. The first way is to rewrite this covariance to $\beta(f, z)\text{Var}(z)$. $\beta(f, z)$ is the efficiency of selection expressed as the regression of fitnesses on characteristics. This coefficient expresses the harshness of the positive selection of railways and the negative selection of mail coaches. The second way of studying the creative destruction is to recognise that our analysis concerns market shares or resource shares rather than the absolute size of our two units. If the whole transportation sector is growing rapidly, the relative destruction of Unit 1 might translate into its absolute growth. The social costs of such 'destruction' are obviously very different from its 'destruction' in absolute terms.

# Schumpeter's Works

## Books

*Capitalism* = *Capitalism, Socialism and Democracy*, 3rd edn., New York: Harper, 1950. This book was originally published in 1942 In the second edition (1947), Schumpeter added a new preface (*S1947b*) and a chapter on "The Consequences of the Second Word War". The third American edition (1950) added the new preface for the British edition as well as the paper "The march into socialism".

*Cycles* = *Business Cycles: A Theoretical, Historical, and Statistical Analysis of the Capitalist Process*, New York and London: McGraw–Hill, 1939. The contents of this large and complex book are distributed across two volumes (with common pagination).

*Cycles Abr.* = *Business Cycles: A Theoretical, Historical, and Statistical Analysis of the Capitalist Process*, abr. edn., New York: McGraw–Hill, 1964. This edition, by Rendigs Fels, omits two thirds of the text of *Cycles* (mainly on statistics and non-American history).

*Development* = *The Theory of Economic Development: An Inquiry into Profits, Capital, Credit, Interest and the Business Cycle*, Cambridge, Mass.: Harvard University Press, 1934. This book was translated by Redvers Opie with the help of Schumpeter. It includes several, but minor, changes when compared with the German edition from 1926. It also omits the section headings that are found in *Entwicklung II* (pp. v–vii).

*Doctrine* = *Economic Doctrine and Method: A Historical Sketch*, London: Allen & Unwin, 1954. Translation by Reinhold Aris of "Epochen der Dogmen- und Methodengeschichte," in Karl Bücher, Joseph A. Schumpeter, and Friedrich von Wieser, "I. Abteilung: Wirtschaft und Wirtschaftswissenschaft" of Max Weber (ed.), *Grundriss der Sozialökonomik*, Tübingen: Mohr, 1914, pp. 19–124.

*Entwicklung I* = *Theorie der wirtschaftlichen Entwicklung*, Leipzig: Duncker & Humblot, 1912. Since Schumpeter wanted the book replaced by *Entwicklung II* and *Development*, the publisher has only recently made it available as a German reprint with a new foreword. Chapter 7 was removed. *S1912b* is an English translation of that chapter and *S1912a* contains a partial translation. *S1912a* also translates samples of the original Chapter 2 that was radically changed. Finally, the radically changed Chapter 6 is included in the translation of *S1910d*.

*Entwicklung II* = *Theorie der wirtschaftlichen Entwicklung: Eine Untersuchung über Unternehmergewinn, Kapital, Kredit, Zins und den Konjunk-*

*turzyklus*, Munich and Leipzig: Duncker & Humblot, 1926. This is a radically revised edition of *Entwicklung I*. The most obvious difference is that Schumpeter deleted Ch. 7 (*S1912b*). Another important difference is that Schumpeter radically rewrote Ch. 2's presentation of innovative entrepreneurs as agents that change mass behaviour (*S1912b*). Furthermore, he reconstructed the old-fashioned analysis of economic crises of Ch. 6 (*S1910d*).

*Geldes* = *Das Wesen des Geldes*, Göttingen: Vandenhoeck & Ruprecht, 1970. Edited by Fritz Mann. This is a transcript of a manuscript developed in the late 1920s and early 1930s. Most of the text seems to have been completed by 1930. The manuscripts for Chapters 13–15 were not available for the German edition. They are presently only available in French translation (*S1930f*).

*History* = *History of Economic Analysis*, London: Allen and Unwin, 1954. Edited by Elizabeth B. Schumpeter. The very complex manuscript was not finished at the time of Schumpeter's death.

*Vergangenheit* = *Vergangenheit und Zukunft der Sozialwissenschaften*, Munich and Leipzig: Duncker & Humblot, 1915. Although this booklet has never been reissued (except for a Japanese translation), it contains interesting views.

*Wesen* = *Das Wesen und der Hauptinhalt der theoretischen Nationalökonomie*, Leipzig: Duncker & Humblot, 1908. Since Schumpeter opposed reprints of this important book, they have only been available since 1970. In the third edition from 1998, the Roman page numbers of the important front matter has been slightly changed. The first edition is used for the present book.

## Collections of papers and letters

*AS* = *Aufsätze zur Soziologie*, Tübingen: Mohr, 1953. Edited by Erich Schneider and Arthur Spiethoff.

*AÖT* = *Aufsätze zur ökonomischen Theorie*, Tübingen: Mohr, 1952. Edited by Erich Schneider and Arthur Spiethoff.

*ATP* = *Aufsätze zur Tagespolitik*, Tübingen: Mohr, 1993. Edited by Christian Seidl and Wolfgang F. Stolper.

*AWP* = *Aufsätze zur Wirtschaftspolitik*, Tübingen: Mohr, 1985. Edited by Wolfgang F. Stolper and Christian Seidl.

*BL* = *Briefe/Letters*, Tübingen: Mohr, 2000. Edited by Ulrich Hedtke and Richard Swedberg.

*BSÖ* = *Beiträge zur Sozialökonomik*, Vienna: Böhlau, 1987. Edited by Stephan Böhm.

*DBA* = *Dogmenhistorische und Biographische Aufsätze*, Tübingen: Mohr, 1954. Edited by Erich Schneider and Arthur Spiethoff.

*EE* = *Essays: On Entrepreneurs, Innovations, Business Cycles, and the Evolution of Capitalism*, New Brunswick, N.J. and London: Transaction, 1989. Edited by Richard Clemence and introduction by Richard Swedberg. The new edition, which is used in the present book, is a repaginated and partly reset version of the 1951 edition (which had the shorter title *Essays on Economic Topics*).

*ESC* = *The Economics and Sociology of Capitalism*, Princeton, N.J.: Princeton University Press, 1991. Edited by Richard Swedberg.

*PR* = *Politische Reden*, Tübingen: Mohr, 1992. Edited by Christian Seidl and Wolfgang F. Stolper.

*RBZ* = "Reden in der Bonner Zeit," *Berliner Debatte Initial*, **8** (3) [1997]: 61–92. Edited by Ulrich Hedtke.

*TGE* = *Ten Great Economists: From Marx to Keynes*, New York: Oxford University Press. Edited by Elizabeth B. Schumpeter.

## Papers and reviews

This is a list of those of Schumpeter's papers and reviews that are cited in the present book. The list covers most of his papers since many papers that are not cited in the main text are classified in Appendix C. Nevertheless, more complete lists should be consulted. A relatively complete list is provided by Massimo Augello (1990, 117–58); but Ulrich Hedtke's (2007a) bibliography covers further items. The years given in the abbreviations for later published manuscripts more or less reflect the time when the manuscripts of posthumously published papers were finished.

*S1901–05* = "Entries in the lecture protocols of the University of Vienna [in German]," see the, not totally identical, listings provided by Ulrich Hedtke and Richard Swedberg ("Einführung," in *BL*, pp. 3–5) and by Kiichiro Yagi (1993, 67–72). Yagi has kindly provided the present author with his underlying data in the form of a typescript by Kurt Mühlberger, archivist at the Archive of the University of Vienna.

*S1905a* = "Die internationale Preisbildung," *Statistische Monatsschrift*, **31** [1905]: 923–8. .

*S1905b* = "Die Methode der Index-Zahlen," *Statistische Monatsschrift*, **31** [1905]: 191–7. .

*S1905c* = "Die Methode der Standard Population," *Statistische Monatsschrift*, **31** [1905]: 188–91. .

*S1906a* = "Über die mathematische Methode der theoretischen Ökonomie," in *AÖT*, pp. 529–48. Original in *Zeitschrift für Volkswirtschaft, Sozialpolitik und Verwaltung*, **15** [1906]: 30–49.

*S1906b* = "Professor Clarks Verteilungslehre," *Zeitschrift für Volkswirtschaft, Sozialpolitik und Verwaltung*, **15** [1906]: 325–33. .

*S1906c* = "Review in German of *Über des Wesen und die Ursachen unserer heutigen Wirtschaftskrisis* by Bernhard Rost and *Zur Lehre von den Wirtschaftskrisen* by Otto Karmin," *Zeitschrift für Volkswirtschaft, Sozialpolitik und Verwaltung*, **15** [1906]: 95–7. .

*S1906d* = "Rudolf Auspitz," in *ATP*, pp. 98–9. Original in *Economic Journal*, **16** [1906]: 309–11.

*S1907a* = "Das Rentenprinzip in der Verteilungslehre," in *AÖT*, pp. 187–265. Original in *Jahrbuch für Gesetzgebung, Verwaltung und Volkswirtschaft*, **31** [1907]: 31–65, 591–634.

*S1907b* = "Review of *Esquisse d'une Sociologie* by M. Émile Waxweiler," *Economic Journal*, **17** [1907]: 109–11. .

*S1908a* = "Bemerkungen über das Zurechnungsproblem," in *AÖT*, pp. 266–319. Original in *Zeitschrift für Volkswirtschaft, Sozialpolitik und Verwaltung*, **18** [1908]: 79–132.

*S1908b* = "Review in German of *Essentials of Economic Theory as Applied to Modern Problems of Industry and Public Policy* by John B. Clark," *Zeitschrift für Volkswirtschaft, Sozialpolitik und Verwaltung*, **17** [1908]: 653–659. .

*S1909a* = "Letter from the Faculty of the University of Vienna to the Austrian Ministry of Education including 'Curriculum Vitae Schumpeters' and 'von Vorlesungen' [in German]," 15 February 1909. Reproduced by Kiichiro Yagi (1993, 73–8).

*S1909b* = "On the concept of social value," in *EE*, pp. 1–20. Original in *Quarterly Journal of Economics*, **23** [1909]: 213–32.

*S1909c* = "Review of 1. Ira Ryner: *On the Crises of 1837, 1847 and 1857 in England, France and the United States*; 2. Minni Throop England: *On Speculation in Relation to the World Prosperity 1897–1902*; 3. W. G. Longworthy Taylor: *The Kinetic Theory of Economic Crises*," *Zeitschrift für Volkswirtschaft, Sozialpolitik und Verwaltung*, **18** [1909]: 683–5. .

*S1910a* = "Die neuere Wirtschaftstheorie in den Vereinigten Staaten," *Jahrbuch für Gesetzgebung, Verwaltung und Volkswirtschaft im Deutschen Reich*, **34** [1910]: 913–963. .

*S1910b* = "How does one study social science?" *Society*, **40** [2003]: 57–63. Translation by Jerry Z. Muller of parts of *Wie studiert man Sozialwissenschaft*, in Schriften des sozialwisenschaftlichen akademischen Vereins in Czernowitz, Pardini: Czernowitz, 1910. The full text is found in *AÖT*, 555–65.

*S1910c* = "Marie Esprit Léon Walras, 1834–1910," in *TGE*, pp. 74–9. Translation by Wolfgang Stolper of *Zeitschrift für Volkswirtschaft, Sozialpolitik und Verwaltung*, **19** [1910]: 397–402.

*S1910d* = "On the nature of economic crises," in Mauro Boianovsky (ed.), *Business Cycle Theory: Selected Texts 1860–1939*, Vol. V, London: Pickering & Chatto, 2005, pp. 5–50. Translation by Alison Lees of "Über das Wesen des Wirtschaftskrisen," *Zeitschrift für Volkswirtschaft, Sozialpolitik*

*und Verwaltung*, **19** [1910]: 271–325.
*S1910e* = "Review of *Manuel d'economie politique* by Vilfredo Pareto," *Archiv für Sozialwissenschaft und Sozialpolitik*, **31** [1910]: 257. .
*S1911* = "Gründungsgewinn in Recht und Wirtschaft," in *ATP*, pp. 124–7. Original in *Zeitschrift für Notariat und freiwillige Gerichtsbarkeit in Österreich*, No 4, 25 January 1911. It is a summary of Schumpeter's lecture for the Viennese Law Society written by a participant.
*S1912a* = "New Translations: *Theorie der wirtschaftlichen Entwicklung*," *American Journal of Economics and Sociology*, **61** [2002]: 405–437. Translations by Marcus Becker and Thorbjørn Knudsen of selected parts of Chapters 2 and 7 of *Entwicklung I*.
*S1912b* = "'The economy as a whole': The seventh chapter to Schumpeter's *Theory of Economic Development* (1912)," *Industry and Innovation*, **9** [2002]: 93–145. Translation by Ursula Backhaus of Chapter 7 of *Entwicklung I*. Also in Jürgen G. Backhaus (ed.), *Joseph A. Schumpeter: Entrepreneurship, Style and Vision*, Boston, Mass.: Kluwer, 2003, pp. 61–116.
*S1913a* = "Eine 'dynamische' Theorie des Kapitalzinses. Eine Entgegung," in *AÖT*, pp. 411–51. Original in *Zeitschrift für Volkswirtschaft, Sozialpolitik und Verwaltung*, **22** [1913]: 599–639. It is a reply on Böhm-Bawerk's criticism of *Entwicklung I*.
*S1913b* = "Meinungsäusserung zur Frage des Werturteils," in *ATP*, pp. 127–8. Original in "Äusserungen zur Werturteilsdiskussion im Ausschuss des Vereins für Socialpolitik", mimeograph, Düsseldorf, 1913. The contribution is reproduced in context in Heino H. Nau (ed.), *Der Werturteilsstreit: Die Äusserungen zur Werturteilsdiskussion im Ausschuss des Vereins für Socialpolitik (1913)*, Marburg: Metropolis-Verlag, 1996, pp. 111–12. Other contributions are by Goldscheid, Neurath, Spann, and Max Weber.
*S1913c* = "Zinsfuss und Geldverfassung," in *AÖT*, pp. 1–28. Original in *Jahrbuch der Gesellschaft österreichischer Volkswirte*, 1913, 38–63.
*S1914a* = "Die 'positive' Methode in der Nationalökonomie," in *AÖT*, pp. 549–52. Original in *Deutsche Litteraturzeitung*, **35** [1914]: 2101–8.
*S1914b* = "Die Wellenbewegung des Wirtschaftslebens," in *BSÖ*, pp. 275–301. Original in *Archiv für Sozialwissenschaft und Sozialpolitik*, **39** [1914]: 1–32.
*S1914c* = "Eugen von Böhm-Bawerk, 1851–1914," in *TGE*, pp. 143–90. Shortened translation by Herbert Zassenhaus of *Zeitschrift für Volkswirtschaft, Sozialpolitik und Verwaltung*, **23** [1914]: 454–528.
*S1914d* = "Railway rate making: discussion," *American Economic Review: Papers and Proceedings*, **4** [1914]: 81–2. .
*S1916a* = "Über das gegenseitige Verhältnis der Völker nach dem Kriege," *Para Pachem Verbands-Mitteilungen*, 1916 [May]: 22–24. Para Pachem was the Austrian Association for the General Agreement of Nations (Österreichischer Verband für allgemeine Völkerverständigung). Link,

www.schumpeter.info/para%20pacem.pdf.
*S1916b* = "The fundamental principle of distribution theory," in Michio Morishima (ed.), *Power or Pure Politics: Joseph A. Schumpeter and Yasuma Takata*, Basingstoke and London: Macmillan, 1998, pp. 1–86. Translation by Cyprian Blamires of *Archiv für Sozialwissenschaft und Sozialpolitik*, **42** [1916]: 1–88.
*S1916–17* = "Memorandum, I–III," in *AWP*, pp. 251–310. Political memoranda produced in 1916–17.
*S1916–18* = "Neue politische Memoranden," Six memoranda from 1916–18, edited by Ulrich Hedtke, 2004. Originally circulated in mimeographed form to clarify issues during World War I. Link, www.schumpeter.info/Memoranden.pdf.
*S1917a* = "Franz Oppenheimers Theorie des Bodenmonopols," in *DBA*, pp. 100–7. Original in *Archiv für Sozialwissenschaft und Sozialpolitik*, **44** [1917]: 495–502. The original title was: "Das Bodenmonopol: Eine Entgegnung auf Dr. Oppenheimers Artikel.".
*S1917b* = "Money and the social product," *International Economic Papers*, **6** [1956]: 148–211. Translation by Arthur Marget of *Archiv für Sozialwissenschaft und Sozialpolitik*, **44** [1917]: 627–715.
*S1918a* = "Karl Marx, der Denker," in *ATP*, pp. 100–3. Original in *Arbeiterwille*, Graz, **29** [1918], 5 May 1918, p. 3.
*S1918b* = "The crisis of the tax state," in *ESC*, pp. 99–140. Translation by Wolfgang Stolper and Richard Musgrave of a booklet in the series *Zeitfragen aus dem Gebiete der Soziologie*, Graz and Leipzig: Leuschner and Lubensky, 1918. The translation was originally published in *International Economic Papers*, 1954.
*S1919a* = "Grundlinien der Finanzpolitik für jetzt und die nächsten drei Jahre," in *AWP*, pp. 344–68. Originally published by Österreichische Staatsdruckerei in Vienna, 17 October 1919.
*S1919b* = "The sociology of imperialisms," in *ESC*, pp. 141–219. Translation by Heinz Norden of *Archiv für Sozialwissenschaft und Sozialpolitik*, **46** [1919]: 1–39, 275–310.
*S1920a* = "Max Weber's work," in *ESC*, pp. 220–9. Translation by Guy Oakes of *Östereichische Volkswirt*, **12** [1920]: 831–4.
*S1920b* = "Sozialistische Möglichkeiten von heute," in *AÖT*, pp. 455–510. Original in *Archiv für Sozialwissenschaft und Sozialpolitik*, **48** [1920]: 305–60.
*S1921* = "Carl Menger, 1840–1921," in *TGE*, pp. 80–90. Translation by Hans Singer of *Zeitschrift für Volkswirtschaft und Sozialpolitik*, N.S., **1** [1921]: 197–206.
*S1923a* = "Angebot," in *ATP*, pp. 132–9. Original in Ludwig Elster, Adolf Weber, and Friedrich von Wieser (eds), *Handwörterbuch der Staatswissenschaften*, Vol. I, Jena: Fischer, 1923, pp. 299–303.
*S1923b* = "Kapital, Nachtrag: Der heutige Stand der Diskussion," in

*ATP*, pp. 150–3. Original in Ludwig Elster, Adolf Weber, and Friedrich von Wieser (eds), *Handwörterbuch der Staatswissenschaften*, Vol. V, Jena: Fischer, 1923, pp. 582–4. The original article was written by Böhm-Bawerk.

*S1924* = "Der Sozialismus in England und bei uns," in *AÖT*, pp. 511–26. Original in *Der Österreichische Volkswirt*, **16** [1924], 295–97, 327–30.

*S1925a* = "Eugen von Böhm-Bawerk," in Henry William Spiegel (ed.), *The Development of Economic Thought: Great Economists in Perspective*, New York: Wiley, 1952, pp. 569–79. Shortened translation by Henry Spiegel of *Neue Österreichische Biographie ab 1815, Grosse Österreicher*, Vol. II, Vienna, 1925, 63–80.

*S1925b* = "Handwritten Curriculum Vitae in German submitted to the Stammbuch (II) der Philosophischen Fakultät der Universität Bonn," 1925. Extracts have been reproduced by Yuichi Shionoya (*Schumpeter and the Idea of Social Science*, Cambridge and New York: Cambridge University Press, 1997, p. 16) and by Richard Swedberg (*Schumpeter: A Biography*, Princeton, N.J.: Princeton University Press, 1991, pp. 11–15).

*S1925c* = "Kreditkontrolle," in *AÖT*, pp. 118–57. Original in *Archiv für Sozialwissenschaft und Sozialpolitik*, **54** [1925]: 289–325.

*S1925d* = "Old and new banking policy," in Yuichi Shionoya and Mark Perlman (eds), *Schumpeter in the History of Ideas*, Ann Arbor, Mich.: University of Michigan Press, 1994, pp. 109–24. Translation from Dutch by C. van Paridon of *Economisch-Statistische Berichten*, **10** [1925], 552–4, 574–7, 600–1.

*S1925e* = "Edgeworth und die neuere Wirtschaftstheorie," in *DBA*, pp. 128–47. Original in *Weltwirtschaftliches Archiv*, **22** [1925]: 183–202.

*S1926a* = "Georg F. Knapp, 1842–1926," in *TGE*, pp. 295–7. Original in *Economic Journal*, **36** [1926]: 512–14.

*S1926b* = "Gustav v. Schmoller und die Probleme von heute," in *DBA*, pp. 148–99. Original in *Schmollers Jahrbuch für Gesetzgebung, Verwaltung und Volkswirtschaft*, **50** [1926]: 337–88.

*S1926c* = "Konjunkturforschung," in *ATP*, pp. 163–73. Original in *Berliner Börsen-Courier*, **58** [1926], 4 April 1926 and 7 April 1926 (as appendices).

*S1927a* = "Die goldene Bremse an der Kreditmaschine," in *AÖT*, pp. 158–84. Original in *Die Kreditwirtschaft, Kölner Vorträge*, Vol. I, Köln, 1927, pp. 80–106.

*S1927b* = "Die Wirtschaftstheorie der Gegenwart in Deutschland," in *DBA*, pp. 255–84. Original in Hans Mayer, Frank A. Fetter, and Richard Reisch (eds), *Die Wirtschaftstheorie der Gegenwart*, Vol. I, Vienna: Springer, 1927, pp. 1–30.

*S1927c* = "Friedrich von Wieser, 1851–1926," in *TGE*, pp. 298–301. Original in *Economic Journal*, **37** [1927]: 328–30.

*S1927d* = "Knut Wicksells mathematische Nationalökonomie," in *DBA*, pp. 241–54. Original in *Archiv für Sozialwissenschaft und Sozialpolitik*, **58**

[1927]: 238–51.
*S1927e* = "Review of *The Social Revolution in Austria* by C. A. Macartney," *Economic Journal*, **37** [1927]: 290–2. .
*S1927f* = "Sombarts Dritter Band," in *DBA*, pp. 220–40. Original in *Schmollers Jahrbuch für Gesetzgebung, Verwaltung und Volkswirtschaft*, **51** [1927]: 349–69.
*S1927g* = "Social classes in an ethnically homogeneous environment," in *ESC*, pp. 230–83. Translation by Heinz Norden of *Archiv für Sozialwissenschaft und Sozialpolitik*, **57** [1927]: 1–67.
*S1927h* = "Cassels Theoretische Sozialökonomik," in *DBA*, pp. 200–19. Original in *Schmollers Jahrbuch für Gesetzgebung, Verwaltung und Volkswirtschaft*, **51** [1927]: 241–60.
*S1927i* = "The explanation of the business cycle," in *EE*, pp. 21–46. Original in *Economica*, **7** [1927]: 286–311.
*S1927j* = "Unternehmerfunktion und Arbeiterinteresse," in *AWP*, pp. 160–73. Original in *Der Arbeitgeber*, **17** [1927]: 166–70.
*S1928a* = "Entrepreneur," in Roger Koppl (ed.), *Austrian Economics and Entrepreneurial Studies*, Amsterdam: Elsevier, 2003, pp. 235–65. Translation by Marcus Becker and Thorbjørn Knudsen of Ludwig Elster, Adolf Weber, and Friedrich von Wieser (eds), *Handwörterbuch der Staatswissenschaften*, Vol. VIII, Jena: Fischer, 1928, pp. 476–87.
*S1928b* = "Individualismus und gebundene Wirtschaft," in *RBZ*, pp. 74–83. Original in *Geschaftliche Mitteilungen für die Mitglieder des Reichsverbandes des Deutschen Gross- und Überseehandels*, No 31/34, October 1928, pp. 61–7.
*S1928c* = "International cartels and their relation to world trade," *Proceedings of the Academy of Political Science*, **12** [1928]: 110–115. .
*S1928d* = "Lohngestaltung und Wirtschaftsentwicklung," in *AWP*, pp. 173–85. Original in *Der Arbeitgeber*, **18** [1928]: 479–82.
*S1928e* = "Staatsreferendar und Staatsassessor," in *AÖT*, pp. 566–83. Original in *Schmollers Jahrbuch für Gesetzgebung, Verwaltung und Volkswirtschaft*, **52** [1928]:703–20.
*S1928f* = "The instability of capitalism," in *EE*, pp. 47–72. Original in *Economic Journal*, **38** [1928]: 361–86.
*S1928–29* = "Finanzwissenschaft, Wintersemester 1928/29," Abstract of Schumpeter's lectures by a student. Produced by Cläre Tisch. Link, `www.schumpeter.info/Tisch.pdf`.
*S1929a* = "Das soziale Antlitz des Deutschen Reiches," in *AS*, pp. 214–25. Original in *Bonner Mitteilungen*, **1** [1929]: 3–14.
*S1929b* = "Der Unternehmer in der Volkswirtschaft von heute," in *AWP*, pp. 226–47. Original in Bernhard Harms (ed.), *Strukturwandlungen der deutschen Volkswirtschaft: Vorlesungen gehalten in der Deutschen Vereinigung für Staatswissenschaftlichen Fortbildung*, 2nd edn, Vol. I, Berlin: Hobbing, 1929, pp. 303–26.

*S1929c* = "Die Wirtschaftslehre und die reformierte Referandarprüfung," in *AÖT*, pp. 584–97. Original in *Schmollers Jahrbuch für Gesetzgebung, Verwaltung und Volkswirtschaft*, **53** [1929]: 637–50.

*S1929d* = "Ökonomie und Psychologie des Unternehmers," in *ATP*, pp. 193–204. Originally as a printed lecture for the Zentralverbandes der deutschen Metallwalzwerks- und Hütten-Industrie, Leipzig, 1929.

*S1930a* = "Böhm-Bawerk, Eugen von (1851–1914)," In Edward R. A. Seligman (ed.), *Encyclopaedia of the Social Sciences*, Vol. II, New York: Macmillan, 1930, pp. 618–19.

*S1930b* = "Money and currency," *Social Research*, **58**: 499–543. Translation by Arthur Marget of *Geldes* (Chs 1–2).

*S1930c* = "Preface to Zeuthen: *Problems of Monopoly and Economic Warfare*," in *ATP*, pp. 108–11. Original in Frederik L. B. Zeuthen, *Problems of Monopoly and Economic Warfare*, London: Routledge, 1930, pp. vii–xiii.

*S1930d* = "Auspitz, Rudolf (1837–1906)," in Edward R. A. Seligman (ed.), *Encyclopaedia of the Social Sciences*, Vol. II, New York: Macmillan, 1930, p. 317.

*S1930e* = "Mitchell's *Business Cycles*," in *EE*, pp. 73–95. Original in *Quarterly Journal of Economics*, **45** [1930]: 150–72.

*S1930f* = *Théorie de la monnaie et de la banque, Vol. II: Theorie appliquée*, L'Harmattan, 2005, pp. 99–221. French translations of the manuscripts of three chapters that were not included in *Geldes*.

*S1931a* = "Das Kapital im wirtschaftlichen Kreislauf und in der wirtschaftlichen Entwicklung," in *ATP*, pp. 204–27. Original in Bernhard Harms (ed.), *Kapital und Kapitalismus: Vorlesungen gehalten in der Deutschen Vereinigung für Staatswissenschaftlichen Fortbildung*, Vol. I, Berlin: Hobbing, 1931, pp. 187–208.

*S1931b* = "Dauerkrise?" in *AWP*, pp. 203–10. Original in *Der Deutsche Volkswirt*, **6** [1931]: 418–21.

*S1931c* = "Recent developments of political economy," in *ESC*, pp. 284–97. Original from 1931 as typescript of a lecture delivered at Kobe University.

*S1931d* = "The 'crisis' in economics—fifty years ago," *Journal of Economic Literature*, **20** [1982]: 1049–59. Loring Allen has edited this manuscript of a lecture held in Tokyo in 1931.

*S1931e* = "The present state of economics, or On systems, schools and methods," *Kokumin Keizai Zasshi* [Journal of Economics and Business Administration], **50**: 679–705. One of three papers from Japan made available by Ulrich Hedtke. Link, `www.schumpeter.info/text2%7E1.htm`.

*S1931f* = "The present state of international commercial policy," *Kokumin Keizai Zasshi* [Journal of Economics and Business Administration], **50**: 481–506. One of three papers from Japan made available by Ulrich Hedtke. Link, `www.schumpeter.info/text2%7E1.htm`.

*S1931g* = "The present world depression: A tentative diagnosis," *American Economic Review: Papers and Proceedings*, **21** [1931]: 179–82. .
*S1931h* = "The theory of the business cycle," *Keizaigaku Ronshu* [Journal of Economics], **4**: 1–18. One of three papers from Japan made available by Ulrich Hedtke. Link, `www.schumpeter.info/text2%7E1.htm`.
*S1932a* = "Charles Rist," *Economic Journal*, **42** [1932]: 333–40. .
*S1932b* = "Das Woher und Wohin unserer Wissenschaft," in *AÖT*, pp. 598–608. Edited by Cläre Tisch after the manuscript of Schumpeter's farewell speach at the University of Bonn, 20 June 1932.
*S1932c* = "Development," *Journal of Economic Literature*, **43** [2005]: 108–120. Translation by Marcus Becker and Thorbjørn Knudsen of the typescript for the unpublished "Festschrift für Emil Lederer: Zu seinem fünfzigsten Geburtstag am 22. Juli 1932".
*S1932d* = "Ladislaus von Bortkiewicz, 1868–1931," in *TGE*, pp. 302–5. Original in *Economic Journal*,. **42** [1932]: 338–40.
*S1932e* = "Review of *Institutes de Science Economique* by G.-H. Bousquet," *Economic Journal*, **42** [1932]: 449–51. .
*S1932f* = "World depression and Franco-German economic relations: a German view," In *ATP*, pp. 227–45. Original in *Lloyds Bank Limited, Monthly Review*, March 1932 (Supplement), 14–15.
*S1933a* = "Der Stand und die nächste Zukunft der Konjunkturforschung," in *ATP*, pp. 112–17. Preface and paper for the Spiethoff Festschrift that was largely organised by Schumpeter. Original in Gustav Clausing (ed.), *Der Stand und die nächste Zukunft der Konjunkturforschung: Festschrift für Arthur Spiethoff*, Munich: Duncker & Humblot, 1933, pp. 263–7.
*S1933b* = "Review of *Essays in Biography* by J. M. Keynes," *Economic Journal*, **43** [1933]: 652–7. .
*S1933c* = "The common sense of econometrics," in *EE*, pp. 100–7. Original in *Econometrica*, **1** [1933]: 5–12.
*S1934a* = "Depressions: Can we learn from past experience?" in *EE*, pp. 108–17. Original in Douglass V. Brown, Edward Chamberlin, Seymour E. Harris, Wassily W. Leontief, Edward S. Mason, Joseph A. Schumpeter, Overton H. Taylor, *The Economics of the Recovery Program*, New York: McGraw–Hill, 1934, pp. 3–21. Also included in Stewart S. Morgan and William H. Thomas (eds), *Opinions and Attitudes in the Twentieth Century*, New York: Nelson, 1934, pp. 324–39.
*S1934b* = "Review of *Economics of Imperfect Competition* by Joan Robinson," in *EE*, pp. 125–33. Original in *Journal of Political Economy*, **42** [1934]: 249–57.
*S1934c* = "The nature and necessity of a price system," in *EE*, pp. 118–24. Original in *Economic Reconstruction*, Report of the Columbia University Commission, New York: Columbia University Press, 1934, pp. 170–6. This book also contains an "Addendum" by Schumpeter, pp. 239–42.

*S1935a* = "Young, Allyn Abbot (1876–1929)," in Edward R. A. Seligman (ed.), *Encyclopaedia of the Social Sciences*, Vol. XV, New York: Macmillan, 1935, pp. 514–15.

*S1935b* = "A theorist's comment on the current business cycle," *Journal of the American Statistical Association, Proceedings*, **30** [1935]: 167–8. .

*S1935c* = "Introduction to the Walras Centennial," *Econometrica*, **3** [1935]: 348. .

*S1935d* = "The analysis of economic change," in *EE*, pp. 134–49. Original in *Review of Economic Statistics*, **17**(4) [1935]: pp. 2–10.

*S1935e* = *Théorie de l'évolution économique: Recherches sur le profit, le crédit, l'intérêt et le cycle de la conjoncture*, Paris: Dalloz. With a very long introduction by François Perroux, which was later published separately in Perroux (1965.).

*S1935f* = "Vorwort zur vierten Auflage," Preface to the fourth German edition of *Theorie der wirtschaftlichen Entwicklung*, Duncker & Humblot: Munich and Leipzig, 1935, pp. xiii–xxi. Partly translated by Yuichi Shionoya, "Schumpeter's preface to the fourth German edition of 'The Theory of Economic Development'," *Journal of Evolutionary Economics*, **14** [2004], 131–42.

*S1936a* = "Can capitalism survive?" in *ESC*, pp. 298–315. Based on transcript of a lecture at the United States Department of Agriculture Graduate School, Washington, D.C., 18 January 1936.

*S1936b* = "Professor Taussig on wages and capital," in *EE*, pp. 150–9. Original in *Explorations in Economics: Notes and Essays Contributed in Honor of F. W. Taussig*, New York and London: McGraw-Hill, 1936, pp. 213–22.

*S1936c* = "Review of *General Theory of Employment, Interest and Money* by John Maynard Keynes," in *EE*, pp. 160–4. Original in *Journal of the American Statistical Association*, **31** [1936]: pp. 791–5.

*S1937* = "Preface to the Japanese edition of emphTheorie der wirtschaftlichen Entwicklung," in *EE*, pp. 165–8. Original published in Japanese translation in Tokyo: Iwanami-shoten, 1937.

*S1940a* = "The influence of protective tariffs on the industrial development of the United States," *Proceedings of the Academy of Political Science*, **19** [1940]: 2–7. .

*S1940b* = "The meaning of rationality in the social sciences," in *ESC*, pp. 316–38. Edited by Richard Swedberg from typescript from 1940 for a book project by Talcott Parsons.

*S1941a* = "Alfred Marshall, 1842–1924," in *TGE*, pp. 91–109. Original in *American Economic Review*, **31** [1941]: pp. 236–48. The original title is "Alfred Marshall's *Principles*: A Semi-Centennial Appraisal".

*S1941b* = "An economic interpretation of our time: The Lowell Lectures," in *ESC*, pp. 339–400. Edited by Richard Swedberg from the typescript for the lecture series in March 1941.

*S1941c* = "Frank William Taussig, 1859–1940," in *TGE*, pp. 191–221. Coauthors: Arthur Cole and Edward Mason. Original in *Quarterly Journal of Economics*, **55** [1941]: pp. 337–63.
*S1942a* = "Karl Marx, 1818–1883," in *TGE*, pp. 3–73. Reprint of *Capitalism*, Part I.
*S1942b* = "Review of *The Theory of Competitive Price* by G. J. Stigler," *American Economic Review*, **32** [1942]: 844–7. .
*S1943* = "Capitalism in the postwar world," in *EE*, pp. 175–88. Original in Seymour E. Harris (ed.), *Postwar Economic Problems*, New York and London: McGraw–Hill, 1943, pp. 113–26.
*S1944a* = "Professor Schumpeter, Austrian Minister, now teaching economic theory here," *Harvard Crimson* [called *Harvard Service News* during the war], 11 April 1944, pp. 1, 4. An imprecise article with extracts from an interview with Schumpeter.
*S1944b* = "Review of *Reflections on the Revolution of Our Time* by H. J. Laski," *American Economic Review*, **34** [1944]: 161–4. .
*S1945* = "Review of *From Economic Theory to Policy* by E. R. Walker," *Journal of Economic History*, **5** [1945]: 94–5. .
*S1946a* = "Keynes and statistics," *Review of Economic Statistics*, **28** [1946]: 194–6. .
*S1946b* = "Capitalism," in *EE*, pp. 189–210. Original in *Encyclopædia Britannica*, 14th edn, Chicago, 1946, Vol. IV, pp. 801–7.
*S1946c* = "Comments on a plan for the study of entrepreneurship," in *ESC*, pp. 406–28. Edited by Richard Swedberg from a typescript from 1946 for Arthur Cole's preparation for the Research Center in Entrepreneurial Studies at Harvard. It was partly published as *S1947c*.
*S1946d* = "John Maynard Keynes, 1883–1946," in *TGE*, pp. 260–91. Original in *American Economic Review*, **36** [1946]: 495–518.
*S1946e* = "Review of *The Road to Serfdom* by F. A. Hayek," *Journal of Political Economy*, **54** [1946]: 269–70. .
*S1946f* = "The decade of the twenties," *American Economic Review: Papers and Proceedings*, **36** [1946]: 1–10. .
*S1946g* = "The future of private enterprise in the face of modern socialistic tendencies," in *ESC*, pp. 401–5. Translation by Michael Prime and David Henderson of *Comment sauvegarder l'entreprise privée*, Montreal: L'Association Professionnelle des Industriels, 1946, pp. 103–8.
*S1947a* = *Rudimentary Mathematics for Economists and Statisticians*, New York and London: McGraw-Hill, 1947. By William L. Crum and Joseph A. Schumpeter. Schumpeter's contribution probably covers pp. 106–9, 129–33, and 159–79. Revised edition of Crum's "Rudimentary Mathematics for Economists and Statisticians," *Quarterly Journal of Economics: Supplement*, **52** [1938], pp. 1–164.
*S1947b* = "Preface," in *Capitalism, Socialism and Democracy*, 2nd edb, New York: Harper, 1947, pp. ix–xii. This preface is not included in the final

edition of *Capitalism.*
*S1947c* = "The creative response in economic history," in *EE,* pp. 221–31. Original in *Journal of Economic History,* **7** [1947]: 149–59. This is a shortened version of *S1946b.*
*S1947d* = "Theoretical problems of economic growth," in *EE,* pp. 232–40. Original in *Journal of Economic History: Supplement,* **7** [1947]: 1–9.
*S1948a* = "Irving Fisher, 1867–1947," in *TGE,* pp. 222–38. Original in *Econometrica,* **16** [1948]: 219–31.
*S1948b* = "Some questions of principle," *Research in the History of Economic Thought and Methodology,* **5** [1987]: 93–116. Edited by Loring Allen from a 1948 manuscript, which provides an alternative to the published version of Part I of *History.*
*S1948c* = "There is still time to stop inflation," *Nation's Business,* **36**(6) [1948]: 33–5, 88–91. .
*S1948d* = "Wage and tax policy in transitional states of society," in *ESC,* pp. 429–37. Edited by Richard Swedberg from the typescript for a lecture series in 1948 at the National Autonomous University of Mexico, Mexico City.
*S1949a* = "American institutions and economic progress," in *ESC,* pp. 438–44. Edited by Loring Allen from the outline of a lecture series at the Walgreen Foundation in Chicago, which was planned to start on 9th January 1950—the day after Schumpeter died.
*S1949b* = "Capitalism, socialism and democracy," in *ATP,* pp. 249–54. Notes from a lecture held in August 1949.
*S1949c* = "Economic theory and entrepreneurial history," in *EE,* pp. 253–71. Original in *Change and the Entrepreneur: Postulates and Patterns for Entrepreneurial History,* Research Center in Entrepreneurial History, Cambridge, Mass.: Harvard University Press, 1949, pp. 63–84.
*S1949d* = "English economists and the state-managed economy," in *EE,* pp. 306–21. Original in *Journal of Political Economy,* **57** [1949]: 371–82.
*S1949e* = "Science and ideology," in *EE,* pp. 272–86. Original in *American Economic Review,* **39** [1949]: 345–59.
*S1949f* = "The *Communist Manifesto* in sociology and economics," in *EE,* pp. 287–305. Original in *Journal of Political Economy,* **57** [1949]: 199–212.
*S1949g* = "The historical approach to the analysis of business cycles," in *EE,* pp. 322–9. Original from 1949 as an unfinished typescript. It was also published in Universities–National Bureau Committee for Economic Research (ed.), *Conference on Business Cycles,* New York: National Bureau of Economic Research, 1951, pp. 149–62.
*S1949h* = "Vilfredo Pareto, 1848–1923," in *TGE,* pp. 110–42. Original in *Quarterly Journal of Economics,* **63** [1949]: 147–73.
*S1950a* = "Review of *Business Cycles and Forecasting* by E. C. Bratt," *Journal of the American Statistical Association,* **45** [1950]: 140–2. .
*S1950b* = "The march into socialism," in *Capitalism,* pp. 415–25. Original

in *American Economic Review: Papers and Proceedings*, **40** [1950]: 446–56.
*S1950c* = "Wesley Clair Mitchell, 1874–1948," in *TGE*, pp. 239–59. Original in *Quarterly Journal of Economics*, **64** [1950]: 139–55.

# Other References

To each bibliographic entry is added a list of the pages on which it is cited. This type indexing as well as several other features of the present book are based on the LaTeX system for the production of complex documents (see Mittelbach and Goossens, 2004).


Aghion, Philippe and Howitt, Peter (1998): *Endogenous Growth Theory*, Cambridge, Mass. and London: MIT Press. See page(s) 284

Alchian, Armen A. (1950): "Uncertainty, evolution and economic theory," *Journal of Political Economy*, **58**: 211–221. See page(s) 259, 260

Aldrich, Howard E. and Ruef, Martin (2006): *Organizations Evolving*, 2nd edn, London: Sage. See page(s) 385

Allen, Peter M. (2007): "Self-organisation in economic systems," in Hanusch and Pyka (2007), pp. 1111–48. See page(s) 428

Allen, Robert Loring (1991): *Opening Doors: The Life and Work of Joseph Schumpeter*, New Brunswick, N.J. and London: Transaction. See page(s) 28, 31, 315, 316, 417, 418, 421

Andersen, Esben Sloth (1993): "Review of *Schumpeter: A Biography* by R. Swedberg," *Journal of Economic Literature*, **31**(4): 1969–1970. See page(s) xiv

Andersen, Esben Sloth (1994): *Evolutionary Economics: Post-Schumpeterian Contributions*, London: Pinter. See page(s) xiii, 118, 379, 427, 428

Andersen, Esben Sloth (2001): "Satiation in an evolutionary model of structural economic dynamics," *Journal of Evolutionary Economics*, **11**: 143–164. See page(s) 291

Andersen, Esben Sloth (2002): "Railroadization as Schumpeter's standard example of capitalist evolution: An evolutionary-ecological account," *Industry and Innovation*, **9**: 41–78. See page(s) 207, 428

Andersen, Esben Sloth (2004a): *Joseph A. Schumpeter: En teori om økonomisk evolution*, Copenhagen: Jurist- og Økonomforbundets Forlag. See page(s) xv

Andersen, Esben Sloth (2004b): "Knowledges, specialization and economic evolution: Modelling the evolving division of human time," in Metcalfe, J. Stanley and Foster, John (eds), *Evolution and Economic Complexity*, Cheltenham and Northampton, Mass.: Elgar, pp. 108–147. See page(s) 286, 290

Andersen, Esben Sloth (2004c): "Population thinking, Price's equation and the analysis of economic evolution," *Evolutionary and Institutional Economics Review*, **1**: 127–148. Also in Witt (2008a), pp. 30–51. See page(s) 391, 436, 438, 442

Andersen, Esben Sloth (2006): "The limits of Schumpeter's *Business Cycles*," *Industry and Innovation*, **13**: 107–116. See page(s) 237

Andersen, Esben Sloth (2007a): Bridging the gap between Schumpeterian competition and evolutionary game theory, Working Paper, Paper for the DRUID Conference on Appropriability, Proximity, Routines and Innovation, Copenhagen, 18–20 June 2007. `www2.druid.dk/conferences/viewpaper.php?id=1595&cf=9`. See page(s) 375

Andersen, Esben Sloth (2007b): "Innovation and demand," in Hanusch and Pyka (2007), pp. 754–765. See page(s) 291

Andersen, Esben Sloth (2008): "Appraising Schumpeter's 'Essence' after 100 years: From Walrasian economics to evolutionary economics," in Puranam, Krishna K. and Jain B, Ravi K. (eds), *Evolutionary Economics*, Hyderabad: Icfai University Press, pp. 75–117. See page(s) 253

Andersen, Esben Sloth (forthcoming): *Joseph A. Schumpeter: A Theory of Social and Economic Evolution*, Basingstoke and New York: Palgrave Macmillan. See page(s) xv

Anderson, Benjamin McAlester (1915): "Schumpeter's dynamic economics," *Political Science Quarterly*, **30**: 645–660. See page(s) 133

Arena, Richard and Dangel-Hagnauer, Cécile (eds) (2002): *The Contribution of Joseph Schumpeter to Economics: Economic Development and Institutional Change*, London and New York: Routledge. See page(s) 18, 464

Arthur, W. Brian (2007): "Complexity and the economy," in Hanusch and Pyka (2007), pp. 1102–10. See page(s) 428

Augello, Massimo M. (1990): *Joseph Alois Schumpeter: A Reference Guide*, Berlin: Springer. See page(s) 417, 421, 425, 449

Bartelsman, Eric J. and Doms, Mark (2000): "Understanding productivity: Lessons from longitudinal microdata," *Journal of Economic Literature*, **38**: 569–594. See page(s) 213, 393

Barton, Nicholas H., Briggs, Derek E.G., Eisen, Jonathan A., Goldstein, David B., and Patel, Nipam H. (2007): *Evolution*, Cold Spring Harbor, N.Y.: Cold Spring Harbor Laboratory Press. See page(s) 374, 435

Baumol, William J. (1959): *Economic Dynamics: An Introduction*, 2nd edn, New York: Macmillan. See page(s) 10, 161

Baumol, William J. (2002): *The Free-Market Innovation Machine: Analyzing the Growth Miracle of Capitalism*, Princeton: Princeton University Press. See page(s) 99, 161

Becker, Marcus C. (ed.) (2008): *Handbook of Organizational Routines*, Cheltenham: Elgar. See page(s) 384

Becker, Marcus C. and Knudsen, Thorbjørn (2003): "The entrepreneur as a crucial juncture in Schumpeter's work: Schumpeter's 1928 handbook entry Entrepreneur," in Koppl, Roger (ed.), *Austrian Economics and Entrepreneurial Studies*, Amsterdam: Elsevier, pp. 199–233. See page(s) 267

Bergson, Henri (1911): *Creative Evolution*, New York: Holt. See page(s) 311

Bharadwaj, Krishna and Kaviraj, Sudipta (eds) (1989): *Perspectives on*

*Capitalism: Marx, Keynes, Schumpeter and Weber*, New Delhi: Sage. See page(s) 418

Birkhoff, George D. and Lewis, Daniel C. (1935): "Stability in causal systems," *Philosophy of Science*, **2**: 304–333. See page(s) 310

Blackmore, John (1995): "Appendix on the Philosophical Society of the University of Vienna 1888–1922: An intellectual link between the period of Boltzmann and Mach and that of the Vienna Circle," in *Ludwig Boltzmann: His Later Life and Philosophy, 1900-1906. Book Two: The Philosopher*, Dordrecht: Kluwer, pp. 267–304. See page(s) 389

Blackmore, John, Itagaki, Ryoichi, and Tanaka, Setsuko (2001): "Alois Höfler: Polymath," in Blackmore, John, Itagaki, Ryoichi, and Tanaka, Setsuko (eds), *Ernst Mach's Vienna 1895–1930: or Phenomenalism as Philosophy of Science*, Dordrecht: Kluwer, pp. 237–276. See page(s) 30

Blaug, Mark (1996): *Economic Theory in Retrospect*, 5th edn, Cambridge: Cambridge University Press. See page(s) 7, 335

Blaug, Mark (2001): "No history of ideas, please, we're economists," *Journal of Economic Perspectives*, **15**: 145–164. See page(s) 343

Böhm-Bawerk, Eugen von (1913): "Eine 'dynamische' Theorie des Kapitalzinses," *Zeitschrift für Volkswirtschaft, Sozialpolitik und Verwaltung*, **22**: 1–62, 640–656. Comments on *Entwicklung I*. See page(s) 107, 132

Böhm-Bawerk, Eugen von (1921): *Kapital und Kapitalzins*, Vol. I, II.1, and II.2, 4th edn, Jena: Fischer. See page(s) 32, 331

Böhm-Bawerk, Eugen von (1949): "Karl Marx and the close of his system," in Sweezy, Paul M. (ed.), *'Karl Marx and the Close of His System' by Eugen von Böhm-Bawerk & 'Böhm-Bawerk's Criticism of Marx' by Rudolf Hilferding*, New York: Kelley, pp. 1–118. See page(s) 125

Bottomore, Tom (1992): *Between Marginalism and Marxism: The Economic Sociology of J. A. Schumpeter*, New York: Harvester Wheatsheaf. See page(s) 107, 417, 419

Bottomore, Tom and Goode, Patrick (eds) (1978): *Austro-Marxism: Texts*, Oxford: Clarendon. See page(s) 105

Boulding, Kenneth E. (1981): *Evolutionary Economics*, Beverly Hills, Cal. and London: Sage. See page(s) 248

Bowles, Samuel (2004): *Microeconomics: Behaviour, Institutions and Evolution*, Princeton, N.J.: Princeton University Press. See page(s) 386

Briefs, Henry W. (1960): *Three Views of Method in Economics*, Washington, D.C.: Georgetown University Press. See page(s) 311

Brooker, Paul (2005): *Leadership in Democracy: From Adaptive Response to Entrepreneurial Initiative*, Basingstoke: Palgrave Macmillan. See page(s) 174

Bös, Dieter and Stolper, Hans-Dieter (eds) (1984): *Schumpeter oder Keynes? Zu Wirtschaftspolitik der neunziger Jahre*, Berlin: Springer. See page(s) 418

Bücher, Karl, Schumpeter, Joseph A., and Wieser, Friedrich von (1914): "I. Abteilung: Wirtschaft und Wirtschaftswissenschaft," in Weber, Max (ed.), *Grundriss der Sozialökonomik*, Tübingen: Mohr. See page(s) 184, 481

Burns, Arthur F. (1934): *Production Trends In The United States Since 1870*, New York: National Bureau of Economic Research. See page(s) 213

Bush, Vannevar (1945): *Science: The Endless Frontier*, Washington, D.C.: National Science Foundation. See page(s) 182

Cantner, Uwe and Krüger, Jens J. (2007): "Empirical tools for the analysis of technological heterogeneity and change: Some basic building blocks of 'evolumetrics'," in Hanusch and Pyka (2007), pp. 493–502. See page(s) 391, 436

Carroll, Glenn R. and Hannan, Michael T. (1999): *The Demography of Corporations and Industries*, Princeton, N.J.: Princeton University Press. See page(s) 385, 393, 428

Cassel, Gustav (1923): *The Theory of Social Economy*, Vol. I, London: Unwin. See page(s) 320

Chamberlin, Edward H. (1951): "The impact of recent monopoly theory on the Schumpeterian system," *Review of Economics and Statistics*, **33**: 133–138. Also in Harris (1951b), pp. 83–88. See page(s) 307

Chandler, Alfred D. (1977): *The Visible Hand: The Managerial Revolution in American Business*, Cambridge, Mass.: Harvard University Press. See page(s) 402

Châteauneuf-Malclès, Anne (2002): "Time and rationality in Schumpeter's construct," in Arena and Dangel-Hagnauer (2002), pp. 106–123. See page(s) 312

Clark, John B. (1898): "The future of economic theory," *Quarterly Journal of Economics*, **13**: 1–14. See page(s) 55

Clark, John B. (1899a): *The Distribution of Wealth: A Theory of Wages, Interest and Profits*, New York and London: Macmillan. See page(s) 55, 56, 57

Clark, John B. (1899b): "Natural divisions in economic theory," *Quarterly Journal of Economics*, **13**: 187–203. See page(s) 55, 57

Clark, John B. (1912): "Review of *Theorie der wirtschaftlichen Entwicklung* by J. A. Schumpeter," *American Economic Review*, **2**: 873–875. See page(s) 133

Clemence, Richard V. and Doody, Francis S. (1950): *The Schumpeterian System*, Cambridge, Mass.: Addison–Wesley. See page(s) 140, 293, 297, 298, 306, 417

Colander, David C. and Landreth, Harry (eds) (1996): *The Coming of Keynesianism to America: Conversations with the Founders of Keynesian Economics*, Cheltenham: Elgar. See page(s) 301

Cole, Arthur H. (1942): "Entrepreneurship as an area of research," *Journal of Economic History: Supplement*, **2**: 118–126. See page(s) 400

Cole, Arthur H. (1946): "An approach to the study of entrepreneurship: A tribute to Edwin F. Gay," *Journal of Economic History: Supplement*, **6**:

1–15. See page(s) 400
Conner, Jeffrey K. and Hartl, Daniel L. (2004): *A Primer of Ecological Genetics*, Sunderland, Mass.: Sinauer. See page(s) 396
Cowan, Robin and Foray, Dominique (2002): "Evolutionary economics and the counterfactual threat: on the nature and role of counterfactual history as an empirical tool in economics," *Journal of Evolutionary Economics*, **12**: 539–562. See page(s) 406
Cyert, Richard M. and March, James G. (1963): *A Behavioural Theory of The Firm*, Englewood Cliffs, N.J.: Prentice-Hall. See page(s) 383
Dawkins, Richard (1991): *The Blind Watchmaker*, Harmondsworth: Penguin. See page(s) 382
Dawson, Cheryl A. (1995): "Gödel's Gabelsberger shorthand," in *Collected Works by Kurt Gödel. Vol. III: Unpublished Essays and Lectures*, New York: Oxford University Press, pp. 7–12. See page(s) 421
DeBresson, Christian (ed.) (1996): *Economic Interdependence and Innovative Activity: An Input–Output Analysis*, Aldershot: Elgar. See page(s) 234
Debreu, Gerard (1959): *Theory of Value: An Axiomatic Analysis of Economic Equilibrium*, New York: Wiley. See page(s) 257
Depew, David J. and Weber, Bruce H. (1995): *Darwinism Evolving: Systems Dynamics and the Genealogy of Natural Selection*, Cambridge, Mass. and London: MIT Press. See page(s) 343
Dobzhansky, Theodosius (1973): "Nothing in biology makes sense except in the light of evolution," *American Biology Teacher*, **35**: 125–129. See page(s) 374
Dopfer, Kurt and Potts, Jason (2008): *The General Theory of Economic Evolution*, London and New York: Routledge. See page(s) 387, 427
Dosi, Giovanni, Nelson, Richard R., and Winter, Sidney G. (eds) (2000): *The Nature and Dynamics of Organizational Capabilities*, Oxford: Oxford University Press. See page(s) 384
Downs, Anthony (1957): *An Economic Theory of Democracy*, New York: Harper. See page(s) 174
Durlauf, Steven N. and Blume, Lawrence E. (eds) (2008): *The New Palgrave Dictionary of Economics*, Vol. I–VIII, 2nd edn, Basingstoke and New York: Palgrave Macmillan. See page(s) 489
Eatwell, John, Milgate, Murray, and Newman, Peter (eds) (1987): *The New Palgrave: A Dictionary of Economics*, Vol. I–IV, London and Basingstoke: Macmillan. See page(s) 466, 489
Eliasson, Gunnar (2007): "On Austrian–Schumpeterian economics and the Swedish growth school," pp. 1055–76. See page(s) 280
Elster, Jon (1978): *Logic and Society: Contradictions and Possible Worlds*, Chichester: Wiley. See page(s) 406
Elster, Jon (1983): *Explaining Technical Change: A Case Study in the Philosophy of Science*, Cambridge: Cambridge University Press. See page(s) 145, 146, 155, 379

Elster, Jon (1989): *Nuts and Bolts for the Social Sciences*, Cambridge: Cambridge University Press. See page(s) 376

Elster, Jon (1998): "A plea for mechanisms," in Hedstrøm, Peter and Swedberg, Richard (eds), *Social Mechanisms: An Analytical Approach to Social Theory*, Cambridge: Cambridge University Press, pp. 45–73. See page(s) 145

Elster, Ludwig, Weber, Adolf, and Wieser, Friedrich von (eds) (1923–29): *Handwörterbuch der Staatswissenschaften*, Vol. I–IX, 4th edn, Jena: Fischer. See page(s) 77, 264

Emmett, Ross (2003): "Exegesis, hermeneutics and interpretation," in Samuels et al. (2003), pp. 523–537. See page(s) 140

Fagerberg, Jan (2003): "Schumpeter and the revival of evolutionary economics: An appraisal of the literature," *Journal of Evolutionary Economics*, **13**: 125–159. See page(s) 2

Fagerberg, Jan, Mowery, David C., and Nelson, Richard R. (eds) (2005): *The Oxford Handbook of Innovation*, Oxford and New York: Oxford University Press. See page(s) 442

Fasol-Boltzmann, Ilse M. (ed.) (1990): *Ludwig Boltzmann: Principien der Naturfilosofi*, Berlin: Springer. See page(s) 421

Fels, Rendigs (1964): "Introduction," in Schumpeter, Joseph A., *Business Cycles: A Theoretical, Historical, and Statistical Analysis of the Capitalist Process*, abr. edn, New York: McGraw–Hill, pp. viii–xiii. See page(s) 8, 140, 196, 232

Fischer, Stanley (1987): "Samuelson, Paul Anthony," in Eatwell et al. (1987), pp. 234–241. See page(s) 39

Fisher, Ronald A. (1934): "Indeterminism and natural selection," *Philosophy of Science*, **1**: 99–117. See page(s) 312

Fisher, Ronald A. (1999): *The Genetical Theory of Natural Selection: A Complete Variorum Edition*, Oxford: Oxford University Press. The first edition was published in 1930. See page(s) 312, 342, 393, 430, 436

Fogel, Robert (1964): *Railroads and American Economic Growth: Essays in Econometric History*, Baltimore, Md.: Johns Hopkins Press. See page(s) 405

Foster, John (1987): *Evolutionary Macroeconomics*, London: Allen & Unwin. See page(s) 239, 387

Frank, Steven A. (1998): *Foundations of Social Evolution*, Princeton, N.J.: Princeton University Press. See page(s) 394, 436, 438, 443

Freeman, Christopher (1982): *The Economics of Industrial Innovation*, 2nd edn, London: Pinter. See page(s) 285

Freeman, Christopher (1990): "Schumpeter's *Business Cycles* revisited," in Heertje, Arnold and Perlman, Mark (eds), *Evolving Technology and Market Structure: Studies in Schumpeterian Economics*, Ann Arbor, Mich.: University of Michigan Press, pp. 17–38. See page(s) 65, 239

Freeman, Christopher (2007): "A Schumpeterian renaissance?" in

Hanusch and Pyka (2007), pp. 130–41. See page(s) 405, 408

Freeman, Christopher and Louçã, Francisco (2001): *As Time Goes By: From the Industrial Revolutions to the Information Revolution*, Oxford: Oxford University Press. See page(s) 209, 405

Frenken, Koen (2007): "Entropy statistics and information theory," in Hanusch and Pyka (2007), pp. 544–55. See page(s) 436

Friedman, Milton (1953): "The methodology of positive economics," in Friedman, Milton (ed.), *Essays in Positive Economics*, Chicago: University of Chicago Press, pp. 3–43. See page(s) 260

Frisch, Ragnar (1933): "Propagation problems and impulse problems in dynamic economics," in *Essays in Honour of Gustav Cassel*, London: Allen & Unwin, pp. 171–205. See page(s) 309

Frisch, Ragnar (1951): "Some personal reminiscences on a great man," *Econometrica*, **19**: 87–91. Also in Harris (1951b), pp. 8–10. See page(s) 390

Frisch, Ragnar (1992): "Statics and dynamics in economic theory," *Structural Change and Economic Dynamics*, **3**: 391–401. Translation of a paper from 1929. See page(s) 318

Futuyma, Douglas J. (1998): *Evolutionary Biology*, 3rd edn, Sunderland, Mass.: Sinauer. See page(s) 314, 435

Galton, Francis (1892): *Hereditary Genius: An Inquiry into its Laws and Consequences*, London: Macmillan. See page(s) 87

Georgescu-Roegen, Nicholas (1971): *The Entropy Law and the Economic Process*, Cambridge, Mass.: Harvard University Press. See page(s) 255, 313

Georgescu-Roegen, Nicholas (1988): "Closing remarks: About economic growth – a variation on a theme by David Hilbert," *Economic Development and Cultural Change: Supplement*, **36**: 291–307. See page(s) 313

Georgescu-Roegen, Nicholas (1992): "Nicolas Georgescu-Roegen about himself," in Szenberg, Michael (ed.), *Eminent Economists: Their Life Philosophies*, Cambridge: Cambridge University Press, pp. 128–159. See page(s) 316

Giersch, Herbert (1984): "The age of Schumpeter," *American Economic Review. Papers and Proceedings*, **74**: 103–109. See page(s) 418

Gigerenzer, Gerd and Selten, Reinhard (eds) (2001): *Bounded Rationality: The Adaptive Toolbox*, Cambridge, Mass. and London: MIT Press. See page(s) 386

Gilfillan, S. Colum (1935): *The Sociology of Invention*, Chicago: Follett. See page(s) 205

Gintis, Herbert (1983): "Review of *Schumpeter's Vision: 'Capitalism, Socialism and Democracy' After 40 Years* edited by A. Heertje," *Journal of Economic Literature*, **21**: 84–86. See page(s) 160

Gintis, Herbert (forthcoming): *The Bounds of Reason: Game Theory and the Unification of the Behavioral Sciences*, Princeton, N.J.: Princeton University Press. See page(s) 386

Goldscheid, Rudolf (1919): "Staatssozialismus oder Staatskapitalismus: ein finanzsoziologischer Beitrag zur Lösung des Staastschuldenproblem," in Hickel, Rudolf (ed.), *Rudolf Goldscheid, Joseph Schumpeter: Die Finanzkrise des Steuerstaats: Beiträge zur politischen Ökonomie der Staatsfinanzen*, Frankfurt: Suhrkamp, pp. 40–252. See page(s) 171

Goodwin, Richard M. (1982): *Essays in Economic Dynamics*, London and Basingstoke: Macmillan. See page(s) 433

Goodwin, Richard M. (1988): Walras and Schumpeter: The Vision Reaffirmed, Conference paper, Conference of the International Schumpeter Society, Siena, Italy, May 24-27, 1988. See page(s) 316

Gould, Stephen Jay (2002): *The Structure of Evolutionary Theory*, Cambridge, Mass. and London: Harvard University Press. See page(s) 14, 314

Haberler, Gottfried (1937): *Prosperity and Depression: A Theoretical Analysis of Cyclical Movements*, Geneva: League of Nations. See page(s) 230

Haberler, Gottfried (1950): "Joseph Alois Schumpeter, 1883–1950," *Quarterly Journal of Economics*, **64**: 333–372. Also in Harris, *Schumpeter*, 1951, 24–47. See page(s) 16, 104, 106, 232, 388

Haberler, Gottfried, Harris, Seymour E., Leontief, Wassily W., and Mason, Edward S. (1951): "Professor Joseph A. Schumpeter: Minute, Faculty of Arts and Sciences, Harvard University," *Review of Economics and Statistics*, **33**: 89–90. Also in Harris (1951b), pp. ix–x. See page(s) 29

Hammerstein, Peter and Hagen, Edward H. (2005): "The second wave of evolutionary economics in biology," *Trends in Ecology and Evolution*, **20**: 604–609. See page(s) 386

Hannan, Michael T. and Freeman, John (1989): *Organizational Ecology*, Cambridge, Mass. and London: Harvard University Press. See page(s) 385, 428

Hannan, Michael T., Pólos, László, and Carroll, Glenn R. (2007): *Logics of Organization Theory: Audiences, Codes, and Ecologies*, Princeton, N.J.: Princeton University Press. See page(s) 385

Hansen, Alvin H. (1938): *Full Recovery or Stagnation?*, New York: Norton. See page(s) 181, 428

Hanusch, Horst (ed.) (1999): *The Legacy of Joseph A. Schumpeter*, Cheltenham and Northampton, Mass.: Elgar. See page(s) 417

Hanusch, Horst and Pyka, Andreas (eds) (2007): *Elgar Companion to Neo-Schumpeterian Economics*, Cheltenham and Northampton, Mass.: Elgar. See page(s) 417, 461, 462, 464, 467, 472, 473, 474, 476, 478, 481

Haralambous, Yannis (1991): "Typesetting old German: Fraktur, Schwabacher, Gotisch and initials," *TUGboat*, **12**: 129–138. `www.tug.org/TUGboat/Articles/tb12-1/tb31hara.pdf`. See page(s) 424

Harris, Seymour E. (1951a): "Introductory remarks," *Review of Economics and Statistics*, **33**: 91–97. Also in Harris (1951b), pp. 1–7. See page(s) 141

Harris, Seymour E. (ed.) (1951b): *Schumpeter: Social Scientist*, Cambridge,

Mass.: Harvard University Press. See page(s) 15, 417

Hébert, Robert F. and Link, Albert N. (2007): *Historical Perspectives on the Entrepreneur*, Boston, Mass.: NOW Publishing. See page(s) 265

Hedtke, Ulrich (2007a): "Joseph Alois Schumpeter: Bibliographie des Gesamtwerkes," www.schumpeter.info/biblio~1.htm. See page(s) 421, 449

Hedtke, Ulrich (2007b): "Joseph Alois Schumpeter: Verzeichnis der deutschsprachigen Briefe," www.schumpeter.info/briefe~1.htm. See page(s) 421

Heertje, Arnold (ed.) (1981): *Schumpeter's Vision: 'Capitalism, Socialism and Democracy' after 40 Years*, Eastbourne and New York: Praeger. See page(s) 160, 477

Heertje, Arnold (2006): *Schumpeter on the Economics of Innovation and the Development of Capitalism*, Cheltenham: Elgar. See page(s) 417

Heilbroner, Robert (1983): *The Worldly Philosophers: The Lives, Times and Ideas of the Great Economic Thinkers*, 2nd edn, Harmondsworth: Penguin. See page(s) 15

Heilbroner, Robert (2000): *The Worldly Philosophers: The Lives, Times and Ideas of the Great Economic Thinkers*, 7th edn, London: Penguin. See page(s) 30

Heiner, Ronald A. (1983): "The origin of predictable behaviour," *American Economic Review*, **73**: 560–595. See page(s) 82

Helburn, Suzanne W. and Bramhall, David F. (eds) (1986): *Marx, Schumpeter, & Keynes: A Centenary of Celebration and Disent*, Armonk, N.Y. and London: Sharpe. See page(s) 418

Helmstädter, Ernst and Perlman, Mark (1996): "Introduction," in Helmstädter, Ernst and Perlman, Mark (eds), *Behavioral Norms, Technological Progress, and Economic Dynamics: Studies in Schumpeterian Economics*, Ann Arbor, Mich.: University of Michigan Press, pp. 1–6. See page(s) 404

Hicks, John R. (1937): "Mr Keynes and the 'classics': A suggested interpretation," *Econometrica*, **5**: 147–159. See page(s) 305

Hicks, John R. (1939): *Value and Capital: An Inquiry into Some Fundamental Principles of Economic Theory*, Oxford: Clarendon. See page(s) 307

Hilferding, Rudolf (1981): *Finance Capital: A Study of the Latest Phase of Capitalist Development*, London and Boston: Routledge & Kegan Paul. See page(s) 128, 164

Hirschman, Albert O. (1958): *The Strategy of Economic Development*, New Haven, Conn.: Yale University Press. See page(s) 234

Hodgson, Geoffrey M. (1993): *Economics and Evolution: Bringing Life Back into Economics*, Cambridge: Polity. See page(s) 329

Hofbauer, Josef and Sigmund, Karl (1998): *Evolutionary Games and Population Dynamics*, Cambridge: Cambridge University Press. See page(s) 380, 428

Höffding, Harald (1905): *Moderne Philosophen*, Leipzig: Reisland. See page(s) 82

Höfler, Alois (ed.) (1899): *Vorreden und Einleitungen zu klassischen Werken der Mechanik: Galilei, Newton, D'Alembert, Lagrange, Kirchhoff, Hertz, Helmholtz*, Veröffentlichungen der Philosophischen Gesellschaft an der Universität Wien, Vol. II, Leipzig: Pfeffer. See page(s) 30

Höfler, Alois (1900): *Studien zur gegenwärtigen Philosophie der Mechanik: Als Nachwort zu Kant's Metaphysische Anfangsgründe der Naturwissenschaft*, Veröffentlichungen der Philosophischen Gesellschaft an der Universität Wien, Vol. IIIb, Leipzig: Pfeffer. See page(s) 30

Hook, Sidney (1943): *Hero in History: A Study in Limitation and Possibility*, New York: Day. See page(s) 77

Hughes, H. Stuart (1977): *Consciousness and Society: The Reorientation of European Social Thought 1890–1930*, 2nd edn, New York: Vintage. See page(s) 76

Jaffé, William (1983): *Essays on Walras*, Cambridge: Cambridge University Press. See page(s) 31

Janik, Allan and Toulmin, Stephen E. (1973): *Wittgenstein's Vienna*, New York: Simon and Schuster. See page(s) 16, 28

Jensen, Hans E. (1987): "New light on J. A. Schumpeter's theory of the history of economics?" *Research in the History of Economic Thought and Methodology*, **5**: 117–148. See page(s) 328

Jensen, Poul E. (1988): "Joseph A. Schumpeter (1883–1950) [in Danish]," in Sørensen, Henning and Fivelsdal, Egil (eds), *Fra Marx til Habermas: Samfundsudvikling og offentlig regulering*, Copenhagen: Nyt fra Samfundsvidenskaberne, pp. 95–112. See page(s) 407

Kamien, Morton I. and Schwarz, Nancy L. (1982): *Market Structure and Innovation*, Cambridge: Cambridge University Press. See page(s) 403

Kendall, Maurice G. (1941): "Review of *Business Cycles* by J. A. Schumpeter," *Journal of the Royal Statistical Society*, **104**: 177–180. The review is signed M. G. K. See page(s) 305

Keynes, John Maynard (1936): *The General Theory of Employment, Interest and Money*, New York: Harcourt Brace. See page(s) 277

Keynes, John Maynard (1971–89): *Collected Writings*, Cambridge: Cambridge University Press. See page(s) 421

Khan, Mohammad Shabbir (1957): *Schumpeter's Theory of Capitalist Development*, Aligarh, India: Muslim University. See page(s) 108, 417

Kilby, Peter (1971): "Hunting the Heffalump," in Kilby, Peter (ed.), *Entrepreneurship and Economic Development*, New York: Free Press, pp. 1–42. See page(s) 400

Kingsland, Sharon E. (1985): *Modeling Nature: Episodes in the History of Population Ecology*, Chicago and London: University of Chicago Press. See page(s) 428

Klamor, Arjo and Leonard, Thomas C. (1994): "So what's an economic metaphor?" in Mirowski, Philip (ed.), *Natural Images in Economic Thought: 'Markets Read in Tooth & Claw'*, Cambridge: Cambridge University Press, pp. 20–51. See page(s) 147

Klausinger, Hansjörg (1990): "The early use of the term 'veil of money' in Schumpeter's monetary writings: A comment on Patinkin and Steiger," *Scandinavian Journal of Economics*, **92**: 617–621. See page(s) 281

Klein, Burton H. (1977): *Dynamic Economics*, Cambridge, Mass.: Harvard University Press. See page(s) 168

Klepper, Steven (1996): "Entry, exit, growth and innovation over the product life cycle," *American Economic Review*, **86**(3): 562–583. See page(s) 393

Klepper, Steven (2002): "The capabilities of new firms and the evolution of the US automobile industry," *Industrial and Corporate Change*, **11**: 645–666. See page(s) 292, 393

Knight, Frank H. (1921): *Risk, Uncertainty, and Profit*, Boston: Hart, Schaffner & Marx. See page(s) 271

Knight, Frank H. (1935): "Statics and dynamics: Some queries regarding the mechanical analogy in economics," pp. 161–185. See page(s) 251

Knudsen, Thorbjørn (2002): "Economic selection theory," *Journal of Evolutionary Economics*, **12**: 443–470. See page(s) 384

Knudsen, Thorbjørn (2004): "General selection theory and economic evolution: the Price equation and the replictor/interactor distinction," *Journal of Economic Methodology*, **11**(2): 147–173. See page(s) 443

Koestler, Arthur (1989): *The Sleepwalkers: A History of Man's Changing Vision of the Universe*, London: Penguin. See page(s) 215

Kondratieff, Nicolai D. (1935): "The long waves in economic life," *Review of Economic Statistics*, **17**: 105–115. Translated by Wolfgang Stolper. See page(s) 212

Kuhn, Thomas S. (1970): *The Structure of Scientific Revolutions*, Chicago and London: University of Chicago Press. See page(s) 328, 336, 342, 370

Kuznets, Simon (1930): *Secular Movements in Production and Prices: Their Nature and Their Bearing Upon Cyclical Fluctuations*, Boston: Houghton Mifflin. See page(s) 213, 428

Kuznets, Simon (1940): "Schumpeter's Business Cycles [review of *Business Cycles* by J. A. Schumpeter]," *American Economic Review*, **30**(2): 257–271. See page(s) 203, 231, 238

Lakatos, Imre (1970): "Falsification and the methodology of scientific research programmes," in Lakatos, Imre and Musgrave, Alan (eds), *Criticism and the Growth of Knowledge*, Cambridge: Cambridge University Press, pp. 91–196. See page(s) 342

Lange, Oscar (1941): "Review of *Business Cycles* by J. A. Schumpeter," *Review of Economic Statistics*, **21**: 190–193. See page(s) 297, 305

Langlois, Richard N. (2007): *Dynamics of Industrial Capitalism: Schumpeter,*

*Chandler, and the New Economy*, London: Routledge. See page(s) 167
Leontief, Wassily (1950): "Joseph A. Schumpeter (1883–1950)," *Econometrica*, **18**: 103–110. See page(s) 28, 32, 40, 41, 302, 323
Leontief, Wassily (1991): "The economy as a circular flow," *Structural Change and Economic Dynamics*, **2**(1): 181–212. Translation of a paper from 1928. See page(s) 234
Lipsey, Richard G., Carlaw, Kenneth I., and Bekar, Clifford T. (2005): *Economic Transformations: General Purpose Technologies and Long-Term Economic Growth*, Oxford: Oxford University Press. See page(s) 209, 405
Louçã, Francisco (2007): *The Years of High Econometrics: A Short History of the Generation that Reinvented Economics*, London: Routledge. See page(s) 303, 304, 310
Lowe, Adolph (1925): "Der gegenwärtige Stand der Konjunkturforschung in Deutschland," in Bonn, Moritz J. and Palyi, Melchior (eds), *Die Wirtschaftswissenschaft nach dem Kriege*, Vol. II, Munich and Leipzig: Dunker & Humblot, pp. 329–377. See page(s) 220
MacDonald, Ronan (1965): "Schumpeter and Max Weber: Central visions and social theories," *Quarterly Journal of Economics*, **79**: 373–396. See page(s) 127
Mach, Ernst (1919): *The Science of Mechanics: A Critical and Historical Account of Its Development*, 4th edn, Chicago: Open Court. See page(s) 31, 344
Machiavelli, Niccolò (1998): *The Prince*, Oxford: Oxford University Press. See page(s) 91
Machlup, Fritz (1951): "Schumpeter's economic methodology," *Review of Economics and Statistics*, **33**: 145–151. Also in Harris (1951b), pp. 95–101. See page(s) 322, 330
Machlup, Fritz (1959): "Statics and dynamics: Kaleidoscopic words," *Southern Economic Journal*, **26**: 91–110. See page(s) 5
Maddison, Angus (1982): *Phases of Capitalist Development*, Oxford and New York: Oxford University Press. See page(s) 209
Maddison, Angus (2003): *The World Economy: Historical Statistics*, Paris: OECD. See page(s) 405
Malerba, Franco (2007): "Schumpeterian patterns of innovation and technological regimes," in Hanusch and Pyka (2007), pp. 344–59. See page(s) 285
Mallock, William H. (1901): *Aristocracy and Evolution: A Study of the Rights, the Origin, and the Social Functions of the Wealthier Classes*, New York: Macmillan. See page(s) 79
Marget, Arthur W. (1951): "Monetary aspects of the Schumpeterian system," *Review of Economics and Statistics*, **33**: 112–121. Also in Harris (1951b), pp. 62–71. See page(s) 281
Marschak, Jacob (1940): "Review of *Business Cycles* by J. A. Schumpeter," *Journal of Political Economy*, **48**: 889–894. See page(s) 219, 231, 305, 310

Marshall, Alfred (1897): "The old generation of economists and the new," *Quarterly Journal of Economics*, **11**: 115–135. See page(s) 24, 303
Marshall, Alfred (1898): "Distribution and exchange," *Economic Journal*, **8**: 37–59. See page(s) 24, 76, 382
Marshall, Alfred (1919): *Industry and Trade: A Study of Industrial Technique and Business Organization; and their Influences on the Conditions of Various Classes and Nations*, London: Macmillan. See page(s) 303
Marshall, Alfred (1961): *Principles of Economics*, variorum edn, London: Macmillan. See page(s) 14, 24, 303
Marx, Karl (1990–92): *Capital: A Critique of Political Economy*, Vol. I–III, Harmondsworth: Penguin. See page(s) 124, 125, 128, 254
Mathews, John A. (2007): "Reflections on Schumpeter's 'lost' seventh chapter to the *Theory of Economic Development*," in Hanusch and Pyka (2007), pp. 78–96. See page(s) 69
Maynard Smith, John (1982): *Evolution and the Theory of Games*, Cambridge: Cambridge University Press. See page(s) 320, 375
Maynard Smith, John and Szathmáry, Eörs (1997): *The Major Transitions in Evolution*, Oxford: Oxford University Press. See page(s) 151
Mayr, Ernst (1976): "Typological versus population thinking," in Mayr, Ernst (ed.), *Evolution and the Diversity of Life: Selected Essays*, Cambridge, Mass. and London: Harvard University Press, pp. 26–29. See page(s) 170, 260
McCraw, Thomas K. (2007): *Prophet of Innovation: Joseph Schumpeter and Creative Destruction*, Cambridge, Mass. and London: Harvard University Press. See page(s) 28, 215, 305, 417, 419
McCrea, Roswell C. (1913): "Schumpeter's economic system," *Quarterly Journal of Economics*, **27**(May): 520–529. See page(s) 67, 132
Medearis, John (2001): *Joseph Schumpeter's Two Theories of Democracy*, Cambridge, Mass. and London: Harvard University Press. See page(s) 174
Meikle, Scott (1995): *Aristotle's Economic Thought*, Oxford: Oxford University Press. See page(s) 123
Merton, Robert K. (1957): "Priorities in scientific discovery: a chapter in the sociology of science," *American Sociological Review*, **22**(6): 635–659. See page(s) 350
Merton, Robert K. (1965): *On the Shoulders of Giants: A Shandean Postscript*, New York: Free Press. See page(s) 350
Messori, Marcello (1998): "The trials and misadventures of Schumpeter's treatise on money," *History of Political Economy*, **29**: 639–673. See page(s) 281
Metcalfe, J. Stanley (1998): *Evolutionary Economics and Creative Destruction*, London and New York: Routledge. See page(s) 232, 443
Metcalfe, J. Stanley (2001): "Evolutionary approaches to population thinking and the problem of growth and development," in Dopfer, Kurt

(ed.), *Evolutionary Economics: Program and Scope*, Vol. Boston, Mass., Kluwer, pp. 141–164. See page(s) 443

Metcalfe, J. Stanley (2002): "Review of *Foundations of Social Evolution* by S. A. Frank," *Journal of Bioeconomics*, **4**: 89–91. See page(s) 394, 443

Metcalfe, J. Stanley (2003): "Industrial growth and the theory of retardation: Precursors of an adaptive evolutionary theory of economic change," *Revue Économique*, **54**: 407–432. See page(s) 213

Metcalfe, J. Stanley (2007): "Replicator dynamics," in Hanusch and Pyka (2007), pp. 440–52. See page(s) 436, 439

Mill, John Stuart (1868): *A System of Logic Ratiocinative and Inductive: Being a Connected View of the Principles of Evidence and the Methods of Scientific Investigation*, Vol. I–II, 7th edn, London: Longmans. See page(s) 55, 142

Mirkowich, Nicholas (1940): "Schumpeter's theory of economic development," *American Economic Review*, **30**: 580. See page(s) 238

Mises, Ludwig von (1978): *Erinnerungen*, Stuttgart and New York: Fischer. See page(s) 105

Mitchell, Brian R. (1981): *European Historical Statistics 1750–1975*, 2nd edn, London and Basingstoke: Macmillan. See page(s) 207

Mitchell, Brian R. (1983): *International Historical Statistics: The Americas and Australasia*, London and Basingstoke: Macmillan. See page(s) 207

Mitchell, Brian R. (1988): *British Historical Statistics*, Cambridge: Cambridge University Press. See page(s) 207, 431

Mitchell, Wesley C. (1925): "Quantitative analysis in economic theory," *American Economic Review*, **15**: 1–12. See page(s) 303, 424

Mitchell, Wesley C. (1969): *Types of Economic Theory: From Mercantilism to Institutionalism*, Vol. I–II, New York: Kelley. See page(s) 44

Mittelbach, Frank and Goossens, Michel (2004): *The LaTeX Companion*, 2nd edn, Boston, Mass.: Addison–Wesley. See page(s) 424, 461

Morgenstern, Oscar (1927): "Review of *Theorie der Wirtschaftlichen Entwicklung* by J. A. Schumpeter," *American Economic Review*, **27**: 281–282. See page(s) 133

Morgenstern, Oscar (1951): "Joseph A. Schumpeter, 1883–1950," *Economic Journal*, **61**: 197–202. See page(s) 41

Morishima, Michio and Catephores, George (1988): "Anti-Say's law versus Say's law: a change in paradigm," in Hanusch, Horst (ed.), *Evolutionary Economics: Applications of Schumpeter's Ideas*, Cambridge: Cambridge University Press, pp. 23–53. See page(s) 65

Moura, Mário da Graça (2002): "Metatheory as a key to understanding: Schumpeter after Shionoya," *Cambridge Journal of Economics*, **26**: 805–821. See page(s) 17

März, Eduard (1991): *Joseph Schumpeter: Scholar, Teacher and Politician*, New Haven, Conn. and London: Yale University Press. See page(s) 67, 417, 419

Muller, Jerry Z. (1999): "Capitalism, socialism, and irony: Understanding

Schumpeter in context," *Critical Review*, **13**: 239–267. See page(s) 157
Nau, Heino Henrich (ed.) (1996): *Der Werturteilsstreit: Die Äusserungen zur Werturteilsdiskussion im Ausschuss des Vereins für Socialpolitik (1913)*, Marburg: Metropolis-Verlag. See page(s) 353
Navratil, K. von (1913): "Review in German of *Theorie der wirtschaftlichen Entwicklung* by J. A. Schumpeter," *Weltwirtschaftliches Archiv*, **1**: 442–445. See page(s) 133
Nelson, Richard R. and Winter, Sidney G. (1982): *An Evolutionary Theory of Economic Change*, Cambridge, Mass. and London: Harvard University Press. See page(s) 255, 284, 291, 292, 375, 380, 383, 427, 443
Niehans, Jürg (1981): "Economics: history, doctrine, science, art," *Kyklos*, **34**: 165–177. See page(s) 300
Nietzsche, Friedrich (1976): *Ecce Homo: How One Becomes What One Is*, Harmondsworth: Penguin. See page(s) 78
Nietzsche, Friedrich (1983): "On the uses and disadvantage of history for life," in *Untimely Meditations*, Cambridge: Cambridge University Press, pp. 57–123. See page(s) 78
North, Douglas C. (1998): "Institutions, organizations, and market competition," in Eliasson, Gunnar and Green, Christopher (eds), *Microfoundations of Economic Growth: A Schumpeterian Perspective*, Ann Arbor, Mich.: University of Michigan Press, pp. 15–25. See page(s) 404
Nowak, Martin A. (2006): *Evolutionary Dynamics: Exploring the Equations of Life*, Cambridge, Mass.: Harvard University Press. See page(s) 380
Oakley, Allen (1990): *Schumpeter's Theory of Capitalist Motion: A Critical Exposition and Reassessment*, Aldershot: Elgar. See page(s) 218, 219, 228, 229, 231, 234, 239, 417
O'Brien, Patric (1977): *The New Economic History of the Railways*, London: Croom Helm. See page(s) 401
Oppenheimer, Franz (1916): *Wert und Kapitalprofit: Neubegründung der objektiven Wertlehre*, Jena: Gustav Fischer. See page(s) 132
Ospovat, Dov (1995): *The Development of Darwin's Theory: Natural History, Natural Theology and Natural Selection, 1883–1859*, paperback edn, Cambridge: Cambridge University Press. See page(s) 123
Oxford University Press (1989): *Oxford English Dictionary*, 2nd edn, Oxford: Oxford University Press. `www.oed.com/`. See page(s) 349
Page, Karen M. and Nowak, Martin A. (2002): "Unifying evolutionary dynamics," *Journal of Theoretical Biology*, **219**: 93–98. See page(s) 435
Pareto, Vilfredo (1935): *The Mind and Society*, New York: Harcourt, Brace & Co. See page(s) 86
Pareto, Vilfredo (1972): *Manual of Political Economy*, London and Basingstoke: Macmillan. See page(s) 85, 193
Parijs, Philippe Van (1981): *Evolutionary Explanation in the Social Sciences: An Emerging Paradigm*, Totowa, N.J.: Rowman and Littlefield. See page(s) 379

Pearson, Karl (1957): *The Grammar of Science*, New York: Meridian. See page(s) 81

Penrose, Edith T. (1952): "Biological analogies in the theory of the firm," *American Economic Review*, **42**: 804–819. See page(s) 261

Penrose, Edith T. (1995): *The Theory of the Growth of the Firm*, 3rd edn, Oxford: Oxford University Press. See page(s) 261

Perez, Carlotta (2002): *Technological Revolutions and Financial Capital: The Dynamic of Bubbles and Golden Ages*, Cheltenham and Northampton, Mass.: Elgar. See page(s) 209, 405

Perez, Carlotta (2007): "Finance and technical change: A neo-Schumpeterian perspective," in Hanusch and Pyka (2007), pp. 775–99. See page(s) 209

Perroux, François (1965): *La pensée économique de Joseph Schumpeter: Les dynamiques du capitalisme*, Genéve: Droz. See page(s) 32, 38, 417, 419

Phelps, Edmund S. (2007): "Foreword," to Frydman, Roman and Goldberg, Michael D., *Imperfect Knowledge Economics: Exchange Rates and Risk*, Princeton, N.J.: Princeton University Press, pp. xii–xx. See page(s) 375

Phelps, Edmund S. (2008): "Toward a model of innovation and performance along the lines of Knight, Keynes, Hayek and M. Polany," in Audretsch, David B., Strom, Robert, and Acs, Zoltan J. (eds), *Entrepreneurship, Growth, and Public Policy*, Cambridge: Cambridge University Press. See page(s) 375

Philippovich, Eugen von (1919–22): *Grundriss der politischen Ökonomie*, Vol. I–III, Tübingen: Mohr. See page(s) 253

Phillips, Almarin (1971): "Technology and industry structure," in Pierce, John R., Fubini, Eugene G., and Haggerty, Patrick E. (eds), *Technology and International Trade*, Washington, D.C.: National Academy of Engineering, pp. 67–87. See page(s) 284

Pianka, Eric R. (1999): *Evolutionary Ecology*, 6th edn, New York: Harper. See page(s) 428, 433

Polanyi, Michael (1962): "The republic of science: Its political and economic theory," *Minerva*, **1**: 54–74. See page(s) 337

Price, George R. (1972a): "Extension of covariance selection mathematics," *Annals of Human Genetics*, **35**: 485–490. See page(s) 394, 436

Price, George R. (1972b): "Fisher's 'Fundamental Theorem' made clear," *Annals of Human Genetics*, **36**: 129–140. See page(s) 394, 436

Price, George R. (1995): "The nature of selection," *Journal of Theoretical Biology*, **175**: 389–396. See page(s) 436

Provine, William B. (1971): *The Origins of Theoretical Population Genetics*, Chicago and London: University of Chicago Press. See page(s) 314, 389

Pyka, Andreas and Fagiolo, Giorgio (2007): "Agent-based modelling: A methodology for neo-Schumpetarian economics," in Hanusch and Pyka (2007), pp. 467–87. See page(s) 428

Raffaelli, Tiziano (2003): *Marshall's Evolutionary Economics*, London and New York: Routledge. See page(s) 25

Redlich, Fritz (1966): "Towards the understanding of an unfortunate legacy," *Kyklos*, **19**: 709–718. See page(s) 265

Reinert, Erik S. (2002): "Schumpeter in the context of the two canons of economic thought," *Industry and Innovation*, **9**: 23–40. See page(s) 27

Reinert, Hugo and Reinert, Erik S. (2006): "Creative destruction in economics: Nietzsche, Sombart and Schumpeter," in Backhaus, Jürgen G. and Drechsler, Wolfgang J. M. (eds), *Friedrich Nietzsche (1844–1900): Economy and Society*, New York: Springer, pp. 55–85. See page(s) 404

Reisman, David (2004): *Schumpeter's Market: Enterprise and Evolution*, Cheltenham and Northampton, Mass.: Elgar. See page(s) 417

Robbins, Lionel (1930): "On a certain ambiguity of the conception of stationary equilibrium," *Economic Journal*, **40**: 194–214. See page(s) 251

Robbins, Lionel (1955): "Schumpeter's 'History of Economic Analysis'," *Quarterly Journal of Economics*, **69**: 1–22. See page(s) 335

Robinson, Joan (1943): "Review of *Capitalism, Socialism and Democracy* by J. A. Schumpeter," *Economic Journal*, **53**: 381–383. See page(s) 160

Rosenberg, Hans (1940): "Review of *Business Cycles* by J. A. Schumpeter," *American Historical Review*, **46**: 96–99. See page(s) 306

Rosenberg, Nathan (1982): *Inside the Black Box: Technology and Economics*, Cambridge: Cambridge University Press. See page(s) 407

Rosenberg, Nathan (1986): "Schumpeter and Marx: How common a vision?" in MacLeod, Roy M. (ed.), *Technology and the Human Prospect: Essays in Honour of Christopher Freeman*, London and Wolfeboro, N.H.: Pinter, pp. 197–213. See page(s) 65

Samuels, Warren J., Biddle, Jeff F., and Davis, John B. (eds) (2003): *A Companion to the History of Economic Thought*, Oxford: Blackwell. See page(s) 336, 466

Samuelson, Paul A. (1943): "Dynamics, statics and the stationary state," *Review of Economic Statistics*, **25**: 58–68. See page(s) 251, 317

Samuelson, Paul A. (1947): *Foundations of Economic Analysis*, Cambridge, Mass.: Harvard University Press. See page(s) 40, 45, 317, 318, 341, 369

Samuelson, Paul A. (1948): *Economics: An Introductory Analysis*, 1st edn, New York and London: McGraw-Hill. See page(s) 341

Samuelson, Paul A. (1951): "Schumpeter as a teacher and economic theorist," *Review of Economics and Statistics*, **33**: 98–103. Also in Harris (1951b), pp. 48–53. See page(s) 305, 325, 399

Samuelson, Paul A. (1981a): "Schumpeter as an economic theorist," in Frisch, Helmut (ed.), *Schumpeterian Economics*, Eastbourne and New York: Praeger, pp. 1–27. See page(s) 1, 215, 300, 369

Samuelson, Paul A. (1981b): "Schumpeter's *Capitalism, Socialism and Democracy*," in Heertje (1981), pp. 1–21. See page(s) 17

Samuelson, Paul A. (1987): "Out of the closet: A program for a Whig history of economic science," *History of Economics Society Bulletin*, **9**: 51–60. See page(s) 301, 323

Samuelson, Paul A. (2003): "Reflections on the Schumpeter I knew well," *Journal of Evolutionary Economics*, **12**: 463–467. See page(s) 155

Sanderson, Stephen K. (1990): *Social Evolutionism: A Critical History*, Cambridge, Mass. and Oxford: Blackwell. See page(s) 3

Saviotti, Paolo P. (2007): "Qualitative change and economic development," in Hanusch and Pyka (2007), pp. 820–39. See page(s) 428

Schefold, Bertram (1986): "Schumpeter as a Walrasian Austrian and Keynes as a classical Marshallian," in Wagener and Drukker (1986), pp. 93–111. See page(s) 65

Schelling, Thomas C. (1978): *Micromotives and Macrobehaviour*, New York and London: Norton. See page(s) 381

Schneider, Erich (1975): *Joseph A. Schumpeter: Life and Work of a Great Social Scientist*, Bureau of Business Research, University of Nebraska–Lincoln. See page(s) 84, 417, 419

Schumpeter, Elizabeth Boody (1950): "Bibliography of the writings of Joseph A. Schumpeter," *Quarterly Journal of Economics*, **64**: 373–384. See page(s) 421, 422

Shackle, George L. S. (1967): *The Years of High Theory: Invention and Tradition in Economic Thought 1926-1939*, Cambridge: Cambridge University Press. See page(s) 301, 307

Shionoya, Yuichi (1990): "The origin of the Schumpeterian research program: A chapter omitted from Schumpeter's *Theory of Economic Development*," *Journal of Institutional and Theoretical Economics*, **146**: 314–327. See page(s) 69, 138

Shionoya, Yuichi (1996): "The sociology of science and Schumpeter's ideology," in Moss, Laurence S. (ed.), *Joseph A. Schumpeter: Historian of Economics*, London and New York: Routledge, pp. 279–316. See page(s) 357

Shionoya, Yuichi (1997): *Schumpeter and the Idea of Social Science*, Cambridge and New York: Cambridge University Press. See page(s) 7, 17, 36, 137, 340, 348, 417

Shionoya, Yuichi (2003): "Review of *The Contribution of Joseph Schumpeter to Economics* edited by R. Arena and C. Dangel-Hagnauer," *History of Political Economy*, **35**: 604–606. See page(s) 18

Shionoya, Yuichi (2004): "Schumpeter's preface to the fourth German edition of 'The Theory of Economic Development'," *Journal of Evolutionary Economics*, **14**: 131–142. See page(s) 191

Shionoya, Yuichi (2005): *The Soul of the German Historical School: Methodological Essays on Schmoller, Weber and Schumpeter*, New York: Springer. See page(s) 18, 37

Silverberg, Gerald and Verspagen, Bart (2003): "Breaking the waves: A Poisson regression approach to Schumpeterian clustering of basic in-

novations," *Cambridge Journal of Economics*, **27**: 671–693. See page(s) 221
Simon, Herbert A. (1982): *Models of Bounded Rationality*, Vol. I–II, Cambridge, Mass. and London: MIT Press. See page(s) 82, 185, 383
Simon, Herbert A. (1991): *Models of My Life*, New York: Basic Books. See page(s) 304
Simon, Herbert A. (1996): *The Sciences of the Artificial*, 3rd edn, Cambridge, Mass. and London: MIT Press. See page(s) 384
Simon, Herbert A. (2002): "Near decomposability and the speed of evolution," *Industrial and Corporate Change*, **11**: 587–599. See page(s) 384
Smith, Adam (1986–99): *An Inquiry into the Nature and Causes of the Wealth of Nations*, Vol. I–II, Oxford: Clarendon. See page(s) 76, 374
Smithies, Arthur (1950): "Memorial: Joseph Alois Schumpeter, 1883–1950," *American Economic Review*, **40**: 628–648. See page(s) 16, 141, 316
Sombart, Werner (1902): *Der moderne Kapitalismus*, Munich and Leipzig: Duncker & Humblot. See page(s) 126
Spencer, Herbert (1896): *The Study of Sociology*, New York: Appleton. See page(s) 77
Spencer, Herbert (1908): "The filiation of ideas," in Duncan, David (ed.), *Life and Letters of Herbert Spencer*, Vol. II, New York: Appleton, pp. 304–365. See page(s) 349
Spiethoff, Arthur (1953): "Business cycles," *International Economic Papers*, **3**: 75–171. See page(s) 200
Spiethoff, Arthur (2002): "Preliminary remarks to a theory of overproduction," in Hagemann, Harald (ed.), *Business Cycle Theory: Selected Texts 1860–1939*, Vol. II, London: Pickering & Chatto, pp. 44–76. See page(s) 116
Sraffa, Piero (1926): "The laws of returns under competitive conditions," *Economic Journal*, **36**: 535–550. See page(s) 292, 301
Staley, Charles E. (1983): "Schumpeter's 1947 course on the history of economic thought," *History of Political Economy*, **15**: 25–37. See page(s) 333
Stigler, George J. (1969): "Does economics have a useful past?" *History of Political Economy*, **1**: 217–230. See page(s) 343
Stigler, Stephen M. (1986): *The History of Statistics: The Measurement of Uncertainty before 1900*, Cambridge, Mass. and London: Harvard University Press. See page(s) 389
Stolper, Wolfgang F. (1994): *Joseph Alois Schumpeter: The Public Life of a Private Man*, Princeton, N.J.: Princeton University Press. See page(s) 305, 310, 311, 375, 417, 418
Streissler, Erich (1973): "To what extent was the Austrian school marginalist?" in Black, R. D. Collinson, Coats, A. W., and Goodwin, Craufurd D. W. (eds), *The Marginal Revolution in Economics: Interpretation and Evaluation*, Durhan, N.C.: Duke University Press, pp. 160–175. See page(s) 61
Streissler, Erich W. (1994): "The influence of German and Austrian eco-

nomics on Joseph Schumpeter," in Shionoya, Yuichi and Perlman, Mark (eds), *Schumpeter in the History of Ideas*, Ann Arbor, Mich.: University of Michigan Press, pp. 13–38. See page(s) 190

Swedberg, Richard (1991): *Schumpeter: A Biography*, Princeton, N.J.: Princeton University Press. See page(s) 28, 190, 417, 419

Sweezy, Paul M. (1942): *The Theory of Capitalist Development: Principles of Marxian Political Economy*, New York: Oxford University Press. See page(s) 252

Sweezy, Paul M. (1943): "Professor Schumpeter's theory of innovation," *Review of Economic Statistics*, **25**: 93–96. See page(s) 242, 250, 251, 252, 263

Tinbergen, Jan (1951): "Schumpeter and quantitative research in economics," *Review of Economics and Statistics*, **33**: 109–111. Also in Harris (1951b), pp. 59–61. See page(s) 305, 390

Tugwell, Rexford G. (ed.) (1924): *The Trend of Economics*, New York: Knopf. See page(s) 303, 424

Universities–National Bureau Committee for Economic Research (ed.) (1951): *Conference on Business Cycles*, New York: National Bureau of Economic Research. See page(s) 324, 325

Veblen, Thorstein (1898a): "Review of *Aristocracy and Evolution* by W. H. Mallock," *Journal of Political Economy*, **6**: 430–435. See page(s) 79

Veblen, Thorstein (1898b): "Why is economics not an evolutionary science?" *Quarterly Journal of Economics*, **12**(July): 373–397. See page(s) 24, 299

Veblen, Thorstein (1899–1900): "The preconceptions of economic science," *Quarterly Journal of Economics*, **13–14**: (13:) 121–150, 396–426, and (14:) 240–69. See page(s) 24

Vecchi, Nicolò De (1995): *Entrepreneurs, Institutions and Economic Change: The Economic Thought of J. A. Schumpeter (1905–1925)*, Aldershot and Brookfield, Vmt.: Elgar. See page(s) 417

Verein für Sozialpolitik (1910): *Verhandlungen des Vereins für Sozialpolitik in Wien 1909*, Leipzig: Duncker & Humblot. See page(s) 353

Viner, Jacob (1954): "Schumpeter's 'History of Economic Analysis'," *American Economic Review*, **44**: 894–910. See page(s) 335

Vromen, Jack J. (1995): *Economic Evolution: An Enquiry Into the Foundations of New Institutional Economics*, London and New York: Routledge. See page(s) 329

Wagener, Hans-Jürgen and Drukker, Jan W. (eds) (1986): *The Economic Law of Motion of Modern Society: A Marx–Keynes–Schumpeter Centennial*, Cambridge: Cambridge University Press. See page(s) 418, 478

Walras, Léon (1954): *Elements of Pure Economics or The Theory of Social Wealth*, London: George Allen and Unwin. See page(s) 53, 60, 114, 115

Weber, Max (1903–06): "Roscher und Knies und die logischen Probleme der historischen Nationalökonomie," in Weber (1968), pp. 1–145. Original in *Schmollers Jahrbuch*. See page(s) 32

Weber, Max (1904): "Die 'Objektivität' sozialwissenschaftlischer und sozialpolitischer Erkenntnis," in Weber (1968), pp. 146–214. Original in *Archiv für Sozialwissenschaft und Sozialpolitik*. See page(s) 32

Weber, Max (1906): "Kritische Studien auf dem Gebite der kulturwissenschaftlichen Logik," in Weber (1968), pp. 215–290. Original in *Archiv für Sozialwissenschaft und Sozialpolitik*. See page(s) 405

Weber, Max (1958): *The Protestant Ethic and the Spirit of Capitalism*, 2nd edn, New York: Scribner. Trans. Talcott Parsons. See page(s) 127

Weber, Max (1968): *Gesammelte Aufsätze zur Wissenschaftslehre*, 3rd edn, Tübingen: Mohr. See page(s) 480, 481

Wicksell, Knut (1936): *Interest and Prices*, London: Macmillan. See page(s) 279, 282

Wicksell, Knut (1954): *Value, Capital and Rent*, London: Allen & Unwin. See page(s) 279

Wieser, Friedrich von (1893): *Natural Value*, London and New York: Macmillan. See page(s) 61

Wieser, Friedrich von (1907): "Arma virumque cano," in Wieser (1929), pp. 335–345. See page(s) 77, 78

Wieser, Friedrich von (1911): "Das Wesen und der Hauptinhalt der theoretischen Nationalökonomie. Kritische Glossen," in Wieser (1929), pp. 10–34. Comments on *Wesen*. See page(s) 41, 304

Wieser, Friedrich von (1914): "Theorie der gesellschaftlichen Wirtschaft," in Bücher et al. (1914), pp. 125–444. See page(s) 77

Wieser, Friedrich von (1929): *Gesammelte Abhandlungen*, Tübingen: Mohr. See page(s) 481

Winter, Sidney G. (1971): "Satisficing, selection and the innovating remnant," *Quarterly Journal of Economics*, **85**(2): 237–261. See page(s) 261

Winter, Sidney G. (1984): "Schumpeterian competition in alternative technological regimes," *Journal of Economic Behavior and Organization*, **5**: 287–320. See page(s) 377

Witt, Ulrich (ed.) (2008a): *Recent Developments In Evolutionary Economics*, Cheltenham and Northampton, Mass.: Elgar. See page(s) 461

Witt, Ulrich (2008b): "What is specific about evolutionary economics?" *Journal of Evolutionary Economics*, **18**: 547–575. See page(s) 376

Witt, Ulrich and Cordes, Christian (2007): "Selection, learning, and Schumpeterian dynamics: A conceptual debate," in Hanusch and Pyka (2007), pp. 316–28. See page(s) 378

Wittgenstein, Ludwig (2001): *Tractatus Logico-Philosophicus*, London and New York: Routledge. See page(s) 42

Wood, J. Cunningham (ed.) (1991): *J. A. Schumpeter: Critical Assessments*, Vol. I–IV, London and New York: Routledge. See page(s) 417

Yagi, Kiichiro (1993): "Schumpeter and the University of Vienna [in Japanese with German documentation]," *Keizai-Ronso Bessatsu* [Research and Study], **5**: 63–83. Special Issue of *Economic Review*, Kyoto University. See page(s) 28

Zambelli, Stefano (2007): "A rocking horse that never rocked: Frisch's 'Propagation Problems and Impulse Problems'," *History of Political Economy*, **39**: 145–166. See page(s) 310

# Index of Schumpeter's Works

The entries cover all mentioning and citation of Schumpeter's works. While monographs and paper collections are ordered by their full titles, Schumpeter's papers are normally found by core words of their titles. The index entries include abbreviations (like *Cycles* and *S1927i*). These abbreviations can be used to look up the information in the Schumpeter bibliography (starting on page 447).

# Index of Persons

The names of some persons have added information in brackets. H = occurrence in the index of Schumpeter's *History of Economic Analysis*. P = entry in one or both of the two editions of the *New Palgrave Dictionary of Economics* (Eatwell et al., 1987; Durlauf and Blume, 2008). Authors of works that are only cited parenthetically are not included (see instead "Other References").

www.ingramcontent.com/pod-product-compliance
Ingram Content Group UK Ltd.
Pitfield, Milton Keynes, MK11 3LW, UK
UKHW041830100926
8556IPUK00002B/100

9 781843 313342

For example, the speed of productivity change depends on the variance of the productivities of the firms of an industry. Since productivity change is not only caused by the competitive selection between firms, it is obvious that this is not a complete explanation. It is not even a full explanation in biology; and Fisher emphasised this fact. Nevertheless, the full operationalisation of his analysis had to await the efforts of George Price (1972b; 1972a).

Through the work of Price and his followers, it has become clear how the mechanism of inter-unit selection can be complemented formally by a mechanism of intra-unit change (see Appendix D.2). The backbone of evolutionary change can be captured by two censuses of the population. At the two points of time, we, for each unit of selection, register its size and the value of its characteristic. Then we, also for each unit, calculate the change in size and the change of the characteristic. Finally, we are able to describe evolutionary change as being composed of two effects. The selection effect is the covariance between the size changes and the characteristics of the first census. The effect of intra-unit change is the average of the size changes of units times their changes of characteristics. Price's Equation demonstrates that the decomposition of evolutionary change into these two effects is an identity. His operationalisation of selection and intra-unit change has been surprisingly fruitful in evolutionary biology (Frank, 1998). With respect to evolutionary economics, Stanley Metcalfe (2002, 90) remarked that "[f]or some years now evolutionary economists have been using the Price equation without realising it". This statistical interpretation of evolutionary change not only helps us to relate to empirical data, but also to define core concepts of evolutionary economics more precisely. A summary of the present approach is found in Table 12.2 on the next page. This table presents an attempt to connect the previous section's theoretical concepts with relatively precise statical procedures. However, the motivations behind these procedures cannot be presented quickly. Instead, they are discussed in the rest of the present section. Further explanations are found in Appendix D.2.

If we return to the case of the productivity studies, it becomes clear that Price's equation seems designed for the study of incumbent firms. Thus it is primarily a tool for the operationalisation of Schumpeter's Mark II model. It serves to decompose the change in average productivity in the effect of the differential growth of firms and the effect of intra-firm change. The selection effect is measured by assuming that all firms have unchanging productivity. The covariance between fixed productivities and size changes can be rewritten as the variance of productivities times the regression coefficient of size changes on fixed characteristics. The regression coefficient can be called the efficiency of selection. If this coefficient is zero, no selection takes place and all the change of productivity comes from intra-firm change. However, if the regression coefficient is cycle: pioneers create the industry and are followed by mass entry; then a shakeout of mass exit is followed by a period of consolidation (Klepper, 1996).

Presently, it does not matter whether or not typical patterns are observed. The important thing is that the collection and analysis of data can follow the rules of vital statistics in general and "corporate demography" in particular. The latter rules have been described by Glenn Carroll and Michael Hannan (1999). Although these researchers no not apply an explicit evolutionary framework, their way of thinking seems very close to that of Schumpeter's Mark I model. In other words, Schumpeter's evolutionary theory can be given a statistical counterpart. This counterpart may demonstrate a need of further theoretical development. For instance, we may distinguish between entry by diversifying firms and entry by new forms. We may also split the latter group according to the prior experience of their founders: some are outsiders and some come from firms within the industry. Empirical evidence from the automobile industry demonstrates that the firms whose founders come from the leading firms of the industry have the largest probability of survival (Klepper, 2002). This result suggests that we should consider a revised version of the Mark I model in which entry is to some extent the result of spin-off and in which new firms to some extent inherit the characteristics of their mother firms.

Although 'vital statistics' (i.e.,statistics of entry and exit) are important for the initial operationalisation of the Mark I model, they are insufficient for a fuller study of this version of the capitalist engine. Furthermore, they cannot capture the capitalist engine Mark II. Since we cannot by a priori reasoning determine which model is predominant, a mix between vital statistics and other types of statistics is necessary. This mix is, for instance, found in modern studies of productivity based on longitudinal microdata. Actually, surveys of longitudinal productivity studies, like that of Bartelsman and Doms (2000), look like compendia of Schumpeter-inspired results. Moreover, it is possible to translate the applied methods of statistical analysis into concepts applied within evolutionary economics. Since the concept of long-term productivity change is notoriously difficult to operationalise and since productivity is at best a proxy for underlying and evolving routines, it is better to analyse the statistics of evolvable characteristics directly. Nevertheless, the problematic assumption that average productivity is an evolving variable may help us to grasp statistical procedures that can be used for the analysis of solid cases of evolution. This procedure can, in several steps, be derived from Fisher's (1999) *Genetical Theory of Natural Selection* from 1930. This book presented what it called "the fundamental theorem of natural selection". According to this "theorem", the speed of the evolution of a characteristic depends on the population's variance with respect to the characteristics.